GEOCHEMISTRY: PATHWAYS AND PROCESSES

Second Edition

SECOND EDITION

GEOCHEMISTRY
Pathways and Processes

Harry Y. McSween, Jr.
University of Tennessee, Knoxville

Steven M. Richardson
Winona State University

Maria E. Uhle
University of Tennessee, Knoxville

Columbia University Press | NEW YORK

Columbia University Press
Publishers Since 1893
New York   Chichester, West Sussex

This book was previously published by Prentice-Hall, Inc.

Library of Congress Cataloging-in-Publication Data

McSween, Harry Y.
    Geochemistry : pathways and processes. — 2nd ed. / Harry Y. McSween, Jr., Steven M.
  Richardson, Maria E. Uhle.
        p.   cm.
    Rev. ed. of: Geochemistry / Steven M. Richardson, c1989.
    Includes bibliographical references and index.
    ISBN 0-231-12440-6
    1. Geochemistry.   I. Richardson, Steven McAfee.   II. Uhle, Maria E.   III. Richardson,
  Steven McAfee. Geochemistry.   IV. Title.
  QE515.R53   2003
  551.9—dc21   2003051638

♾ Columbia University Press books are printed on permanent and durable acid-free paper.

Printed in the United States of America
10 9 8 7 6 5 4 3 2 1

For Sue, Cathy, and Mike

CONTENTS

FOUR   HOW TO HANDLE SOLUTIONS   56

FIVE   DIAGENESIS: A STUDY IN KINETICS   79

FIFTEEN   STRETCHING OUR HORIZONS: COSMOCHEMISTRY   313

PREFACE TO THE SECOND EDITION

The various operations of Nature, and the changes which take place in the several substances around us, are so much better understood by an attention to the laws of chemistry, that in every walk of life the chemist has a manifest advantage.

Samuel Parkes (1816), *The Chemical Catechism*

The modern geologist with no knowledge of geochemistry will be severely limited. Indeed, geochemistry now pervades the discipline—providing the basis for measurement of geologic time, allowing critical insights into the Earth's inaccessible interior, aiding in the exploration for economic resources, understanding how we are altering our environment, and unraveling the complex workings of geochemical systems on the Earth and its neighboring planets. Geochemical reasoning reveals both processes and pathways, as the book's title implies.

Our book is about chemistry, written expressly for geologists. To be more specific, it is a text we hope will be used to introduce advanced undergraduate and graduate students to the basic principles of geochemistry. It may even inspire some to explore further and to become geochemists themselves. Whether that happens or not, our major objective is to help geology students gain that manifest advantage noted in the quote above by showing that concepts from chemistry have a place in all corners of geology. The ideas in this book should be useful to all practicing geologists, from petrologists to paleontologists, from geophysicists to astrogeologists.

Students exposed to a new discipline should expect to spend some time assimilating its vocabulary and theoretical underpinnings. It isn't long, though, before most students begin to ask, "How can I *use* it?" Geologists, commonly being rather practical people, reach this point perhaps earlier than most scientists. In this book, therefore, we have tried not to just talk about geochemical theory, but to show how its principles can be used to solve problems. We have integrated into each chapter a number of worked problems, solved step by step to

demonstrate the way that ideas can be put to practical use. In most cases we have devised problems that are based on published research, so that the student can relate abstract principles to the real concerns of geologists. Working the problems at the end of each chapter will further reinforce what has been covered in the text.

This attention to problem solving reflects our belief that geochemistry, beyond the most rudimentary level, is quantitative. To answer real questions about the chemical behavior of the Earth, a geologist must be prepared to manipulate quantitative data and perform calculations. We address this philosophy directly at the end of chapter 1. It should be understood at the outset, however, that most questions at the level of this book can be answered by anyone with a standard undergraduate background in the sciences. We assume only that the student has had a year of college-level chemistry and a year of calculus. In this second edition, we have added a chapter (chapter 2) to help students put a geological frame around fundamental concepts in chemistry. As in the earlier edition, we have outlined more advanced methods for handling a few specialized math concepts, such as partial derivatives, in an appendix. With these aids, most routine approaches to solving geochemical problems should already be within the student's reach.

As the book's title suggests, we have tried from the beginning to balance the traditional equilibrium perspective with observations from a kinetic viewpoint, emphasizing both pathways and processes in geochemistry. We have focused initially on processes in which temperature and pressure are nearly constant. After an abbreviated introduction to the laws of thermodynamics and to fundamental equations for flow and diffusive transport, we spend the first half of the book investigating diagenesis, weathering, and solution chemistry in natural waters. We have made significant changes in this second edition of the book by including new chapters on mineral structure and bonding and on organic geochemistry. These concepts are also integrated into other chapters. In the second half of the book, we return for a closer look at thermodynamics and kinetics as they apply to systems undergoing changes in temperature and pressure during magmatism and metamorphism. Stable and radiogenic isotopes are treated near the end of the book, where their applications to many of the previous topics are emphasized. The concept of geochemical systems, introduced in chapter 1, is illustrated in chapters 8, 12, and 15, as we emphasize, respectively, the ocean-atmosphere, the core-mantle-crust, and the planets as systems. In this second edition, we have made special efforts to update examples and worked problems to reflect new discoveries in the field.

The division of geochemical topics outlined above has served us well in the classroom. Among other advantages, it spreads out the burden of learning basic theory and allows the instructor to address practical applications of geochemistry before students' eyes begin to glaze over. For those who might want to leap directly to applicable equations rather than follow our derivations, we have placed key equations in boxes. At the other extreme, we hope that instructors and curious students are intrigued by the occasional advanced or offbeat ideas that we have included as highlighted boxes in each chapter. They are intended to provoke discussion and further investigation.

This book is a distillation of what we ourselves have learned from our own teachers and from our many colleagues and friends in geochemistry; the book itself is (eloquent, we hope) testimony to their influences. The appearance of the second edition results largely from the unflagging encouragement and contagious enthusiasm of Holly Hodder, Prentice-Hall editor of the first edition. We have also greatly benefited from the counsel of our new Columbia University Press editor, Robin Smith. We hope that geology students will find this revised edition stimulating, accessible, and useful.

CHAPTER ONE

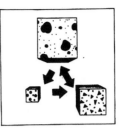

INTRODUCING CONCEPTS IN GEOCHEMICAL SYSTEMS

OVERVIEW

In this introductory chapter, our goal is to examine the range of problems that interest geochemists and to compare two fundamental approaches to geochemistry. Thermodynamics and kinetics are complementary ways of viewing chemical changes that may take place in nature. By learning to use both approaches, you will come to appreciate the similarities among geochemical processes and be able to follow the many pathways of change. To develop some skills in using the tools of geochemistry, we also examine the limitations of thermodynamics and kinetics, and discuss practical considerations in problem solving.

WHAT IS GEOCHEMISTRY?

In studying the Earth, geologists use a number of borrowed tools. Some of these come from physicists and mathematicians, others were developed by biologists, and still others by chemists. When we use the tools of chemistry, we are looking at the world as geochemists. From this broad perspective, the distinction between geochemistry and some other disciplines in geology, such as metamorphic petrology or crystallography, is a fuzzy and artificial one.

Historical Overview

The roots of geochemistry belong to both geology and chemistry. Many of the practical observations of Agricola (Georg Bauer), Nicolas Steno, and other Renaissance geologists helped to expand knowledge of the behavior of the elements and their occurrence in nature. As we see in more detail in chapter 2, these observations formed the basis from which both modern chemistry and geology grew in the late eighteenth century. The writings of Antoine Lavoisier and his contemporaries, during the age when modern chemistry began to take shape, were filled with conjectures about the oceans and atmosphere, soils, rocks, and the processes that modify them. In the course of separating and characterizing the properties of the elements, Lavoisier, Humphry Davy, John Dalton, and others also contributed to the growing debate among geologists about the composition of the Earth. Similarly, Romé de L'Isle and René-Just Haüy made the first modern observations of crystals, leading very quickly to advances in mineralogy and structural chemistry.

The term *geochemistry* was apparently used for the first time by the Swiss chemist C. F. Schönbein in 1838. The emergence of geochemistry as a separate discipline, however, came with the establishment of major laboratories at the U.S. Geological Survey in 1884; the Carnegie Institution of Washington, D.C., in 1904; and in several European countries, notably Norway and the Soviet Union, between roughly 1910 and 1925. From these laboratories came some of the first systematic surveys of rock and mineral compositions and some of the first experimental studies in which the thermodynamic conditions of mineral stability were investigated. F. W. Clarke's massive work, *The Data of Geochemistry*, published in 1909 and revised several times over the next 16 years, summarized analytical work performed at the U.S. Geological Survey and elsewhere, allowing geologists to estimate the average composition of the crust. The phase rule, which was first suggested through theoretical work by J. Willard Gibbs of Yale University in the 1880s, was applied to field studies of metamorphic rocks by European petrologists B. Roozeboom and P. Eskola, who thus established the chemical basis for the metamorphic facies concept. Simultaneously, Arthur L. Day and others at the Carnegie Institution of Washington Geophysical Laboratory began a program of experimentation focused on the processes that generate igneous rocks.

During the twentieth century, the course of geochemistry has been guided by several technological advances. The first of these was the discovery by Max von Laue (1912) that the internal arrangement of atoms in a crystalline substance can serve as a diffraction grating to scatter a beam of X-rays. Shortly after this discovery, William L. Bragg used this technique to determine the structure of halite. In the 1920s, Victor M. Goldschmidt and his associates at the University of Oslo determined the structures of a large number of common minerals, and from these structures, formulated principles regarding the distribution of elements in natural compounds. Their papers, published in the series *Geochemische Verteilungsgesetze der Elemente,* were perhaps the greatest contributions to geochemistry in their time.

During the 1930s and into the years encompassing World War II, new steel alloys were developed that permitted experimental petrologists to investigate processes at very high pressures for the first time. Percy Bridgeman of Harvard, Norman L. Bowen of the Carnegie Geophysical Laboratory, and a growing army of colleagues built apparatus with which they could synthesize mineral assemblages at temperatures and pressures similar to those found in the lower crust and upper mantle. With these technological advances, geochemists were finally able to investigate the chemistry of inaccessible portions of the Earth.

The use of radioactive isotopes in geochronology began at the end of the nineteenth century, following the dramatic discoveries of Ernest Rutherford, Henri Becquerel, and Marie and Pierre Curie. The greatest influence of nuclear chemistry on geology, however, came with the development of modern, high-precision mass spectrometers in the late 1930s. Isotope chemistry rapidly gained visibility through the careful mass spectroscopic measurements of Alfred O. C. Nier of the University of Minnesota; between 1936 and 1939, he determined the isotopic compositions of 25 elements. Nier's creation in 1947 of a simple but highly precise mass spectrometer brought the technology within the budgetary reach of many geochemistry labs and opened a field that, even today, is still a rapidly growing area of geochemistry. Beginning with the work of Harold Urey of the University of Chicago and his students in the late 1940s, stable isotopes have become standard tools of economic geology and environmental geochemistry. In recent years, precise measurements of radiogenic isotopes have also proved increasingly valuable in tracing chemical pathways in the Earth.

Advances in instrumentation brought explosive growth to other analytical corners of geochemistry during the last half of the twentieth century. Geochemists made use of increasingly sophisticated detectors and electronic signal processing systems to perform chemical analyses at lower and lower detection levels, opening the new subfield of trace element studies. As we show in chapter 12, this has made it possible to refine our understanding of the evolution of the Earth's mantle and crust and of many fundamental petrologic processes. The electron microprobe, developed independently by R. Castaing and A. Guinier in France and I. B. Borovsky in Russia in the late 1940s, became commercially available to geochemists in the 1960s. Because the new instrument made it possible to perform analyses on the scale of microns, geochemists were able to study not only microscopic samples but also subtle chemical gradients within larger samples. The 1970s and 1980s, as a result, saw increasing interest in geochemical kinetics.

Finally, as exploration for and exploitation of petroleum became a major focus for geology in the early decades of the twentieth century, the field of organic geochemistry slowly gained importance. In the 1930s, Russia's "father of geochemistry," V. I. Vernadsky, was among the first to speculate on the genetic relationships among sedimentary organic matter, petroleum, and hydrocarbon gases. Alfred Treibs confirmed the biological origin of oil in the 1930s. Others, including Wallace Pratt of Humble Oil Company, correctly deduced that oil is formed by natural cracking of large organic molecules and noted that local differences in source material and diagenetic conditions create a distinctive chemical "fingerprint" for each petroleum reservoir. In the latter half of the century, organic geochemists turned their attention to environmental contaminants, redirecting the petroleum geochemists' methods to trace and control industrial organic matter in soils and groundwater and reconstruct ancient environments.

The growth of geochemistry has been paralleled by the development of cosmochemistry, and many scientists have spent time in both disciplines. From the uniformitarian point of view, the study of terrestrial pathways and processes in geochemistry is merely an introduction to the broader search for principles that govern the general behavior of planetary materials. In the 1920s Victor M. Goldschmidt, for example, refined many of his notions about the affinity of elements for metallic, sulfidic, or siliceous substances by studying meteorites. (You will read more about this in chapter 2.) In more recent years, our understanding of crust-mantle differentiation in the early Earth has been enhanced by studies of lunar samples.

This historical sketch has unavoidably bypassed many significant events and people in the growth of geochemistry. It should be apparent from even this short treatment, however, that the boundary between geology and chemistry has been porous. For the generations of investigators who have worked at the boundary during the past two centuries, it has been an exciting and challenging time. This period has also been marked by recurring cycles of interest in data gathering and the growth of basic theory, to which both chemists and geologists have contributed. Each advance in technology has invited geochemists to scrutinize the Earth more closely, expanded the range of information available to us, and led to a new interpretation of fundamental processes.

Beginning Your Study of Geochemistry

In this book, we explore many ideas that geology shares with chemistry. In doing so, we offer a perspective on geology that may be new to you. You don't need to be a chemist to use the techniques of geochemistry, just as you don't need to be a physicist to understand the constraints of seismic data or a biologist to benefit from the wealth of evolutionary insights from paleontology. However, the language and methods of geochemistry are different from much that is familiar in standard geology. You will have to pick up a new vocabulary and become familiar with some fundamental concepts of chemistry to appreciate the role geochemistry plays in interpreting geological events and environments. We start to introduce some of these concepts in this chapter and the next, and introduce others as we go along. As we explore environments from magma chambers to pelagic sediment columns, we hope that your understanding of other disciplines in geology will also be enriched.

As a student of geology, you have already been introduced to a number of unifying concepts. Plate tectonics, for example, provides a physical framework in which seemingly independent events such as arc volcanism, earthquakes, and high-pressure metamorphism can be interpreted as the products of large-scale movements in the lithosphere. As geologists fine-tune the theory of plate tectonics, its potential for letting us in on broader secrets of how the Earth works is even more tantalizing than its promise of showing us how the individual parts behave.

In the same way, as we try to answer the question "What is geochemistry?" in this book, we look for unifying concepts by discovering how the tools of geochemistry can help us interpret a wide variety of geologic environments. In chapter 3, for example, we introduce the concept of chemical potential and demonstrate its role in characterizing directions of change in chemical systems. We then encounter chemical potential in various guises in later chapters as we examine topics in chemical weathering, diagenesis, stable isotope fractionation, magma crystallization, and condensation of gases in the early solar system, to mention a few. In this way, you will come to recognize that chemical potential is a common strand in the fabric of geochemistry. We offer you the continuing challenge of recognizing other unifying concepts as we proceed through this book.

A glance at the Table of Contents shows you that geochemistry virtually reaches into all areas of geology. Because this book is meant to explore broadly applicable ideas in geochemistry, we do not concentrate on one corner of geology exclusively. Instead, we choose problems that are of interest to many kinds of geologists and help you find ways to approach and (we hope) solve them.

GEOCHEMICAL VARIABLES

Most geochemical processes can be described in rather broad, qualitative terms if we are presenting them for a lay audience or we are trying to establish a "big picture" of how things work. As a rule, however, qualitative arguments in geochemistry are believable only if they are backed up by more rigorous, quantitative analyses. At a minimum, it is important to begin any investigation of a geochemical system by identifying a set of properties (*variables*) that may account for potentially significant changes in the system.

Let's begin by defining two types of variables, which we will refer to as *extensive* and *intensive*. The first of these is a measure of the size or extent of the system in which we are interested. For example, mass is an extensive variable because (all other properties being held constant) the mass of a system is a function of its size. If we choose to examine a block of galena (PbS) that is 20 cm on a side, we expect to find that it has a different mass than a block that measures only 5 cm on a side. Volume is another familiar extensive property, clearly a measure of the size of a system. It should be obvious from these examples that extensive quantities are additive. That is, the mass of galena in the two blocks described above is equal to the sum of their individual masses. Volumes are also additive.

Intensive variables, in contrast, have values that are independent of the size of the system under consideration. Two of the most commonly described intensive properties are temperature and pressure. To verify that these are independent of the size of a system, imagine a block of galena that is uniformly at room temperature (~25°C). Experience tells us that if we do nothing to the block except cut it in half, we will then have two blocks of galena, but each will still be at 25°C. A similar thought experiment will confirm that pressure is also independent of the dimensions of a system, provided that it too is uniform throughout the system.

It is useful to conceive many geochemical problems in terms of intensive rather than extensive properties. By doing so, we free ourselves from considering the size of a system and think simply about its behavior. For example, if we want to determine whether a mineral assemblage consisting of quartz and olivine is more or less stable than an assemblage of pyroxene, the answer will be most useful if it is cast in terms of energy per unit of mass or volume.

The conversion from extensive to intensive properties involves normalizing to some convenient measure of the size of the system. For example, we may choose to describe a block of galena by its mass (an extensive property) normalized by its volume (another extensive property). The resulting quantity, density, is the same regardless of which block we choose to consider, and is therefore an intensive property of the system. We might choose to normalize instead by multiplying the mass of the block by 6.02×10^{23} (Avogadro's number: the number of formula units per mole of any substance, and in this case an appropriate scale factor) and then dividing by the number of PbS formula units in it. The result would be the *molecular weight* of galena, another intensive property. In general, then, the ratio of two extensive variables is an intensive variable. When we start to consider energy functions, in chapter 3, we will begin using an overscore on symbols (as in $\Delta \bar{G}_r$) to indicate that they are molar quantities, and therefore intensive. In that chapter and beyond, it will be useful to recognize that the derivative of an extensive variable with respect to either an extensive or an intensive variable is also intensive.

GEOCHEMICAL SYSTEMS

Several times we have used the word *system* in this chapter. By this, we mean the portion of the universe that is of interest for a particular problem, whether it is as small as an individual clay particle or as large as the Solar System.

Because the results of geochemical studies will usually depend sensitively on how a system is defined, it is always important to begin the study by defining the boundaries of the system. There are three generic types of systems, each defined by the condition of its boundaries. Systems are *isolated* if they cannot exchange either matter or energy with the universe beyond their limits. A perfectly insulated thermos bottle is an example of such a system. Geochemists often make the simplifying assumption that systems are isolated because it is easier to keep track of the state of a system both physically and mathematically if they can imagine a well-defined wall around it. In chapter 3, we also show that the theoretical existence of isolated systems enables us to put limits on the behavior of fundamental thermodynamic functions. In the real universe, however, truly isolated systems cannot exist because there are no perfect insulators to prevent the exchange of energy across system boundaries.

In nature, systems can either be *closed* or *open*. A closed system is one that can exchange energy, but not matter, across its boundaries. An open system can exchange both energy and matter. This distinction is actually somewhat artificial, although it is practical in most cases. For example, provided that the beds confining an aquifer have an acceptably low permeability, we may consider the aquifer to be a closed system if we are studying only the relatively rapid chemical reactions within it. If, however, we are considering a long-term toxic waste disposal problem, our definition of "acceptably low" permeability would probably be different. The same aquifer might more realistically be considered an open system. Often, then, the definition of a system will have to include an assessment of the rates of transfer across its boundaries. So long as the system is contained within limits of acceptably low transfer rates, we can usually justify treating it as either a closed or an isolated system. Thus, time is an important consideration when we are defining systems.

THERMODYNAMICS AND KINETICS

There are two different sets of conditions under which we might study a geochemical system. First, we might consider a system in a state of *equilibrium*, a consequence of which is that observable properties of the system undergo no change with time. (We offer a more exact definition in terms of thermodynamic conditions in chapter 3.) If a system is at equilibrium, we are generally not concerned with (and can't readily determine) how it reached equilibrium. The tools of thermodynamics are not useful for asking questions about a system's evolutionary history. However, with the appropriate equations, we *are* able to estimate what a system at equilibrium would look like under any environmental conditions we choose. The methods of thermodynamics allow us to use environmental properties such as temperature and pressure, plus the system's bulk composition, to predict which substances will be stable and in what relative amounts they will be present. In this way, the thermodynamic approach to a geochemical system can help us measure its stability and predict the direction in which it will change if its environmental parameters change.

The alternative to studying a system in terms of thermodynamic conditions is to examine the *kinetics* of the system. That is, we can study each of the pathways along which the system may evolve between states of thermodynamic equilibrium and determine the rates of change of system properties along those pathways. Most geochemical systems can change between equilibrium states along a variety of pathways, some of which are more efficient than others. In a kinetic study, the task is often to determine which of the competing pathways is dominant.

As geochemists, we have an interest in determining not only what should happen in real geologic systems (the thermodynamic answer) but also how it is most likely to happen (the kinetic answer). The two are intimately linked. While the focus of thermodynamics is on the end states of a system—before and after a change has taken place—a kinetic treatment of geochemical systems sheds light on what happens between the end states. Where possible, it is smart to keep both approaches in mind.

AN EXAMPLE: COMPARING THERMODYNAMIC AND KINETIC APPROACHES

Figure 1.1a is a schematic drawing of a hypothetical system that we can use to illustrate these two approaches. It is a closed system consisting of five subsystems or *reservoirs* (indicated by boxes) that are connected by *pathways* (labeled by arrows). To make it less abstract, we have redrawn this schematic in figure 1.1b so that you can see the real system it is meant to represent: the floor plan of a small house without exterior doors or windows (unrealistic, perhaps, but the absence of exterior outlets is what makes this a closed system). We can use this familiar setting to demonstrate many of the principles that apply to dynamic geochemical systems.

Assume that you, the owner of the house, have invited several of your friends to a party, and that they are all hungry. As we take our first look at the party, people are distributed unevenly in the living room, the dining room, the kitchen, the bedroom, and the bathroom. Your guests are free to move from one room to another, provided that they use doorways and do not break holes through the walls, but there are some rules of population dynamics (which we will describe shortly) that govern the rates at which they move.

From a thermodynamic point of view, the system we have described is initially in a state (i.e., hungry) that we would have a hard time describing mathematically,

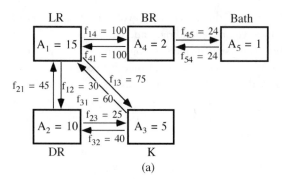

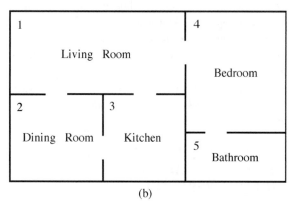

FIG. 1.1. (a) Schematic diagram for a house with five rooms. The number of people in room i is given by A_i, and their rates of movement between rooms by f_{ij}, where i is the room being vacated and j is a destination. The units of f_{ij} are people hour^{-1}. (b) Floor plan for the house described schematically in (a).

but which is clearly different from the final state that you, as host, would like to approach (i.e., well fed). It is also clear that if you were to begin serving food to your friends, they would gradually move from the hungry state to the well-fed state. If people were as well behaved as chemical systems, we ought to be able to use an approach analogous to thermodynamics to calculate the "energy" difference between the two states—just how hungry are your guests?—and therefore predict the amount of food you will have to provide to let everyone go home feeling satisfied. The mathematics necessary to do this calculation might involve writing a differential equation for each room, expressing variations in some function W (a "fullness" function) in terms of variations in amounts of ham, cole slaw, and beer:

$$dW = (\partial W/\partial \mathrm{Ham})d\mathrm{Ham} + (\partial W/\partial \mathrm{Slaw})d\mathrm{Slaw}$$
$$+ (\partial W/\partial \mathrm{Beer})d\mathrm{Beer}. \qquad (1.1)$$

This equation is in the form of a total differential, described in appendix A. It will become familiar in later chapters.

Thermodynamics, then, allows us to recognize that the initial hungry state of your guests is less stable than the final well-fed state, and provides a measure of how much change to expect during the party. Even with an adequate description of a system, however, thermodynamics can give no guarantee that the real world will reach the equilibrium configuration that we calculate on paper. Despite our best thermodynamic calculations, for example, the people at our party may remain indefinitely in a hungry state. Maybe they don't like your cooking, or the large dog in the corner of your kitchen discourages them.

In the geologic realm, minerals or mineral assemblages may appear to be stable simply because we observe them over times that are short compared with the rates of reactions that alter them. Such materials are said to be *metastable*. Granites and limestones are metastable at the Earth's surface, even though we know them to be less stable than clays and salt-bearing groundwaters. The aragonitic tests that are typically excreted by modern shelled organisms are less stable than calcite, yet aragonite is only slowly converted to the calcite that thermodynamics tells us should be present. Diamonds, stable only under the high-pressure conditions of the mantle, do not spontaneously restructure themselves to form graphite, even over very large spans of time. In each of these cases, the reaction paths between the metastable and stable states involve intermediate steps that are difficult to perform and therefore take place very slowly or not at all. We refer to these intermediate steps as *kinetic barriers* to reaction. Given an adequate description of the compositional and environmental constraints on a system, thermodynamics can tell us what that system *should* look like if it is given infinite time in which to overcome kinetic barriers.

We encounter many geologically important systems that are in transition between equilibrium states, yet restrained by kinetic barriers. Is it possible to apply thermodynamic reasoning to these systems? It is common to find that very large systems, such as the ocean, are in a state of *dynamic balance* or *steady state,* in which environmental conditions differ from one part of the system to another, while the overall system may persist indefinitely with no net change in its state. Under these conditions, thermodynamics remains a useful approach, although it should be clear that no system-wide state of thermodynamic equilibrium can be described.

To return to the example in figure 1.1, let us assume that randomly moving guests will nibble food whenever they can find it. Further, suppose that people in any particular room at the party always move around more rapidly within the room than they do from one room to another. If that is so, then the proportion of hungry to moderately well-fed people at different places within that room will be fairly uniform any time we peek into it. Different rooms, however, may have different proportions of hungry and well-fed people, depending on the size of the room and whether it has food in it. Each of the rooms, in other words, is always in a local equilibrium state that differs from the states of the other rooms, and the entire house is a mosaic of equilibrium states. If we knew how the "thermodynamics" of people works, then we could apply "fullness" functions such as equation 1.1 within rooms to help describe how hungry the guests in them were at any time. Furthermore, despite the fact that the rooms are not in equilibrium with each other, consideration of the mosaic provides valuable information about the direction of overall change in the house. (Note: Every example has its limitations. In this case, the statistics of small numbers make it difficult to take this observation literally. The more people we invite to the party, however, the more uniform the population is in any room.)

This is the basis of the concept of local equilibrium that is often used in metamorphic petrology to discuss mineral assemblages that are well defined on the scale of a thin section, but which may differ from one hand sample to another across an outcrop. Within small parts of the system, rates of reaction may often be rapid enough that equilibrium is maintained and thermodynamics can give useful information about the existence and relative abundance of minerals in which we are interested. Provided that we can agree on the definition of the mineral assemblages and therefore the sizes of the regions of mosaic equilibrium, we can use thermodynamics to analyze the regions individually and to predict directions of change in the system as a whole.

The point is that thermodynamics remains a useful part of our tool kit even if we examine a system that is far from equilibrium. We can't use thermodynamics, however, to answer questions such as how rapidly the party will approach its well-fed state, or how people may be distributed throughout the house at any instant in time.

In the world of geochemistry, we may be able to describe deviations from equilibrium by applying thermodynamics, but we find ourselves incapable of predicting which of many pathways a system may follow toward equilibrium or how rapidly it gets there. Those are kinetic problems.

Suppose we address one of the kinetic problems for the party system: how are people distributed among the rooms as a function of time? To answer the question, let's choose one room at random and look at the operations that will cause its population to change with time. We do this by taking an inventory of the rates of input to and output from the room, and comparing them. We can write this as a differential equation:

$$dA_i/dt = \text{total input} - \text{total output}$$
$$= \sum f_{ji} - \sum f_{ij}. \qquad (1.2)$$

Here we are using the symbol A_i to indicate the number of people in the room we have chosen (room i, where i is a number from one to five). The quantities labeled f are *fluxes;* that is, numbers of people moving from one room to another per unit of time. The order of subscripts on f is such that the first one identifies where people are moving from, and the second tells us where they are moving to. This notation is used to label the pathways in figure 1.1a. Room j in each case is some room other than i. The summations in equation 1.2 are over all values of i.

We have to write an expression of this type for each room in the example (for dA_1/dt, dA_2/dt, and so forth). If we define the initial values for each A_i and all f_{ij}, then we can solve these *flux equations* simultaneously for any time t. The result will be a room-by-room population census as the party proceeds. This step is often intimidating, but only occasionally impossible, especially with appropriate software for numerically solving sets of differential equations.

Armed with new analytical insight, let's reexamine the concept of a steady state for our hypothetical system. It can now be seen to be a condition in which each of the flux equations 1.2 is equal to zero. If there is such a state, the population in each room will not change with time. Notice that this does not imply that the fluxes are zero, but that the inputs and outputs from each room are balanced. That is, steady state may also be defined as a condition in which the time derivatives df_{ij}/dt are equal to zero. Is this confusing? Consider worked problem 1.1.

Worked Problem 1.1

In figure 1.1a, we have shown numerical values for each of the nonzero fluxes in the party example. How can we determine, using these constant values for f_{ij}, whether the five flux equations for this system describe a system in steady state?

The flux equations are

$$dA_1/dt = f_{21} + f_{31} + f_{41} - f_{12} - f_{13} - f_{14} = 45 + 60 + 100$$
$$- 30 - 75 - 100 = 0$$

$$dA_2/dt = f_{12} + f_{32} - f_{21} - f_{23} = 30 + 40 - 45 - 25 = 0$$

$$dA_3/dt = f_{13} + f_{23} - f_{31} - f_{32} = 75 + 25 - 60 - 40 = 0$$

$$dA_4/dt = f_{14} + f_{54} - f_{41} - f_{45} = 100 + 24 - 100 - 24 = 0$$

$$dA_5/dt = f_{45} - f_{54} = 24 - 24 = 0$$

Because each of the time derivatives is equal to zero, the reservoir contents, A_i, must be fixed at the values shown in figure 1.1a. Therefore, although people are free to move between rooms, the distribution of party guests is in a steady state. If we solved the five differential equations simultaneously, we would see that each population census in time looks like the previous one: no change.

There is no guarantee that all systems have a steady state, particularly if they are open. A system may continually accumulate or lose mass through time (a dissolving crystal is one such system). At one time, it was believed that the salinity of the oceans was a unidirectional function of time, and that the progressive accumulation of salt in seawater could be used to infer the age of the oceans. In that view, the ocean is not at or approaching a steady state. It is also possible that a system may have more than one steady state. Systems that involve interacting fluxes for large numbers of species, or that have an oscillatory behavior (due, perhaps, to the passage of seasons), often have multiple steady states.

What might cause a system to move away from a steady state? In our party example, we can define some laws of population dynamics to describe the rates at which people move from room to room. These "laws" are a set of empirical statements that tell us how the fluxes f_{ij} in the set of equations we are developing change with time. It is important to do this because fluxes are rarely constant in natural systems. In this problem, for example, it is fair to assume that the rate at which people leave any given room to move to another is directly proportional to the number of people in the room. The more

crowded the room becomes, the more likely people are to leave it. That is,

$$dA_i^{out}/dt = \sum f_{ij} = \sum k_{ij}A_i = A_i\sum k_{ij}, \tag{1.3}$$

in which the quantities k_{ij} are simple *rate constants* expressing migrations from room i to room j and the summations are over all values of j. Thus, the rate at which people leave the living room (dA_1^{out}/dt) is equal to the number of people in the room (A_1) times the sum of k_{12}, k_{13}, k_{14}, and k_{15}, the rate constants that govern movement out of the living room. The k_{ij} have units of inverse time. This "law" assumes *first-order* kinetics, which is justifiable in many geochemical problems as well. The rate of removal of magnesium from seawater at midocean ridges appears to be directly proportional to the Mg^{2+} concentration of seawater, for example. Similarly, the rate of formation of halite evaporates is probably also a first-order function of the concentration of NaCl in the oceans.

By introducing rate laws such as the first-order expression applied here, we recognize that fluxes, in general, are not constant in systems away from their steady state. If we assume a first-order law, $f_{ij} = k_{ij}A_i$ in the party system, this means that we could change the fluxes of people from one room to another either by changing the population in one or more of the rooms or by varying the rate constants k_{ij}. For example, without abandoning the simple first-order law or changing the rate constants, we could instantly move several people to the dining room as a way of modeling the effect of a one-time impulse, such as bringing out more food. Equation 1.3 predicts that overcrowding would make the fluxes out of the dining room higher for a while, and thus that the party would begin relaxing again to its steady state. If we chose instead to model the gradual effects as people get full and mellow, however, we would need to replace the rate constants with time-variant functions somehow linked to the "thermodynamic" fullness function we described earlier. This would make the set of kinetic equations 1.2 much trickier to solve, although probably more realistic.

It is not appropriate to assume that all geologic processes follow simple rate laws. Many do not. Complicated rate laws may be appropriate for systems that exhibit seasonal variability or in which the rate of removal of material from reservoir i is a thermodynamic function of A_i. Complex processes by which ions detach from the surfaces of crystals, for example, may cause rates of solubility to be highly nonlinear functions of undersaturation. In most cases, determination of appropriate rate laws and constants is a difficult business at best. For this reason, you will find that many potentially interesting kinetic problems have been discussed in schematic terms by geochemists, but solved only in a qualitative or greatly simplified form.

NOTES ON PROBLEM SOLVING

In this chapter, we have outlined the range of problems that interest geochemists and introduced in broad terms the thermodynamic and kinetic approaches to them. In the remaining chapters, we look in detail at selected geochemical topics and develop specific techniques for analyzing them. As you progress, reflect on this introduction. In it, we have tried to emphasize basic principles that you should always bear in mind as you study a geochemical question.

Remember that predictions made from thermodynamic or kinetic calculations cannot be any better than our description of the system allows them to be. In the frivolous example we have been using, we cannot expect thermodynamics to comment on events outside of the house, or on the state of hunger being felt by guests lurking in undescribed closets. We also cannot comment on the relative efficiency of kinetic pathways we have failed to include in the model, such as extra doors between rooms. As we have just shown, this leads us to make strategic compromises, simplifying models to make them solvable while keeping them as true to nature as we can.

This limitation may seem obvious, but it is surprisingly easy to overlook. If we are examining a system consisting of metamorphosed shales, for example, and forget to include in our description any chemical reactions that produce biotite, there is no way that we can expect thermodynamics to predict its existence. This omission has a serious effect. For a mineral such as biotite, which may be a volumetrically significant phase in some metamorphosed shales, this error in description will render meaningless any thermodynamic analysis of the system as a whole. If we decide to omit a minor phase like apatite, however, the effect on thermodynamic calculations should be minimal. Thermodynamics will not be able to comment on its stability, but our overall description of the system should not be seriously in error. Between these two extremes, it may not be easy to decide how much

simplification is too much. Is it important to include ilmenite and pyrite? The answer may not be clear, but you should always ask the question.

It is just as easy to forget that this limitation applies to kinetic studies as well. Because geochemical reactions commonly proceed along several parallel pathways, a valid kinetic treatment depends critically on how well we have identified and characterized those pathways. Most systems of even moderate complexity, however, include pathways that are easy to underestimate or to overlook completely. In chapter 8, for example, we consider the effect of hydrothermal circulation at midoceanic ridge crests on the mass balance of calcium and magnesium in the oceans. Until the mid-1970s, few data suggested that this was a significant pathway in the geochemical cycle for magnesium. Kinetic models, therefore, regularly overestimated the importance of other pathways, such as glauconite formation in marine sediments.

It is clear, then, that the credibility of any statements we make regarding the appearance of phases, the stability of a system, or rates of change depends in large part on how carefully we have described the system *before* applying the tools of geochemistry. This may be the single greatest source of difficulty you will experience in applying geochemical concepts to real-world problems.

Remember also that there are few concepts that are the exclusive property of igneous or sedimentary petrologists, of oceanographers or soil chemists, of economic geologists or planetary geologists. The breadth of geochemistry can be overwhelming. However, breadth can also be advantageous when we try to solve practical problems. Familiarity with the data and methodology used in one area of geochemistry will often help you see alternative solutions to a problem in another area. Many of the fundamental quantities of thermodynamics and kinetics should gradually become familiar as they appear in different contexts. Where possible, we try to point out areas of obvious overlap. Do not hesitate to do as we have done in this first chapter, though, and step completely out of the realm of geochemistry to play with an idea. A simple and apparently unrelated thought problem, such as our party example, can sometimes suggest a new way to approach a geochemical question.

Finally, we encourage you to develop the simplest geochemical models that will adequately answer questions posed about the Earth. All too often, problems appear impossible to solve because we have made them unnecessarily complex and treat them as abstract mathematical exercises rather than as investigations of the real Earth. In most cases, a careful selection of variables and a set of physically justifiable simplifying assumptions can make a problem tractable without sacrificing much accuracy. What you already know as a geologist about the Earth's properties and behavior should always serve as a guide in making such choices.

Having said this, we also emphasize that mathematics is an essential tool for the quantitative work of geochemistry. Now is a good time to familiarize yourself with the contents of the appendices at the end of this book, before we begin our first excursion into thermodynamics. The first of these appendices is an abbreviated overview of mathematical concepts, emphasizing methods that may not have been part of your first year of college-level calculus. Among these concepts are principles of partial differentiation and an introduction to numerical methods for fitting functions to data and finding roots. Refer to appendix A as often as necessary to make the mathematical portions of this book more palatable.

Some introductory texts include an appendix with preferred data needed in problem solving. We have decided not to provide data tables, except when we need an abbreviated table in a chapter. Instead, appendix B is a reference list of standard sources for geochemical data with which you should become familiar. Yes, this is less convenient than a compact data table would have been. We hope, though, that you will quickly become accustomed to finding and evaluating data for yourself, and that our list will be more useful in the long run than a table might have been.

As a slight concession in the opposite direction, we have listed the values of selected fundamental constants and conversion factors in appendix C. The literature of geochemistry, particularly articles written prior to the 1990s, uses a rich mixture of measurement units. We have made no attempt to favor one system over another in this book. Our practical opinion is that because data are reported in several conventional forms, you should be conversant with them all and should know how to shift readily from one set of units to another.

SUGGESTED READINGS

These first references illustrate the breadth of geochemistry. They also provide a wealth of data that may be most useful to professionals.

Fairbridge, R. W., ed. 1972. *The Encyclopedia of Geochemistry and Environmental Sciences.* New York: Van Nostrand Reinhold. (A frequently referenced source book for geochemical data.)

Goldschmidt, V. M. 1954. *Geochemistry.* Oxford: Clarendon Press. (This book is now dated, but remains a classic.)

Mason, B., and C. Moore. 1982. *Principles of Geochemistry.* New York: Wiley. (Probably the most readable survey of the field of geochemistry now available.)

Wedepohl, K. H., ed. 1969–1978. *Handbook of Geochemistry.* Berlin: Springer-Verlag. (A five-volume reference set with ring-bound pages that are designed for convenient updating. A definitive source book on the occurrence and natural chemistry of the elements.)

PROBLEMS

(1.1) You are already accustomed to describing geologic materials and environments in terms of measurable variables such as temperature or mass. Make a list of 10 such variables and indicate which ones are intensive and which extensive.

(1.2) Consider the mass balance and the pathways for change in a residual soil profile.
- (a) Sketch a diagram similar to figure 1.1a, and indicate the way in which major species, such as Fe, Ca, or organic matter, may be redistributed with time.
- (b) Write a complete set of schematic flux equations to describe mass balance in your system, following the example of worked problem 1.1.

(1.3) It has been said that diagenetic reactions generally occur in an open system, whereas metamorphic reactions more commonly take place in a closed system. Explain what this geochemical difference means and suggest a reason for it.

(1.4) Consider the party problem again. What might you have to do to include each of the following extra factors in the problem?
(a) You run out of food while guests are still hungry.
(b) Guests are free to enter and leave the party.
(c) The bathroom door is locked.
(d) Half of the guests try to avoid being in the same room as the other half, but everyone wants to spend some time in the kitchen.

The following problems for math review test your skills with concepts we will be using from time to time.

(1.5) Write the derivative with respect to t for each of the following expressions:
(a) $\ln (t)^3$
(b) $\exp (2\pi a/\alpha t)$ (π, a, and α are constants)
(c) $\sin (\alpha t^2)$ (α is a constant)
(d) $\log_{10} (\alpha/t)$ (α is a constant)
(e) $t \ln t + (1 - t) \ln (1 - t)$
(f) $\Sigma \exp (-E_i/Rt)$ (R and all E_i are constants)
(g) $\int_0^{\alpha t} \sin (t - z)dz$ (α and z are constants)

(1.6) Write down the total differential, df, for each of the following expressions:
(a) $f(x, y, z) = x^2 + y^2 + z^2$
(b) $f(x, y, z) = \sin (xyz)$
(c) $f(x, y, z) = x^3y - xy^2$

(1.7) Determine which of the following expressions are exact differentials:
(a) $df = (y + z)dx + (z + y)dy + (x + y)dz$
(b) $df = 2xy\, dx + 2yz^2\, dy - (1 - x^2 - 2y^2z)dz$
(c) $df = z\, dx + x\, dy + y\, dz$
(d) $df = (x^3 + 3x^2y^2z)dx + (y + 2x^3yz)dy + (x^3y^2)dz$

(1.8) Consider the two differentials, $df = y\, dx - x\, dy$ and $df = y\, dx + x\, dy$. One of them is exact; the other is not. Evaluate the line integral of each over the following paths described by Cartesian coordinates (x, y):
(a) A straight line from $(1, 1)$ to $(2, 2)$
(b) A straight line from $(1, 1)$ to $(2, 1)$, then another from $(2, 1)$ to $(2, 2)$
(c) A closed loop consisting of straight lines from $(1, 1)$ to $(2, 1)$ to $(2, 2)$ and then back to $(1, 1)$

(1.9) Change units on the following quantities as directed:
(a) Convert 1.5×10^3 bars to GPa
(b) Convert 3.5 g cm^{-3} to g bar cal^{-1}
(c) Convert 1.987 cal deg^{-1} to kJ deg^{-1}
(d) Convert 5.0×10^{-13} cm^2 sec^{-1} to m$^2(10^6$ yr$)^{-1}$

23 V	24 Cr	25 Mn
41 Nb	42 Mo	43 Tc
73 Ta	74 W	75 Re

CHAPTER TWO

HOW ELEMENTS BEHAVE

OVERVIEW

Our goal in this chapter is to refresh concepts that you may have first encountered in a chemistry course and to put them in a geologic context, where they will be useful for the chapters to come. We begin with a review of atomic structure and the periodic properties of the elements, focusing carefully on the electronic structure of the atom. This leads us to consider the nature of chemical bonds and the effects of bond type on the properties of compounds. These should be familiar topics, but they may look different through the lens of geochemistry.

ELEMENTS, ATOMS, AND THE STRUCTURE OF MATTER

Elements and the Periodic Table

Long before the seventeenth century, alchemists determined that most pure substances can be broken down, sometimes with difficulty, into simpler substances. They reasoned, therefore, that all matter must ultimately be made of a few kinds of material, commonly called *elements*. Following the lead of Aristotle, they identified these as *earth, air, fire,* and *water*. Without stretching too far, we can recognize a parallel to what chemists today call the three *states* of matter (solid, liquid, and gas) and

energy. These are useful distinctions to make, but unfortunately, medieval alchemists had difficulty explaining how the Aristotelian elements could combine to form everyday materials. Two key concepts were missing: the idea that matter is composed of individual particles (*atoms*) and the notion that *compounds* are made of atoms held together electrostatically.

Philosophers claim that the atomic theory had its roots in ancient Greece. The sage Democritus declared that the motions we observe in the natural world can only be possible if matter consists of an infinite number of infinitesimal particles separated by void. This was not a scientific hypothesis, however, but a proposition for debate. Democritus had neither the means nor the inclination to test his idea. Still, for more than 2000 years, scholars argued about whether matter was infinitely divisible or made of tiny, discrete particles. The issue was resolved only when scientists finally recognized the difference between chemical compounds, on the one hand, and *alloys* or *mixtures* on the other.

Through the seventeenth and eighteenth centuries, scientists gradually convinced themselves that the Aristotelian *earth,* in particular, was not a fundamental substance. Analytical techniques improved and empirical rules began to emerge. Robert Boyle, for example, declared that a substance cannot be an element if it loses weight during a chemical change, thus clarifying the distinction between pure elements and compounds. By the 1790s, systematic purification had yielded roughly a third of the elements known today. Most of these are metals: copper, gold, silver, lead, zinc, iron, magnesium, mercury, and tungsten, to name a few.

Geologists might argue, however, that the greatest accomplishment of that era was the isolation of non-metallic elements. In particular, the discovery of oxygen in the 1780s (attributed independently and with great controversy to Joseph Priestley, Karl Scheele, and Antoine Lavoisier) clarified the mysterious relationship between metallic elements and the clearly nonmetallic minerals that constitute most of the Earth's crust. The most common minerals were revealed as complex oxides of metallic elements. This realization, coming in the midst of an explosion of national investments in mining and metallurgy in Europe, was perhaps the first major step toward modern geochemistry. It was also a marked change in direction for science at large. Lavoisier, Joseph Louis Proust, Jeremias Benjamin Richter, Joseph-Louis Gay-Lussac, and others established by experimentation that oxygen and other gaseous elements combine in fixed proportions to form compounds. Water, for example, is always 11.2% hydrogen and 88.8% oxygen by weight.

Fixed proportions suggest a systematic arrangement of discrete particles. In 1805, John Dalton proposed four postulates:

1. Atoms of elements are the basic particles of matter. They are indivisible and cannot be destroyed.
2. Atoms of a given element are identical, having the same weight and the same chemical properties.
3. Atoms of different elements combine with one another in simple whole-number ratios to form *molecules* of compounds.
4. Atoms of different elements may combine in more than one small whole-number ratio to form more than one compound.

These postulates have now been revised, particularly in light of the discovery of radioactivity in the late 1800s. We now realize, for example, that energy and matter are interchangeable, and that atoms can be "smashed" into smaller particles. Still, for most chemical purposes, the postulates have proved correct.

Once a rational basis for the atomic theory had been established, the number of newly isolated elements began to grow. Chemists began recognizing relationships among the elements. Johann Dobereiner, in 1829, noted several triads of elements with similar chemical properties. In each triad, the atomic mass of one element was nearly equal to the average of the atomic masses of the other two. Chlorine, bromine, and iodine, for example, are all corrosive, colored, diatomic gases; the atomic mass of bromine is halfway between the masses of chlorine and iodine. Over the next 40 years, scientists documented other patterns when they arranged the elements in order of increasing atomic mass. Finally, in 1869, Russian chemist Dmitri Mendeleev made a careful reevaluation of what was known about the elements, corrected some erroneous values, and proposed the system known today as the *periodic table* (fig. 2.1).

We return several times in this chapter to examine the periodic properties of elements and to offer explanations for their behavior in the Earth. For a first look, consider figure 2.2, in which we compare the melting and boiling points of elements in the first three main rows of the periodic table (omitting for now the elements scandium through zinc: the first *transition series*). The trend toward increasing values near the center of each row is unmistakable. Mendeleev was able to predict the properties of germanium, as yet undiscovered in the 1860s, by using trends like these.

Although the periodic table was developed empirically, its structure suggests that atoms of each element are built according to regular architectural rules, thereby offering some hope for figuring out what those rules are. By the end of the nineteenth century, chemists understood that atoms are made of three fundamental types of particles. The heaviest of these, *protons* and *neutrons*, are tightly packed into the *nucleus* of the atom, where they constitute very little of the atom's volume but virtually all of its mass. The lightest particles, *electrons*, orbit the nucleus in an "electron cloud" that is largely empty space. Protons and electrons are opposing halves of an electrical system that controls most of the atom's chemical properties. By convention, each proton is said to have a unit of positive charge, and each electron has an equal unit of negative charge.

The Atomic Nucleus and Isotopes

Three numbers, describing the abundance and type of nuclear particles in an atom, are in common use. The first of these, symbolized by the letter Z, counts the number of protons and is known as the *atomic number*. This number identifies which element the atom represents

IA																	8A
1 H	2A											3A	4A	5A	6A	7A	2 He
3 Li	4 Be											5 B	6 C	7 N	8 O	9 F	10 Ne
11 Na	12 Mg	3B	4B	5B	6B	7B		8B		1B	2B	13 Al	14 Si	15 P	16 S	17 Cl	18 Ar
19 K	20 Ca	21 Sc	22 Ti	23 V	24 Cr	25 Mn	26 Fe	27 Co	28 Ni	29 Cu	30 Zn	31 Ga	32 Ge	33 As	34 Se	35 Br	36 Kr
37 Rb	38 Sr	39 Y	40 Zr	41 Nb	42 Mo	43 Tc	44 Ru	45 Rh	46 Pd	47 Ag	48 Cd	49 In	50 Sn	51 Sb	52 Te	53 I	54 Xe
55 Cs	56 Ba	57 La*	72 Hf	73 Ta	74 W	75 Re	76 Os	77 Ir	78 Pt	79 Au	80 Hg	81 Tl	82 Pb	83 Bi	84 Po	85 At	86 Rn
87 Fr	88 Ra	89 Ac+	104 Db	105 Jl	106 Rf	[107] Bh	[108] Hn	[109] Mt									

*Lanthanide series		58 Ce	59 Pr	60 Nd	61 Pm	62 Sm	63 Eu	64 Gd	65 Tb	66 Dy	67 Ho	68 Er	69 Tm	70 Yb	71 Lu
+Actinide series		90 Th	91 Pa	92 U	93 Np	94 Pu	95 Am	96 Cm	97 Bk	98 Cf	99 Es	100 Fm	101 Md	102 No	103 Lr

FIG. 2.1. The elements are described in this periodic table by their atomic number and symbol. The names dubnium (Db), joliotium (Jl), rutherfordium (Rf), bohrium (Bh), hahnium (Hn) and meitnerium (Mt) for elements 104–109 have been proposed but not yet formally approved.

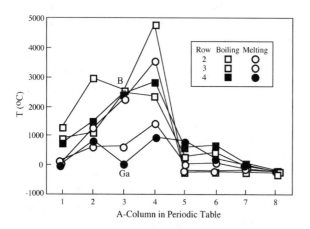

FIG. 2.2. Mendeleev suggested that the elements could be tabulated in a way that emphasizes trends in chemical properties. For example, with some exceptions (boron's boiling point and gallium's melting point, in column 3), melting and boiling points increase toward the middle of each row in the periodic table.

(any atom with $Z = 8$, for example, is an atom of oxygen, one with $Z = 16$ is sulfur, and so forth). Because an electrically neutral atom must have equal numbers of protons and electrons, Z also tells us the number of electrons in an uncharged atom. The second useful number is the neutron number, symbolized by the letter N. Within limits imposed by the attractive and repulsive forces among nuclear particles, atoms of any element may have different numbers of neutrons. Atoms for which Z is the same but N is different are called *isotopes*.

The numbers Z and N are sufficient to describe atoms, but chemists find it convenient to use a third quantity, A, equal to the sum of Z and N. Because protons and neutrons differ in mass by only a factor of one part in 1836, we assign each the same arbitrary unit mass. The quantity A, therefore, represents the total mass of the nucleus. Because electrons are so light, A is very nearly the mass of the entire atom. From a practical perspective, it is easier to measure the total mass of an atom than to count its neutrons, so A is a more useful number than N. In shorthand form, we convey our knowledge about a

WHAT'S IN A NAME?

A beginning pianist often finds that the greatest obstacle to progress is musical notation. Dotted quarter notes, the bass clef, and all the strange comments in Italian are literally a foreign language that must be mastered by any musician. Beginning chemists often have a similar problem with the periodic table, their musical staff. For each element, there are atomic masses, oxidation states, ionic radii, and a host of other things to remember. The easiest, perhaps, is the element's symbol, but even there a student chemist may be dazzled by strange conventions.

The classical elements, most of which are metals, were named so long ago that there is no record of where their names came from. The words *iron, lead, gold,* and *tin,* for example, are obscure Germanic names. They were good enough for common folk, but too plebian for alchemists, who preferred the more "scientific" sound of the Latin names *ferrum, plumbum, aurium,* and *stannum.* From these we get the symbols Fe, Pb, Au, and Sn. Many modern names have a Latin or Greek flavor as well, particularly those that were given as the periodic table went through a season of explosive growth during the early nineteenth century. Carl Mosander, for example, was so annoyed by

the difficulty of isolating the element lanthanum that he gave it a name derived from the Greek verb *lanthanein* ("to escape notice"). French chemist Lecoq de Boisbaudran had a similar experience with dysprosium, for which he crafted a name from the Greek word *dysprositos* ("hard to get at").

A geochemist may find it amusing to track down elements named for mining districts. In its first appearance, for example, copper was called *aes Cyprium*—literally, "metal from Cyprus." Magnesium was named for the Magnesia district in Thessaly. The grand prize, however, has to go to a black mineral collected from a granite pegmatite near Stockholm in 1794 by the Finnish mineralogist Johan Gadolin. He named it yttria, after the nearby village of Ytterby. In 1843, Mosander isolated three new elements from yttria, calling them yttrium, erbium, and terbium. Almost forty years later, Swiss chemist Jean Charles de Marignac isolated yet another element, ytterbium, from the same ore. To complete the package, Marignac isolated one more element in 1880 and named it gadolinium for the original discoverer of the mineral (which is now known as gadolinite). Quite a yield from one little mining district!

---

particular atom (or *nuclide,* as it is called when the focus is on nuclear properties) by writing its chemical symbol, a familiar alternative for Z, preceded by a superscript that specifies A. For a nuclide consisting of 8 protons and 8 neutrons, for example, we write ^{16}O. For the one with 26 protons and 30 neutrons, we write ^{56}Fe.

The existence of isotopes was not suspected during the nineteenth century, but it accounts for some of the confusion among chemists who were trying to make sense of periodic properties. Chlorine, for example, has two common isotopes: ^{35}Cl and ^{37}Cl. Because ^{35}Cl is roughly three times as abundant as ^{37}Cl, the weighted average mass of the two in a natural chlorine sample is close to 35.5. A natural sample of *any* element, in fact, will yield a nonintegral atomic mass for this reason. (Atomic mass is measured in *atomic mass units* [amu], equal to 1/12 of the mass of a ^{12}C atom, or 1.6605×10^{-24} g.)

Worked Problem 2.1

A geochemist analyzes a very large number of minerals containing lead and finds four isotopes in the following relative abundances:

^{204}Pb (mass = 203.973 amu) 1.4%

^{206}Pb (mass = 205.974 amu) 24.1%

^{207}Pb (mass = 206.976 amu) 22.1%

^{208}Pb (mass = 207.977 amu) 52.4%

Except for ^{12}C, the masses of nuclides are never integral. Notice, for example, that the mass of ^{204}Pb is slightly less than 204 amu. This is because a small fraction of the mass of any atom is actually in the binding energy that holds its nucleus together. (Recall Albert Einstein's famous equation, $e = mc^2$.) Given the nuclide masses in parentheses, then, what is the atomic mass of lead in nature, as implied by these analyses?

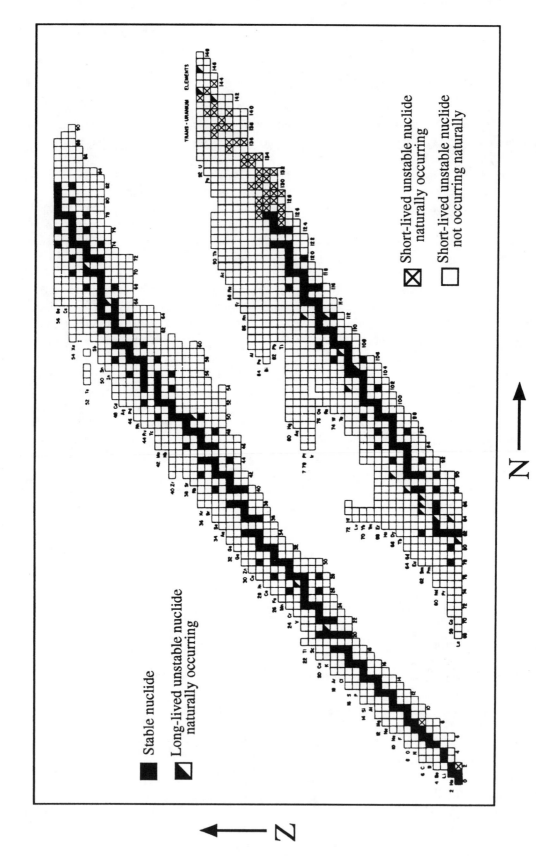

FIG. 2.3. The stable nuclides, shown in black, define a narrow band within a wider band of unstable nuclides in this plot of protons (*Z*) versus neutrons (*N*). In general, elements with *Z* or *N* even have more stable nuclides than those with *Z* or *N* odd.

To calculate, multiply the mass of each nuclide by its proportion in the sample and sum the results:

$$203.973 \times 0.014 + 205.974 \times 0.241 + 206.976 \times 0.221 + 207.977 \times 0.524 = 207.217 \text{ amu.}$$

TABLE 2.1. Abundance of Stable Nuclides

	N (even)	N (odd)	Total
Z (even)	159	53	212
Z (odd)	50	4	54

Roughly 1700 nuclides are known. Of these, only ~260 are stable. The rest have nuclei that can disintegrate spontaneously to produce subatomic particles and leave a nuclide in which Z or N has changed. Figure 2.3, a plot of the number of neutrons versus protons in each of the known nuclides, illustrates this point. Stable nuclides only occur within a thin diagonal band, in which N is slightly greater than Z, flanked on both sides by unstable *radioactive* nuclides. Careful inspection of a segment of this band (fig. 2.4) shows that, in most cases, the stable nuclides have an even number of neutrons and protons. Stable nuclides with either Z or N as odd numbers are much less common, and isotopes with both Z and N odd are generally unstable. This distribution is shown numerically in table 2.1.

Most of the known radioactive isotopes do not occur in nature, although all have been produced artificially in nuclear reactors. Some of the "missing" isotopes may have occurred naturally in the distant past, but have such high decay rates that they have long since become extinct. The radioactive isotopes of greatest interest to geochemists generally decay very slowly or are replenished continually by natural nuclear reactions.

Geochemists study both stable and radioactive isotopes in the Earth. In chapter 13, we look carefully at ways in which stable isotopes are fractionated in nature and how we can measure their relative abundances to figure out which processes may have affected geologic samples. Stable isotopes can also be used as tracers to infer chemical pathways in the Earth or to follow the progress of reactions. Radioactive isotopes, the topic of chapter 14, can also be used as tracers. Their primary value, however, is in dating geologic samples. Radiometric dating methods, suggested by Ernest Rutherford in the nineteenth century and put in useful form by Bertram Boltwood and Arthur Holmes early in the twentieth, are now the basis for all "absolute" ages in geology.

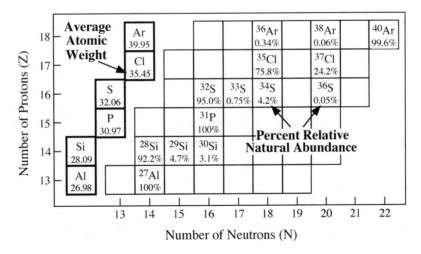

FIG. 2.4. Expanded segment of the nuclide chart of figure 2.3, showing stable nuclides and their percentages of relative natural abundances. The average atomic weight for each element (shown in the boxes to the left of the figure) is the sum of the weights of the various isotopes times their respective relative abundances.

The Basis for Chemical Bonds: The Electron Cloud

An atom is electrically independent of its neighbors if the number of electrons around it is the same as the number of protons in its nucleus. If not, then an excess positive charge on one atom must be balanced by negative charges elsewhere. This is the fundamental basis for *bonding,* the force that holds atoms together to form compounds.

Why should an atom ever have "too many" or "too few" electrons? Even as chemists at the start of the twentieth century gained confidence in the predictive capacity of the periodic table, *stoichiometry* (the rules governing how many atoms of each element belong in a compound) remained a stubborn mystery. It was easy to see, for example, that the composition of hydrides (and other simple compounds) follows a simple pattern across each row of the periodic table. Lithium combines with one hydrogen atom to form LiH, beryllium forms BeH_2, boron makes BH_3, and carbon CH_4. From there to the end of the row, the combining ratio drops: NH_3, H_2O, and HF. There is no neon hydride. In the next row, we see NaH, MgH_2, AlH_3, SiH_4, PH_3, H_2S, and HCl. There is no argon hydride. As the transition elements are added in successive rows, the pattern becomes less obvious, but is still there. Why? In fact, why are *any* of the periodic properties periodic?

The answer lies in the distribution of electrons around the nucleus. Chemists began to glimpse the structure of the "cloud" of orbiting electrons in 1913, when Niels Bohr first successfully attempted to explain why atomic spectra consist of discrete lines instead of a continuous blur (fig. 2.5). Building on theoretical advances by Max Planck and Albert Einstein in the previous decade, he made the novel assumption that the angular momentum of an orbiting electron can only have certain fixed values. The orbital energy associated with any electron, similarly, cannot vary continuously but takes only discrete *quantum* values. As a result, only specific orbits are possible. In mathematical terms, Bohr postulated that the angular momentum of the electron ($m_e v r$) must be equal to $nh/2\pi$, where m_e is the mass of the electron, v is its linear velocity, r is its distance from the nucleus, h is Planck's constant, and n takes positive integral values from 1 to infinity. Because n can only increase in integral steps, r defines a set of spherical shells at fixed radial distances from the nucleus.

Clever though this approach was, chemical physicists soon recognized that an electron cannot be described correctly as if it were a tiny planet orbiting a nuclear star. A better model uses the language of wave equations developed to describe electromagnetic energy. In 1926, Erwin Schrödinger proposed such a model incorporating not only the single quantum number as Bohr had suggested, but three others as well. Schrödinger's equations describe geometrical charge distributions (atomic orbitals) that are unique for each combination of quantum numbers. The electrical charge that corresponds to each electron is not ascribed to an orbiting particle, as in Bohr's model. Instead, you can think of each electron as being spread over the entire orbital—a complex three-dimensional figure.

The *principal quantum number, n,* taking integral values from 1 to infinity, describes the effective volume of an orbital—commonly calculated as the volume in which there is a 95% chance of finding the electron at any instant. This is similar to Bohr's definition of *n*. Chemists

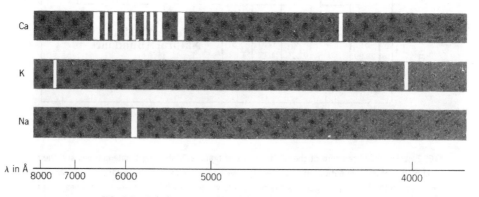

FIG. 2.5. Emission spectra for calcium, potassium, and sodium.

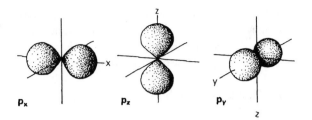

FIG. 2.6. The three p orbitals in any shell have figure-eight shapes, oriented along Cartesian x, y, or z axes centered on the atomic nucleus.

commonly use the word *shell* to refer to all orbitals with the same value of n, because each increasing value of n defines a layer of electron density that is farther from the nucleus than the last n.

The *orbital angular momentum quantum number, l,* determines the shape of the region occupied by an electron. Depending on the value of n, a particular electron can have values of l that range from zero to $n - 1$. If $l = 0$, the orbital described by Schrödinger's equations is spherical and is given the shorthand symbol s. If $l = 1$, the orbital looks like one of those in figure 2.6 and is called a p orbital. Orbitals with $l = 2$ (d orbitals) are shown in figure 2.7. The $l = 3$ orbitals (f orbitals) have shapes that are too complicated to illustrate easily here.

The third quantum number, m_l, describes the orientation of the electron orbital relative to an arbitrary direction. Because an external magnetic field (such as might be induced by a neighboring atom) provides a convenient reference direction, m_l is usually called the *magnetic orbital quantum number.* It can take any integral value from $-l$ to l.

The fourth quantum number, m_s, does not describe an orbital itself, but imagines the electron as a particle within the orbital spinning around its own polar axis, much like the Earth does. In doing so, it becomes a tiny magnet with a north and a south pole. The *magnetic spin quantum number,* m_s, can be either positive or negative, depending on whether the electron's magnetic north pole points up or down relative to an outside magnetic reference.

Each orbital therefore can contain no more than two electrons, with opposite spin quantum numbers. This rule, which affects the order in which electrons may fill orbitals, is known as the *Pauli exclusion principle.* The simplest atom, hydrogen, has one electron in the orbital closest to the nucleus ($n = 1$). With each increase in atomic number, atoms gain an electron in the unfilled orbital with the lowest energy level. We code electron orbitals with a label composed of the numerical value of n and a letter corresponding to the value of l, so the lowest orbital is called the 1s orbital. It can hold up to two electrons, as shown in table 2.2. The atom with atomic number 3, lithium, has two electrons in the 1s orbital and a third in 2s, which has the next lowest energy level. It takes another 7 electrons to fill the $n = 2$ shell with s and p electrons, and 18 more to fill the $n = 3$ shell with s, p, and d electrons.

You can extend the table to calculate the number of s, p, d, and f orbitals in the $n = 4$ shell. The order in which orbitals beyond 3p fill, however, is not what you might expect from this exercise. As illustrated in figure 2.8, complex interactions among electrons and the nucleus reduce the energy associated with each orbital as atomic number increases. In neutral potassium and calcium atoms, the first to have >18 electrons, the 4s orbitals have lower energy levels than the 3d orbitals, so the nineteenth and twentieth electrons fill them instead. With increasing atomic number, electrons fill the 3d, 4p, 5s, 4d, and 5p orbitals. This order could only be anticipated by calculating the relative energy levels of

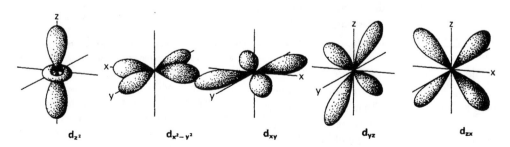

FIG. 2.7. Three of the d orbitals (d_{xy}, d_{yz}, and d_{zx}) in any shell have four lobes, oriented between the Cartesian axes. A fourth ($d_{x^2-y^2}$) also has four lobes, but along the x and y axes. The fifth (d_{z^2}) has lobes parallel to the z axis and a ring of charge density in the x-y plane.

TABLE 2.2. Configuration of Electrons in Orbitals

	n	l	m_l	m_s
1s orbital	1	0	0	+
	1	0	0	−
2s orbital	2	0	0	+
	2	0	0	−
2p orbitals	2	1	−1	+
	2	1	−1	−
	2	1	0	+
	2	1	0	−
	2	1	1	+
	2	1	1	−
3s orbital	3	0	0	+
	3	0	0	−
3p orbitals	3	1	−1	+
	3	1	−1	−
	3	1	0	+
	3	1	0	−
	3	1	1	+
	3	1	1	−
3d orbitals	3	2	−2	+
	3	2	−2	−
	3	2	−1	+
	3	2	−1	−
	3	2	0	+
	3	2	0	−
	3	2	1	+
	3	2	1	−
	3	2	2	+
	3	2	2	−

Another reason is that an atom's electronic configuration determines its affinity for atoms of other elements. By studying the periodic properties of the elements, we can begin to understand how they are distributed in the Earth.

Size, Charge, and Stability

The outermost electrons in an atom determine its effective size, or *atomic radius*, as well as much of its behavior in bonding. These electrons are sometimes called the *valence* electrons. Figure 2.9 illustrates two trends with increasing atomic number. First, as s and p electrons are added across a single row of the periodic table, atomic radius decreases. The simple explanation for this is that the positive charge of the nucleus continues to increase as atomic number increases. Electrons in the same shell cannot effectively shield each other from the pull of the nucleus, so valence electrons are drawn in as nuclear charge increases, shrinking atoms from left to right across a row. (Shielding effects among electrons in d and f or-

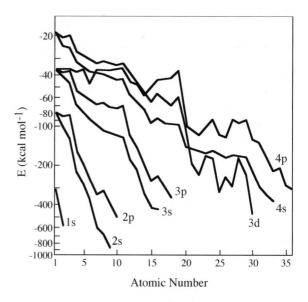

FIG. 2.8. As atomic number increases, electrons are increasingly shielded from the charge of the nucleus by the presence of other electrons. As a result, the energy (E) necessary to stabilize an electron in each orbital generally decreases with increasing atomic number. This rule is made more complex by interactions between the orbitals themselves. Energy levels in the 3d orbitals, for example, increase across much of row 3 in the periodic table. Potassium and calcium, which in their ground state have enough electrons to enter 3d orbitals, instead place them in 4s orbitals, which have a lower energy.

successive orbitals, as was done in producing figure 2.8. Once recognized, however, the order provides insight into the layout of the periodic table. If you compare table 2.2 with the periodic table, you will see an interesting pattern. Each row in the periodic table begins with a new s orbital (n increases by 1 and $l = 0$), and ends as the last p orbital for that value of n is filled. The s and p orbitals, in other words, define the basic outline of the periodic table. Atoms whose last-filled orbitals are d or f orbitals make up the *transition* elements that occupy the central block in rows 4 and beyond.

Why should a geochemist care about this level of detail? One reason is that electronic configuration offers a way to interpret bonding and, with it, the driving force for chemical reactions. Site occupancies in minerals and mineral structures themselves are best understood when we predict the shapes and orientations of electron orbitals. We explore this general topic later in this chapter.

Radius Increases ↓

Radius Decreases →

0.37 H																	– He
1.23 Li	0.89 Be											0.80 B	0.77 C	0.74 N	0.74 O	0.72 F	– Ne
1.57 Na	1.36 Mg											1.25 Al	1.17 Si	1.10 P	1.04 S	0.99 Cl	– Ar
2.03 K	1.74 Ca	1.44 Sc	1.32 Ti	1.22 V	1.17 Cr	1.17 Mn	1.17 Fe	1.16 Co	1.15 Ni	1.17 Cu	1.25 Zn	1.25 Ga	1.22 Ge	1.21 As	1.17 Se	1.14 Br	– Kr
2.16 Rb	1.91 Sr	1.62 Y	1.45 Zr	1.34 Nb	1.29 Mo	– Tc	1.24 Ru	1.25 Rh	1.28 Pd	1.34 Ag	1.41 Cd	1.50 In	1.41 Sn	1.41 Sb	1.37 Te	1.33 I	– Xe
2.35 Cs	1.98 Ba	1.69 La*	1.44 Hf	1.34 Ta	1.30 W	1.28 Re	1.26 Os	1.26 Ir	1.29 Pt	1.34 Au	1.44 Hg	1.55 Tl	1.54 Pb	1.52 Bi	1.53 Po	– At	– Rn
– Fr	– Ra	– Ac+	– Db	– Jl	– Rf	– Bh	– Hn	– Mt									

*Lanthanide series

1.65 Ce	1.65 Pr	1.64 Nd	– Pm	1.66 Sm	1.85 Eu	1.61 Gd	1.59 Tb	1.59 Dy	1.58 Ho	1.57 Er	1.56 Tm	1.70 Yb	1.56 Lu

+Actinide series

1.65 Th	– Pa	1.42 U	– Np	– Pu	– Am	– Cm	– Bk	– Cf	– Es	– Fm	– Md	– No	– Lr

FIG. 2.9. Atomic radii, indicated here in Ångstroms (Å), decrease across each row in the periodic table as nuclear charge increases. Each added shell of electrons, however, is shielded from the nucleus by inner electrons. Atomic radii therefore increase from one row to the next. Chemists generally determine the size of atoms by studying the dimensions of their coordination sites in crystal structures (discussed later in this chapter). It is difficult to measure atomic radii for elements that are rare in nature (the actinides, for example) or that do not bond with other elements to form crystals (the noble elements). For this reason, no radii are reported for those elements in this figure.

bitals are more complex, so this trend is less consistent across the transition elements.) Increasing atomic number beyond the end of a row, however, means adding electrons to a new shell, farther from the nucleus. As a result, elements in the next row in the periodic table have larger atomic radii. Within a single column of the periodic table, therefore, atomic radius increases with increasing atomic number. This second trend is also apparent in figure 2.9.

The measure of the energy needed to form a positively charged *ion* (a *cation*) by removing electrons from an atom is called *ionization potential* (IP), plotted in figure 2.10. When electrons are closest to the nucleus, they are hardest to remove with energy supplied from the outside. For this reason, it is easier to produce cations of atoms toward the bottom or the left side of the periodic table than from atoms toward the top or the rightmost columns.

IP is greatest in the right hand column of the periodic table, among the group of elements known as the *noble*

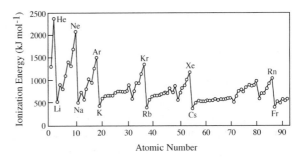

FIG. 2.10. The energy required to remove one electron from an atom increases across each row of the periodic table, but decreases slightly from one row to the next because successive shells of electrons are held less tightly by the positive charge of the nucleus. The quantity plotted here is the first ionization potential. A plot of the second or third ionization potential would be similar to this one, although removal of a second or third electron requires more energy.

gases (He, Ne, Ar, Kr, Xe, and Rn). In this group, all appropriate s and p orbitals are filled. This configuration, known as a complete outer shell, is particularly stable. Atoms on the left side of the periodic table form cations readily, losing enough electrons that their outer shell looks like the noble gas at the end of the previous row. In the second row, for example, lithium easily loses one electron to become Li^+, beryllium becomes Be^{2+}, and boron becomes B^{3+}, in each case ending up with an outer shell that looks like that of neutral helium.

The converse of ionization potential is *electron affinity* (EA), the amount of energy released if we succeed in adding a valence electron to a neutral atom to create a negatively charged *anion*. Like IP, EA also increases to the right in the periodic table. The highest affinities are among the halogens (F, Cl, Br, and I), which only need to gain a single electron to reach a noble gas configuration. These and the four elements to their immediate left (O, S, Se, and Te) are the only ones that commonly form anions. For all other elements, EA is very small or negative, indicating that they gain little or no stability by gaining electrons.

Ionization changes the size of an atom. If it loses electrons, the net positive excess charge in its nucleus draws the remaining electrons in tighter. The overall trend in cation radii, therefore, is the same as for atomic radii: they increase toward the bottom and decrease toward the right side of the periodic table. Gaining electrons has the opposite effect, so anions are larger than their neutral counterparts. Figure 2.11 illustrates how ionic radii vary with atomic number.

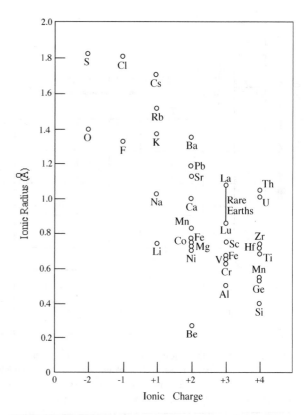

FIG. 2.11. As electrons are removed from an atom, the remaining electrons are drawn more tightly by the charge of the nucleus, so that ionic radius decreases with increasing positive charge. Compare, for example, the radii of Fe^{2+} and Fe^{3+} or Mn^{2+} and Mn^{4+}. Also, compare each of the ionic radii in this figure with the atomic radii in figure 2.9.

Elemental Associations

Trends among the periodic properties suggest logical ways to group elements. For example, it is customary to speak of all elements in the leftmost column of the periodic table as *alkali metals* and those in the second column as *alkaline earth metals*. The rightmost column we have already identified as the *noble gases;* the column to its left contains the *halogens*. In each of these groups, elements have the same configuration of valence electrons. The alkali metals (Li, Na, K, Rb, Cs, and Fr), for example, each have a lone electron in the outermost shell and therefore readily form cations with a +1 charge. Chemists also speak of the *transition metals,* a group of elements in the middle of rows 4, 5, and 6, across which the d electron orbitals are gradually filled. Similarly, the

lanthanides and *actinides* in rows 6 and 7 are groups across which the f orbitals are filled. Each of these groups, labeled in figure 2.12, contains elements that, by virtue of their common electron configuration, behave in similar ways during chemical reactions and therefore form similar compounds.

Appreciation for the periodic properties of the elements has cast light on many geochemical mysteries. One of the most fundamental observations that geologists make about the Earth, for example, is that it is a differentiated body with a core, mantle, and crust that are chemically distinct. Why? Is there a rational way to explain why such elements as potassium, calcium, and strontium are concentrated in the crust and not in the core? Or why copper is almost invariably found in sulfide ores rather than in oxides?

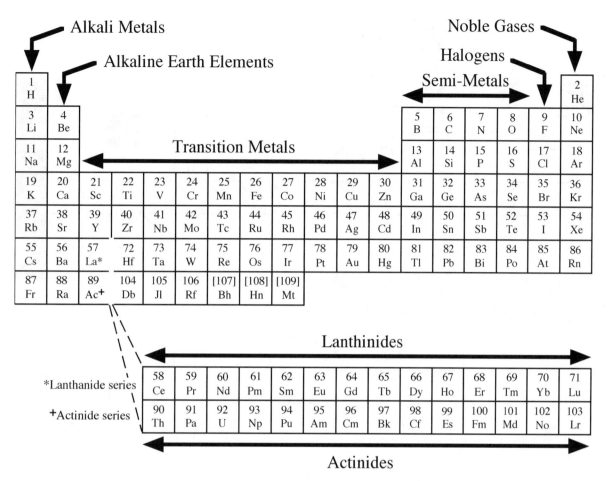

FIG. 2.12. Elements with similar properties fill similar roles in nature and are therefore commonly recognized as members of groups, as indicated in this figure. The lanthanides are often referred to as rare earth elements (REE).

Early in the twentieth century, Victor Goldschmidt was prompted by a study of differentiated meteorites to propose a practical scheme for grouping elements according to their mode of occurrence in nature. His analyses of three kinds of materials—silicates, sulfides, and metals—suggested that most elements have a greater affinity for one of these three materials than for the other two. Calcium, for example, can be isolated as a pure metal with great difficulty, and its rare sulfide, oldhamite (CaS), occurs in some meteorites. Calcium is most common, however, in silicate minerals. Gold, however, is almost invariably found as a native metal.

Goldschmidt's classification, summarized below, ultimately included a fourth type of material not found in abundance in meteorites:

1. *Lithophile* elements (from the Greek *lithos,* meaning "rock") are those that readily form compounds with oxygen. The most common oxygen-based minerals in the Earth are silicates, but lithophile elements also dominate in oxides and other "stony" minerals.
2. *Chalcophile* elements are those that most easily form sulfides. The Greek root for the name of this group is *chalcos,* meaning "copper," a prominent element in this category.
3. *Siderophile* elements prefer to form metallic alloys rather than stoichiometric compounds. The most abundant siderophile element is iron (in Greek, *sideros*), for which this group is named.
4. *Atmophile* elements are commonly found as gases, whether in the atmosphere or in the Earth.

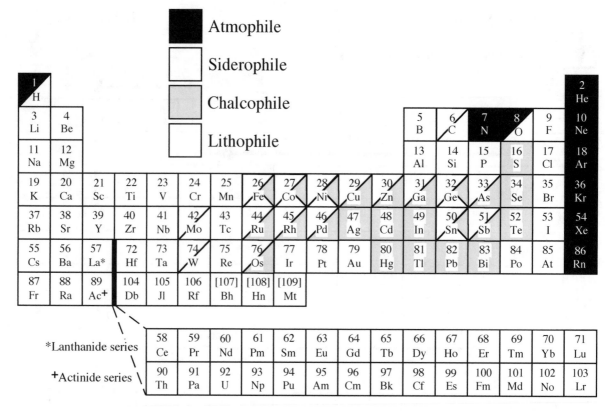

FIG. 2.13. Goldschmidt's classification of the elements emphasizes their affinities with three fundamental types of Earth materials (sulfides, silicates, or native metals) or with the atmosphere. Several elements—the most abundant being iron—have affinities with more than one group.

As figure 2.13 suggests, these four categories overlap quite a bit. Tin, for example, is commonly found as an oxide (the mineral cassiterite, SnO_2) and less often as a sulfide (stannite, Cu_2FeSnS_4). Carbon, as graphite or diamond, occurs as a pure element and it alloys readily with other metals, as in the manufacture of steel, but it is also common as a lithophile element in carbonates. Iron forms abundant silicate and sulfide minerals, but it is also the major constituent of the Earth's metallic core. Nevertheless, most elements belong primarily in one group, and Goldschmidt's classification is a guide to geochemical behavior.

Why does this geochemical classification work? Again, it is based loosely on the periodic properties of elements and, in particular, on those properties that control bonding between elements. We pursue this topic in the final sections of this chapter.

BONDING

Perspectives on Bonding

A *bond* exists between atoms or groups of atoms when the forces between them are sufficient to create an atomic association that is stable enough for a chemist to consider it an independent entity. This definition gives us a lot of room for subjective judgment about what "stable enough" and "independent" mean, and it covers a range of situations in which the types and strength of forces differ. In general, chemists find this flexibility to be an advantage. Students of geochemistry, however, can be led into uncertainty about the role of individual atoms in natural materials. In crystalline solids, for example, the "independent entity" is not a freestanding molecule but an extended periodic array of atoms in which bonds may range subtly both in strength and in character. One source of confusion is the apparently sharp distinction between *ionic* and *covalent* bonds that many students learn in basic chemistry courses. Briefly and in the simplest terms, you may have learned that an ionic bond is one in which one or more electrons is transferred from a cation to an anion; a covalent bond is one in which electrons are shared between adjacent atoms. In both cases, the goal is for each atom to end up with a noble gas configuration in its outer shell of electrons. In fact, the difference between ionic and covalent bonding is more a matter of degree than a clear distinction.

A common way to determine whether a bond is ionic or covalent is based on *electronegativity* (EN). EN is proportional to the sum of IP and EA. Because affinity is difficult to measure, however, EN is usually calculated directly from bond energies. Figure 2.14 summarizes EN values for many elements.

The greater the EN difference between atoms, the more likely it is that electrons will be transferred from one atom to the other, forming an ionic bond. Typically, a bond is considered ionic if the EN difference in it is >2.1. For example, when a chlorine atom (EN = 3.16) is adjacent to a sodium atom (EN = 0.93), it easily draws the lone electron from sodium to itself. This is apparent in the electronegativity difference (ΔEN) of 2.23 between the atoms. The Mg-O and Ca-O bonds that are common in rock-forming silicates are also ionic (ΔEN = 2.13 and 2.44, respectively). Si-O and Al-O bonds in the same minerals have ΔEN = 1.54 and 1.83, however, and are therefore covalent, by this definition.

Nobel Prize-winning chemist Linus Pauling suggested an alternative way to describe a bond by calculating its percentage of ionic character with the empirical formula:

$$p = 16|EN_a - EN_b| + 3.5|EN_a - EN_b|^2,$$

in which a and b are atomic species to be bonded. For Na-Cl, we find that $p = 54.7\%$ and for Ca-O, $p = 59.9\%$, whereas for Si-O, $p = 32.9\%$ and for Al-O, $p = 41.0\%$. If we look at bonding in this way, we see "ionic" and "covalent" as relative terms on a continuum of bond types. This offers a more realistic perspective on variations in bonding, but can present conceptual difficulties when we consider mineral structures. It is customary, for example, for geochemists to talk about the "ionic radius" of Mg^{2+}, as we have earlier in this chapter. If an Mg-O bond is described in Pauling's terms, however, $p = 49.9\%$. Roughly half of the electron charge transferred from the Mg^{2+} ion, therefore, is actually shared with the O^{2-} ion, so it is difficult to conceive of either ion as an independent entity with a clearly defined radius. The standard practice of referring to "ions" in mineral structures as if they were the same as free ions in aqueous solution is clearly misleading.

The percentage that Pauling's formula estimates can be thought of in another way: as the *polarity* of the bond. Geochemists who deal with gases or aqueous solutions find that a knowledge of polarity helps to explain solubility, reactivity, and other key properties. Those who deal with solid materials can use the same sort of information to interpret optical, thermal, and electric properties. Polarity can greatly affect the boiling point and conductivity

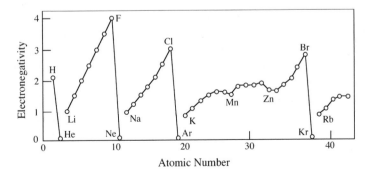

FIG. 2.14. Electronegativity generally increases with atomic number, within a single row of the periodic table. It may be thought of as a measure of the ease with which an atom can add electrons to complete its outer shell.

of a substance, for example, and the way it combines with other substances, as in H_2O (see the box on this topic, later in the chapter).

Structural Implications of Bonding

If bonds in a substance have a highly ionic character, it is convenient to view the atoms like marbles arranged according to size and electrical charge. Each ion must be surrounded by oppositely charged ions, so that electrical charge is balanced locally, but the number of surrounding ions depends on their relative sizes. This is the case because ionic bonding leads to the most stable structure when the ions are packed together as closely as possible. In the halite (NaCl) structure, for example, each sodium atom is at the center of a cluster of six chloride ions and each chloride ion is surrounded by six sodium ions. The positive charge on each sodium ion is distributed evenly over six Na-Cl bonds, and so is the negative charge on each chloride ion. But why six ions?

Figure 2.15a shows a perspective drawing of the halite structure, with cations and anions drawn as spheres. The distance between ions, which touch each other in the actual structure, has been exaggerated for ease of viewing. Figure 2.15b isolates a portion of the structure: a cation at the geometric center of an octahedral array of anions. In figure 2.15c, we have removed the top and bottom anions for clarity, and in figure 2.15d, we have rotated that view to be viewed directly from above. In this final perspective, we have restored interatomic distances to show the cation-anion contact.

If we were to decrease the ionic radius ratio r^+/r^- somehow, it should be apparent that the distance d between the anions would gradually shrink to zero, at which point they would be touching. At that point, the length of each side of the square connecting their centers would be $2r^+$. By the Pythagorean Theorem, it would also be equal to $\sqrt{2}(r^+ + r^-)$. Algebraic manipulation of the two expressions reveals that the value of r^+/r^- would have become $\sqrt{2} - 1 = 0.414$. We could not decrease r^+/r^- further without either allowing the anions to overlap or creating free space between the anions and the cation, so that they no longer touched. The first of these conditions is self-limiting, because the valence electrons of the anions repel each other, preventing further overlap. The second condition is far more likely. In fact, it is not uncommon for a small cation to rattle around in an oversized site sur-

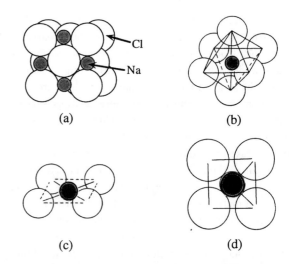

FIG. 2.15. (a) The structure of sodium chloride—the mineral halite—is based on a cubic face-centered lattice. Na^+ and Cl^- ions alternate along all three coordinate axes. As a result, each ion is surrounded by six ions of the other element. (b) A single Na^+ ion and the six Cl^- ions surrounding it, "exploded" to illustrate the octahedral arrangement of anions. (c) The top and bottom Cl^- ions have been removed for clarity, revealing a square planar array of Cl^- ions around the Na^+. (d) A view straight down on the planar arrangement.

rounded by anions that touch each other. In such cases, however, the cation-anion bonds are longer and thus weaker than they are when the anions are farther apart. This can be seen, for example, in the abnormally low melting points of lithium halides (table 2.3).

For ionic compounds with $r^+/r^- < 0.414$, fourfold (tetrahedral) coordination is more likely than the sixfold (octahedral) coordination of the NaCl structure, because the cation no longer rattles around in an oversized site. We have to be cautious about applying geometric arguments for such compounds, however, because cations tend to form bonds with a more covalent character when they are in fourfold coordination. Unlike purely ionic bonds, which are electrostatic and therefore nondirectional, covalent bonds are directional. As we show shortly, this adds further complexity to the study of structure.

What if we increase r^+/r^-, though, instead of decreasing it? If the central cation becomes large enough, it is eventually possible to create a more stable site by surrounding it with eight, rather than six anions. In prob-

TABLE 2.3. Melting Points of Some Alkali Halides

	Melting Point (°C)	r^+/r^-		Melting Point (°C)	r^+/r^-		Melting Point (°C)	r^+/r^-
CsBr	627	0.89	CsI	621	0.79	CsCl	638	0.96
RbBr	681	0.80	RbI	638	0.71	RbCl	717	0.86
KBr	748	0.70	KI	693	0.63	KCl	768	0.76
NaBr	740	0.52	NaI	653	0.46	NaCl	803	0.56
LiBr	535	0.39	LiI	450	0.35	LiCl	606	0.42

Ionic radii are from Shannon (1976).
Cesium ions are in eightfold, rather than sixfold, coordination in halide compounds.

lem 2.9, at the end of this chapter, you are challenged to show that this arrangement is preferred when $r^+/r^- > \sqrt{3} - 1 = 0.732$. As indicated in table 2.3, this is the case for each of the cesium halides.

We can anticipate the coordination number and geometric arrangement of anions around any ion in an ionic solid, then, by referring to the limiting ratios r^+/r^- shown in table 2.4. These coordination guidelines are only approximate, because ions are not like the rigid marbles we have assumed in this discussion. Instead, ions expand or shrink slightly to fit their environment. This makes it difficult to apply the guidelines when r^+/r^- is close to one of the limiting values, as is the case for RbBr and RbCl, which have the halite structure rather than the predicted cesium halide structure. Still, it is surprising how well simple radius ratios can help us anticipate a coordination number even for groups of ions with relatively little ionic character.

Another approach to understanding the geometric implications of bonding builds on *valence-bond theory*, an extension of the quantum model described earlier in this chapter. According to valence-bond theory, a chemical bond occurs when electron orbitals in adjacent atoms overlap, creating new valence electron orbitals that encompass both atoms. Only certain orbitals are allowed to overlap, however. The geometric arrangement of those orbitals determines how atoms cluster around each other.

To explore briefly how this theory works, let's look back at the quantum model for a single atom. Recall that for each combination of n, l, and m_l, there are two possible values of m_s, the magnetic spin quantum number. Two electrons with the same n, l, and m_l, but opposite values of m_s are said to be *paired*. Each shell, for example, can hold up to six electrons in the orbitals labeled p_x, p_y, and p_z in figure 2.6. Because the second electron in an orbital occupies a slightly higher energy level than the first, electrons find an advantage to being distributed evenly among the three orbitals. If an atom doesn't have enough electrons to fill all three orbitals completely, then it accepts only one electron each in p_x, p_y, and p_z before adding a second electron to any of them. Why is all of this important? Because the electrons in the unpaired orbitals are the ones that can overlap between atoms to form bonds with a covalent character.

A neutral oxygen atom, for example, has eight electrons. Two of them fill the s orbital in the first shell, another two enter the s orbital in the second shell, and the other four are in second shell p orbitals. Two are paired in $2p_x$, one is in $2p_y$, and one in $2p_z$. Oxygen is therefore divalent; covalent bonds can form at right angles, along the partially occupied p_y and p_z orbitals. Nitrogen, which has one fewer electron than oxygen, is trivalent; its 2p electrons are unpaired—one each in p_x, p_y, and p_z. In a molecule such as NH_3, mutually perpendicular covalent bonds form along each orbital lobe.

Valence-bond theory predicts molecular structures, then, by examining the number and arrangement of unpaired electrons in each atom. It also takes into account the energy level of each orbital and their orientations as they overlap. The union of new valence orbitals

TABLE 2.4. Coordination Relationships in Ionic Compounds

Radius Ratio r^+/r^-	Arrangement of Anions	Coordination Number
0.15–0.22	Corners of a triangle	3
0.22–0.41	Corners of a tetrahedron	4
0.41–0.73	Corners of an octahedron	6
0.73–1.0	Corners of a cube	8
1.0	Corners and edges of a cube	12

constitutes a new entity, a *bonding orbital,* that has a lower energy and is therefore more favored than the valence orbitals considered independently. In a molecule of H_2, for example, each atom contributes an unpaired 1s electron, and their orbitals overlap to form s-s bonding orbitals. The H_2 molecule is stable because the pair of bonding orbitals has a lower potential energy than the combination of two partially occupied valence orbitals. This will be a useful perspective to recall in chapter 3, where we introduce the concept of Gibbs free energy and explore the driving force for chemical reactions.

There are many ways to form bonding orbitals, and a discussion of them is beyond the scope of this chapter. The following example emphasizes the usefulness of valence-bond theory in geochemistry—perhaps enough to encourage you to make a more rigorous study of the topic.

Worked Problem 2.2

Silicon, a key element in rock-forming minerals, poses an interesting conceptual challenge in valence-bond theory. It has only two electrons in partially filled orbitals—silicon's configuration is $1s^2\ 2s^2\ 2p_x^2\ 2p_y^2\ 2p_z^2\ 3s^2\ 3p_x^1\ 3p_y^1$—so, by analogy with what we have just illustrated about oxygen and nitrogen, silicon should be divalent. In fact, however, it has a formal valence of +4, as if the atom had two more unpaired electrons. Why?

The solution to this paradox lies in the formation of *hybrid valence orbitals* (fig. 2.16). In compounds of silicon, an electron is "promoted" from an s orbital to a vacant p orbital, giving the outer shell the potential structure $3s^1\ 3p_x^1\ 3p_y^1\ 3p_z^1$. It takes energy to redistribute electrons in this way, however, and this energy is obtained in part by making each of the four unpaired electrons equivalent to the others. The new entities are sp^3 *hybrid* orbitals, oriented toward the four corners of a regular tetrahedron. As these form bonding orbitals to oxygen atoms, they become the basis for the SiO_4 tetrahedral unit that is common to all silicate minerals.

We might have tried to predict this tetrahedral arrangement of bonds by comparing ionic radii. The ionic radius of Si^{4+} is 0.26 Å; the radius of O^{2-} is 1.35 Å. A quick check of table 2.3 indicates that their ratio (= 0.19) is only slightly smaller than expected for fourfold coordination. Although this result is reassuring, it is somewhat misleading. Using r^+/r^- ignores the fact that the Si-O bond is highly covalent and that the Si^{4+} ion, therefore, cannot be considered as spherically symmetrical. Valence-bond theory is a more appropriate tool in this case.

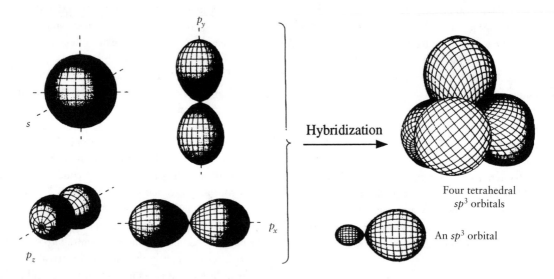

FIG. 2.16. Bonding electrons in a silicon atom are in $3p_x$ and $3p_y$ orbitals. There are no electrons in $3p_z$. The paired electrons in 3s are not available for bonding. If the s and p orbitals are combined, then each of the four electrons becomes a bonding electron in a hybrid sp^3 orbital. Each hybrid orbital has two lobes, one of which is larger than the other. The four large lobes point toward the corners of a tetrahedron.

So far, we have emphasized interactions between pairs of atoms. In extended structures, such as crystals and most organic molecules, however, each atom is affected not only by its nearest neighbors but also by those farther away. Because electrons are generally bound closely to their parent atoms, these distant forces are minimal. In some materials, though, extended molecular orbitals may include several atoms. For example, the benzene (C_6H_6) molecule (fig. 2.19a) is described classically as a ring of carbon atoms held together by covalent single and double bonds. As the structural diagrams in figure 2.19b indicate, there are five different ways to arrange the six bonds in the ring. Valence-bond theory suggests that each of them is equally probable and that the true

WATER IS A STRANGE SUBSTANCE

It is virtually impossible to find a geochemical environment in which water does not play a role, whether at low or high temperature, on the Earth's surface, or deep in the crust. Although water is a familiar substance, it has unusual properties that can only be understood by studying hybrid bonding orbitals.

A neutral oxygen atom contains unpaired electrons in $2p_y$ and $2p_z$ orbitals. When these bond to hydrogen atoms, the result should be a pair of H-O bonds at right angles. Instead, interelectronic repulsion between the filled $2s^2$ and $2p_x^2$ orbitals favors the formation of sp^3 hybrid orbitals such as those discussed in worked problem 2.2. The oxygen atom in a water molecule therefore finds itself at the center of a tetrahedral arrangement of hybrid orbitals. Two of these contain lone pairs of electrons; the other two bond to hydrogen atoms.

A water molecule, therefore, is asymmetrical. The center of positive charge is offset quite a bit from the center of negative charge. Because H-O bonds lie on one side of the oxygen atom, the water molecule has a strong dipole moment and can exert an orientation effect on nearby charged particles. As a result, water is an excellent solvent for ionic substances. In a sodium chloride solution, for example, each Na^+ ion is surrounded by a hydration tetrahedron of water molecules, each one with its negative end drawn toward the Na^+ ion by a weak ion-dipole interaction (fig. 2.17). Each Cl^- ion attracts the positive end of a water dipole and is also enclosed in a hydration sphere. Sodium and chloride ions are thus shielded from each other and prevented from precipitating. We investigate solubility more fully in chapters 4 and 7.

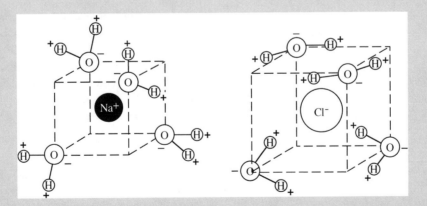

FIG. 2.17. Visualize a hydrated Na^+ or Cl^- ion at the center of a cube with water molecules at four of the corners. The negative end of each water dipole is attracted to the Na^+ ion, so the hydrogen ions at the positive end point outward. A Cl^- ion attracts the positive end of the water dipole, so that the hydrogen ions point inward.

Water molecules also attract each other through dipole-dipole interactions that we refer to as *hydrogen bonds*. The attraction is greatest in ice, in which each hydrogen atom is drawn electrostatically to a lone electron pair on a neighboring water molecule (fig. 2.18). The result is an open network of molecules that has a dramatically lower density than liquid water. This attraction also accounts for water's unusually high surface tension, boiling point, and melting point, as well as many other distinctive properties we will explore elsewhere in this book.

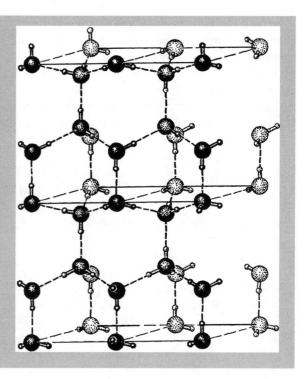

FIG. 2.18. When water freezes, sp³ orbitals on adjacent H_2O units align so that there is one hydrogen atom along each oxygen-oxygen axis. The hydrogen bonds (dashed lines) between H_2O units thus make an electrostatic connection between hydrogen ions on one H_2O molecule and paired (nonbonding) electrons on the other molecule. These weak bonds are nevertheless strong enough to support the open crystal structure of ice.

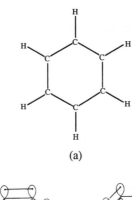

(a)

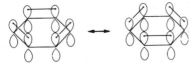

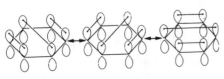

(b)

structure of benzene, therefore, is a resonant mixture of all five. Some of the bonding electrons, in other words, do not belong to specific pairs of carbon atoms but to the entire ring.

Taken to an extreme, this multiatom example suggests a parallel to the popular description of bonding in metals. We observed earlier that the distinction between ionic and covalent bonding is not sharp, but a matter of degree. Metallic bonding is yet another mode along a continuum of atomic interactions, a mode in which electrons are even more free to roam from their parent atoms

FIG. 2.19. (a) Bonding electrons in a benzene molecule bind six carbon atoms in a ring and connect a hydrogen ion to each one. The remaining bonding electron from each carbon atom occupies a p orbital perpendicular to the plane of the ring. (b) At any instant, electrons in the projecting p orbitals (technically, π orbitals) can form stable pairs (hence a "second" or "double" bond) in any of five ways. The benzene structure is a resonant mixture of these equally probable arrangements.

than in covalently bonded materials. In a pure metal, all atoms are the same size, so the coordination number for any one of them is 12 (see table 2.4). Statistically, therefore, each atom should have 12 equivalent bonds with its neighbors. None of the metals, however, has 12 unpaired electrons or can generate that many hybrid orbitals. This apparent problem can be resolved if we consider the bonding orbitals to be an extensive, nondirected resonant network, often described as an electron "gas" that permeates the structure. As in the benzene molecule, valence electrons belong to the entire structure, not to specific atoms. As a result, metals have a higher electrical and thermal conductivity than nonmetals.

This simplistic "free electron" model for metallic bonding is limited in ways that would become apparent if we studied the electrical potential across a metallic structure carefully. The quantum nature of the structure makes some energy levels accessible to the free electrons and prohibits others. Even at this basic level, however, the concept of nonlocalized bonding offers a useful starting point for understanding the behavior of metals.

Yet another variation on this theme of nonlocalized bonding is *Van der Waals bonding*, which may also be described as an extended, nondirectional field of forces between atoms or molecular units. Geologists may know of Van der Waals bonding primarily through their familiarity with the structure of graphite, which consists of stacked sheets of covalently bonded carbon (fig. 2.20). Overlapping sp^2 hybrid orbitals shape the hexagonal array of atoms within a sheet. Electrons in these orbitals are tightly bound to pairs of atoms. Electrons in the remaining unhybridized p orbital on each carbon atom stabilize the sheet by being shared more extensively within the array. Because all of these interatomic forces are confined within sheets, each sheet can be thought of as a separate "molecule" of graphite. The unhybridized p orbitals, however, are perpendicular to the sheets and thus contribute to local charge polarization, analogous to the polarity on a water molecule. This polarization provides the weak cohesion—the Van der Waals attraction—that binds sheets together in crystals of graphite.

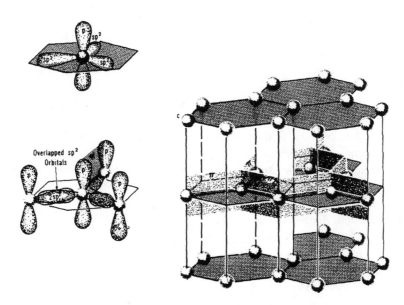

FIG. 2.20. Carbon atoms within a sheet of the graphite structure are bonded by electrons in overlapping sp^2 hybrid orbitals and by electrons shared among p orbitals perpendicular to the sheet, as in benzene (see fig. 2.19). The weak Van der Waals attraction between sheets is due to local charge polarization on those p orbitals.

TABLE 2.5. Properties Associated with Interatomic Bonds

Property	Type of Bond			
	Ionic	Covalent	Metallic	Van der Waals
Interatomic force is . . .	nondirected	spatially directed	nondirected	nondirected
Bonds are mechanically . . .	strong, making hard crystals	strong, making hard crystals	variable; gliding is common	weak, making soft crystals
In response to temperature change, phases have . . .	fairly high melting point, low coefficients of expansion, ions in melt	high melting point, low coefficients of expansion, molecules in melt	variable melting points	low melting points, large coefficients of expansion
In an electrical field, phases are . . .	moderate insulators, conducting by ion transport in melt	insulators in solid and in melt	conductors, by electron transport	insulators in solid and in melt
In response to light, phases . . .	have properties similar to individual ions, and therefore also similar in solution	have high refractive indices; absorption very different in solution or gas	are opaque; properties similar in melt	have properties similar to individual molecules, and therefore also similar in solution

After Evans (1964).

Retrospective on Bonding

Unlike other treatments of chemical bonding that you are likely to have seen, we have tried to call attention to similarities among bond types rather than differences. As so often is true in science, the simple categories you may have learned to recognize in early chemistry or mineralogy courses have blurred edges. Ionic, covalent, metallic, and Van der Waals bonds constitute a continuum of forces that hold atoms together. It is particularly important to emphasize this, because the structures of the most abundant geologic materials include bonds from more than one of the standard categories, as well as some that cannot be placed easily in any of them. As a result, many geologic materials do not behave the way you may have learned to expect ideal ionic, covalent, or metallic substances to behave.

We do not mean to imply that the distinctions among the bond types are meaningless. As table 2.5 indicates, there are some fundamental differences between compounds based primarily on each of these types of bonds. These differences are apparent across broad categories of Earth materials that we consider in later chapters.

SUMMARY

In this chapter, we have established a vocabulary for the chemical language we use throughout this book, with-out straying too far from our peculiar interest in the chemistry of the Earth. With luck, much of what you have just read is a review of familiar material. We hope, though, that we have succeeded in presenting it from a novel perspective. The structure of atoms, particularly the configuration of their electrons, controls the ways that they combine to form natural materials. The type and strength of chemical bonds in a substance determine the amount of energy needed to involve it in reactions. The masses, sizes, and electrical charges on atoms influence the rates at which they react with each other. All of these are topics we explore in the chapters ahead.

SUGGESTED READINGS

Ahrens, L. H. 1965. *Distribution of the Elements in our Planet.* New York: McGraw-Hill. (This concise paperback, now dated, is still one of the best overviews of the principles affecting element distribution in the Earth.)

Barrett, J. 1970. *Introduction to Atomic and Molecular Structure.* New York: Wiley. (A rigorous introduction to the topic, without the heavy reliance on mathematics that can make quantum chemistry inaccessible for nonspecialists.)

Faure, G. 1991. *Principles and Applications of Inorganic Geochemistry.* New York: Macmillan. (A well-written, broad introduction to geochemistry, with particular attention to nuclear chemistry.)

McMurray, J. and R. C.Fay. 1995. *Chemistry.* Englewood Cliffs: Prentice-Hall. (Every geochemistry student should have a basic comprehensive chemistry text; we consider this one to be the best.)

Pauling, L. 1960. *The Nature of the Chemical Bond,* 3rd ed. Ithaca: Cornell University Press. (This is the classical treatise on bonding, surprisingly easy to follow.)

Van Melsen, A.G. 1960. *From Atomos to Atom.* New York: Harper and Brothers. (For students who are curious about the history of science, this book is a highly readable account of the evolution of atomic theory.)

The following sources were referenced in this chapter. The reader may wish to examine them for further details.

Evans, R. C. 1964. *An Introduction to Crystal Chemistry.* Cambridge: Cambridge University Press.

Shannon, R. D. 1976. Revised effective ionic radii and systematic studies of interatomic distances in halides and chalcogenides. *Acta Crystallographica, Section A* 32:751–767.

PROBLEMS

(2.1) Magnesium has three naturally occurring isotopes: ^{24}Mg (78.99% abundance, mass = 23.985 amu), ^{25}Mg (10% abundance, mass = 24.986 amu), and ^{26}Mg (10% abundance, mass = 25.983amu). What is the average atomic weight of magnesium?

(2.2) The average atomic weight of silicon is 28.086 amu. It has three stable isotopes, two of which are ^{28}Si (92.23% abundance, mass = 27.977 amu) and ^{30}Si (3.10% abundance, mass = 29.974 amu). What is the mass of the third stable isotope?

(2.3) ^{14}C, important in radiometric dating, has the same atomic mass number as ^{14}N. How is this possible?

(2.4) Arrange the following bonds in order of their increasing ionic character, using Pauling's formula: Be-O, C-O, N-O, O-O, Si-O, Se-O. Repeat, using DEN values. How do the two sets of results compare?

(2.5) What electrically neutral atoms have each of following the ground state configurations?
(a) $1s^2 2s^2 2p^6 3s^2 3p^6 4s^2$
(b) $1s^2 2s^2 2p^6 3s^2 3p^6 4s^2 3d^6$
(c) $1s^2 2s^2 2p^6 3s^2 3p^6 4s^2 3d^{10} 4p^5$

(2.6) What is the ground state configuration for each of the following ions?
(a) Mg^{2+}
(b) F^-
(c) Cu^+
(d) Mn^{3+}
(e) S^{2-}

(2.7) Predict the coordination number for each element in the following compounds or complexes:
(a) CuF_2
(b) SO_3
(c) CO_2
(d) MnO^{2+}

(2.8) Considering their EN values, explain why potassium, strontium, and aluminum are classified as lithophile rather than chalcophile elements in Goldschmidt's classification.

(2.9) Show algebraically that eightfold coordination for a cation is likely only if the ionic radius ratio $r^+/r^- > \sqrt{3} - 1 = 0.732$.

(2.10) According to one empirical set of definitions, atoms joined by covalent bonds form molecules; those joined by ionic or metallic bonds do not. What might be the justification for these definitions? Are they valid? What are their limitations?

(2.11) Graphite will conduct an electrical current applied parallel to its sheet structure, but will not conduct a current perpendicular to the sheets. Why?

CHAPTER THREE

A FIRST LOOK AT THERMODYNAMIC EQUILIBRIUM

OVERVIEW

In this chapter, we introduce the foundations of thermodynamics. A major goal is to establish what is meant by the concept of equilibrium, described informally in the context of the party example in chapter 1. The idea of equilibrium is an outgrowth of our understanding of three laws of nature, which describe the relationship between heat and work and identify a sense of direction for change in natural systems. These laws are the heart of the subject of thermodynamics, which we apply to a variety of geochemical problems in later chapters.

To reach our goal, we need to examine some familiar concepts like temperature, heat, and work more closely. It is also necessary to introduce new quantities such as entropy, enthalpy, and chemical potential. The search for a definition of equilibrium also reveals a set of four fundamental equations that describe potential changes in the energy of a system in terms of temperature, pressure, volume, entropy, and composition.

TEMPERATURE AND EQUATIONS OF STATE

It is part of our everyday experience to arrange items on the basis of how hot they are. In a colloquial sense,

our notion of temperature is associated with this ordered arrangement, so that we speak of items having higher or lower temperatures than other items, depending on how "hot" or "cool" they feel to our senses. In the more rigorous terms we must use as geochemists, though, this casual definition of temperature is inadequate for two reasons. First of all, a practical temperature scale should have some mathematical basis, so that changes in temperature can be related to other continuous changes in system properties. It is hard to use the concept of temperature in a predictive way if we have to rely on disjoint, subjective observations. Second, and more important, it is hard to separate this common perception of temperature from the more elusive notion of "heat," the quantity that is transferred from one body to another to cause changes in temperature. It would be helpful to develop a definition of temperature that does not depend on our understanding of heat, but instead relies on more familiar thermodynamic properties.

Imagine two closed systems, each of which is homogeneous; that is, each system consists of a single substance with continuous physical properties—from now on, we will call this kind of substance a *phase*—that is not undergoing any chemical reactions. If this condition is met, the thermodynamic state of either system can be completely described by defining the values of any two of its intensive properties. These might be pressure and viscosity, for example, or acoustic velocity and molar volume. To study the concept of temperature, let's examine the relationship between pressure and molar volume, using the symbols p and $\bar{v}$ in one system, and P and $\bar{V}$ in the other. (Refer back to chapter 1 to confirm that molar volume, the ratio of volume to the number of moles in the system, is indeed an intensive property. Notice, also, that we are now beginning to use the convention of an overscore to identify molar quantities, as described in chapter 1.)

If we place the two containers in contact, changes of pressure and molar volume will take place spontaneously in each system until, if we wait long enough, no further changes occur. When the two systems reach that point, they are in *thermal equilibrium.* Let's now measure their pressures and molar volumes and label them P_1, $\bar{V}_1$ and p_1, $\bar{v}_1$. There is no reason to expect that P_1 will necessarily be the same as p_1 or that $\bar{V}_1$ will be equal to $\bar{v}_1$. In general, the systems will not have identical properties.

If we separate the systems for a moment, we will find it possible to change one of them so that it is described by new values P_2 and $\bar{V}_2$ that still lead to thermal equilibrium with the system at p_1 and $\bar{v}_1$. There are, in fact, an infinite number of possible combinations of P and $\bar{V}$ that satisfy this condition. These combinations define a curve on a graph of P versus $\bar{V}$ (fig. 3.1a). Each combination of P and $\bar{V}$ describes a state of the system in thermal equilibrium with p_1 and $\bar{v}_1$. All points on the curve are, therefore, also in equilibrium with each other. This observation has been called the zeroth law of thermodynamics: "If A is in equilibrium with B and B is in

equilibrium with C, then A is in equilibrium with C." Notice—this is important—that figure 3.1a could describe either system. We could just as well have varied pressure and molar volume in the *other* system and found an infinite number of values of p and $\bar{v}$ that lead to thermal equilibrium with the system at $P_1, \bar{V}_1$. These values would define a curve on a graph of p versus $\bar{v}$, shown in figure 3.1b.

Curves like these are called *isothermals.* What we have just demonstrated is that states of a system have the same temperature if they lie on an isothermal. Furthermore, two systems have the same temperature if their states on their respective isothermals are in thermal equilibrium. These two statements, together, define what we mean by "temperature." You could express these two statements algebraically by saying that the each of isotherms in our example can be described by a function:

$$f(p, \bar{v}) = t \ \text{ or } \ F(P, \bar{V}) = T,$$

and that the systems are in thermal equilibrium if $t = T$. Functions like these, which define the interrelationships among intensive properties of a system, are called *equations* or *functions of state.*

Because many real systems behave in roughly similar ways, several standard forms of the equation of state have been developed. We find, for example, that many gases at low pressure can be described adequately by the ideal gas law:

$$P\bar{V} = RT,$$

in which R is the gas constant, equal to 1.987 cal mol^{-1} K^{-1}. Other gases, particularly those containing many atoms per molecule or those at high pressure, are characterized more appropriately by expressions like the Van der Waals equation:

$$(P + a/\bar{V}^2)(\bar{V} - b) = RT,$$

in which a and b are empirical constants. We apply these and other equations of state in chapter 4.

Our notion of temperature, therefore, depends on the conventions we choose to follow in writing equations of state. A practical temperature scale can be devised simply by choosing a well-studied reference system (a *thermometer*), writing an arbitrary function that remains constant (that is, generates isothermals) for various states of the system in thermal equilibrium, and agreeing on a convenient way to number selected isothermals.

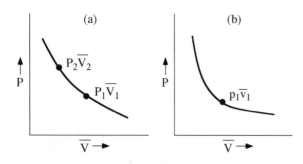

FIG. 3.1. (a) All states of a system for which $f(P, V) = T$ are said to be in thermal equilibrium. A line connecting all such states is called an *isothermal.* (b) If another system contains a set of states lying on the isothermal $f(p, v) = t$, and if $t = T$, then the two systems are in thermal equilibrium.

WORK

In general terms, *work* is performed whenever an object is moved by the application of a force. An infinitesimal amount of work dw, is therefore described by writing:

$$dw = F\,dx,$$

in which F is a generalized force and dx is an infinitesimal displacement. By convention, we define dw so that it is positive if work is performed by a system on its surroundings, and negative if the environment performs work on the system. We know from experience that an object may be influenced simultaneously by several forces, however, so it is more useful to write this equation as:

$$dw = \sum_i F_i\,dx_i.$$

The forces may be hydrostatic pressure, P, or may be directed pressure, surface tension, or electrical or magnetic potential gradients. All forms of work, though, are equivalent, so the total work performed on or by a system can be calculated by including all possible force terms in the equation for dw. The thermodynamic relationships that build on the equation do not depend on the identity of the forces involved.

This description of work is broader than it often needs to be in practice. Work is performed in most geochemical systems when a volume change dV is generated by application of a hydrostatic pressure:

$$dw = P\,dV.$$

This is the equation we most often encounter, and geochemists commonly speak as if the only work that matters is pressure-volume work. Usually, the errors introduced by this simplification are small. It is always wise, however, to examine each new problem to see whether work due to other forces is significant. Later chapters of this book discuss some conditions in which it is necessary to consider other forces.

Note that we have defined work with a differential equation. The total amount of work performed in a process is the integral of that equation between the initial and final states of the system. Because forces in geologic environments rarely remain constant as a system evolves, the integral becomes extremely difficult to evaluate if we do not specify that the process is a slow one. The work performed by a gas expanding violently, as in a volcanic eruption, is hard to estimate, because pressure is not the same at all places in the gas and the gas is not expanding in mechanical equilibrium with its surroundings.

THE FIRST LAW OF THERMODYNAMICS

During the 1840s, a series of fundamental experiments were performed in England by the chemist James Joule. In each of them, a volume of water was placed in an insulated container, and work was performed on it from the outside. Some of these experiments are illustrated in figure 3.2. The paddle wheel, iron blocks, and other mechanical devices are considered to be parts of the insulated container. The temperature of the water was monitored during the experiments, and Joule reported the surprising result that a specific amount of work performed on an insulated system, by any process, always results in the same change of temperature in the system.

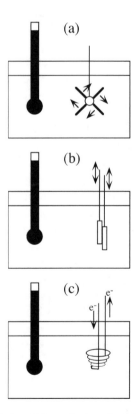

FIG. 3.2. English chemist James Joule performed experiments to relate mechanical work to heat. (a) A paddlewheel is rotated and a measurable amount of work is performed on an enclosed water bath. (b) Two blocks of iron are rubbed together. (c) Electrical work is performed through an immersion heater.

What changed inside the container, and how do we explain Joule's results? In physics, work is often introduced in the context of a potential energy function. For example, a block of lead may be lifted from the floor to a tabletop by applying an appropriate force to it. When this happens, we recognize not only that an amount of work has been expended on the block, but also that its potential energy has increased. Usually, problems of this type assume that no change takes place in the internal state of the block (that is, its temperature, pressure, and composition remain the same as the block is lifted), so the change in potential energy is just a measure of the change in the block's position. When Joule performed work on the insulated containers of water, however, it was not the position of the water that changed, but its internal state, as measured by a change in its temperature. The energy function used to describe this situation is called the *internal energy* (symbolized by E) of the system, and Joule's results can be expressed by writing

$$dE = -dw. \qquad (3.1)$$

Thus, there are two ways to change the internal energy of a system. First, as we implied when considering the meaning of temperature, energy can pass directly through system walls if those walls are noninsulating. Energy transferred by this first mode is called *heat*. If Joule's containers had been noninsulators, he could have produced temperature changes without performing any work, simply by lighting a fire under them. Second, as Joule demonstrated, the system and its surroundings can perform work on each other.

To investigate these two modes of energy transfer, we need to distinguish between two types of walls that can surround a closed system. What we have spoken of loosely as an "insulated" wall is more properly called an *adiabatic* wall. (The Greek roots of the word *adiabatic* mean, appropriately, "not able to go through.") *Adiabatic processes,* like those in Joule's experiments, do not involve any transfer of heat between a system and its surroundings. Perfect adiabatic processes are seldom seen in the real world, but geochemists often simplify natural environments by assuming that they are adiabatic. A *nonadiabatic* wall, however, allows the passage of heat. It is possible to perform work on a system that is bounded by either type of wall.

Equation 3.1 is an extremely useful statement, referred to as the *First Law of Thermodynamics*. In words,

it tells us that the work done on a system by an adiabatic process is equal to the increase in its internal energy, a function of the state of the system. We may conclude further that if a system is isolated, rather than simply closed behind an adiabatic wall, no work can be performed on it from the outside, and its internal energy must remain constant.

Equation 3.1 can be expanded to say that for any system, a change in internal energy is equal to the sum of the heat gained (dq) and the work performed on the system ($-dw$):

$$dE = dq - dw. \qquad (3.2)$$

Notice that dE is a small addition to the amount of internal energy already in the system. Because this is so, the total change in internal energy during a geochemical process is equal to the sum of all increments dE:

$$dE = E_{final} - E_{initial} = \Delta E.$$

The value of ΔE, in other words, does not depend on how the system evolves between its end states. In fact, if the system were to evolve along a path that eventually returns it to its initial state, there would be no net change whatever in its internal energy. This is an important property of functions of state, described in more mathematical detail in appendix A.

In contrast, the values of dq and dw describe the amount of heat or work expended across system boundaries, rather than increments in the amount of heat or work "already in the system." When we talk about heat or work, in other words, the emphasis is on the process of energy *transfer,* not the state of the system. For this reason, the integral of dq or dw depends on which path the system follows from its initial to its final state. Heat and work, therefore, are not functions of state.

Worked Problem 3.1

Consider the system illustrated in figure 3.3, whose initial state is described by a pressure P_1 and a temperature T_1. This might be, for example, a portion of the atmosphere. Recall that the equation of state that we use to define isothermals for this system tells us what the molar volume, $\bar{V}_1$, under these conditions is. For ease of calculation, assume that the equation of state for this system is the ideal gas law, $\bar{V} = RT/P$. Compare two ways in which the system might slowly evolve to a new state in which $P = P_2$ and $T = T_2$. In the first process, let pressure increase slowly from P_1 to P_2 while the temperature remains constant, then

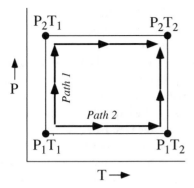

FIG. 3.3. As a system's pressure and temperature are adjusted from P_1T_1 to P_2T_2, the work performed depends on the path that the system follows.

let temperature increase from T_1 to T_2 at constant pressure P_2. In the second process, let temperature increase first at constant pressure P_1, and then let pressure increase from P_1 to P_2. (If this were, in fact, an atmospheric problem, the isobaric segments of these two paths might correspond to rapid surface warming on a sunny day, and the isothermal segments might reflect the passage of a frontal system.) Is the amount of work done on these two paths the same?

This problem is similar to problem 1.8 at the end of chapter 1. If the only work performed on the system is due to pressure-volume changes, then the work along each of the paths is defined by the integral of $Pd\bar{V}$. For path 1,

$$w_1 = \oint_{P_1T_1}^{P_2T_1} Pd\bar{V} + \oint_{P_2T_1}^{P_2T_2} Pd\bar{V}.$$

Along path 2,

$$w_2 = \oint_{P_1T_1}^{P_1T_2} Pd\bar{V} + \oint_{P_1T_2}^{P_2T_2} Pd\bar{V}.$$

To evaluate these four integrals, find an expression for $d\bar{V}$ by expressing the equation of state as a total differential of $\bar{V}(T, P)$:

$$d\bar{V} = (\partial\bar{V}/\partial T)_P \, dT + (\partial\bar{V}/\partial P)_T \, dP = (R/P)dT - (RT/P^2)dP,$$

and substitute the result into w_1 and w_2 above. The result, after integration, is that the work performed on path 1 is:

$$w_1 = R[(T_2 - T_1) + T_1 \ln(P_1/P_2)]$$

but the work performed on path 2 is:

$$w_2 = R[(T_2 - T_1) + T_2 \ln(P_1/P_2)],$$

which is clearly different. Similarly, if we had a function for dq, the heat gained, we could integrate it to show that q_1 and q_2, the amounts of heat expended along these paths, must also be different. We are about to do just that.

ENTROPY AND THE SECOND LAW OF THERMODYNAMICS

Taken by itself, equation 3.2 tells us that heat and work are equivalent means for changing the internal energy of a system. In developing that expression, however, we have engaged in a little sleight-of-hand that may have led you, quite incorrectly, to another conclusion as well. By stating that heat and work are equivalent modes of energy transfer, the First Law may have left you with the impression that heat and work can be freely exchanged for one another. This is not the case, as a few commonplace examples will show.

A glass of ice water left on the kitchen counter gradually gains heat from its surroundings, so that the water warms up and the room temperature drops ever so slightly. In the process, there has been an exchange of heat from the warm room to the relatively cool water. As Joule showed, the same transfer could have been accomplished by having the room perform work on the ice water. It is clearly impossible, however, for a glass of water at room temperature to cool spontaneously and begin to freeze, although we can certainly remove heat by transferring it first from the water to a refrigeration system. In a similar way, a lava flow gradually solidifies by transferring heat to the atmosphere, but this natural process cannot reverse itself either. It is impossible to melt rock by transferring heat to it directly from cold air.

If we were shown films of either of these events, we would have no trouble recognizing whether they were being run forwards or backwards. Equation 3.2, however, does not provide us with a means of making this determination of direction theoretically. On the basis of experience with natural processes, therefore, we are driven to formulate a *Second Law of Thermodynamics*. In its simplest form, first stated by Rudolf Clausius in the middle of the nineteenth century, the Second Law says that heat cannot spontaneously pass from a cool body to a hotter one. Another way of stating this, which may also be useful, is that any natural process involving a transfer of energy is inefficient, with the result that a certain amount is irreversibly converted into heat that cannot be involved in further exchanges. The Second Law, therefore, is a recognition that natural processes have a sense of direction.

Two difficult concepts are embedded in what we have just said. One is *spontaneity* and the other is *reversibility*. A *spontaneous* change is one that, under the right

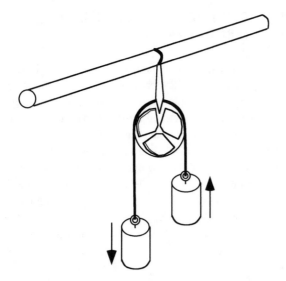

FIG. 3.4. Two weights are connected by a rope that passes over a frictionless pulley. If one weight is infinitesimally heavier than the other, then it will perform the maximum possible amount of work (the work of lifting the other weight) as it falls. The smaller the weight difference between them is, the greater the amount of work and the more nearly reversible it is.

conditions, can be made to perform work. The heavy weight in figure 3.4 can perform the work of raising the lighter weight as it falls, so that this change is spontaneous. If the lighter weight raised the heavier one, we would recognize the change as nonspontaneous and we would assume that some change outside the system had probably caused it. When we face a conceptual challenge in telling whether a change is spontaneous, it is usually because we haven't defined the bounds of the system clearly. Notice also that a spontaneous change doesn't have to *perform* work; it just has to be capable of it. The heavy weight will fall spontaneously whether it lifts the lighter weight or not.

This leads us to consider reversibility. A change is *reversible* if it does the maximum amount of possible work. The falling weight in figure 3.4, for example, will be undergoing a reversible change if it lifts an identical weight and loses no energy to friction in the pulley and rope that connects it to the other weight. Experience tells us that reversibility is an unattainable ideal, of course, but that doesn't stop engineers from designing counterweighted elevators to use as little extra energy as possible. It also doesn't stop geochemists from considering gradual changes in systems that are never far from equilibrium as

if they were reversible. The conceptual challenge arises because true reversibility is beyond our experience with the natural world.

Another perspective on the Second Law, then, is that it tells us that nature favors spontaneous change, and that maximum work is never performed in real-world processes. Just as the internal energy function was introduced to present the First Law in a quantitative fashion, we need to define a thermodynamic function that quantifies the sense of direction and systematic inefficiency that the Second Law identifies in natural processes. This new function, known as *entropy* and given the symbol S, is defined in differential form by:

$$dS = (dq/T)_{rev}, \qquad (3.3)$$

in which dq is an infinitesimal amount of heat gained by a body at temperature T in a reversible process. It is a measure of the degree to which a system has lost heat and therefore some of its capacity to do work. Many people find entropy an elusive concept to master, so it is best to become familiar with the idea by discussing examples of its use and properties.

Figure 3.5 illustrates a potentially reversible path for a system enclosed in a nonadiabatic wall. Suppose that the system expands isothermally from state A to state B. In doing so, it performs an amount of work on its surroundings that can be calculated by the integral of PdV. Graphically, this integral is represented by the area AA′B′B. Because this is an isothermal process, state A and state B must be in equilibrium with each other,

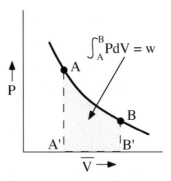

FIG. 3.5. The work performed by isothermal compression from state A to state B is equal to the area AA′B′B. Because $dE = 0$, an equivalent negative amount of heat is gained by the system. If this change could be reversed precisely, the net change in entropy dS would be zero.

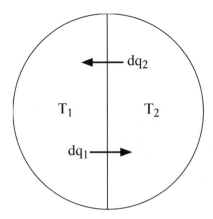

FIG. 3.6. Two bodies in an isolated system are separated by a nonadiabatic wall and may, therefore, exchange quantities of heat dq_1 and dq_2 as they approach thermal equilibrium.

which is another way of saying that the internal energy of the system must remain constant along the path between them. Therefore, the work performed by the system must be balanced by an equivalent amount of heat gained, according to equation 3.2. Suppose, now, that it were possible to compress the system isothermally, thus reversing along the path B → A without any interference due to friction or other real-world forces. We would find that the work performed *on* the system (the heat *lost* by the system) would be numerically equal to the work on the forward path, but with a negative sign. That is, the net change in entropy around the closed path is:

$$dS = dq/T - dq/T = 0.$$

The entropy function for a system following a reversible pathway, therefore, is a function of state, just like the internal energy function.

Next, consider an isolated system in which there are two bodies separated by a nonadiabatic wall (fig. 3.6). Let the two bodies initially be at different temperatures T_1 and T_2. If we allow them to approach thermal equilibrium, there must be a transfer of heat dq_2 from the body at temperature T_2. Because the total system is isolated, an equal amount of heat dq_1 must be gained by the body at temperature T_1. The net change in entropy for the system is:

$$dS_{sys} = dS_1 + dS_2,$$

or

$$dS_{sys} = dq_1\,([1/T_1] - [1/T_2]).$$

Remember, though, that the Second Law tells us that heat can only pass spontaneously from a hot body to a cooler one, so this result is only valid if $T_2 > T_1$. Therefore, the entropy of an isolated system can only increase as it approaches internal equilibrium:

$$dS_{sys} > 0.$$

Finally, look at the closed system illustrated in figure 3.7. It is similar to the previous system in all respects, except that is bounded by a nonadiabatic wall, so that the two bodies can exchange amounts of heat dq_1' and dq_2' with the world outside. As they approach thermal equilibrium, the heat exchanged by each is given by:

$$(dq_1)_{total} = dq_1 + dq_1',$$

and

$$(dq_2)_{total} = dq_2 + dq_2'.$$

Any heat exchanged *internally* must still show up either in one body or the other, so as before:

$$dq_1 = -dq_2.$$

The net change in system entropy is, therefore,

$$dS_{sys} = (dq_1)_{total} + (dq_2)_{total}$$
$$= (dq_1'/T_1) + (dq_2'/T_2) + dq_1([1/T_1] - [1/T_2]).$$

It has already been shown that the Second Law requires the last term of this expression to be positive. Therefore,

$$dS_{sys} > (dq_1'/T_1) + (dq_2'/T_2).$$

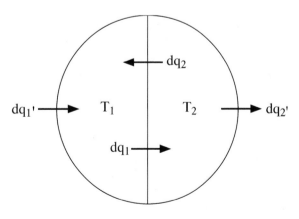

FIG. 3.7. The system in figure 3.6 is allowed to exchange heat with its surroundings. The net change in internal entropy depends on the values of dq_1' and dq_2'.

This result is more open ended than the previous one. Although the entropy change due to *internal* heat exchanges in any system will always be positive, it says that we have no way of predicting whether the *overall* dS_{sys} of a nonisolated system will be positive unless we also have information about dq_1', dq_2', and the temperature in the world outside the system.

To appreciate this ambiguity, think back to the difference between placing a glass of ice water on the kitchen counter and placing it in the freezer. In both cases, internal heat exchange results in an increase in system entropy. When the glass is refrigerated, however, (dq_1'/T_1) and (dq_2'/T_2) become potentially large negative quantities, with the result that the change due to internal processes is overwhelmed and the entropy within the glass decreases. The Second Law, therefore, does not rule out the possibility that thermodynamic processes can reverse direction. It does tell us, though, that a spontaneous change can *only* reverse direction if heat is lost to some external system. A local decrease in entropy must be accompanied by an even larger increase in entropy in the world at large. When we have difficulty with this notion, as we commented earlier, it is usually because we have not been clear about defining the size and boundaries of the system. Therefore, except for those cases in which a system evolves sufficiently slowly that we are justified in approximating its path as a reversible one, its change in entropy must be defined not by equation 3.3, but by:

$$dS > (dq/T)_{irrev}. \tag{3.4}$$

Entropy and Disorder

Entropy is commonly described in elementary texts as a measure of the "disorder" in a system. For those who learned about entropy through statistical mechanics, following the approach pioneered by Nobel physicist Max Planck in the early 1900s, this may make sense. This is sometimes less than satisfying for people who address thermodynamics as we have, because the mathematical definition in equations 3.3 and 3.4 doesn't seem to address entropy in those terms. To see why it is valid to think of entropy as an expression of disorder, consider this problem.

Worked Problem 3.2

Two adjacent containers are separated by a removable wall. Into one of them, we introduce 1 mole of nitrogen gas. We fill the other with 1 mole of argon. If the wall is removed, we expect that the two gases will mix spontaneously rather than remaining in their separate ends of the system. It would require considerable work to separate the nitrogen and argon again. (See problem 3.5 at the end of this chapter.) By randomizing the positions of gas molecules and contributing to a loss of order in the system, we have caused an increase in entropy.

Suppose that the mixing takes place isothermally, and that the only work involved is mechanical. Assume also, for the sake of simplicity, that both argon and nitrogen are ideal gases. How much does the entropy increase in the mixing process? First, write equation 3.2 as:

$$dq = dE + PdV.$$

Because we agreed to carry out the experiment at constant temperature, dE must be equal to zero. Using the equation of state for an ideal gas, we can see that dq now becomes:

$$dq = (nRT/V)dV,$$

from which:

$$dS = dq/T = (nR/V)dV.$$

(Notice that this problem is cast in terms of *volume*, rather than *molar volume*, because we need to keep track of the number of moles, n.) We can think of the current problem as one in which each of the two gases has been allowed to expand from its initial volume (V_1 or V_2) into the larger volume $V_1 + V_2$. For each, then, the entropy change due to isothermal expansion is:

$$\Delta S_i = \Delta n_i R \ln([V_1 + V_2]/V_i),$$

and the combined entropy change is:

$$\Delta S_1 + \Delta S_2 = n_1 R \ln([V_1 + V_2]/V_1) + n_2 R \ln([V_1 + V_2]/V_2).$$

This can be simplified by defining the *mole fraction*, X_i, of either gas as $X_i = n_i/(n_1 + n_2) = V_i/(V_1 + V_2)$; thus,

$$\Delta \bar{S} = -R(X_1 \ln X_1 + X_2 \ln X_2),$$

where the bar over $\Delta \bar{S}$ is our standard indication that it has been normalized by $n_1 + n_2$ and is now a molar quantity. The entropy of mixing will always be a positive quantity, because X_1 and X_2 are always <1.0. The answer to our question, therefore, is that entropy increases by:

$$2 \times \Delta \bar{S} = 2 \times -(1.987)(0.5 \ln 0.5 + 0.5 \ln 0.5)$$

$$= 2.76 \text{ cal K}^{-1} \text{ or } 11.55 \text{ J K}^{-1}.$$

REPRISE: THE INTERNAL ENERGY FUNCTION MADE USEFUL

With the First and Second laws in hand, we can now return to the subject of equilibrium, which was defined informally at the beginning of this chapter as a state in

ENTROPY AND DISORDER: WORDS OF CAUTION

The description of entropy as a measure of disorder is probably the most often used and abused concept in thermodynamics. In many applications, the idea of "disorder" is coupled with the inference from the Second Law that entropy should increase through time. Like all brief descriptions of complex ideas, though, the statement that "entropy measures system disorder" must be treated with some care. It is common practice among chemical physicists to estimate the entropy of a system by calculating its statistical degrees of freedom—computing, in effect, the number of different ways that atoms can be configured in the system, given its bulk conditions of temperature, pressure, and other intensive parameters. A less rigorous description of disorder, however, can lead to very misleading conclusions about the entropy of a system and thus the course of its evolution.

Creationists, for example, have grasped at the idea of entropy as disorder as a vindication of their belief that biologic evolution is impossible. In their view, "higher" organisms are more ordered than the more primitive organisms from which biologists presume that they evolved. Because this interpretation presents evolution as a historical progression from a less ordered to a more ordered state, it describes an apparent violation of the Second Law. There are so many flaws in this argument that it is difficult to know where to begin to discuss it, and we will mention only a few.

The first is the question of the meaning of "order" as applied to organisms or families of organisms. This is far from a semantic problem, because it is by no means clear that higher organisms are more ordered than primitive ones, from the point of view of thermo-dynamics. The biologic concept of order, although potentially connected to thermodynamics in the realm of biochemistry, is based largely on functional complexity, not equations for energy utilization. The situation becomes even murkier when applied to groups of organisms, rather than to individuals.

Furthermore, from a thermodynamic perspective, living organisms cannot be viewed as isolated systems, separated from their inorganic surroundings. As we have tried to emphasize, the definition of system boundaries is crucial if we are to interpret system changes by the Second Law. A refrigerator, for example, might be mistakenly seen to violate the Second Law if we failed to recognize that entropy in the kitchen around it increases even as the contents of the refrigerator become more ordered.

Finally, we note that there may be instances in which entropy and disorder in the macroscopic sense can be decoupled, even in a properly defined isolated system. For example, if two liquids are mixed in an adiabatic container, we might expect that they will mix to a random state at equilibrium. We know, however, that other system properties may commonly preclude a random mixture. Oil and vinegar in a salad dressing unmix readily, as a manifestation of differences in their bonding properties. If the attractive force between similar ions or molecules is greater than that between dissimilar ones, then complete random mixing would imply an increase, rather than a decrease, in both internal energy and entropy. A macroscopically ordered system, in other words, can easily be more stable than a disordered one without shaking our faith in the Second Law of Thermodynamics.

---

which no change is taking place. The First Law restates this condition as one in which dE is equal to zero. From the Second Law, we find that entropy is always maximized during the approach to equilibrium, and that dS becomes equal to zero once equilibrium is reached.

The results expressed in equations 3.3 and 3.4 can be used to recast equation 3.2 as:

$$dE \leq TdS - dw.$$

Under most geologic conditions, mechanical (pressure-volume) work is the only significant contribution to dw, and it is reasonable to substitute PdV for dw:

$$dE \leq TdS - PdV. \tag{3.5}$$

This is a practical form of equation 3.2, although only correct under the conditions just discussed. In most geologic environments, we assume that change takes place very slowly, so that systems can be regarded as following nearly reversible paths. Generally, therefore, the inequality in equation 3.5 can be disregarded, even though it is strictly necessary.

Because much of the remaining discussion of thermodynamics in this book derives from equation 3.5, we should recognize three other relationships that follow directly from it and from the fact that the internal energy function is a function of state. First, equation 3.5 is a differential form of $E = E(S, V)$. Assuming that changes take place reversibly, it can be written as a total differential:

$$dE(S, V) = (\partial E/\partial S)_V \, dS + (\partial E/\partial V)_S \, dV.$$

Comparison with equation 3.5 yields the two statements:

$$T = (\partial E/\partial S)_V,$$

and

$$P = -(\partial E/\partial V)_S.$$

In other words, the familiar variables temperature and pressure can be seen as expressions of the manner in which the internal energy of a system responds to changes in entropy (under constant volume conditions) or volume (under adiabatic conditions). Second, because $E(S, V)$ is a function of state, $dE(S, V)$ is a perfect differential (perfect differentials are reviewed in appendix A). That is,

$$(\partial^2 E/\partial S \partial V) = (\partial^2 E/\partial V \partial S),$$

or

$$(\partial T/\partial V)_S = -(\partial P/\partial S)_V. \tag{3.6}$$

This last equation is known as one of the *Maxwell relationships*. These and other similar expressions to be developed shortly can be used to investigate the interactions of S, P, V, and T. Some of these will be discussed more fully in chapter 9, when we look in more detail at the effects of changing temperature or pressure in the geologic environment.

AUXILIARY FUNCTIONS OF STATE

Equation 3.5 is adequate for solving most thermodynamic problems. In many situations, however, it is possible to use equations of greater practical interest, which can be derived by imposing environmental constraints on the problem.

Enthalpy

The first of these equations can be derived by writing equation 3.5 as:

$$dE + PdV \leq TdS.$$

If we restrict ourselves by looking only at processes that take place at constant pressure, this is the same as writing:

$$d(E + PV) \leq TdS,$$

because $d(PV) = PdV + VdP$, which equals PdV if $dP = 0$.

The quantity $E + PV$ is a new function, called *enthalpy* and commonly given the symbol H. It is useful as a measure of heat exchanged under isobaric conditions, where $dH = dq$. Reactions that evolve heat, and therefore have a negative change in enthalpy, are *exothermic*. Those that result in an increase in enthalpy are *endothermic*.

The left side of this last equation can be expanded to reveal a differential form for dH:

$$dH = d(E + PV) = dE + PdV + VdP,$$

or

$$dH \leq TdS + VdP. \tag{3.7}$$

As we did with equation 3.5, we can compare the total differential $dH(S, P)$ and equation 3.7 under reversible conditions to recognize that:

$$T = (\partial H/\partial S)_P,$$

and

$$V = (\partial H/\partial P)_S.$$

Enthalpy, like internal energy and entropy, can be shown to be a function of state, because it experiences no net change during a reversible cycle of reaction paths. This makes it possible, among other things, to determine the amount of heat that would be exchanged in geologically important reactions, even when those reactions may be too sluggish to be studied directly at the low temperatures at which they take place in nature. The careful determination of such values is the business of *calorimetry*.

Because H is a function of state, it is also possible to extract one more Maxwell relation like equation 3.6 from its cross-partial derivatives (also reviewed in appendix A):

$$(\partial T/\partial P)_S = (\partial V/\partial S)_P. \tag{3.8}$$

The Helmholtz Function

By analogy with the way we introduced enthalpy, we can discover another useful function by writing equation 3.5 as:

$$dE - TdS \leq -PdV.$$

Under isothermal conditions, this expression is equivalent to:

$$d(E - TS) \leq -PdV.$$

The function $F = E - TS$ is best referred to as the *Helmholtz function*, although you may see it referred to elsewhere as *Helmholtz free energy* or the *work function*.

Unfortunately, a variety of symbols have been used for the internal energy, enthalpy, and Helmholtz functions, as well as Gibbs free energy, which is discussed shortly. This has led to some confusion in the literature. In the United States, F is commonly used to designate Gibbs free energy (for example, in publications of the National Bureau of Standards). The International Union of Pure and Applied Chemistry, however, has recommended that G be the standard symbol for Gibbs free energy. This usage, if not yet standard, is at least widespread. In the United States, those who use F for Gibbs free energy generally use the symbol A for the Helmholtz function. To complete the confusion, internal energy, which we have identified with the symbol E, is frequently referred to as U, to avoid confusing it with total (internal plus potential plus kinetic) energy. Always be sure you know which symbols you are using. Nicolas Vanserg has written an excellent article on the subject, which is listed among the references at the end of this chapter (Vanserg 1958).

Because it might be mistaken for Gibbs free energy, which is ultimately a more useful function in geochemistry, it is best to avoid the term *Helmholtz free energy*. The name *work function*, however, is fairly informative. The integral of dF at constant temperature is equal to the work performed on a system. This expression has its greatest application in mechanical engineering.

It is easy to derive the differential form dF:

$$dF = d(E - TS) = dE - TdS - SdT,$$

or

$$dF \leq -SdT - PdV. \tag{3.9}$$

The total differential $dF(T, V)$, when compared with equation 3.9, yields the useful expressions:

$$S = -(\partial F / \partial T)_V,$$

and

$$P = -(\partial F / \partial V)_T.$$

Because F is a function of state and $dF(T, V)$ is therefore a perfect differential, we also gain another Maxwell relation:

$$(\partial S / \partial V)_T = (\partial P / \partial T)_V. \tag{3.10}$$

Although the Helmholtz function is rarely used in geochemistry, the Maxwell relationship equation 3.10 is quite widely used. When we return for a second look at thermodynamics in chapter 9, equation 3.10 will be discussed as the basis of the *Clapeyron equation*, a means for describing pressure-temperature relationships in geochemistry.

Gibbs Free Energy

The most frequently used thermodynamic quantity in geochemistry can be derived by writing equation 3.5 in the form:

$$dE - TdS + PdV < 0.$$

Under conditions in which both temperature and pressure are held constant, this expression becomes:

$$d(E - TS + PV) < 0.$$

In the now familiar fashion in which we have already defined H and F, we designate the quantity $E - TS + PV$ with the symbol G and call it the *Gibbs free energy*, in honor of the Josiah Willard Gibbs, a chemistry professor at Yale University, who wrote a classic series of papers in the 1870s in which virtually all of the fundamental equations of modern thermodynamics appeared for the first time. Colloquially, among geochemists, it is common to speak simply of "free energy."

The differential dG can be written:

$$dG = d(E - TS + PV) = dE - TdS - SdT + PdV + VdP,$$

or

$$dG \leq -SdT + VdP. \tag{3.11}$$

As with the previous fundamental equations, we can apply our knowledge of the total differential $dG(T, P)$ of this state function to find that:

$$-S = (\partial G/\partial T)_P \, , \ V = (\partial G/\partial P)_T,$$

and

$$-(\partial S/\partial P)_T = (\partial V/\partial T)_P. \tag{3.12}$$

It can be seen from equation 3.11 that the Gibbs free energy is the first of our fundamental equations written solely in terms of the differentials of intensive parameters. This and the ease with which both temperature and pressure are usually measured contributes to the great practical utility of this function.

However, what is "free" about the Gibbs free energy? Consider the intermediate step in equation 3.11. At constant temperature and pressure, this reduces to:

$$dG = dE - TdS + PdV.$$

Substituting $dE = dq - dw$, we have:

$$dG = dq - dw - TdS + PdV.$$

If the quantity of heat dq is transferred to the system isothermally and any changes are reversible, then $dq = TdS$ and $dw = dw_{rev}$, so we can rewrite this last expression as:

$$-dG = dw_{rev} - PdV.$$

The decrease in free energy of a system undergoing a reversible change at constant temperature and pressure, therefore, is equal to the nonmechanical (i.e., not pressure-volume) work that can be done by the system. If the condition of reversibility is relaxed, then:

$$-dG > dw_{irrev} - PdV.$$

In either case, the change in G during a process is a measure of the portion of the system's internal energy that is "free" to perform nonmechanical work.

The free energy function provides a valuable practical criterion for equilibrium. At constant temperature, the net change in free energy associated with a change from state 1 to state 2 of a system can be calculated by integrating dG:

$$\int_1^2 dG = \int_1^2 d(E - TS + PV) = \int_1^2 d(H - TS),$$

from which:

$$\Delta G = \Delta H - T\Delta S, \tag{3.13}$$

where $\Delta G = G_2 - G_1$, $\Delta H = H_2 - H_1$, and $\Delta S = S_2 - S_1$.

Energy is available to produce a spontaneous change in a system as long as ΔG is negative. According to equation 3.13, this can be accomplished under any circumstances in which $\Delta H - T\Delta S$ is negative. Most exothermic changes, therefore, are spontaneous. Endothermic processes (those in which ΔH is positive) can also be spontaneous, but only if they are associated with a large positive change in entropy. This possibility was not appreciated at first by chemists. In fact, in 1879 the French thermodynamicist Marcelin Berthelot used the term *affinity*, defined by $A = -\Delta H$, as a measure of the direction of positive change. According to his reasoning, a spontaneous chemical reaction could only occur if $A > 0$. If an endothermic reaction turned out to be spontaneous, he assumed that some unobserved mechanical work must have been done on the system. As chemists became familiar with Gibbs's papers on thermodynamics, however, it became clear that Berthelot and others had misunderstood the concept of entropy. Affinity was a theoretical blind alley.

In summary, a thermodynamic change proceeds as long as there can be a further decrease in free energy. Once free energy has been minimized (that is, when $dG = 0$), the system has attained equilibrium.

Worked Problem 3.3

To illustrate how enthalpy, entropy, and temperature are related to G, consider what happens when ice or snow sublimates at constant temperature. Those who live in northern climates will recognize this as a common midwinter phenomenon. Figure 3.8 is a schematic representation of the way in which various energy functions change as functions of the proportions of water vapor and ice in a closed system. (Strictly speaking, snow does not sublimate into a closed system, but into the open atmosphere. On a still night, close to the ground, however, this is not a bad approximation.) Notice that the enthalpy of the system increases linearly as vaporization takes place. This can be rationalized by noting we have to add heat to the system to break molecular bonds in the ice. If enthalpy were the only factor involved in the process, therefore, the system would be most stable when it is 100% solid, because that is where H is minimized. Entropy, however, is maximized if the system is 100% water vapor, because the degree of system randomness is greatest there. It can be shown from statistical arguments, however, that entropy is a logarithmic function of the proportion of vapor, rising most rapidly when the vapor fraction in the system is low. Therefore the free energy of the system, $G = H - TS$, is less than the free energy of the pure solid until a substantial amount of

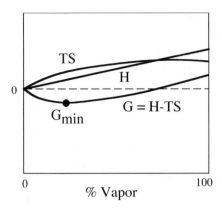

FIG. 3.8. The quantity $G = H - TS$ varies with the amount of vapor as sublimation takes place in an enclosed container. The free energy when vapor and solid are in equilibrium is G_{min}. (Modified from Denbigh 1968.)

vapor has been produced. Sublimation occurs spontaneously. At some intermediate vapor fraction, $H - TS$ reaches a minimum, and solid and vapor are in equilibrium at G_{min}. If the vapor fraction is increased further, the difference $H - TS$ exceeds G_{min} again, and condensation occurs spontaneously. This is the source of the beautiful hoar frost that develops on tree branches and other exposed surfaces on still winter mornings.

CLEANING UP THE ACT: CONVENTIONS FOR *E, H, F, G,* AND *S*

Except in the abstract sense that we have just used G in worked problem 3.3, there is no way we can talk about absolute amounts of energy in a system. To say that a beaker of reagents contains 100 kJ of enthalpy or 55 kcal of free energy is meaningless. Instead, we compare the value of each of the energy functions to its value in a mixture of pure elements under specified temperature and pressure conditions (a *standard state*). For example, we would measure the enthalpy of a quantity of NaCl at 298 K relative to the enthalpies of pure sodium and pure chlorine at the same temperature and at one atmosphere of pressure, using the symbol ΔH_f^0. The superscript 0 indicates that this is a standard state value and the subscript f indicates that the reference standard is a mixture of pure elements. By convention, the ΔH_f^0 or ΔG_f^0 for a pure element at any temperature is equal to zero. The same notation is used for internal energy and the Helmholtz function as well, although they are less frequently encountered in geochemistry.

In our discussions so far, we have perhaps given the impression that whatever thermodynamic data we might need to solve a particular problem will be readily available. The references in appendix B do, in fact, include data for a large number of geologic materials. We have largely ignored the thorny problem of where they come from, however, and where we turn for data on materials that have not yet appeared in the tables. As an example of how thermodynamic data are acquired, we now show how solution calorimetry is used to measure enthalpy of formation. In chapter 10, we show how to extract thermodynamic data from phase diagrams.

It is not necessary (or even possible) to make a direct determination of ΔH_f^0 for most individual phases, because we can rarely generate the phases from their constituent elements. Instead, we measure the heat evolved or absorbed as the substance we are interested in is produced by a specific reaction between other phases for which we already have enthalpy data. Because most reactions of geologic interest are abysmally slow at low temperatures, even this enthalpy change may be almost impossible to measure directly. However, because enthalpy is an extensive variable, we can employ some sleight-of-hand: we can measure the heat lost or gained when the products and reactants are each dissolved in separate experiments in some solvent at low temperature, and then add the heats of solution for the various products and reactants to obtain an equivalent value for the heat of reaction.

Worked Problem 3.4

Consider the following laboratory exercise. We wish to determine the molar enthalpy for the reaction:

$$2MgO + SiO_2 \rightleftharpoons Mg_2SiO_4,$$
periclase quartz   forsterite

to which we assign the value $\Delta \bar{H}_1$. This can be determined by measuring the heats of solution for periclase, quartz, and forsterite in HF at some modest temperature:

$$2MgO + SiO_2 + HF \rightleftharpoons (2MgO, SiO_2)_{solution},$$

which gives $\Delta \bar{H}_2$, and:

$$Mg_2SiO_4 + HF \rightleftharpoons (2MgO, SiO_2)_{solution},$$

which gives $\Delta \bar{H}_3$. If the solutions are identical, we can see that the first of these equations is mathematically equivalent to the second minus the third, so that:

$$\Delta \bar{H}_1 = \Delta \bar{H}_2 - \Delta \bar{H}_3.$$

In this way, solution calorimetry can be used to determine the enthalpy of formation for forsterite from its constituent oxides at low temperature. As you might expect, however, solvents suitable for silicate minerals (such as hydrofluoric acid or molten lead borate) are generally corrosive and hazardous to handle. Consequently, these experiments require considerable skill and specialized equipment. Most geochemists refer to published collections of calorimetric data rather than making the measurements themselves.

What we have just described is not a measurement of $\Delta \bar{H}_f^0$, because the reference comparison was to constituent oxides, not pure elements. If we wanted to determine $\Delta \bar{H}_f^0$, we would search for tabulated data for reactions forming the oxides from pure elements (typically measured by some method other than solution calorimetry). Again, recalling that enthalpies are additive, we could recognize that $\Delta \bar{H}_f^0$ for MgO is defined by the reaction:

$$2Mg + O_2 \rightleftarrows 2MgO,$$

and $\Delta \bar{H}_f^0$ for SiO_2 is defined by:

$$Si + O_2 \rightleftarrows SiO_2.$$

Label these two values $\Delta \bar{H}_4$ and $\Delta \bar{H}_5$. Therefore, for the enthalpy $\Delta \bar{H}_6$ of the net reaction:

$$2Mg + 2 O_2 + Si \rightleftarrows Mg_2SiO_4,$$

we obtain:

$$\Delta \bar{H}_6 = \Delta \bar{H}_2 - \Delta \bar{H}_3 + \Delta \bar{H}_4 + \Delta \bar{H}_5.$$

The value $\Delta \bar{H}_6$ in this case is the enthalpy of formation for forsterite derived from those for the elements, $\Delta \bar{H}_f^0$.

Although most thermodynamic functions are best defined as relative quantities like ΔH_f^0, entropy is a major exception. The convention observed most commonly is an outgrowth of the *Third Law of Thermodynamics,* which can be stated, in paraphrase from Lewis and Randall (1961): If the entropy of each element in a perfect crystalline state is defined as zero at the absolute zero of temperature, then every substance has a nonnegative entropy; at absolute zero, the entropy of all perfect crystalline substances becomes zero. This statement, which has been tested in a very large number of experiments, provides the rationale for choosing the absolute temperature scale (i.e., temperature in Kelvins) as our standard for scientific use. For now, the significance of the Third Law is that it defines a state in which the *absolute* or *third law entropy* is zero. Because this state is the same for all materials, it makes sense to speak of S^0, rather than ΔS_f^0.

With this new perspective, return for a moment to equation 3.13. It should now be clear that the symbol Δ has a different meaning in this context. In fact, although it is rarely done, it would be less confusing to write equation 3.13 as:

$$\Delta(\Delta G) = \Delta(\Delta H) - T\Delta S,$$

in which the deltas outside the parentheses and on S refer to a change in state (that is, a change in these values during some reaction), and the deltas inside the parentheses refer to values of G or H relative to some reference state. We examine this concept more fully in later chapters.

Although we have not felt it necessary to prove, it should be apparent that each of the functions E, H, F, G, and S is an extensive property of a system. The amount of energy or entropy in a system, therefore, depends on the size of the system. In most cases, this is an unfortunate restriction, because we either don't know or don't care how large a natural system may be. For this reason, it is customary to normalize each of the functions by dividing them by the total number of moles of material in the system, thus making each of them an intensive property.

COMPOSITION AS A VARIABLE

Up to now, the functions we have considered all assume that a system is chemically homogeneous. Most systems of geochemical interest, however, consist of more than one phase. The problem most commonly faced by geochemists is that the bulk composition of a system can be packaged in a very large number of ways, so that it is generally impossible to tell by inspection whether the assemblage of phases actually found in a system is the most likely one. To answer questions dealing with the stability of multiphase systems, we need to write a separate set of equations for E, H, F, and G for each phase and apply criteria for solving them simultaneously. We will do this job in two steps.

Components

To describe the possible variations in the compositions and proportions of phases, it is necessary to define a set of *thermodynamic components* that satisfy the following rules:

1. The set of components must be sufficient to describe all of the compositional variations allowable in the system.

2. Each of the components must vary independently in the system.

As long as these criteria are met, the specific set of components chosen to describe a system is arbitrary, although there may often be practical reasons for choosing one set rather than another.

These rules set very stringent restrictions on the way components can be chosen, so it is a good idea to spend time examining them carefully. First, notice that the solid phases we encounter most often in geochemical situations have compositions that are either fixed or are variable only within bounds allowed by stoichiometry and crystal structure. This means that if we are asked to find components *for a single mineral*, they must be defined in such a way that they can be added to or subtracted from the mineral without destroying its identity. A system consisting only of rhombohedral Ca-Mg carbonates, for example, cannot be described by entities such as Ca^{2+}, MgO, or CO_2, because none of them can be independently added to or subtracted from the system without violating its crystal chemistry. The most obvious, although not unique, choice of components in this case would be $CaCO_3$ and $MgCO_3$.

When a system consists of more than one phase, it is common to find that some components selected for individual phases are redundant in the system as a whole. This occurs because it is possible to write stoichiometric relationships that express a component in one phase as some combination of those in other phases. For each stoichiometric equation, therefore, it is possible to remove one phase component from the list of system components and thus to arrive at the independent variables required by rule 2. It is also possible, and often desirable, to choose system components that cannot serve as components for any of the individual phases in isolation. Such a choice is compatible with the selection rules if the amount of the component in the system as a whole can be varied by changing proportions of individual phases. Petrologists usually refer to nonaluminous pyroxenes, for example, in terms of the components $MgSiO_3$, $FeSiO_3$, and $CaSiO_3$, even though $CaSiO_3$ cannot serve as a component for any pyroxene considered by itself.

Worked Problem 3.5

Olivine and orthopyroxene are both common minerals in basic igneous rocks. For the purposes of this problem, assume that

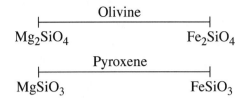

FIG. 3.9. Olivine and pyroxene solid solutions can each be represented by two end-member components.

olivine's composition varies only between Mg_2SiO_4 (Fo) and Fe_2SiO_4 (Fa), and that orthopyroxene is a solid solution between $MgSiO_3$ (En) and $FeSiO_3$ (Fs). What components might be used to describe olivine and orthopyroxene individually, and how might we select components for an ultramafic rock consisting of both minerals?

The simplest mineral components are the end-member compositions themselves. The end-member compositions are completely independent of one another in each mineral and, as can be seen at a glance in figure 3.9, any mineral composition in either solid solution can be formed by some linear combination of the end members. Compositions corresponding to Fo, Fa, En, and Fs, therefore, satisfy our selection rules.

A choice of FeO or MgO would *not* be valid, because neither one can be added to or subtracted from olivine or orthopyroxene unless we also add or subtract a stoichiometric amount of SiO_2. Changing FeO or MgO alone would produce

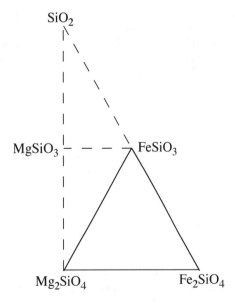

FIG. 3.10. If we choose to describe a system containing olivines *and* pyroxenes, we need three components. Phase compositions lying outside the triangle of system components require negative amounts of one or more components.

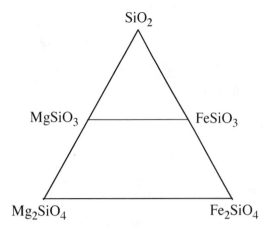

FIG. 3.11. An alternative selection of components for the system in figure 3.10. All olivine and pyroxene compositions now lie within the triangle.

system compositions that lie off of the solid solution lines in figure 3.9.

If olivine and orthopyroxene are not isolated phases but are constituents of a rock, however, we need to choose a different set of components. The mineral components are still appropriate, but one of them is now redundant. We can eliminate it by writing a stoichiometric relationship involving the other three. For example:

$$MgSiO_3 = 0.5Mg_2SiO_4 + FeSiO_3 - 0.5Fe_2SiO_4.$$

Notice that this is only a mathematical relationship among abstract quantities, not necessarily a chemical reaction among phases. We have tried to emphasize this by using an equal sign rather than an arrow. The solid triangle in figure 3.10 illustrates the system as defined by the components on the right side.

As required by the negative amount of Fe_2SiO_4 in the equation above, magnesium-bearing orthopyroxenes have compositions outside of the triangle defined by the system components. There is nothing wrong with this representation, but it is awkward for most petrologic applications. A more conventional selection of components is shown in figure 3.11. The mineral compositions at the ends of the olivine solid solution are still retained as system components, but the third component, SiO_2, does not correspond to either mineral in the system.

It is very important to recognize that components are an abstract means of characterizing a system. They do not need to correspond to substances that can be found in nature or manufactured in a laboratory. Orthopyroxenes, for example, are frequently described by the components $MgSiO_3$ and $FeSiO_3$, even though there is no natural

phase with the composition $FeSiO_3$. Monatomic components such as F, S, or O are also legitimate, even though fluorine, sulfur, and oxygen invariably occur as molecules containing two or more atoms. For some petrological applications, it makes sense to use such components as $CaMg_{-1}$, which clearly do not exist as real substances. In fact, the carbonates discussed above can be characterized quite well by $MgCO_3$ and $CaMg_{-1}$, as can be seen from the stoichiometric relationship:

$$MgCO_3 + CaMg_{-1} = CaCO_3.$$

Components of this type, known as *exchange operators,* have been used to great advantage in describing many metamorphic rocks (Thompson et al. 1982; Ferry 1982). Because components are abstract constructions, we are not required to use them in positive amounts. Fe_2O_3, for example, can be described by the components Fe_3O_4 and Fe, even though we need to add a negative amount of Fe to Fe_3O_4 to do the job:

$$3Fe_3O_4 - Fe = 4Fe_2O_3.$$

Worked Problem 3.6

The ratio K/Na in coexisting pairs of alkali feldspars and alkali dioctahedral micas can be used to infer pressure and temperature conditions for rocks in which they were formed. We see how this is done in chapter 9. For now, let's ask what selection of components might be most helpful if we were interested in K $\rightleftarrows$ Na exchange reactions between these two minerals.

The exchange operator KNa_{-1} is a good choice for a system component in this case. Feldspar compositions can be generated from:

$$NaAlSi_3O_8 + x\,KNa_{-1} = K_xNa_{1-x}AlSi_3O_8,$$

for any value of x between 0 and 1. Similarly, mica compositions can be derived from:

$$NaAl_3Si_3O_{10}(OH)_2 + yKNa_{-1}$$
$$= K_yNa_{1-y}Al_3Si_3O_{10}(OH)_2.$$

If we also select the two sodium end-member compositions as components, then all possible compositional variations in the system can be described. Figure 3.12a shows one way of illustrating this selection. A more subtle diagram, using the same set of components, is presented in figure 3.12b.

This is not a contrived example. We have selected it to emphasize that chemical components are abstract mathematical entities, but it is also meant to illustrate how a very common class of geochemical problems can be reconceived and simplified by choosing components creatively.

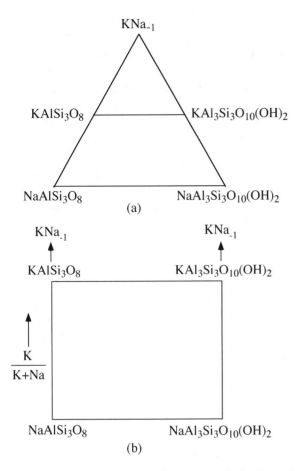

FIG. 3.12. (a) Alkali micas and feldspars can be described by a component set that includes KNa_{-1}. (b) An orthogonal version of the diagram in (a). Its vertical edges both point to KNa_{-1}.

CHANGES IN *E, H, F,* AND *G* DUE TO COMPOSITION

Consider an open system containing only magnesian olivine, a pure phase. The internal energy of the system is equal to $E = \bar{E}_{Fo} n_{Fo}$, where $\bar{E}_{Fo}$ is the molar internal energy for pure forsterite and n_{Fo} is the number of moles of the forsterite component in the system. Suppose that it were possible to add a certain number of moles of the fayalite component, n_{Fa}, to the one-phase system without causing any increase in energy as a result of the mixing itself. Because E is an extensive property, the internal energy of the phase would then be equal to:

$$E = \bar{E}_{Fo} n_{Fo} + \bar{E}_{Fa} n_{Fa}.$$

Further infinitesimal changes in the amounts of forsterite or fayalite in the system would result in a small change in E:

$$dE = \bar{E}_{Fo} dn_{Fo} + \bar{E}_{Fa} dn_{Fa}.$$

This differential equation can be written in the form of a total differential, from which it can be seen that:

$$\bar{E}_{Fo} = (\partial E / \partial n_{Fo}) n_{Fa},$$

and

$$\bar{E}_{Fa} = (\partial E / \partial n_{Fa}) n_{Fo}.$$

This is an idealized process, of course, chosen to demonstrate that the internal energy of a phase can be changed—in addition to the ways we have already discussed—by varying its composition. It is more realistic to recognize that the mixing process as described involves increases in both entropy and volume. Notice also that the total internal energy of the phase, *E*, is *not* a molar quantity, because it has not been divided by the total number of moles in the system. The relationships $dE_{olivine} = \bar{E}_{Fo} dn$ or $dE_{olivine} = \bar{E}_{Fa} dn$ can only be valid if the olivine is either pure forsterite or pure fayalite. In between, as we see in later chapters, dE takes a nonlinear and generally complicated form. The quantities labeled $\bar{E}_{Fo}$ and $\bar{E}_{Fa}$ above are more useful if written with these restrictions in mind:

$$E_i = (\partial E / \partial n_i)_{S,V,n_{j \neq i}}. \qquad (3.14)$$

We have now identified a new thermodynamic quantity, the *partial molar internal energy,* which describes the way in which total internal energy for a phase responds to a change in the amount of component *i* in the phase, all other quantities being equal. You may think of it, if you like, as a chemical "pressure" or a force for energy change in response to composition, in the same way that pressure is a force for energy change in response to volume. To emphasize the importance of this new function, it has been given the symbol μ_i and is called the *chemical potential* of component *i* in the phase.

The internal energy of a phase, then, is:

$$E = E(S, V, n_1, n_2, \ldots, n_j).$$

We can rewrite equation 3.5 to include the newfound chemical potential terms:

$$\begin{aligned} dE \leq TdS - PdV + \mu_1 dn_1 + \mu_2 dn_2 + \ldots \\ + \mu_j dn_j \leq TdS - PdV + \sum \mu_i dn_i. \end{aligned} \qquad (3.15a)$$

In the same way, it can be shown that the auxiliary functions H, F, and G are also functions of composition:

$$dH \leq TdS + VdP + \mu_1 dn_1 + \mu_2 dn_2 + \ldots$$
$$+ \mu_j dn_j \leq TdS + VdP + \sum \mu_i dn_i, \quad (3.15b)$$

$$dF \leq -SdT - PdV + \mu_1 dn_1 + \mu_2 dn_2 + \ldots$$
$$+ \mu_j dn_j \leq -SdT - PdV + \sum \mu_i dn_i, \quad (3.15c)$$

$$dG \leq -SdT + VdP + \mu_1 dn_1 + \mu_2 dn_2 + \ldots$$
$$+ \mu_j dn_j \leq -SdT + VdP + \sum \mu_i dn_i. \quad (3.15d)$$

Chemical potential, therefore, can be defined in several equivalent ways:

$$\mu_i = (\partial E/\partial n_i)_{S,V,n_{j \neq i}} \quad (3.16a)$$

$$= (\partial H/\partial n_i)_{S,P,n_{j \neq i}} \quad (3.16b)$$

$$= (\partial F/\partial n_i)_{T,V,n_{j \neq i}} \quad (3.16c)$$

$$= (\partial G/\partial n_i)_{T,P,n_{j \neq i}} \quad (3.16d)$$

and can be correctly described as the *partial molar enthalpy*, the *partial molar Helmholtz function*, or the *partial molar free energy*, provided that the proper variables are held constant, as indicated in equations 3.16a–d.

CONDITIONS FOR HETEROGENEOUS EQUILIBRIUM

We are now within reach of a fundamental goal for this chapter. Having examined the various ways in which internal energy is affected by changes in temperature, pressure, or composition, we may now ask: what conditions must be met if a system containing several phases is in internal equilibrium? This is a circumstance usually referred to as *heterogeneous equilibrium*.

A system consisting of several phases can be characterized by writing an equation in the form of equation 3.15a for each individual phase:

$$dE_\Phi \leq T_\Phi dS_\Phi - P_\Phi dV_\Phi + \sum (\mu_{i\Phi} dn_{i\Phi}),$$

where we are using the subscript Φ to identify properties with the individual phase. For example, if the system contained phases A, B, and C, we would write an equation for dE_A, another for dE_B, and a third for dE_C. The final term of each equation is the sum of the products $\mu_i dn_i$ for each system component (1 through i) for the individual phase.

It is easiest to examine the equilibrium condition among phases if we consider only the simple (and geochemically unlikely) situation in which system components correspond one-for-one with components of the individual phases. We also consider only the equilibrium conditions for an isolated system. We are doing this only for the sake of simplicity, however. We would arrive at the same conclusions if we took on the more difficult challenge of a closed or open system, or if we considered a set of system components that differ from the set of all phase components.

Because the system is isolated, we know that its extensive parameters must be fixed. That is,

$$dE_{sys} = \sum dE_\Phi = 0$$
$$dS_{sys} = \sum dS_\Phi = 0$$
$$dV_{sys} = \sum dV_\Phi = 0$$
$$dn_{1,sys} = \sum dn_{1,\Phi} = 0$$
$$dn_{2,sys} = \sum dn_{2,\Phi} = 0$$
$$\vdots$$
$$dn_{i,sys} = \sum dn_{i,\Phi} = 0.$$

There is one $dn_{i,sys}$ equation for each component in the system. Despite these equations, there is no constraint that prevents the relative values of E, S, V, or the various n's from readjusting themselves, as long as their totals remain zero. That is, the various individual values of E_Φ, V_Φ, and the $n_{i,\Phi}$ are not independent. Whatever leaves one phase in the system must show up in at least one of the others. On the other hand, there *are* constraints on the *intensive* parameters in the system. To see what they are, let's write an equation for the total internal energy change for the system, dE_{sys}:

$$dE_{sys} = \sum dE_\Phi = 0 = \sum (T_\Phi dS_\Phi) - \sum (P_\Phi dV_\Phi)$$
$$+ \sum \left(\sum (\mu_{i,\Phi} dn_{i\Phi}) \right).$$

The only way to guarantee that $\sum dE_{sys} = 0$ is for each of the terms on the right side of this equation to equal zero. We have already agreed that dS, dV, and each of the dn_i might differ from phase to phase. It would be a remarkable coincidence if T_Φ, P_Φ, and each of the $\mu_{i\Phi}$ could also vary among phases in such a way that $\sum (T_\Phi dS_\Phi)$, $\sum (P_\Phi dV_\Phi)$, and $\sum (\sum (\mu_{i\Phi} dn_{i\Phi}))$ were always equal to zero. Fortunately, we do not need to rely on coincidence. Unlike extensive properties, intensive properties are not free to vary among phases in equilibrium. We showed earlier in this chapter that this is true for temperature and pressure; at equilibrium

$$T_A = T_B = \ldots = T_\Phi$$

$$P_A = P_B = \ldots = P_\Phi.$$

It should be evident now that the same is true for chemical potentials; at equilibrium

$$\mu_{1A} = \mu_{1B} = \ldots = \mu_{1\Phi}$$

$$\mu_{2A} = \mu_{2B} = \ldots = \mu_{2\Phi}$$

$$\vdots$$

$$\mu_{iA} = \mu_{iB} = \ldots = \mu_{i\Phi}.$$

Because this conclusion is so crucial in geochemistry, we emphasize it again: *At equilibrium, the chemical potential of any component must be the same in all phases in a system.* To be sure that these general results are clear, let's examine heterogeneous equilibrium in a specific system.

Worked Problem 3.7

An experimental igneous petrologist, working in the laboratory, has produced a run product which consists of quartz (qz) and a pyroxene (cpx) intermediate in composition between $FeSiO_3$ and $MgSiO_3$. Assuming that the two minerals were formed in equilibrium, what conditions must have been satisfied?

To answer this question, first choose a set of components for the system. Several selections are possible, some of which we considered in an earlier problem. This time, let's choose the end-member mineral compositions $FeSiO_3$, $MgSiO_3$, and SiO_2. Heterogeneous equilibrium then requires that the intensive parameters be:

$$T_{qz} = T_{cpx} = T$$

$$P_{qz} = P_{cpx} = P$$

$$\mu_{SiO_2,qz} = \mu_{SiO_2,cpx} = \mu_{SiO_2}$$

$$\mu_{FeSiO_3,qz} = \mu_{FeSiO_3,cpx} = \mu_{FeSiO_3}$$

$$\mu_{MgSiO_3,qz} = \mu_{MgSiO_3,cpx} = \mu_{MgSiO_3}.$$

Notice that the chemical potentials of $FeSiO_3$ and $MgSiO_3$ are defined in quartz, despite the fact that they are not components of the mineral quartz itself, and μ_{SiO_2} is defined in pyroxene, although SiO_2 is not a component of pyroxene. All three are *system* components. The chemical potential of any component is a measure of the way in which the energy of a phase changes if we change the amount of that component in the phase. For example, if chemical potentials are defined by equation 3.16d, the constraint on $FeSiO_3$ should be read to mean that at constant temperature and pressure, the derivative of free energy with

respect to the mole fraction of $FeSiO_3$ is identical in pyroxene and in quartz. If it were possible to add the same infinitesimal amount of $FeSiO_3$ to each, the free energies of the two phases would each change by the same amount:

$$dG = \mu_{FeSiO_3} \, dn_{FeSiO_3}.$$

The Gibbs-Duhem Equation

One final, very useful relationship can be derived from this discussion of chemical potentials. Consider equation 3.15a for a phase that is in equilibrium with other phases around it. It is possible to write an integrated form of equation 3.15a:

$$E = TS - PV + \sum \mu_i n_i.$$

Therefore, the differential energy change, dE, that takes place if the system is allowed to leave its equilibrium state by making small changes in any intensive or extensive properties is:

$$dE = TdS + SdT - PdV - VdP + \sum \mu_i dn_i + \sum n_i d\mu_i.$$

To see how the intensive properties alone are interrelated, subtract equation 3.15a, which was written in terms of variations in extensive parameters alone, from this equation to get:

$$0 = SdT - VdP + \sum n_i d\mu_i. \qquad (3.17)$$

This expression is known as the *Gibbs-Duhem* equation. It tells us that there can be no net gradients in intensive parameters at equilibrium. Of more specific interest in geochemical problems, at constant temperature and pressure, there can be no net gradient in internal energy as a result of composition in a phase in internal equilibrium. This can be seen, in a way, as another way of stating the conclusions from our discussion of heterogeneous equilibrium.

SUMMARY

What is equilibrium? Our discussion of the three laws of thermodynamics has led us to discover several characteristics that answer this question. First, systems in equilibrium must be at the same temperature: this is the condition of *thermal equilibrium*. Second, provided that work is exclusively defined as the integral of PdV, there can be no pressure gradient between systems in equilibrium: this is the condition of *mechanical equilibrium*. Finally, the chemical potential of each component must be the same in all phases at equilibrium. This is the least obvious of the conditions we have discussed, and is the focus of many discussions in subsequent chapters.

We have also repackaged the internal energy function in several ways and examined the role of entropy, which will always be maximized during the approach to equilibrium. To demonstrate that each of the equilibrium criteria is a direct consequence of the three laws, we used the internal energy function, $E(S, V, n)$, as the basis of our discussion. As we progress, though, we will quickly abandon E in favor of the Gibbs free energy. Any of the energy functions, however, can be used to clarify the conditions of equilibrium.

With the ideas we have introduced in this chapter, we are now ready to begin exploring geologic environments. This has only been a first look at thermodynamics, however. We return to the topic many times.

SUGGESTED READINGS

There are many good introductory texts in thermodynamics, although most are written for chemists rather than for geochemists. Among the best, or at least most widely used, texts for geochemists are listed here, along with excellent but challenging articles and historical accounts.

Denbigh, K. 1968. *The Principles of Chemical Equilibrium.* London: Cambridge University Press. (Heavy going, but almost anything you want to know is in here somewhere.)

Ferry, J. M., ed. 1982. *Characterization of Metamorphism through Mineral Equilibria.* Reviews in Mineralogy 10. Washington, D.C.: Mineralogical Society of America. (An enlightening selection of articles designed for classroom use. Chapters 1–4 are particularly relevant to our first discussion of thermodynamics. Be prepared to stretch.)

Fraser, D. G., ed. 1977 *Thermodynamics in Geology.* Dordrecht, Holland: D. Reidel. (Topical chapters showing both fundamentals and applications of thermodynamics.)

Goldstein, M., and I. F. Goldstein. 1978. *How We Know: An Exploration of the Scientific Process.* New York: Plenum. (The historical development of our notions of heat and temperature is discussed in chapter 4 of this thought-provoking book.)

Greenwood, H. J., ed. 1977. *Short Course in Application of Thermodynamics to Petrology and Ore Deposits.* Mineralogical Association of Canada. (A very readable book, with many sections of interest to economic geologists.)

Kern, R., and A. Weisbrod. 1967. *Thermodynamics for Geologists.* San Francisco: Freeman, Cooper. (Perhaps the simplest treatment of thermodynamics available for the geologist, although now a bit dated.)

Lewis, G. N., and M. Randall. 1961. *Thermodynamics*, revised by K. S. Pitzer and L. Brewer. New York: McGraw-Hill. (A classic text still in widespread use.)

Nordstrom, D. K., and J. Munoz. 1985. *Geochemical Thermodynamics.* Menlo Park: Benjamin Cummings. (An excellent text for the serious student.)

Smith, E. B. 1982. *Basic Chemical Thermodynamics*, 3rd ed. New York: Oxford University Press. (A remarkably clear text for the serious beginner.)

Thompson, J. B., J. Laird, and A. B. Thompson. 1982. Reactions in amphibolite, greenschist, and blueschist. *Journal of Petrology* 23:1–27. (A good study for the motivated student. Reactions in amphibole-bearing assemblages are discussed in terms of exchange operators.)

Vanserg, N. 1958. Mathmanship. *The American Scientist* 46: 94A–98A. (This article does not deal with thermodynamics directly, but is a commentary on the use of symbols. Its author was better known by his real name, Hugh McKinstry, professor of geology at Harvard for many years.)

PROBLEMS

(3.1) Refer to figures 3.9 and 3.10. Notice that by adding or subtracting SiO_2, we change the proportions of olivine and pyroxene in a rock, but not their compositions. Changing the amount of any other system component in figure 3.9 or 3.10, however, alters both the compositions and the proportions of minerals. Can you find another set of components that includes one which, when varied, changes the compositions but not the proportions of olivine and pyroxene?

(3.2) Show that the work done by 1 mole of a gas obeying the Van der Waals equation during isothermal expansion from V_1 to V_2 is equal to $w = RT \ln([\bar{V}_2 - b]/[\bar{V}_1 - b] + a([1/\bar{V}_2] - [1/\bar{V}_1])$.

(3.3) How much work can be obtained from the isothermal, reversible expansion of one mole of chlorine gas from 1 to 50 liters at 0°C, assuming ideal gas behavior? What if chlorine behaves as a Van der Waals gas? Use the result of problem 3.2; $a = 6.493$ l^2 atm mol^{-2}, $b = 0.05622$ l mol^{-1}.

(3.4) What is the maximum work that can be obtained by expanding 10 grams of helium, an ideal gas, from 10 to 50 liters at 25°C? Express your answer in (a) calories, (b) joules.

(3.5) Suppose that a mixture of two inert gases is divided into two containers separated by a wall. Imagine a microscopic valve placed in the wall that allows molecules of one gas to move into container A and molecules of the other gas to move into container B, but does not allow either gas to move in the opposite direction. Such an imaginary device is called *Maxwell's demon*. How would such a device violate the Second Law of Thermodynamics?

(3.6) What conditions must be satisfied if hematite, magnetite, and pyrite are in equilibrium in an ore assemblage?

(3.7) Construct a triangular diagram for the alkali feldspar-mica system similar to figure 3.12a, but in which you swap the positions of $NaAl_3Si_3O_{10}(OH)_2$ and $KAlSi_3O_8$. What component must now take the place of KNa_{-1} at the top corner of the diagram? Where would this component have been plotted on figure 3.12a? Where does KNa_{-1} plot on your diagram?

(3.8) Using the Maxwell relations, verify that $-(\partial V/\partial T)_P /(\partial P/\partial T)_V = (\partial V/\partial P)_T$.

CHAPTER FOUR

HOW TO HANDLE SOLUTIONS

OVERVIEW

This is a chapter about solutions, important to us because most geological materials have variable compositions. They are, in fact, mixtures at the submicroscopic scale between idealized end-member substances such as albite, water, grossular, dolomite, or carbon dioxide, which lose their molecular identities in the mixture. To see how thermodynamics can be used to predict the equilibrium state in a system dominated by phases with variable compositions, we examine first the structure of solutions of solids, liquids, and gases, and discuss ways in which this architecture is reflected in mole fractions of end-member components.

At the end of chapter 3, we developed thermodynamic equations that describe the state of equilibrium for a heterogeneous system. Among these equations are constraints implying that the compositions of phases in equilibrium are controlled by the system's overall drive toward a minimum ΔG. As we study the structure of solutions in this chapter, we look explicitly at the compositions of phases in equilibrium and apply the thermodynamic principles from chapter 3. Specifically, we begin to develop equations that relate mole fractions and other mixing parameters (which we introduce) to ΔG.

The results of this investigation can be applied to many problems in the realm of geochemistry. We begin applying

them in this chapter by studying aqueous fluids and the phenomenon of solubility. This study not only presents an opportunity to use the equations we develop, but also prepares the way for our later discussions of the oceans, diagenesis, and weathering reactions.

WHAT IS A SOLUTION?

Most phases of geochemical interest do not have fixed compositions. With the conspicuous exception of quartz, each of the major rock-forming minerals is a solid solution between two or more end-member molecules. The range of allowable compositions in any solution is dictated by rules of structural chemistry that consider both crystal architecture and the size and electronic configuration of ions. For many minerals, these rules permit liberal replacement of iron by magnesium or calcium, sodium by potassium, or silicon by aluminum. At first glance, the degree of flexibility that we observe in mineral compositions looks random. It seems to spoil any hope that we might be able to apply thermodynamic principles to geologic materials. Liquids and gases follow even fewer structural guidelines and have even more variable compositions. Can the principles of equilibrium thermodynamics we have just explored in chapter 3 make sense in the more complicated world of these *real* phases? To answer, we first need to know more about solutions.

Crystalline Solid Solutions

Most geologically important materials are crystalline solids; that is, solids in which atoms are arranged in a periodic three-dimensional array that maintains its gross structural identity over fairly large distances. A "fairly large" distance is not easily defined, however. If we see a mineral grain in thin section that has continuous optical properties, it is clearly large enough. Even submicroscopic grains are judged to be crystalline if their structures are continuous over distances sufficient to yield an X-ray diffraction pattern. We can get reliable results from thermodynamic calculations for most geochemical purposes if the materials we study are crystalline at this scale. We can still apply principles of thermodynamics at smaller scales, but the rules become more complex. In very fine-grained crystalline aggregates, for example, surface properties make a major contribution to free energy. In highly stressed materials, too, dislocations or defects constitute a significant volume fraction of the material and call for special methods. Finally, intimately intergrown materials, such as those in figure 4.1, in which two or more phases are structurally compatible and randomly intermixed, are difficult to model. We take a look at some of these special cases in chapter 10, but for now we stick with crystalline materials that have uniform properties over distances of thousands of unit cell repeats.

The rules that dictate where atoms sit in a crystalline substance are quite flexible. Two major factors govern which atoms may occupy a given site: their ionic charges and their sizes. Charge is important because of the need to obtain electrical neutrality over the structure. Size is important because it influences the degree of "overlap" between the orbitals of valence electrons of the ion and those that surround it. Recall our discussion of these principles in chapter 2 and scrutinize figure 2.11 again. You will find that several ions available to a growing crystal will satisfy these constraints. In addition, the charge balance requirement can be satisfied by the formation of randomly distributed vacancies or by inserting ions into defects or into positions that are normally unoccupied. Local charge imbalances caused when an inappropriately charged ion occupies a site, therefore, can be averaged out over the structure as a whole. As a result, minerals are most properly viewed as crystalline solutions in which many competing ions substitute freely for one another. We must consider both the bulk composition of a solution and the way in which it is mixed as we describe it with thermodynamics.

To illustrate, let us compare two possible mixing schemes, using the clinopyroxene solid solution $CaMgSi_2O_6$ (diopside)-$NaAlSi_2O_6$ (jadeite) as an example. First, take a look at figure 4.2 as we do a quick overview of pyroxene structural chemistry. The atoms in silicates are

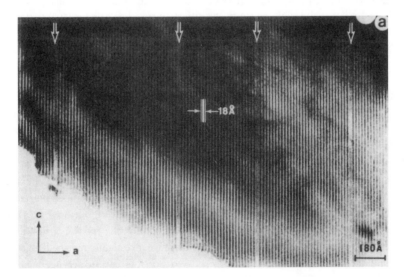

FIG. 4.1. High-resolution transmission electron microscope (HRTEM) image of an intergrowth of ortho- and clinopyroxenes. This material is crystalline, but the identity of the structural unit changes after a random, rather small number of repeats (arrows). Thermodynamic characterization of such material is very difficult.

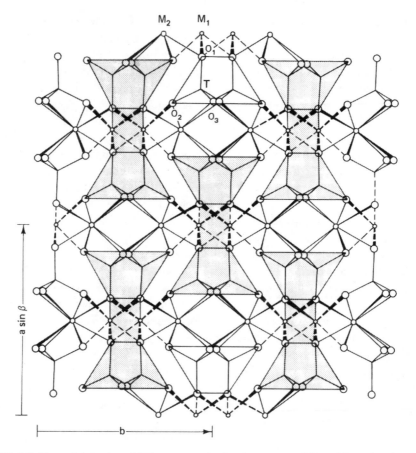

FIG. 4.2. The crystal structure of C2/c pyroxene, showing the geometry of M1 and M2 cation sites, each of which is surrounded by six oxygen atoms, and their relationship to tetrahedral (T) sites, which are largely occupied by silicon atoms. This diagram, a projection onto the (100) crystallographic plane, is drawn to emphasize the geometry of cation sites. (Modified from Cameron and Papike 1981.)

arranged so that each silicon is surrounded by four oxygens to form a tetrahedral unit with a net electrical charge of −4. These are connected to each other and to other atoms by bonds that are largely ionic in nature. The manner in which the units are linked together, and the identity of other atoms that occupy sites between them, determine the mineral's structure and establish its bulk chemical identity. (Review our discussion of coordination polyhedra in chapter 2.) In a pyroxene, SiO_4 tetrahedra are linked by corners to form long single chains. Diopside and jadeite are C2/c pyroxenes, in which cations occupy two other types of sites, designated M1 and M2, between these chains. M1 is surrounded by six oxygens and is nearly a regular octahedron. It is occupied by the relatively small cations Mg^{2+} and Al^{3+}. The

M2 site is eight-coordinated, 20–25% larger, and irregular in shape. Ca^{2+} and Na^+, relatively large cations, prefer this site. For a detailed but very readable description of pyroxene crystal chemistry, see the articles by Cameron and Papike (1980, 1981).

So much for the general context. Now, if we consider only the pure end-member pyroxenes, we should have no trouble figuring out how sites are occupied, because there is only one type of cation available for each site. Once we investigate intermediate compositions, however, two possible cation arrangements suggest themselves. The first, called *molecular mixing*, assumes that charge balance is always maintained over very short distances. That is, calcium and magnesium ions substitute for sodium and aluminum ions as a coupled pair: each time we find a Ca^{2+} in an M2 site, there is a Mg^{2+} in an adjacent M1 site. The end-member pyroxenes mix as discrete molecules and complete short-range order results. In this case, the mole fraction of $CaMgSi_2O_6$ in the solid solution ($X_{CaMgSi_2O_6,cpx}$) is equal to the proportion of the diopside molecule present. Until recently, molecular mixing was widely assumed by petrologists.

The other possibility, known as *mixing on sites,* assumes that Ca^{2+} and Na^+ are randomly distributed on M2 sites, unaffected by whether adjacent M1 sites are filled with Mg^{2+} or Al^{3+} ions. This style of mixing may produce local charge imbalances, but the average stoichiometry is satisfied over long distances even in the absence of short-range order. Because each of the sites, according to this model, behaves independently of the others, we must now calculate the mole fraction $X_{CaMgSi_2O_6,cpx}$ as the product of $X_{Ca,M2,cpx}$ and $X_{Mg,M1,cpx}$. This follows from a basic principle of statistics: if the probability of finding a Mg^{2+} ion in M1 is some value x and the probability of finding a Ca^{2+} in M2 is y, then the combined probability of finding both ions in the structure is xy. Thus, if 80% of the ions in the M1 site are Mg^{2+} and 80% of the M2 atoms are Ca^{2+}, $X_{CaMgSi_2O_6,cpx}$ is equal to 0.64. If the tetrahedral site were partially occupied by something other than silicon, we would have to multiply by $X_{Si,T,cpx}$ as well.

How can we tell which of these two mixing models is the correct one? It is logical to infer that the degree of randomness among atoms distributed on M1 and M2 affects the molar entropy of a phase, and therefore each of the energy functions E, H, F, and G. In theory, then, it should be possible to use these two extreme models to calculate the relative stabilities of clinopyroxene solutions and see which one matches what we see in the natural world. To do this requires calculating bond energies from electrostatic equations and detailed crystallographic data. Large-scale computer models exist for this purpose, but are successful only with very simple structures. Ronald Cohen (1986), however, has done the next best thing. By evaluating a large body of calorimetric data and phase equilibrium studies for aluminous clinopyroxenes, he concluded that the macroscopic behavior of pyroxene solid solutions provides no observational justification for molecular mixing. His conclusions suggest that other common silicate solutions (feldspars, micas, amphiboles) behave similarly. Unless crystallographic evidence suggests otherwise, therefore, we may assume that mixing takes place on sites.

This is a significant finding. It is not the end of the discussion about how to model clinopyroxenes, however. We have assumed that each of the cations occupies only one type of site, for example. In fact, this is not the case. Although Mg^{2+} prefers the smaller M1 site, it may also be present in M2. Fe^{2+} and Mn^{2+} may also be found in either M1 or M2. Other possible complications include the stabilizing effect felt by certain transition metal cations as a result of site distortions (more about this in chapter 12) and the degree of covalency of bonds in the structure (refer again to chapter 2). Because we generally have no way of testing each of these effects directly, thermodynamic interpretation of crystalline solutions is an empirical process, based largely on the macroscopic behavior of materials. Models of site occupancy rely strongly on observations made by spectroscopic and X-ray diffraction methods, and on our understanding of the rules of crystal chemistry that govern element partitioning. These still leave considerable uncertainty, so most calculations can only be performed by making simplifying assumptions.

Worked Problem 4.1

To illustrate how we can calculate mole fractions in a disordered mineral structure, consider the following analysis of a C2/c omphacite (pyroxene), which was first reported by Clark et al. (1969). The abundances are recorded as numbers of cations per six oxygens in the clinopyroxene unit cell. (The analysis in Clark et al. [1969] also includes 0.002 Ti^{3+}, which we have omitted for simplicity. Including Ti^{3+} would change the contents of the M1 site slightly.)

Atom	Abundance
Si	1.995
Al	0.238
Fe^{3+}	0.123
Fe^{2+}	0.116
Mg	0.582
Ca	0.583
Na	0.325

To estimate the site occupancy, we first assume that all tetrahedral sites are filled with either Si or Al (the only two atoms in the analysis that regularly occupy tetrahedral sites). Because the ratio of tetrahedral cations to oxygens in a stoichiometric pyroxene should be 2/6, we see that 0.005 Al must be added to the 1.995 Si to fill the site. An alternative, which might also be justifiable, is to assume that Si is the only tetrahedral cation, and simply normalize the Si in the analysis to 2.00.

Of the remaining cations, we may assume that Ca and Na (both of which are large) are restricted to the M2 site, whereas

Al and Fe^{3+} (both relatively small) prefer the more compact M1 site. X-ray structure refinement by Clark et al. indicates, further, that the ratio Fe/Mg in M1 is 0.40 and in M2 is 0.54. Using these pieces of information, we proceed as follows:

1. The amount of Fe^{2+} in M1 is some unknown quantity that we call x. The amount of Mg in M1, also unknown, we call y. We know, however, that $(Fe^{3+} + x)/y = 0.4$.

2. The amount of Fe^{2+} in M2 is unknown, but must be equal to $0.116 - x$, just as the amount of Mg in M2 must be $0.582 - y$. We know, therefore, that $(0.116 - x)/(0.582 - y) = 0.54$.

3. If we solve the two equations in x and y simultaneously, we conclude that $x = 0.092$ and $y = 0.538$.

4. The total cation abundance in M1 ($\sum M1$) is, therefore,

$$\sum M1 = (Al) + (Fe^{3+}) + x + y = 0.233 + 0.123 + 0.092 + 0.538 = 0.986.$$

The total cation abundance in M2 ($\sum M2$), similarly, is:

$$\sum M2 = (Ca) + (Na) + (0.116 - x) + (0.582 - y) = 0.583 + 0.325 + 0.024 + 0.044 = 0.976.$$

5. To calculate the mole fractions on each site, we normalize their respective contents by $1/\sum M1$ or $1/\sum M2$ to get:

M1		M2	
Cation	$X_{,M1}$	Cation	$X_{,M2}$
Al	0.236	Ca	0.598
Fe^{3+}	0.125	Na	0.333
Fe^{2+}	0.093	Fe^{2+}	0.025
Mg	0.546	Mg	0.045

One way to model this pyroxene is to choose likely molecular end members such as $NaAlSi_2O_6$ (jadeite), $CaMgSi_2O_6$ (diopside), $NaFe^{3+}Si_2O_6$ (aegirine), among others. We might then calculate mole fractions of each on the assumption that an M1 site occupied by Al is always adjacent to an M2 site containing Na, an M2 occupied by Na is either adjacent to an Al or an Fe^{3+} in M1, and so forth. We have just argued that molecular mixing is generally a poor way to proceed, however. Instead, if we wish to consider a reaction in which the $NaAlSi_2O_6$ component of this pyroxene is involved, we calculate its mole fraction as the cumulative probability of finding both Al and Na at random on their respective sites:

$$X_{Al,M1,cpx} X_{Na,M2,cpx} = (.236)(.333) = 0.079.$$

Likewise,

$$X_{Aeg,cpx} = X_{Fe^{3+},M1,cpx} X_{Na,M2,cpx} = (.125)(.333) = 0.042,$$

and

$$X_{Di,cpx} = X_{Mg,M1,cpx} X_{Ca,M2,cpx} = (.546)(.598) = 0.327.$$

Amorphous Solid Solutions

Glasses and other amorphous geological materials have structures that differ from crystalline materials only in their degree of coherency. An amorphous solid has no long-range structural continuity but may still consist of numerous very small-scale ordered domains. Many of the concerns we had about mixing in crystalline solids are therefore also valid for amorphous materials. Their greatest importance for geochemists, however, is that they provide a conceptual bridge toward our understanding of liquids.

Melt Solutions

The only geologically significant melt systems that we understand in any detail are silicate magmas. Our knowledge of silicate melt structures has been acquired primarily since the late 1970s, with the help of improved X-ray diffraction and spectroscopic methods. Geochemists now understand that silicon and oxygen atoms continue to associate in SiO_4 tetrahedral units beyond the melting point, and that the units tend to form polymers. Depending on the composition of the melt, these units may be chain polymers similar to the ones found in pyroxenes, or they may be branched structures of various sizes—the remnants (or precursors) of sheet and framework crystalline structures.

The degree of polymerization depends on temperature, but is most strongly influenced by the proportion of *network-formers* such as silicon and aluminum and *network-modifiers* such as Fe^{2+}, Mg^{2+}, Ca^{2+}, K^+, and Na^+. In silica-rich melts, the highly covalent nature of the Si-O bond tends to stabilize structures in which a large number of oxygen atoms are shared between SiO_4 units. As a result, these melts are highly viscous and have low electrical conductivities. Adding even a small proportion of Na_2O or another network modifier, however, reduces viscosity drastically and increases melt conductivity. As in crystalline silicates, the bonds between oxygen and the larger cations are ionic. These break more easily than the Si-O bonds, thus reducing the melt's overall tendency to form large, tenacious polymeric structures. Water also acts as a network modifier in silica-rich melts, apparently by reacting with the bridging oxygens in a polymer. This reaction may be described generically by:

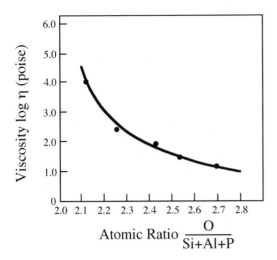

FIG. 4.3. Relationship between melt viscosity and composition at 1150°C. The variable on the horizontal axis is the atomic ratio of oxygen atoms to network-formers (Si + Al + P). The filled circles represent compositions of natural magmas.

$$H_2O + \overset{\mid}{\underset{\mid}{-Si}}-O-\overset{\mid}{\underset{\mid}{Si}}- \rightleftarrows 2(-\overset{\mid}{\underset{\mid}{Si}}-OH).$$

Figure 4.3 illustrates the degree to which melt structure influences thermodynamic properties by showing the qualitative relationship between viscosity and composition.

Worked Problem 4.2

Before much of the modern experimental work to examine to relationship between melt structure and viscosity was begun, Jan Bottinga and Daniel Weill (1972) developed a simple empirical method for predicting melt viscosity. They noted that the natural logarithm of viscosity (η) may be approximated by a function of composition that is linear within restricted intervals of SiO_2 content. Specifically, they showed that the empirical coefficients D_i in the equation:

$$\ln \eta = \sum X_i D_i$$

are constants at a given temperature and a given value of X_{SiO_2}.

In the table below are values of D_i at 1400°C for melts with two different silica contents, calculated statistically by Bottinga and Weill from the measured viscosities of a large number of silicate melts. Notice from the arithmetic signs on D_i that some melt components tend to increase viscosity (that is, they act as network-formers), whereas others tend to decrease it (by acting as network-modifiers). Notice also that some components, such as FeO, behave differently in silica-rich melts than in silica-poor ones.

Component	$.75 < X_{SiO_2} < .81$ D_i	$.55 < X_{SiO_2} < .65$ D_i
SiO_2	10.50	9.25
TiO_2	−3.61	−4.26
FeO	6.17	−6.64
MnO	−6.41	−5.41
MgO	−2.23	−4.27
CaO	−3.61	−5.54
$MgAl_2O_4$	−4.82	−1.71
$CaAl_2O_4$	−1.74	−0.22
$NaAlO_2$	15.1	7.48
$KAlO_2$	15.1	7.48

To illustrate this method, we consider two different melt compositions, a granite and a diabase. For each, we have tabulated a bulk analysis in weight percentage of oxides and a recalculated analysis in mole percentage. In the recalculation, all iron has been converted to FeO, phosphorus has been added to SiO_2, water has been omitted (a serious omission, but at least we can compare anhydrous melt viscosities this way), and the total has been adjusted to 100 mole percent.

Component	Granite Wt. %	Diabase Wt. %
SiO_2	70.18	50.48
TiO_2	0.39	1.45
Al_2O_3	14.47	15.34
Fe_2O_3	1.57	3.84
FeO	1.78	7.78
MnO	0.12	0.20
MgO	0.88	5.79
CaO	1.99	8.94
Na_2O	3.48	3.07
K_2O	4.11	0.97
H_2O	0.84	1.89
P_2O_5	0.19	0.25
	Mole %	Mole %
SiO_2	80.12	58.45
TiO_2	0.38	1.26
FeO	1.83	10.83
MnO	0.11	0.12
MgO	1.02	9.95
CaO	0.00	4.76
$MgAl_2O_4$	0.47	0.00
$CaAl_2O_4$	2.41	6.28
$NaAlO_2$	7.68	6.85
$KAlO_2$	5.97	1.43

For the granite,

$$\ln \eta = \sum X_i D_i = (.8013)(10.5) + (.0038)(-3.61)$$
$$+ (.0183)(6.17) + (.0011)(-6.41)$$
$$+ (.0102)(-2.23) + (.0047)(-4.82)$$
$$+ (.0241)(-1.74) + (.0768)(15.1)$$
$$+ (.0597)(15.1) = 10.48$$

$$\eta = 3.56 \times 10^4 \text{ poise} = 3.56 \times 10^3 \text{ kg m}^{-1} \text{ sec}^{-1}.$$

For the diabase,

$$\ln \eta = \sum X_i D_i = (.5845)(9.25) + (.0126)(-4.26)$$
$$+ (.1083)(-6.64) + (.0019)(-5.41)$$
$$+ (.0995)(-4.27) + (.0476)(-5.54)$$
$$+ (.0628)(-0.22) + (.0685)(7.48)$$
$$+ (.0143)(7.48) = 4.54$$

$$\eta = 9.37 \times 10^1 \text{ poise} = 9.37 \text{ kg m}^{-1} \text{ sec}^{-1}.$$

To an even greater extent than is true for crystalline solids, our limited understanding of melt structures makes it impossible to calculate thermodynamic quantities from models that consider the potential energy contributions of individual atoms. Enthalpies and free energies of formation for melts are determined on macroscopic systems, and therefore represent an average of the contributions from an extremely large number of local structures. Thermodynamic models that make simple assumptions about how mixing takes place between melts of different end-member compositions are surprisingly successful, but tend to be more descriptive than predictive. We discuss these more fully in chapter 9.

Electrolyte Solutions

An *electrolyte solution* is one in which the dissolved species (the *solute*) are present in the form of ions in a host fluid that consists primarily of molecular species (the *solvent*). In the geological world, these are usually aqueous fluids, although CO_2-rich solvents may be increasingly important with increasing depth in the crust and upper mantle. Electrolyte solutions do not exhibit the large-scale periodic structures found in either melts or solids. The ions and solvent molecules, however, cannot generally be treated as a mechanical mixture of independent species. Electrostatic interactions among solute and solvent species place limits on the concentration of free solute ions and can produce complexes that influence the thermodynamic behavior of a solution. These effects are most obvious as the concentration of dissolved species increases and require a rather complicated mathematical treatment. Many of the later sections of this chapter are devoted to understanding the behavior of aqueous electrolyte solutions.

Gas Mixtures

Gases at low pressure mix nearly as independent molecules. In this way, they are perhaps the simplest of geologic materials and therefore the best to study initially. Except for transient species in high-energy environments, most gas species are electrically neutral. Their neutrality and the relatively large distances between molecules at low pressure allow us to treat atmospheric and most volcanic gases as mechanical mixtures. As such gas mixtures interact with other geological materials (say, in weathering reactions), the thermodynamic influence of each of the individual gas species is directly proportional to its abundance in the mixture. Only at elevated pressures do gas molecules interfere with each other and begin to alter the thermodynamic behavior of the mixture.

SOLUTIONS THAT BEHAVE IDEALLY

For each type of solution we have considered in this brief discussion, there is some range of composition, pressure, and temperature over which the end-members mix as nearly independent species. For many, this range is distressingly narrow, but it provides us with a simplified system from which we can extrapolate to examine the behavior of "real" solutions. When the end-member constituents of solutions act nearly as if they were independent, we refer to their behavior as *ideal*.

In this section and the following one on nonideal solutions, we begin with a discussion of gases. This is done partly because gases are found in every geological environment, but largely because their behavior is the easiest to model.

We have already seen (equation 3.11) that the Gibbs free energy for a one-component system can be written as $G(T, P)$. In differential form:

$$dG = -SdT + VdP.$$

Furthermore, dG can be recognized as $dG = nd\bar{G} = nd\mu$. Suppose we are interested in the difference in chemical potential between some reference state for the system (for which we will continue to use the superscript 0) and another state in which temperature or pressure is different. To extract this information, we must integrate dG between the two states:

$$\int_{\bar{G}^0}^{\bar{G}} nd\bar{G} = n\int_{\mu^0}^{\mu} d\mu = -\int_{T^0}^{T} SdT + \int_{P^0}^{P} VdP.$$

The integration, unfortunately, is harder than it looks. We can move the variable n outside of the integrals for $\bar{G}$ and μ because the amount of material in a system is not a function of its molar free energy. We could not do the same with S or V, however, because they are, respectively, functions of T and P. To make the problem solvable, we need to supply $S(T)$ and $V(P)$. The first of these is difficult enough to define that we will avoid it for now by declaring that we are only interested in isothermal processes. In this chapter, therefore, the integral of SdT is equal to zero. Later, in chapter 9, we consider temperature variations.

The pressure-volume integral is easier to handle. The function $V(P)$ is obtained from an equation of state, the most familiar of which is the ideal gas equation: $V = nRT/P$. The integral equation for an isothermal ideal gas, therefore, reduces to:

$$n\int_{\mu^0}^{\mu} d\mu = nRT\int_{P^0}^{P} dP.$$

If we do the integration, we see that:

$$\mu - \mu^0 = RT \ln(P/P^0).$$

It is standard practice to choose the reference pressure as unity, so that this equation takes the form

$$\mu = \mu^0 + RT \ln P. \qquad (4.1)$$

Notice that because P in equation 4.1 really stands for P/P^0, it is a dimensionless number. We have now derived an expression for the chemical potential of a single component in a one-component ideal gas at a specified temperature and pressure.

Next, consider a phase that is a multicomponent mixture of ideal gases. There must be a stoichiometric equation that describes all of the potential interactions between gas species in the mixture:

$$v_{j+1}A_{j+1} + \ldots + v_{i-1}A_{i-1} + v_iA_i$$
$$= v_1A_1 + v_2A_2 + \ldots + v_jA_j.$$

This is a standard equation that you have seen many times, but the notation may look a little strange, so let's pause to explain it. We keep track of each gas species by associating it with a unique subscript. Subscripts 1 through j refer to product species and $j + 1$ through i refer to reactants. The quantities v_1 through v_i are stoichiometric coefficients and $A_1, A_2, A_3, \ldots, A_i$ are the individual gases. If the gas were a two-component system of carbon and oxygen species, for example, and if we were to ignore all possible species except for CO_2, CO, and O_2, then the stoichiometric equation would be

$$CO + \tfrac{1}{2}O_2 = CO_2.$$

The species CO, O_2, and CO_2 here are A_1, A_2, and A_3, and the stoichiometric coefficients are $v_1 = v_2 = 1$ and $v_3 = 0.5$. We use this notation in stoichiometric equations throughout the rest of this book.

Returning now to the gas mixture, we find that the expanded form of equation 3.11 to describe variations in free energy looks like:

$$dG = -SdT + VdP + \sum_{k=1}^{j} \mu_k dn_k - \sum_{k=j+1}^{i} \mu_k dn_k, \qquad (4.2)$$

in which the chemical potentials are properly defined as partial molar free energies (equation 3.16d):

$$\mu_i = (\partial G/\delta n_i)_{T,P,n_{j\neq i}} = \bar{G}_i.$$

The summation terms in equation 4.2 do not contradict equation 3.15d, where positive and negative compositional effects on free energy were combined implicitly in a single term. Here, we separate them explicitly merely to emphasize that quantities associated with product species (1 through j) are meant to be added to the free energy of the system; those associated with reactant species ($j + 1$ through i) are subtracted.

If you have been following this discussion closely, it might bother you that we said we would use the subscript i in stoichiometric equations to identify real chemical species and have now reverted to the convention of using i to identify components. As we have emphasized before, the relationship between species and components is one of the hardest concepts in this book to understand, so we encourage you to re-examine the final sections of chapter 3 if you are confused. The practice of using i in equation 4.2 to count species implies that there are stoichiometric equations that relate species compositions to system components. For example, if we were describing the two-component system of carbon and oxygen species it would be fair to incorporate μ_{CO}, μ_{O_2}, and μ_{CO_2} in equation 4.2, even though only two of the quantities CO, O_2, and CO_2 can be system components. The parallel to the stoichiometric equation $CO + \tfrac{1}{2}O_2 = CO_2$ is a Gibbs-Duhem equation (3.17), which assures us that at equilibrium $\mu_{CO} + 0.5\mu_{O_2} = \mu_{CO_2}$. The individual quantities dn_i in equation 4.2, in other words, are not independent, but are tied to the stoichiometric coefficients v_i and related to each other by:

$$(dn_1/v_1) = (dn_2/v_2) = \ldots = (dn_j/v_j) = -(dn_{j+1}/v_{j+1})$$
$$= \ldots = -(dn_{i-1}/v_{i-1}) = -(dn_i/v_i) = d\zeta.$$

The important point here is that changes in the amount of each species in the system are proportional by a common factor to their stoichiometric coefficients in the reaction. That factor, $d\zeta$, is a convenient measure of how far a reaction among species has proceeded.

With the insight that $v_i = dn_i/d\zeta$, we can rewrite equation 4.2 at constant T and P as:

$$dG = (v_1\mu_1 + v_2\mu_2 + \ldots + v_j\mu_j - v_{j+1}\mu_{j+1} - \ldots$$
$$- v_{i-1}\mu_{i-1} - v_i\mu_i)d\zeta$$
$$= \left(\sum_{k=1}^{j} v_k\mu_k - \sum_{k=j+1}^{i} v_k\mu_k\right)d\zeta$$
$$= \Delta\bar{G}_r d\zeta.$$

We have just derived a very useful quantity,

$$\Delta\bar{G}_r = dG/d\zeta , \qquad (4.3)$$

known as the *free energy of reaction* or the *chemical reaction potential*. At equilibrium, it has the value zero. If $\Delta\bar{G}_r < 0$, then the reaction proceeds to favor the product species; if $\Delta\bar{G}_r > 0$, then the reaction is reversed to favor reactant species. This is a quantitative explanation of LeChatlier's Principle, which states that an excess of species involved on one side of a reaction causes the reaction to proceed toward the opposite side.

Because this is still a discussion of ideal gas mixtures, the equation of state for the gas phase remains $V = nRT/P$, but the total amount of material in the system, n, is now equal to the sum of the amounts of species, n_i. The pressure exerted by any individual species in the mixture is, therefore, a *partial pressure,* defined by:

$$P_i = n_i RT/V = Pn_i/n = PX_i.$$

For each species in a gas mixture, then, we can write an expression like equation 4.1:

$$\mu_i = \mu_i^0 + RT \ln P_i. \qquad (4.4)$$

For the mixture as a whole, equations 4.3 and 4.4 tell us that the free energy of reaction can be divided into two parts, one of which refers to the free energy in some standard state (often one in which solutions with variable compositions consist of a limiting end member) and the other of which tells us how free energy changes as a result of compositional deviations from the standard state. In summary, the free energy of reaction is written as:

$$\Delta\bar{G}_r = \sum v_i\mu_i$$
$$= \sum v_i\mu_i^0 + RT \sum \ln P_i$$
$$= \Delta\bar{G}^0 + RT \sum \ln P_i^{v_i} \qquad (4.5)$$
$$= \Delta\bar{G}^0 + RT \ln(P_1^{v_1}P_2^{v_2} \ldots P_j^{v_j}/P_{j+1}^{v_{j+1}} \ldots P_{i-1}^{v_{i-1}}P_i^{v_i}).$$

If the gas phase is in internal (homogeneous) equilibrium, then $\Delta\bar{G}_r = 0$ and:

$$-\Delta\bar{G}^0/RT = \ln(P_1^{v_1}P_2^{v_2} \ldots P_j^{v_j}/P_{j+1}^{v_{j+1}} \ldots P_{i-1}^{v_{i-1}}P_i^{v_i}),$$

or

$$e^{-\Delta\bar{G}^0/RT} = K_{eq}$$
$$= (P_1^{v_1}P_2^{v_2} \ldots P_j^{v_j}/P_{j+1}^{v_{j+1}} \ldots P_{i-1}^{v_{i-1}}P_i^{v_i}). \qquad (4.6)$$

The new term, K_{eq}, is called the *equilibrium constant*, and equation 4.6 is the first answer to our question: "What controls the compositions of phases at equilibrium?"

It is time to apply what we have discussed to a specific example.

Worked Problem 4.3

The Soviet Venera 7 planetary probe measured the surface temperature on Venus to be 748 K and determined that the total pressure of the atmosphere is 90 bar. The Pioneer Venus mission, which parachuted four probes to the surface in 1978, determined that the lower atmosphere consists of 96% CO_2 and contains 20 ppm of CO. Other gases in the atmosphere (primarily N_2) have a marginal effect on chemical equilibria in the C-O system. Given these data and assuming that CO_2, CO, and O_2 behave as ideal gases, what should be the partial pressure of O_2 at the surface of Venus?

The first step is to write a balanced chemical reaction among the species:

$$CO + \tfrac{1}{2}O_2 \rightleftarrows CO_2.$$

Equation 4.6 for this example, then, should be written:

$$K_{eq} = e^{-\Delta\bar{G}^0/RT} = P_{CO_2}/(P_{CO}P_{O_2}^{1/2}).$$

We can calculate the value of $\Delta\bar{G}^0$ from the standard molar free energies of formation of each of the reactant and product species from the elements at 475°C (748 K):

$$\Delta\bar{G}^0 = \Delta\bar{G}_{f,CO_2} - \Delta\bar{G}_{f,CO} - \tfrac{1}{2}(\Delta\bar{G}_{f,O_2})$$
$$= -94.516 - (-42.494) - \tfrac{1}{2}(0.0)$$
$$= -52.022 \text{ kcal mol}^{-1}.$$

The value of K_{eq}, therefore, is:

$$K_{eq} = \exp[52.022 \text{ kcal mol}^{-1}/(0.001987 \text{ kcal deg}^{-1} \text{ mol}^{-1})$$
$$(748 \text{ K})] = 1.59 \times 10^{15}.$$

From this and the expression for K_{eq}, we calculate that:

$$P_{O_2} = (P_{CO_2}/K_{eq}P_{CO})^2 =$$
$$(86.4 \text{ bar}/(1.59 \times 10^{15})(.00002)(90 \text{ bar}))^2 =$$
$$9.11 \times 10^{-22} \text{ bar}.$$

This quick "back of the envelope" answer is consistent with what we know from observations: the amount of free oxygen in

Venus's atmosphere is below the detection limits of our planetary probes. Except as an example of how to use K_{eq}, however, you should not take this numerical result very seriously. Several other factors would need to be considered if we were to attempt a rigorous calculation, including equilibria involving solid silicate and carbonate minerals in the Venusian crust.

With only small modifications, this derivation for gaseous systems can yield a similar expression that can be used for ideal liquid or solid solutions. If we stipulate that both pressure and temperature are constant for these solutions, then:

$$dP_i = d(PX_i) = P\,dX_i.$$

The equation analogous to 4.4 is therefore:

$$\mu_i = \mu_i^* + RT \ln X_i, \tag{4.7}$$

and the free energy of reaction is:

$$\Delta \bar{G}_r = \Delta \bar{G}^* + RT \ln [X_1^{v_1} X_2^{v_2} \ldots X_j^{v_j} / (X_{j+1}^{v_{j+1}} \ldots X_{i-1}^{v_{i-1}} X_i^{v_i})]. \tag{4.8}$$

The superscript * on both μ^* and $\Delta \bar{G}^*$ is a reminder that each is a function of temperature and pressure, unlike μ^0 and $\Delta \bar{G}^0$, which depend only on temperature. When we discuss the effects of pressure on mineral equilibria in chapter 9, we will see that it is necessary to add an extra term to the free energy equations for solid and liquid phases to allow for compressibility. For gases, that adjustment is made in the $RT \ln K_{eq}$ term.

Because equations 4.5 and 4.8 are similar in form, they can be combined to formulate a general equation relating the compositions of several ideal solutions or gas mixtures in a geochemical system to their molar free energies:

$$\Delta \bar{G}_r = (\Delta \bar{G}^0 + \Delta \bar{G}^*) + RT \ln [X_1^{v_1} X_2^{v_2} \ldots X_j^{v_j} P_1^{v_1} P_2^{v_2} \ldots P_j^{v_j} / (X_{j+1}^{v_{j+1}} \ldots X_{i-1}^{v_{i-1}} X_i^{v_i} P_{j+1}^{v_{j+1}} \ldots P_{i-1}^{v_{i-1}} P_i^{v_i})]. \tag{4.9}$$

Each of the free energies, mole fractions, and partial pressures should be understood to carry a subscript to identify the phase to which they refer. For example, the mole fraction of $CaAl_2Si_2O_8$ in plagioclase should be shown as $X_{CaAl_2Si_2O_8,Pl}$. Except where they are necessary to avoid ambiguity, however, we usually follow standard practice and omit phase subscripts. Let's try a problem involving both a gas phase and condensed phases to see how equation 4.9 works in practice.

Worked Problem 4.4

During greenschist facies metamorphism (~400°C), rocks containing crysotile, calcite, and quartz commonly react to form tremolite, releasing both CO_2 and water. Assume, for the moment, that calcite is pure $CaCO_3$, that the other two solids are pure magnesian end members of crystalline solutions, and that water and CO_2 mix as an ideal gas phase. This reaction is strongly affected by total pressure, and we have not yet discussed adjustments for pressure. Suppose, however, that you tried to simulate this reaction by heating crysotile, calcite, and quartz to 400°C in an open crucible. What would $\Delta \bar{G}_r$ be under these circumstances?

The balanced reaction for this example is:

$$5\,Mg_3Si_2O_5(OH)_4 + 6\,CaCO_3 + 14\,SiO_2$$
$$\rightleftarrows 3\,Ca_2Mg_5Si_8O_{22}(\dot{O}H)_2 + 6\,CO_2 + 7\,H_2O.$$

Data relevant to this problem are available in Helgeson (1969):

	Tremolite	CO_2	H_2O	Crysotile	Calcite	Quartz
$\Delta \bar{H}^0$	−2894140	−87451	−54610	1012150	−279267	−212480 cal mol⁻¹
$\bar{S}^0$	258.8	65.91	51.97	118.55	41.65	20.87 cal deg⁻¹ mol⁻¹

The partial pressure of CO_2 in the atmosphere is ~3.2×10^{-4} atm, and the partial pressure of H_2O, although variable, is ~1×10^{-2} atm. If the crucible is open to the atmosphere and mixing is rapid, then these partial pressures should remain roughly constant.

The molar free energy of reaction can be calculated from:

$$\Delta \bar{G}_r = \Delta \bar{H}^0 - T\Delta \bar{S}^0 + RT \ln K_{eq},$$

in which:

$$K_{eq} = \frac{X_{Ca_2Mg_5Si_8O_{22}(OH),trem}^3 P_{CO_2}^6 P_{H_2O}^7}{(X_{Mg_3Si_2O_5(OH)_4,cry}^5 X_{CaCO_3,cc}^6 X_{SiO_2,qz}^{14})}.$$

From the data above, we calculate:

$$\Delta \bar{H}^0 = 3(-2894140) + 6(-87451) + 7(-54610)$$
$$- 5(-1012150) - 6(-279267) - 14(-212480)$$
$$= 121256 \text{ cal mol}^{-1}$$

$$\Delta \bar{S}^0 = 3(258.8) + 6(65.91) + 7(51.97) - 5(118.55)$$
$$- 6(41.65) - 14(20.87) = 399.32 \text{ cal deg}^{-1} \text{ mol}^{-1}.$$

Because we assumed that the solid phases have end-member compositions, each of the mole fractions is equal to one. To reach the final answer, then, we calculate:

$$\Delta \bar{G}_r = 121256 - 673(399.32) + 1.987(673)[6 \ln(3.2 \times 10^{-4}) + 7 \ln(1 \times 10^{-2})] = -255161 \text{ cal mol}^{-1}.$$

The reaction, therefore, favors the products under the conditions specified. We should expect CO_2 and water vapor to be released from our open crucible.

Suppose that we performed the same experiment, but used an impure calcite with a composition $(Ca_{0.8}Mn_{0.2}CO_3)$. How would you expect the free energy of reaction to change, assuming that Ca-Mn-calcite behaves as an ideal solid solution?

The expressions for $\Delta \bar{G}_r$ and K_{eq} remain the same, as do the values for $\Delta \bar{H}^0$ and $\Delta \bar{S}^0$. The mole fraction of $CaCO_3$ in calcite is no longer 1.0, however, so the value of $(-6 \ln X_{CaCO_3})$ is no longer zero. We find, now, that:

$$\Delta \bar{G}_r = 121256 - 673(399.32) + 1.987(673)$$
$$(6 \ln(3.2 \times 10^{-4}) + 7 \ln(1 \times 10^{-2}) - 6 \ln(0.8)$$
$$= -253371 \text{ cal mol}^{-1}.$$

The reaction still favors the products, but $\Delta \bar{G}_r$ has increased by 1790 cal mol^{-1}.

SOLUTIONS THAT BEHAVE NONIDEALLY

In many cases, it is appropriate to assume that real gases follow the ideal gas equation of state. More generally, however, interactions between molecules force us to modify the ideal gas law by adding correction terms to adjust for nonideality. Before we worry about the precise nature of these adjustments, let us illustrate how nonideality complicates matters by writing the corrected equation of state as a generic power function:

$$PV = RT + BP + CP^2 + DP^3 + EP^4 + \ldots,$$

in which the coefficients B, C, D, and so forth are empirical functions of T alone. Assuming isothermal conditions and following the procedure by which we derived equation 4.1, we now get:

$$\mu = \mu^0 + RT \ln P + B(P - 1) + C(P^2 - 1)/2 + \ldots.$$

This is a clumsy equation to handle, particularly if we are headed for a generalized form of equation 4.9, and it is only valid for the empirical equation of state we chose. The way we avoid this problem is to repackage all the terms containing P and introduce a new variable, f, so that:

$$\mu = \mu^0 + RT \ln(f/f^0). \tag{4.10}$$

This new quantity is known as *fugacity*, and it will serve in each of the equations as a "corrected" pressure; that is, as a pressure which has been adjusted for the effects of nonideality. Notice that equation 4.10 is identical in form to equation 4.1. The reference fugacity, f^0, used for comparison is usually taken to be the fugacity of the

pure substance at 1 atmosphere total pressure and at a specified temperature. (A variety of standard states are possible. For any given problem, the one we choose is partly governed by convenience and partly by convention. In chapter 9, we discuss a number of alternate standard states.) For a pure (one-component) gas, $f^0 = P^0 = 1$ atm. The ratio f/f^0 is given the special name *activity*, for which we use the symbol a. The activity of a gas is, therefore, a dimensionless value, numerically equal to its fugacity.

We have certainly made the integrated power function look simpler, but this is an illusion because we have only redefined variables. We have not improved our understanding of nonideal gases unless we can evaluate fugacity. To do this, we differentiate equation 4.10 with respect to P at constant temperature:

$$(\partial \mu / \partial P)_T = RT (\partial \ln a / \partial P)_T.$$

(Remember that μ^0 is a function of T only, so that $(\partial \mu^0/\partial P)_T$ is equal to zero.) This can be recast another way by recalling from equation 3.11 that $(\partial \mu/\partial P)_T = (\partial \bar{G}/\partial P)_T = \bar{V}$, so that:

$$RT \, d \ln a = \bar{V} dP.$$

Subtract the quantity $RT \, d \ln P$ from both sides to get:

$$RT \, d \ln(a/P) = \bar{V} dP - RT \, d \ln P$$
$$= (\bar{V} - [RT/P]) dP,$$

and then integrate the result between the limits zero and P. This procedure yields the expression:

$$\ln a = \ln P + \int_0^P (\bar{V}/RT - 1/P) dP,$$

which is just what we were looking for. The adjustment for nonideality is expressed as an integral that can be evaluated by replacing $\bar{V}$ with an appropriate equation of state, $\bar{V}(P)$. Notice that if $\bar{V}(P) = RT/P$ (the ideal gas equation), then the integral has the value zero at any pressure. The quantity:

$$\gamma = \exp\left[\int_0^P (\bar{V}/RT - 1/P) dP\right], \tag{4.11}$$

defines the *activity coefficient*, γ, which is equal to the ratio a/P.

Activities of species in nonideal gas mixtures are defined the same way we followed in deriving equation 4.5, except that all partial pressures P_i must now be

ACTIVITY COEFFICIENTS AND EQUATIONS OF STATE

One good way to illustrate the nonideal behavior of real gases with increasing pressure is to plot values of pressure against the experimentally-determined quantity $P\bar{V}/RT$, which serves as a measure of compressibility. For an ideal gas, $P\bar{V}/RT$ will, of course, be equal to 1 at all pressures; values >1 indicate that a gas is more compressible than an ideal gas; values >1 indicate that it is less compressible. In figure 4.4, we show the behavior of molecular hydrogen and oxygen on this type of plot.

At low pressures, most gases occupy less volume than we would expect from the ideal gas equation of state, suggesting that attractive forces between gas molecules reduce the effective mean distance between them. In 1879, the Dutch physicist Van der Waals recognized this phenomenon and adjusted the equation of state by adding a term $a/\bar{V}^2$ to the observed pressure, where a is an empirically determined constant for the gas. This modification correctly describes the deviation shown in figure 4.4 at very low pressures, although research has shown that a is not a constant, in fact, but depends on both pressure and temperature.

This correction for intermolecular attractions, however, predicts that real gases will become even more compressible as pressure is increased. This is clearly not the case. Consequently, the full form of Van der Waals' equation includes a second empirical adjustment on the molar volume:

$$(P + [a/\bar{V}^2])(\bar{V} - b) = RT.$$

As pressure increases, an increasingly significant volume in the gas is occupied by the molecules themselves. The factor b, therefore, can be thought of as a measure of the *excluded volume* that already contains molecules. The remaining volume, in which molecules are free to move, is much less than the total volume, and pressure is higher than it would be for an ideal gas. The full equation, therefore, reflects a balance between attractive and repulsive intermolecular forces that affect the molar volume of a real gas.

To calculate an activity coefficient using the Van der Waals equation of state, we need to rearrange it in terms of $\bar{V}$:

$$\bar{V}^3 - \bar{V}^2(b + RT/P) + \bar{V}(a/P) - ab/P = 0.$$

This cubic equation can be shown to have three real roots at temperatures and pressures below the critical point, of which the largest root is the true molar volume for the gas. (We do not define the critical point until chapter 10. For now, think of it as a condition in temperature and pressure beyond which there is no physical distinction between liquid and vapor states. Virtually all near-surface environments are below the critical point.) It is this solution that should replace $\bar{V}$ in equation 4.11. In practice, equation 4.11 is solved numerically by computing molar volumes iteratively over the range of pressures from zero to P as part of the algorithm that solves the integral. This is a reasonably straightforward and reliable procedure at moderate temperatures and pressures.

At the more elevated pressures at which many igneous or metamorphic reactions may involve gas phases, the Van der Waals equation of state is less

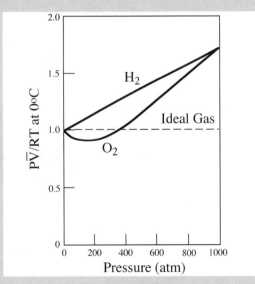

FIG. 4.4. Total pressure versus $P\bar{V}/RT$ for H_2 and O_2. A gas for which $P\bar{V}/RT < 1$ is more compressible than an ideal gas; if $P\bar{V}/RT > 1$, the gas is less compressible.

successful. Redlich and Kwong (1949) found that they could only predict the nonideal behavior of gases at very high pressures by modifying the attractive force term. Their equation,

$$P = RT/(\bar{V} - b) - a/([\bar{V} + b]V\sqrt{T}),$$

has been widely used in petrology, and has been modified further in many studies. Good reviews of these modifications, and of the use of Redlich-Kwong type equations of state in calculating activity coefficients, can be found in Holloway (1977) and Kerrick and Jacobs (1981).

multiplied by appropriate activity coefficients γ_i. The resulting equilibrium constant is given by:

$$K_{eq} = e^{-\Delta\bar{G}^*/RT} = [a_1^{v_1} a_2^{v_2} \ldots a_j^{v_j} / \\
(a_{j+1}^{v_{j+1}} \ldots a_{i-1}^{v_{i-1}} a_i^{v_i})][P_1^{v_1} P_2^{v_2} \ldots P_j^{v_j} / \\
(P_{j+1}^{v_{j+1}} \ldots P_{i-1}^{v_{i-1}} P_i^{v_i})][\gamma_1^{v_1} \gamma_2^{v_2} \ldots \gamma_j^{v_j} / \\
(\gamma_{j+1}^{v_{j+1}} \ldots \gamma_{i-1}^{v_{i-1}} \gamma_i^{v_i})].$$

(4.12)

In a similar way, nonideal liquid or crystalline solutions can be treated by modifying equation 4.9. As with gases, activity is defined as the ratio f/f^0. Most species in liquid or solid solution, however, have negligible vapor pressures at 1 atmosphere total pressure, so that f^0 is rarely equal to one. Furthermore, because fugacities in solution are so small, it is impractical to use the fugacity ratio to measure quantities like the activity of Fe_2SiO_4 in a magma. Instead, it is common to write:

$$K_{eq} = e^{-\Delta\bar{G}^*/RT} = [a_1^{v_1} a_2^{v_2} \ldots a_j^{v_j} / \\
(a_{j+1}^{v_{j+1}} \ldots a_{i-1}^{v_{i-1}} a_i^{v_i})][X_1^{v_1} X_2^{v_2} \ldots X_j^{v_j} / \\
(X_{j+1}^{v_{j+1}} \ldots X_{i-1}^{v_{i-1}} X_i^{v_i})][\gamma_1^{v_1} \gamma_2^{v_2} \ldots \gamma_j^{v_j} / \\
(\gamma_{j+1}^{v_{j+1}} \ldots \gamma_{i-1}^{v_{i-1}} \gamma_i^{v_i})].$$

(4.13)

in which we declare that $a_i = \gamma_i X_i$. In practice, it is common to specify, further, that a_i is equal to 1 for a pure substance at any chosen temperature. More specifically,

$$a_i \rightarrow X_i \text{ as } X_i \rightarrow 1.$$

Following this convention and applying equation 4.13, the activity of $MgSiO_3$, for example, would be equal to 1 in pure enstatite ($X_{MgSiO_3} = 1$). Other conventions are also common, however. Aqueous geochemists are more comfortable with compositions expressed in molal units (moles per kilogram of solution), rather than in mole fractions. Therefore, in very dilute aqueous solutions, we usually define the activity of solute species by the relation $a_i = \gamma_i m_i$, in which m_i is the molality of species i. This produces an equation like equation 4.13 in which

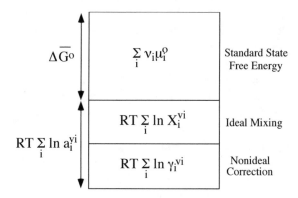

FIG. 4.5. Schematic representation of the various contributions to the free energy of a nonideal solution.

the X_i quantities are replaced by m_i. We generally apply this expression by specifying a standard state in which $m_i = 1$ and then extrapolating to infinite dilution, where:

$$a_i \rightarrow m_i \text{ as } m_i \rightarrow 0.$$

We discuss this standard state more fully in chapter 9.

According to either of these conventions and others that we have not mentioned, solutions approach ideal behavior ($\gamma_i \rightarrow 1$) as their compositions approach a single-component end member. The greatest difficulty arises when we have to consider nonideal solutions that are far from a pure end-member composition. We consider that problem for nonelectrolyte solutions and examine standard states for activity more fully in chapter 9. Figure 4.5 summarizes graphically the various contributions to ΔG_r we have discussed so far.

ACTIVITY IN ELECTROLYTE SOLUTIONS

Species in aqueous solutions such as seawater or hydrothermal fluids present an unusual challenge. Although natural fluids commonly contain some associated (that is, uncharged) species, it is much more common to find free ions. In a typical problem, for example, we may be asked to evaluate the solubility of sodium sulfate in seawater. Ignoring the fact that crystalline sodium sulfate usually takes the form of a hydrate, $Na_2SO_4 \cdot H_2O$, the relevant chemical reaction is:

$$Na_2SO_4 \rightleftarrows 2Na^+ + SO_4^{2-},$$

which we always write with the solid phase as a reactant and fully dissociated ions as products. For this type of reaction, we refer to the equilibrium constant as an equilibrium *solubility product constant, K_{sp}*:

$$K_{sp} = (a_{Na^+})^2 a_{SO_4 2}/a_{Na_2SO_4}.$$

(The activity of Na_2SO_4 is unity, because it is a pure solid phase.)

In general, ionic species in solution do not behave ideally unless they are very dilute, because ions tend to interact electrostatically. They also generally associate with water molecules to produce "hydration spheres" in which the chemical potential of H_2O is different from that in pure water. (Look again at figure 2.17 and accompanying discussion.) The result is that the free energies of both the solvent (H_2O) and the solute (dissolved ions) differ from their standard states, and their activities differ from their concentrations.

Unfortunately, there is no way to measure the activity of a dissolved ion independently. Because charge balance must always be maintained, we cannot vary the concentration of a cation such as Na^+ without also adjusting the anions in solution. It is impossible, for example, to determine how much of the potential free energy change during evaporation of seawater is due to increasing a_{Na^+} and how much is the result of parallel increases in a_{Cl^-}, $a_{SO_4^{2-}}$, and other anion activities.

The Mean Salt Method

Several approaches have been taken to solve this problem. One is to define the *mean ionic activity*, given by:

$$a_\pm = (a_+^{v_+} a_-^{v_-})^{1/v},$$

in which a_+ and a_- are the individual activities of cations and anions, respectively, and the stoichiometric coefficients $v = v_+ + v_-$ count the number of ions formed per dissociated molecule of solute. For example, the dissociation reaction for $MgCl_2$ is:

$$MgCl_2 \rightleftarrows Mg^{2+} + 2Cl^-,$$

so $v_+ = 1$, $v_- = 2$, and $v = 3$. The mean ionic activity of $MgCl_2$ in aqueous solution is therefore equal to:

$$a_\pm (MgCl_2) = (a_{Mg^{2+}} a_{Cl^-}^2)^{1/3}.$$

Mean activities, unlike the activities of charged ions, are readily measurable. From these, it is possible to calculate mean ion activity coefficients by using the relationship:

$$\gamma_\pm = a_\pm/m_\pm.$$

The quantity $m_\pm$ is the *mean ionic molality*, which is related to the molal concentration of total (nondissociated solute), m, by:

$$m_\pm = m(v_+^{v_+} v_-^{v_-})^{1/v}. \tag{4.14}$$

Again, for $MgCl_2$ in aqueous solution,

$$m_\pm(MgCl_2) = m_{MgCl2}([1]^1[2]^2)^{1/3}.$$

It is customary to calculate *individual ion activity coefficients*, γ_+ and γ_-, by referring to a standard univalent electrolyte in a solution of the same effective concentration as the one we are studying. Potassium chloride is a common standard electrolyte for this purpose because it has been determined experimentally that $\gamma_{K^+} = \gamma_{Cl^-}$, so that:

$$\gamma_\pm(KCl) = (\gamma_{K^+} \gamma_{Cl^-})^{1/2} = \gamma_{K^+} = \gamma_{Cl^-}.$$

If we wanted to calculate an individual ion activity coefficient for Mg^{2+} by this *mean salt method*, then, we would assume that the value of γ_{Cl^-} in an $MgCl_2$ solution is the same as the value of $\gamma_\pm$ (KCl) that we measured in a KCl solution. In other words,

$$\gamma_\pm(MgCl_2) = (\gamma_{Mg^{2+}}[\gamma_{Cl^-}]^2)^{1/3} = (\gamma_{Mg^{2+}} \gamma_{\pm_{(KCl)}}^2)^{1/3},$$

and, therefore:

$$\gamma_{Mg^{2+}} = \gamma_\pm(MgCl_2)^3/\gamma_\pm(KCl)^2.$$

Geochemists use the mean salt method to estimate activity coefficients for anions as well as cations. Because the basis for the method is an empirical comparison to a well studied electrolyte such as KCl, and because the comparison is always done between solutions with the same effective concentration, the method is reliable over a very wide range of conditions.

The Debye-Hückel Method

Because the mean salt method relies on the use of $\gamma_\pm$ from simple electrolyte solutions, you need access to measurements on a large number of electrolytes over a wide range of concentrations. Fortunately, there are extensive tables and graphs of experimentally derived activity coefficients in the chemical literature (see the highlighted discussion of nonideality in natural waters for one example). For those who prefer to generate their own values, an alternative nonempirical method for calculating ion activity coefficients in dilute solutions is provided by Debye-Hückel theory, which attempts to calculate the effect of electrostatic interactions among ions on their free energies of formation. The disadvantage of the theory is that it is only reliable in dilute solutions. Nevertheless, because very many solutions of geologic interest fall within its range of reliability, geochemists find the Debye-Hückel method extremely useful.

To evaluate the cumulative effect of attractive and repulsive forces, we define first a charge-weighted function of species concentration known as *ionic strength* (I):

$$I = \tfrac{1}{2} \sum m_i z_i^2.$$

In this way, we recognize that ions in solution are influenced by the total electrostatic field around them and that polyvalent ions ($|z| > 1$) exert more electrostatic force on their neighbors that monovalent ions. A 1 m solution of $MgSO_4$, for example, has an ionic strength of 4, whereas a 1 m solution of $NaCl$ has an ionic strength of only 1.

In its most commonly used form, the Debye-Hückel equation takes into account not only the ionic strength of the electrolyte solution, but also the effective size of the hydrated ion of interest, thus estimating the number of neighboring ions with which it can come into contact. Values of the size parameter, å, for representative ionic species, are given in table 4.1, along with values of empirical parameters A and B, which are functions only of pressure and temperature. We calculate the activity coefficient for ion i from:

$$\log_{10} \gamma_i = -A z_i^2 \sqrt{I}/(1 + B\,å\,\sqrt{I}).$$

Pytkowicz (1983) has discussed the theoretical justification for this equation and a critique of its use in aqueous geochemistry.

Worked Problem 4.5

What is the activity coefficient for Ca^{2+} in a 0.05 m aqueous solution of $CaCl_2$ at 25°C? How closely do answers obtained by the mean salt method and the Debye-Hückel equation agree?

To calculate $\gamma_{Ca^{2+}}$ by the Debye-Hückel method, we first determine the ionic strength of the solution:

$$I = \tfrac{1}{2}\sum m_i z_i^2 = \tfrac{1}{2}([0.05][2]^2 + [0.1][-1]^2) = 0.15.$$

Then, using the data in table 4.1, we find that:

$$\log_{10}\gamma_{Ca^{2+}} = -A z_i^2 \sqrt{I}/(1 + B\,å\,\sqrt{I})$$

$$= -(0.5085)(2)^2 \sqrt{(0.15)}/[1 + (0.3281 \times 10^8)$$
$$(6 \times 10^{-8})\sqrt{(0.15)}]$$

$$= -0.447.$$

So,

$$\gamma_{Ca^{2+}} = 0.357.$$

How does this compare with $\gamma_{Ca^{2+}}$ calculated by the mean salt method? The mean ionic activity coefficient $\gamma_{\pm CaCl_2}$ in 0.05 m

TABLE 4.1. Values of Constants for Use in the Debye-Hückel Equation

Temperature °C	A	B	å × 10⁸	Ion
0	0.4883	0.3241	2.5	Rb^+, Cs^+, NH_4^+, Tl^+, Ag^+
5	0.4921	0.3249	3.0	K^+, CL^-, Br^-, I^-, NO_3^-
10	0.4960	0.3258	3.5	OH^-, F^-, HS^-, BrO_3^-, IO_4^-, MnO_4^-
15	0.5000	0.3262	4.0–4.5	Na^+, HCO_3^-, $H_2PO_4^-$, HSO_3^-, SO_4^{2-}, SeO_4^{2-}, CrO_4^{2-}, HPO_4^{2-}, PO_4^{3-}
20	0.5042	0.3273	4.5	Pb^{2+}, CO_3^{2-}, SO_3^{2-}, MoO_4^{2-}
25	0.5085	0.3281	5.0	Sr^{2+}, Ba^{2+}, Ra^{2+}, Cd^{2+}, Hg^{2+}, S^{2-}, WO_4^{2-}
30	0.5130	0.3290	6	Li^+, Ca^{2+}, Cu^{2+}, Zn^{2+}, Sn^{2+}, Mn^{2+}, Fe^{2+}, Ni^{2+}, Co^{2+}
35	0.5175	0.3297	8	Mg^{2+}, Be^{2+}
40	0.5221	0.3305	9	H^+, Al^{3+}, Cr^{3+}, trivalent rare earths
45	0.5271	0.3314	11	Th^{4+}, Zr^{4+}, Ce^{4+}, Sn^{4+}
50	0.5319	0.3321		
55	0.5371	0.3329		
60	0.5425	0.3338		

Data from Garrels and Christ (1982).

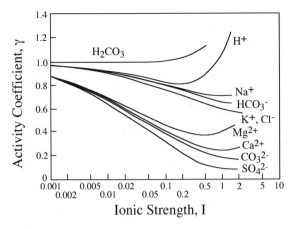

FIG. 4.6. Activity coefficients for common aqueous species as a function of ionic strength.

aqueous solution has been reported by Goldberg and Nuttall (1978) to be 0.5773. To use the mean salt method, we need to know the mean ion activity coefficient for KCl at the same effective concentration as the $CaCl_2$ solution; that is, at the same ionic strength. A KCl solution of ionic strength 0.15 is 0.15 molal. Robinson and Stokes (1949) report that $\gamma_{\pm KCl} = 0.744$ in 0.15 m aqueous KCl. You can verify this, although with less precision, from the graph in figure 4.6. From this value, we calculate that:

$$\gamma_{\pm CaCl_2} = (\gamma_{Ca^{2+}}[\gamma_{Cl^-}]^2)^{1/3} = (\gamma_{Ca^{2+}}[\gamma_{\pm KCl}]^2)^{1/3}.$$

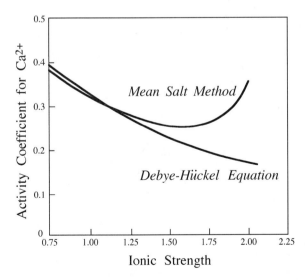

FIG. 4.7. Ion activity coefficient for Ca^{2+} as a function of ionic strength, calculated from the Debye-Hückel equation and by the mean salt method. Notice the deviations at higher ionic strengths.

Therefore,

$$\gamma_{Ca^{2+}} = (\gamma_{\pm CaCl_2})^3/[\gamma_{\pm KCl}]^2) = (0.577)^3/(0.649)^2 = 0.347.$$

This is very close to the 0.357 value calculated by Debye-Hückel theory. In figure 4.7, we have repeated this pair of calculations for a range of ionic strengths from 0.05 to 3.0. It is apparent that the methods agree at ionic strengths less than ~0.3, but diverge strongly in more concentrated solutions, as Debye-Hückel theory fails to model changes in the structure of the solute adequately. Other calculation schemes have been used to model highly concentrated solutions, particularly at elevated temperatures (see, for example, Harvie et al. [1984], or the Helgeson references in appendix B), but most researchers continue to use values obtained by some variant of the mean salt method in those situations.

SOLUBILITY

Many of the interesting questions facing aqueous geochemists have to do with mineral solubility. How much sulfate is in groundwater that is in equilibrium with gypsum beds? How much fluoride can be carried by an ore-forming hydrothermal fluid? What sequence of minerals should be expected to form during evaporation of seawater? What processes can cause deposition of ore minerals from migrating fluids? How are the compositions of residual (zonal) soils affected by the chemistry of soil water? How might a chemical leak from a waste disposal site affect the stability of surrounding rocks and soils? To address these questions and others like them in coming chapters, we must first look at broad conditions that affect solubility.

In worked problems 4.6–4.10, we will make successive refinements in our approach to a typical problem, illustrating by stages the major thermodynamic factors influencing solubility.

Worked Problem 4.6

What is the solubility of barite in water at 25°C? The simplest answer can be calculated from data for the molar free energies of formation of species in the reaction:

$$BaSO_4 \rightleftarrows Ba^{2+} + SO_4^{2-}.$$

We find the following values in Parker et al. (1971):

	$BaSO_4$	Ba^{2+}	SO_4^{2-}
$\Delta \bar{G}_f$ (kcal mol^{-1})	−325.6	−134.02	−177.97

NONIDEALITY IN NATURAL WATERS

The water in streams and the aquifers that we typically draw on for irrigation and public water supplies is in fact a dilute electrolyte solution. People and plants are sensitive to small variations in composition. Some electrolytes affect cosmetic qualities of water such as taste and smell, and others can have profound effects on health. Inorganic solutes can also interact with pipes and machinery, producing corrosion or scale. For this reason, the composition of natural water in most parts of the world is monitored quite closely.

As we discuss in greater detail in chapters 5 and 7, the composition of natural waters is controlled largely by weathering reactions. The dominant cations in surface and ground waters (Ca^{2+}, Mg^{2+}, Na^+, and K^+) and the dominant anions (HCO_3^-, Cl^-, and SO_4^{2-}) are derived from silicate and carbonate minerals and the oxidation of metal sulfides. Hydrogeologists commonly trace the chemical history of water from an aquifer or reservoir by studying the relative abundances of these ions, expressed as concentration ratios in either molalities (moles/kg) or equivalencies (moles × charge/liter). Using diagrams such as those in figure 4.8, for example, we can distinguish between waters that have equilibrated with limestone (relatively Ca^{2+}, Mg^{2+}, and HCO_3^- rich) and those that are more likely to have equilibrated with shale (relatively Na^+, and K^+ rich). These are useful qualitative indicators of where a water sample has been.

The more quantitative thermodynamic and kinetic analyses that we introduce in chapters 5 and 7 will

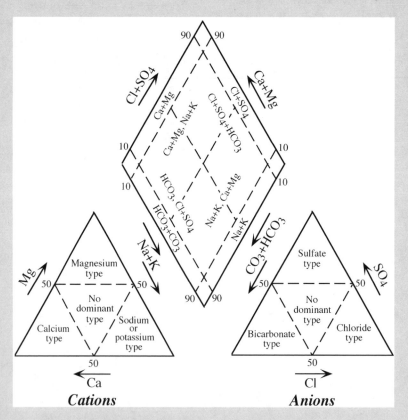

FIG. 4.8. Classification diagram for anion and cation facies in terms of major ion percentages. Groundwater types are designated according to the domain in which they occur on diagram segments. (After Freeze and Cherry 1979.)

require ion activities, rather than concentrations. You can use either the mean salt method or the Debye-Hückel method to calculate ion activity coefficients. To decide which method is most likely to give a reliable result for a given water sample, calculate its ionic strength. Except when some other ion is present in significant amounts, this is equal to:

$$I = \tfrac{1}{2}[m_{Na^+} + m_{K^+} + 4m_{Ca^{2+}} + 4m_{Mg^{2+}} + m_{HCO_3^-} + m_{Cl^-} + 4m_{SO_4^{2-}}].$$

As table 4.2 suggests, the ionic strengths of natural waters range from about 1.0×10^{-3} in rivers and lakes to as much as 1×10^{-1} in groundwater. Oilfield brines and saline lakes have significantly higher ionic strengths. The ionic strength of seawater is 7×10^{-1}. Because the Debye-Hückel method is unreliable above ionic strengths of 1.0×10^{-1} (see figure 4.7), therefore, you should use the mean salt method for any solution more saline than groundwater.

TABLE 4.2. Analyses of Groundwater and River Waters

Ion	Source						
	1	2	3	4	5	6	7
K^+	1.0	9.0	0.6	0.2	0.2	1.5	1.6
Na^+	7.9	37.0	2.0	0.3	0.9	6.5	6.6
Ca^{2+}	56.0	60.0	4.8	2.1	1.7	20.1	16.6
Mg^{2+}	12.0	60.0	1.5	0.1	0.4	4.9	4.3
HCO_3^-	160.0	417.0	24.0	4.9	1.6	71.4	66.2
Cl^-	12.0	27.0	0.6	—[1]	0.5	7.0	7.6
SO_4^{2-}	53.0	96.0	1.1	2.1	6.2	14.9	9.7
I	0.0065	0.0146	0.0006	0.0002	0.0003	0.0026	0.0022

Values in mg/l. Sources are: 1, limestone aquifer, Florida (Goff 1971); 2, dolomite aquifer, Manitoba (Langmuir 1971); 3, granitic glacial sand aquifer, Northwest Ontario (Bottomley 1974); 4, streams draining greenstone, South Cascade Mountains, Oregon (Reynolds and Johnson 1972); 5, streams draining granite, metamorphics, and glacial till, Hubbard Brook watershed, New Hampshire (Likens et al. 1977), analysis also contains trace amounts of NO_3^-; 6, average river waters, corrected for pollution, North America (Meybeck 1979); 7, average river waters, corrected for pollution, Asia (Meybeck 1979).

[1]No measurement.

The standard molar free energy of reaction is calculated by:

$$\Delta \bar{G}_r = -134.02 + (-177.97) - (-325.6)$$
$$= 13.61 \text{ kcal mol}^{-1}.$$

If the fluid is saturated with respect to $BaSO_4$, then the reaction above is perfectly balanced and $\Delta \bar{G}_r = 0$. Therefore,

$$\log K_{sp} = -\Delta \bar{G}_r^0 / RT = -13.61/([298][0.001987])$$
$$= -9.98,$$

$$K_{sp} = 1.04 \times 10^{-10}.$$

If we assume that the solution behaves ideally, then the solubility of barite can be calculated by recognizing that charge balance in the fluid requires that $m_{Ba^{2+}} = m_{SO_4^{2-}}$, and that:

$$m_{Ba^{2+}} \cdot m_{SO_4^{2-}} = K_{sp}.$$

From this, we get:

$$m_{Ba^{2+}} = m_{SO_4^{2-}} = 1.02 \times 10^{-5} \text{ mol kg}^{-1}.$$

The result in worked problem 4.6 is quite close to the value of 1.06×10^{-5} mol kg^{-1} measured in experiments by Blount (1977), even though we assumed the fluid was ideal. This agreement may be fortuitous, however, because we should anticipate some nonideal behavior. Let's do the problem again.

Worked Problem 4.7

What are the activity coefficients for Ba^{2+} and SO_4^{2-} in a solution saturated with respect to barite, and how much does the calculated solubility change if we allow for nonideal behavior?

The easiest way to solve this problem is to design a numerical algorithm and iterate toward a solution by computer. In outline form, the procedure looks like this:

1. Using the ideal solubility as a first guess, calculate the ionic strength of a saturated solution.

2. Calculate activity coefficients for Ba^{2+} and SO_4^{2-} from the Debye-Hückel equation, using constants in table 4.1.
3. Using the activity coefficients and the value of K derived from $\Delta \bar{G}_r^0$, calculate improved values of $m_{Ba^{2+}}$ and $m_{SO_4^{2-}}$.
4. Repeat steps 1 through 3 until $m_{Ba^{2+}}$ and $m_{SO_4^{2-}}$ do not change in successive iterations by more than some acceptably small amount.

When we did this calculation, it took four cycles to satisfy the condition $|m_{Ba^{2+}} (\text{old}) - m_{Ba^{2+}} (\text{new})| < 10^{-9}$, which is a calculation error of less than one part in 10^4 (much better than is justified by the uncertainties in $\Delta \bar{G}_r^0$). Notice that charge balance constraints require that $m_{Ba^{2+}} = m_{SO_4^{2-}}$.

Iteration	I	$\gamma_{Ba^{2+}}$	$\gamma_{SO_4^{2-}}$	$\gamma_\pm$	$m_{Ba^{2+}}$
1	2.41×10^{-5}	0.9775	0.9774	0.9774	1.0443×10^{-5}
2	4.18×10^{-5}	0.9705	0.9704	0.9705	1.0519×10^{-5}
3	4.21×10^{-5}	0.9704	0.9703	0.9704	1.0520×10^{-5}
4	4.21×10^{-5}	0.9704	0.9703	0.9704	1.0520×10^{-5}

As we suspected in worked problem 4.6, a solution saturated with $BaSO_4$ is nearly ideal. By taking the slight nonideality into account, however, the computed solubility increased to within 1% error of the value measured by Blount (1977).

The Ionic Strength Effect

The reason that there is progressive improvement with each iteration in worked problem 4.7 is that each cycle calculates the ionic strength with an improved set of activity coefficients. Suppose, however, that there were other ions in solution as well as Ba^{2+} and SO_4^{2-}. Natural fluids, in fact, rarely contain only a single electrolyte in solution. Most hydrothermal fluids or formation waters are dominated by NaCl. To calculate the ionic strength correctly, we must consider not only the amount of Ba^{2+} and SO_4^{2-} but also include the concentrations of all other ions in solution. This, in turn, leads to an adjustment in the solubility of $BaSO_4$.

Worked Problem 4.8

What effect do other solutes have on the solubility of barite? In particular, what would be the solubility of barite in a solution at 25°C that is already 0.2 m in NaCl? Assuming that there is little tendency for Ba^{2+} to associate with Cl^- (Sucha et al. 1975) or for Na^+ to associate with SO_4^{2-} (Elgquist and Wedborg 1978), we should expect that barite solubility changes only as the result of increased ionic strength. To verify this, we repeat the calculation of worked problem 4.7, this time calculating ionic strength in each iteration by $I = \frac{1}{2}(m_{Ba^{2+}} \times 4 + m_{SO_4^{2-}} \times 4 + m_{Na^+} + m_{Cl^-})$.

I	$\gamma_{Ba^{2+}}$	$\gamma_{SO_4^{2-}}$	$\gamma_\pm$	$m_{Ba^{2+}}$
0.2000	0.2986	0.2671	0.2824	3.615×10^{-5}

These values, once again, compare favorably with Blount (1977), who measured barite solubility in 0.2-m $NaCl$-H_2O at 25°C and found $m_{Ba^{2+}} = 3.7 \times 10^{-5}$ mol kg^{-1} and $\gamma_\pm = 0.2754$.

The Common Ion Effect

In many natural fluids, the solubility of minerals like barite increases with ionic strength, as we saw in worked problem 4.8. This is not the case, however, if ions produced by dissociation of the solute are also present from some other source in solution. For example, if barite is in equilibrium with a solution that contains both NaCl and Na_2SO_4, the charge balance constraint on the fluid becomes:

$$2m_{Ba^{2+}} + m_{Na^+} = 2m_{SO_4^{2-}} + m_{Cl^-}.$$

Sulfate is a *common ion* in $BaSO_4$ and Na_2SO_4. Extra SO_4^{2-} ions in solution, whether they come from Na_2SO_4 or some other source, force the reaction:

$$BaSO_4 \rightleftarrows Ba^{2+} + SO_4^{2-}$$

toward the left (remember equation 4.3?), decreasing the solubility of barite. To demonstrate this common ion effect, we offer the following problem.

Worked Problem 4.9

How does the addition of successive amounts of Na_2SO_4 to an aqueous fluid affect the solubility of barite?

The molal concentration of Ba^{2+} can be calculated from this expression and the equilibrium constant, K:

$$K = m_{Ba^{2+}} m_{SO_4^{2-}} \gamma_{Ba^{2+}} \gamma_{SO_4^{2-}}$$
$$= m_{Ba^{2+}} \gamma_{Ba^{2+}} \gamma_{SO_4^{2-}} (2m_{Ba^{2+}} + m_{Na^+} - m_{Cl^-})/2$$
$$m_{Ba^{2+}} = 2K/(\gamma_{Ba^{2+}} \gamma_{SO_4^{2-}} [2m_{Ba^{2+}} + m_{Na^+} - m_{Cl^-}]).$$

This equation can be solved by the same numerical procedure we applied in worked problems 4.7 and 4.8. We can simplify the calculation in this case, because the solubility of barite has been shown to be quite low. We can assume that all SO_4^{2-} in solution is due to dissociation of Na_2SO_4, as long as we examine only solutions of moderately high $m_{SO_4^{2-}}$. The charge balance constraint, in other words, is approximately:

$$m_{Na^+} = 2m_{SO_4^{2-}} + m_{Cl^-}.$$

The ionic strength calculated with this simplification will differ only very slightly from the "true" ionic strength.

To illustrate the effect of a common ion (SO_4^{2-} in this case), we have calculated $m_{Ba^{2+}}$ in a variety of $NaCl$-Na_2SO_4-H_2O solutions with ionic strength equal to 0.2.

$m_{Ba^{2+}}$	$m_{SO_4^{2-}}$	m_{Na^+}	m_{Cl^-}
3.91×10^{-6}	3.33×10^{-4}	2.00×10^{-1}	1.99×10^{-1}
1.31×10^{-6}	1.00×10^{-3}	1.99×10^{-1}	1.97×10^{-1}
6.53×10^{-7}	2.00×10^{-3}	1.98×10^{-1}	1.94×10^{-1}
1.96×10^{-7}	6.67×10^{-3}	1.93×10^{-1}	1.80×10^{-1}
9.80×10^{-8}	1.33×10^{-2}	1.86×10^{-1}	1.60×10^{-1}
6.53×10^{-8}	2.00×10^{-2}	1.80×10^{-1}	1.40×10^{-1}

As expected, the addition of sulfate, even in very small amounts, dramatically lowers the solubility. Barite will precipitate if the product $m_{Ba^{2+}} m_{SO_4^{2-}}$ equals or exceeds K_{eq} (1.31×10^{-9} at ionic strength 0.2). With increasing amounts of sulfate in solution, this condition is met with vanishingly small amounts of dissolved Ba^{2+}. The same effect would be observed if we were to provide barium ions from some source other than barite (for example, from barium chloride).

Complex Species

In stream waters (ionic strength < 0.01 on average) and in most surficial environments on the continents, dissolved species are highly dissociated. As ionic strength increases, however, electrical interactions between ions also increase. The result is that many ions in concentrated solutions exist in groups or clusters known as *complexes*. For ease of discussion, these may be divided into three broad classes: *ion pairs, coordination complexes,* and *chelates.* By convention (not universally observed, unfortunately), the stability of any complex is expressed by a *stability constant, K_{stab},* which is written for the reaction forming a complex species from simpler dissolved species. Thus, for example, Mg^{2+} and SO_4^{2-} ions can associate to form the neutral ion pair $MgSO_4^0$ by the reaction

$$Mg^{2+} + SO_4^{2-} \rightleftarrows MgSO_4^0,$$

for which $K_{stab} = a_{MgSO_4^0}/(a_{Mg^{2+}} a_{SO_4^{2-}}) = 5.62 \times 10^{-3}$ at 25°C. The larger a stability constant, the more stable the complex it describes.

The distinctions among different types of complexes are not easy to define and, for thermodynamic purposes, not very important. Ion pairs are characterized by weak bonding; hence, small stability constants. The dominant complex species in seawater and most brines fall into this category. In chapter 8, we examine the effect of ion pair formation on the chemistry of the oceans.

Coordination complexes involve a more rigid structure of *ligands* (usually anions or neutral species) around a central atom. This well-defined structure contributes to greater stability. Transition metals, such as copper, vanadium, and uranium have been shown to exist in aqueous solution primarily as coordination complexes like $Cu(H_2O)_6^{2+}$, in which the negatively charged ends of several polar water molecules are attracted to the metal cation to form a "hydration sphere" like the one we illustrated in figure 2.17. Other polar molecules also commonly form coordination complexes with metal cations. One that may be familiar to you from analytical chemistry is ammonia, which combines readily with Cu^{2+} in solution to form the bright blue complex $Cu(NH_3)_4^{2+}$. The stability constant for this species is 2.13×10^{14}.

Molecules like water, ammonia, Cl^-, and OH^-, which form simple coordination complexes because they can only attach to one metal cation at a time, are called *unidentate* ligands. Other species, usually organic molecules or ions, are called *multidentate* or *polydentate* ligands. Chelates, which form around these, have larger, more complicated structures that include several metal cations. Chelates differ from coordination complexes only in their degree of structure. Some, like the hexadentate anion of ethylenediaminetetraacetic acid (mercifully known as EDTA) are widely used by analytical chemists to scavenge metals from very dilute aqueous solutions.

In nature, the vast number of humic and fulvic acids and other dissolved organic species also serve as chelating agents. In general, the larger and more complicated these are, the more stable the complexes they form. Because they have very high stability constants and can coordinate several cations at once, chelating agents can contribute significantly to the chemistry of transition metals and of cations such as Al^{3+} in natural waters, even if both they and the metal cations are in very low concentration.

The formation of complex species increases the solubility of electrolytes by reducing the effective concentration of free ions in solution. In a sense, we can think of ions bound up in complexes as having become electrostatically shielded from oppositely charged ions and therefore unable to combine with them. Because the activity product of free ions is lowered, equilibria that involve those ions favor the further dissolution of solid material. Thus, although the activity product of free ions at saturation will still be equal to K_{sp}, the total concentration of species in solution will be much higher than in a solution without complexing.

Worked Problem 4.10

Calcium and sulfate ions can associate in aqueous solution to form a neutral ion pair, $CaSO_4^0$. This is one of several ion pairs that affect the concentration of species in seawater, as we shall see in chapter 8. How does it affect the solubility of gypsum?

The solubility product constant, K_{sp}, for gypsum is calculated from the free energy of the reaction:

$$CaSO_4 \cdot 2H_2O \rightleftarrows Ca^{2+} + SO_4^{2-} + 2\,H_2O.$$

In reasonably dilute solutions, the activity of water can be assumed to be unity. Gypsum, a pure solid, is in its standard state and also has unit activity. K_{sp}, then, is equal to $a_{Ca^{2+}}\, a_{SO_4^{2-}}$. At 25°C, it has the value 2.45×10^{-5}.

The stability constant, K_{stab}, for $CaSO_4^0$ is derived by experiment for the reaction:

$$Ca^{2+} + SO_4^{2-} \rightleftarrows CaSO_4^0.$$

$$K_{stab} = a_{CaSO_4^0}/(a_{Ca^{2+}}\,a_{SO_4^{2-}}) = 2.09 \times 10^2$$

at 25°C. By combining the two expressions for K_{sp} and K_{stab} and rearranging terms, we can calculate:

$$a_{CaSO_4^0} = K_{stab}K_{sp} = 5.09 \times 10^{-3}.$$

Because $CaSO_4^0$ is a neutral species, its concentration in solution will be nearly the same as its activity. The concentration of the ion pair, therefore, is ~5 mmol kg^{-1}. It can be shown, using the iterative procedure in worked problem 4.7, that the concentration of Ca^{2+} in the absence of complexing is ~10 mmol kg^{-1}. The total calcium concentration in solution is thus ~15 mmol kg^{-1}, or ~50% higher than the concentration of Ca^{2+} alone.

SUMMARY

We see now that the structure and stoichiometry of solutions play a key role in determining how they act in chemical reactions. The free energy of any reaction can be expressed in terms of a standard state free energy and a contribution due to mixing of end-member species in solutions. Relatively few solutions of geochemical interest are ideal; in most, interactions among dissolved species or among atoms on different structural sites produce significant departures from ideality. In many cases, however, we gain valuable first impressions of a system by assuming ideal mixing and then adding successive corrections to describe its actual behavior.

We discuss solution models more fully in chapters 9 and 10, where, in particular, we address the problem of calculating activity coefficients for nonideal systems of

petrologic interest. In the next few chapters, however, we make use of the Debye-Hückel equation and the mean salt method, which yield the activity coefficients we need to explore dilute aqueous solutions. Solubility equilibria will dominate our discussion of the oceans, of chemical weathering, and of diagenesis.

SUGGESTED READINGS

Several excellent textbooks are available for the student interested in the thermodynamic treatment of solutions. The ones listed here are a selection of some that the authors have found particularly useful.

Drever, J. I. 1997. *The Geochemistry of Natural Waters*. New York: Simon and Schuster. (This is a very good book for the student who is new to aqueous geochemistry. There are quite a few carefully written case studies to show the use of geochemistry in the real world.)

Garrels, R. M., and C. L. Christ. 1982. *Solutions, Minerals, and Equilibria*. San Francisco: Freeman, Cooper. (Perhaps the most often used text in its field.)

Powell, R. 1978. *Equilibrium Thermodynamics in Petrology*. London: Harper and Row. (A good text for classroom use. Chapters 4 and 5 consider the calculation of mole fractions in crystalline and liquid solutions.)

Pytkowicz, R. M. 1983. *Equilibria, Nonequilibria, and Natural Waters*, vol. I. New York: Wiley. (An exhaustive treatment of the theoretical basis for the thermodynamics of aqueous solutions. Chapter 5 has more than 100 pages devoted to the determination of activity coefficients.)

Robinson, R. A., and R. H. Stokes. 1968. *Electrolyte Solutions*. London: Butterworths. (A reference for chemists rather than geologists, but widely quoted by aqueous geochemists. Highly theoretical, but not difficult to follow.)

Stumm, W., and J. J. Morgan. 1995. *Aquatic Chemistry*. New York: Wiley. (A detailed, yet highly readable text with an emphasis on the chemical behavior of natural waters. Chapter 6, dealing with complexes, is particularly well written.)

The following articles were referenced in this chapter. An interested student may wish to explore them more thoroughly.

Blount, C. W. 1977. Barite solubilities and thermodynamic quantities up to 300°C and 1400 bars. *American Mineralogist* 62:942–957.

Bottinga, J., and D. F. Weill. 1972. The viscosity of magmatic silicate liquids: a model for calculation. *American Journal of Science* 272:438–475.

Bottomley, D. 1974. *Influence of Hydrology and Weathering on the Water Chemistry of a Small Precambrian Shield Watershed*. Unpublished MSc. Thesis, University of Waterloo.

Cameron, M., and J. J. Papike. 1980. Crystal chemistry of silicate pyroxenes. *Mineralogical Society of America Reviews in Mineralogy* 7:5–92.

Cameron, M., and J. J. Papike. 1981. Structural and chemical variations in pyroxenes. *American Mineralogist* 66:1–50.

Clark, J., D. E. Appleman, and J. J. Papike. 1969. Crystal-chemical characterization of clinopyroxenes based on eight new structure refinements. *Mineralogical Society of America Special Paper* 2:31–50.

Cohen, R. E. 1986. Thermodynamic solution properties of aluminous clinopyroxenes: Non-linear least-squares refinements. *Geochimica et Cosmochimica Acta* 50:563–576.

Elgquist, B., and M. Wedborg. 1978. Stability constants of $NaSO_4^-$, $MgSO_4$, MgF^+, $MgCl^+$ ion pairs at the ionic strength of seawater by potentiometry. *Marine Chemistry* 6:243–252.

Freeze, R. A., and J. A. Cherry. 1979. *Groundwater.* Englewood Cliffs: Prentice-Hall.

Goff, K.J. 1971. *Hydrology and Chemistry of the Shoal Lakes Basin, Interlake Area, Manitoba.* Unpublished Master's thesis, University of Manitoba.

Goldberg, R. N., and R. L. Nuttall. 1978. Evaluated activity and osmotic coefficients for aqueous solutions: The alkaline earth halides. *Journal of Physical and Chemical Reference Data* 7:263.

Harvie, C. E., N. Moller, and J. H. Weare. 1984. The prediction of mineral solubilities in natural waters: The Na-K-Mg-Ca-H-Cl-SO_4-OH-HCO_3-CO_3-CO_2-H_2O system to high ionic strengths at 25°C. *Geochimica et Cosmochimica Acta* 48: 723–752.

Helgeson, H. C. 1969. Thermodynamics of hydrothermal systems at elevated temperatures and pressures. *American Journal of Science* 267:729–804.

Holloway, J. R. 1977. Fugacity and activity of molecular species in supercritical fluids. In D. G. Fraser, ed. *Thermodynamics in Geology*, pp. 161–180. Boston: D. Reidel.

Kerrick, D. M., and G. K. Jacobs. 1981. A modified Redlich-Kwong equation for H_2O, CO_2, and H_2O-CO_2 mixtures at elevated pressures and temperatures. *American Journal of Science* 281:735–767.

Langmuir, D. 1971 The chemistry of some carbonate groundwaters in Central Pennsylvania. *Geochimica et Cosmochimica Acta* 35:1023–1045.

Likens, G. E., F. H. Bormann, R. S. Pierce, J. S. Eaton, and N. M. Johnson. 1977. *Biogeochemistry of a Forested Ecosystem.* New York: Springer-Verlag.

Livingstone, D. A. 1963. *Chemical Composition of Rivers and Lakes.* U.S. Geological Survey Professional Paper 440G.

Meybeck, M. 1979. Concentrations des eaux fluviales en éléments majeurs et apports en solution aux océans. *Revues de Géologie Dynamique et de Géographie Physique* 23: 215–246.

Parker, V. B., D. D. Wagman, and W. H. Evans. 1971. *Selected Values of Chemical Thermodynamic Properties: Tables for the Alkaline Earth Elements (Elements 92 through 97 in the Standard Order of Arrangement).* Technical Note 270-6. Washington, D.C.: U.S. National Bureau of Standards.

Redlich, O., and J.N.S. Kwong. 1949. An equation of state: Fugacities of gaseous solutions. *Chemical Review* 44: 233–244.

Reynolds, R. C., and N. M. Johnson. 1972. Chemical weathering in the temperate glacial environment of the Northern Cascade Mountains. *Geochimica et Cosmochimica Acta* 36: 537–544.

Robinson, R. A., and R. H. Stokes. 1949. Tables of osmotic and activity coefficients of electrolytes in aqueous solutions at 25°C. *Transactions of the Faraday Society* 45:612.

Sucha, L., J. Cadek, K. Hrabek, and J. Vesely. 1975. The stability of the chloro complexes of magnesium and of the alkaline earth metals at elevated temperatures. *Collected Czechoslovakian Chemical Communications* 40:2020–2024.

PROBLEMS

(4.1) Use the Debye-Hückel equation to calculate values for the ion activity coefficients for Cl^- and Al^{3+} at 25°C in aqueous solutions with ionic strengths ranging from 0.01 to 0.2. Plot these on a graph of γ versus log I. What physical differences between these two ions account for the differences in their activity coefficients?

(4.2) Using the procedure introduced in worked problem 4.5, calculate the solubility of gypsum in water at 25°C.

(4.3) To produce a solution that is saturated with respect to fluorite at 25°C, you need to dissolve 6.8×10^{-3} g of CaF_2 per 0.25 liter of pure water. Assuming that CaF_2 is 100% dissociated, and that the solution behaves ideally, calculate the K_{sp} for fluorite. What is the free energy for the dissociation reaction?

(4.4) The K_{sp} for celestite is 7.6×10^{-7} at 25°C. How many grams of $SrSO_4$ can you dissolve in a liter of 0.1 m $Sr(NO_3)_2$? Assume that the activity coefficients for all dissolved species are unity, and that no complex species are formed in solution.

(4.5) Repeat the calculation in problem 4.4, again assuming that complex species are absent, but correcting for the ionic strength effect by calculating activity coefficients with the Debye-Hückel equation.

(4.6) The following analysis of water from Lake Nipissing in Ontario was reported by Livingstone (1963):

HCO_3^-	26.2 ppm	SO_4^{2-}	8.5 ppm
Cl^-	1.0	NO_3^-	1.33
Ca^{2+}	9.0	Mg^{2+}	3.6
Na^+	3.8		

What is the ionic strength of this water? (Recall that concentrations in ppm are equal to concentrations in mmol kg^{-1} multiplied by formula weight.)

(4.7) Calculate the mole fractions of forsterite and fayalite in an olivine with the composition $Mg_{0.7}Fe_{1.3}SiO_4$.

(4.8) A garnet has been analyzed by electron microprobe and found to have the following composition:

SiO_2	39.08 wt %
Al_2O_3	22.66
FeO	29.98
MnO	1.61
MgO	6.67

Calculate the mole fractions of almandine, pyrope, and spessartine in this garnet.

(4.9) Calculate the mole fractions of enstatite, ferrosilite, diopside, $CaCrAlSiO_6$ (Cr-CATS), and $Mg_{1.5}AlSi_{1.5}O_6$ (pyrope) in the orthopyroxene analysis below. Assume complete mixing on sites in such a way that all tetrahedral sites are filled by either Si or Al; M1 sites by Al, Cr, Fe^{3+}, Fe^{2+}, or Mg; and M2 sites by Ca, Fe^{2+}, or Mg. Assume also that the ratio of X_{Fe} to X_{Mg} is that same in M1 and M2 sites. (This problem, modified from Powell 1978, requires a fair amount of effort.)

SiO_2	57.73 wt %	Cr_2O_3	0.46 wt %
Al_2O_3	0.95	FeO	3.87
Fe_2O_3	0.42	MgO	36.73
CaO	0.23		

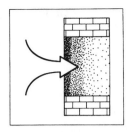

CHAPTER FIVE

DIAGENESIS

A Study in Kinetics

OVERVIEW

Diagenesis embraces all of the changes that may take place in a sediment following deposition, except for those due to metamorphism or to weathering at the Earth's surface. Intellectual battles have been waged over these two environmental limits to diagenesis. Rather than join these battles, we focus on some of the geochemical processes that affect sediments after burial and consider some of the pathways along which they may change.

Diagenetic changes take place slowly at low temperature. The assemblages we see in sediment, therefore, usually represent some transition between stable states. In previous chapters, we began to study thermodynamic principles that allow us to interpret those stable states. This chapter, instead, focuses exclusively on building our understanding of kinetic methods—the best tools we have for examining the transition between stable states. We develop a set of fundamental equations that describe the transport of chemical species by diffusion and advection, and consider ways to describe dissolution and precipitation reactions from a kinetic perspective. Our goal is a general diagenetic equation that summarizes the various processes controlling the redistribution of chemical species in sediments after burial.

WHAT IS DIAGENESIS?

Many diagenetic processes are associated with lithification. Among these are compaction, cementation, recrystallization, and growth of new (authigenic) minerals. Through these mechanisms, unconsolidated sediment undergoes a general reduction in porosity and develops a secondary framework that converts it to solid rock. Other changes also occur, often independent of the progress of lithification. Chief among these are the many processes that modify buried organic matter, which we consider in chapter 6.

Each of these diagenetic processes commonly takes place between ~20°C and 300°C and occurs close enough to the surface that the total pressure is less than ~1 kbar. Almost invariably, diagenetic processes also involve the participation of fluids in the interstitial spaces between sediment particles. In freshly buried sediments, this fluid has the same bulk composition as the waters from which the sediments were deposited. As time proceeds, these *formation waters* accumulate and transmit the products of reactions within the sediment column. The thermodynamic principles we introduced in earlier chapters are sometimes useful in examining these reactions. In most cases, however, sediments and fluids undergoing diagenesis are highly nonuniform; that is, their compositions vary from place to place within a distance of a few centimeters to a few meters. Because of this variation, it is possible to extract kinetic information by studying concentration-distance profiles in a diagenetic environment.

KINETIC FACTORS IN DIAGENESIS

Transport of material can take place in either of two ways: by *diffusion* or *advection*. When we focus on diffusive transport, we are considering the dispersion of ions or molecules through a medium that is not itself moving. The path followed by any single particle is random, but a net movement of material may result if there is a gradient in some intensive property such as temperature or chemical potential across the system. In contrast, advective transport takes place when the ions or molecules are carried in a medium, like an intergranular fluid, that is moving. The driving force in this case is simply a gradient in hydrostatic pressure.

In this section, we develop a general expression to illustrate how the concentration of any mobile species in a diagenetic environment may change as the result of diffusive or advective transport, or as the result of chemical reactions between the fluid and solid phases in a sedimentary column.

Diffusion

In chapter 3, we introduced thermodynamic functions that describe the behavior of systems in bulk, rather than developing a statistical method for describing systems at the atomic level. This phenomenological approach addresses the thermodynamic properties of systems without acknowledging the interactions among atoms, or even their existence. In the same way, nineteenth-century scientists such as Joseph Fourier and Georg Ohm described the transport of heat and electrical charge by assuming that these entities travel through a continuous medium rather than one consisting of discrete atoms. This assumption leads to a description of transport properties by differential equations, just as it did when we adopted this perspective for thermodynamics. Diffusive transport was first described in this way in the 1850s by Adolf Fick, a German chemist.

According to *Fick's First Law*, the flux (J) of material along a composition gradient at any specific time, t, is directly proportional to the magnitude of the gradient.

That is, if the ions or molecules in which we are interested move along a direction, x, parallel to a gradient in concentration, c, then:

$$J = -D(\partial c / \partial x)_t. \tag{5.1}$$

The quantity D is called the *diffusion coefficient,* and is commonly tabulated in units of $cm^2 \ sec^{-1}$. Equation 5.1 satisfies the empirical observation that the flux drops to zero in the absence of a concentration gradient. It is applicable in any system experiencing diffusive transport, but is most appropriate when the system of interest is in a steady state; that is, when $(\partial c / \partial t)_x$ is a constant. (Recall our brief introduction to the idea of a steady state in the party example of chapter 1.) Equation 5.1 is similar to Ohm's law or to fundamental heat flow equations in that D can be shown experimentally to be independent of the magnitude of $(\partial c / \partial x)_t$. Because all of the quantities in equation 5.1 are positive numbers, we must place a negative sign on the right hand side to indicate that material diffuses in the direction of decreasing concentration.

If the concentration at a given point in the system is changing with time, equation 5.1 still holds, but it is not the most convenient way to describe diffusive transport. To derive a more appropriate differential equation, consider the schematic sediment column in figure 5.1, in which we have measured the flux of some species at two depths x_1 and x_2, separated by a distance Δx, and have found J_1 to be different from J_2. So long as Δx is small, we can describe the relationship between J_1 and J_2 by writing:

$$J_1 = J_2 - \Delta x (\partial J / \partial x).$$

Because the fluxes at the two depths differ, the concentration of the species of interest in the sediment volume between them must change during any time interval we choose. If we assume a unit cross-sectional area for the column, then the affected volume is $1 \times \Delta x$ and the change can be written as:

$$(J_1 - J_2) = \Delta x (\partial c / \partial t) = -\Delta x (\partial J / \partial x).$$

We can now substitute this result into equation 5.1 to obtain

$$\frac{\partial c}{\partial t} = \frac{\partial}{\partial x}\left(D \frac{\partial c}{\partial x}\right). \tag{5.2}$$

This equation is *Fick's Second Law,* sometimes referred to as a *continuity equation,* because it arises directly from the need to conserve matter during transport.

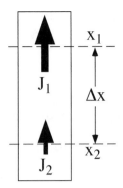

FIG. 5.1. Schematic sediment column with a unit cross-sectional area. The flux J_1 of some chemical species at depth x_1 is greater than the flux J_2 at depth x_2. This means that concentration of the species between x_1 and x_2 must vary as a function of time.

As they are presented here, Fick's laws describe diffusion in a straight line and assume that the medium for diffusion is isotropic. In most diagenetic environments, where diffusion takes place in pore waters, this formulation is adequate. Lateral variations in the composition of interstitial waters in a sediment, in most cases, are much less pronounced than changes with depth, so the driving force for diffusion is primarily one dimensional. In general, however, diffusion takes place in three dimensions, (x, y, z), and equation 5.2 is more properly written as:

$$\frac{\partial c}{\partial t} = \frac{\partial}{\partial x}\left(D\frac{\partial c}{\partial x}\right)_{y,z} + \frac{\partial}{\partial y}\left(D\frac{\partial c}{\partial y}\right)_{x,z} + \frac{\partial}{\partial z}\left(D\frac{\partial c}{\partial z}\right)_{x,y}.$$

In this chapter, we deal exclusively with the one-dimensional form. We also generally assume that the diffusion coefficient, D, does not vary from place to place. These simplifications make mathematical modeling much easier, and in most cases we lose very little accuracy by adopting them.

If D is not a function of distance, then Fick's Second Law can be written more simply as:

$$\frac{\partial c}{\partial t} = D\left(\frac{\partial^2 c}{\partial x^2}\right). \tag{5.3}$$

Solutions for this equation in a nonsteady state system (the most general situation) involve determining concentration as a function of distance and time, $c(x, t)$, for a given set of initial and boundary conditions. Even in this simplified form of Fick's Second Law, the task is often quite difficult. Many of the most useful solutions are described in Crank (1975). The two following problems

illustrate common situations in diagenetic environments of interest to geochemists.

Worked Problem 5.1

Suppose a potentially mobile chemical species is deposited in a very thin layer of sediment and subsequently buried. Can we describe how it becomes redistributed in the surrounding sediments during diagenesis?

Geochemists commonly encounter problems like this when they are asked to evaluate a site where heavy metals, arsenic, or similar toxic materials from a contaminant spill have entered soils. Imagine, for example, that copper-rich mud from a mining operation has accidentally been flushed into a lake, where it settles in a thin layer and is then quickly covered with other sediment. As a result, there is soluble copper in an infinitesimally thin sediment layer at some distance x_0 below the sediment-water interface. At the start of the problem, this marker horizon contains a quantity, a, of Cu^{2+} that is not found elsewhere in the column.

The solution to Fick's Second Law under these conditions is:

$$c(x, t) = \frac{\alpha}{2\sqrt{\pi D t}} \exp\left(\frac{x^2}{4Dt}\right),$$

where x is a distance either above or below the reference horizon x_0. Unless you want to practice your skills at solving differential equations, you can look up solutions like this in Crank (1975) or any other reference in which common solutions are tabulated. This particular one, called the *thin-film* solution, is most valuable as the basis for solving more complex problems. We can verify that it is the correct solution for this problem, because it can be differentiated to yield equation 5.2 again and because it satisfies the initial conditions we specified:

for $|x| > 0$, $c(x, t) \to 0$ as $t \to 0$,

and

for $x = 0$, $c(x_0, t) \to \infty$ as $t \to 0$,

in such a way that the total amount of the mobile species is fixed:

$$\int_{-\infty}^{\infty} c(x, t)\, dx = \alpha.$$

Suppose, then, that the marker horizon contains 50 mg of Cu^{2+} per cm^2, and assume that the diffusion coefficient for Cu^{2+} in water is 5×10^{-6} cm^2 sec^{-1}. If diffusion proceeds outward from the marker horizon for 2 months, what will be the concentration of copper in pore waters 1 cm away? Applying the thin-film solution, we find that:

$$c(1\ cm, 5.1 \times 10^6\ sec) = 2.74\ mg\ cm^{-3}$$

$$= 2.74 \times 10^3\ mg\ kg^{-1}$$

$$= 2740\ ppm.$$

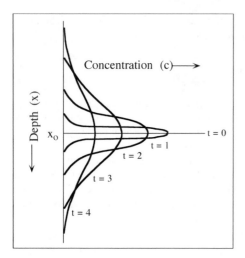

FIG. 5.2. Diffusion away from a thin-film source creates a symmetrical concentration profile that flattens and stretches out with time t. The areas under the curves are identical because the total amount of diffusing material is constant.

The total mass of diffusing material in a thin-film problem is typically very small and is concentrated initially in a layer with "zero" thickness. In a real setting, of course, the source layer has a finite thickness. For that case, we would need to apply a slightly different solution to equation 5.2.

We can examine the general consequences of thin-film geometry by studying figure 5.2, in which we have plotted $c(x, t)$ against x at successive times after diffusion outward from the marker horizon has begun. The area under the $c(x, t)$ curve remains constant (because the total amount of material is fixed), but the diffusing species spreads out symmetrically above and below the source layer as time advances. Referring to the solution in worked problem 5.1, we can see that the concentration at $x = 0$, $c(x_0, t)$, decreases at a rate proportional to $1/\sqrt{t}$. We also see that the curve has inflection points (where $\partial^2 c/\partial x^2 = 0$) above and below x_0, and that these separate a zone of solute depletion near the source bed from zones of solute enrichment at a distance. One of the other questions we might ask, then, is How far will material have diffused by some time t? This is a difficult question to answer, because $c(x, t)$ never reaches zero at any distance once diffusion has begun. One practical measure of the limit to e^{-x^2} is the point at which it decays to $1/e$ (~0.36) times its value at x_0. In this example, that distance is equal to $x_p = 2\sqrt{Dt}$. For the numerical example

in worked problem 5.1, $x_p = 10$ cm. This result is often generalized as a rule called the *Bose-Einstein relation*, describing the penetration distance over which a diffusing species has traveled:

$$x_p = k\sqrt{Dt}. \tag{5.4}$$

The size of the proportionality constant, k, may vary from one problem to another, depending on the geometry of the system and how strictly we want to define the practical limits of the $c(x, t)$ curve.

Now let's try another common problem, more complicated than the first.

Worked Problem 5.2

Suppose that a diffusing species is initially concentrated in one rather thick layer or a body of water, and then begins to travel across a boundary into a second layer. How does the concentration of diffusing ions vary in the second layer as a function of position and time?

Consider the situation illustrated in figure 5.3a. Here, lake water containing Cu^{2+} at a concentration $c = c'$ overlies a sediment column that initially contains no copper. This is the configuration we might expect if, for example, acid mine drainage were diverted into a previously clean lake, so that the lake water was suddenly copper-rich while the pore water in the underlying sediment was still clean. The boundary between the two water volumes is at the lake bed, where $x = x_0 = 0$. Verify for yourself that the initial conditions are:

for $x > 0$ at $t = 0$, $c = 0$ (in the sediment),

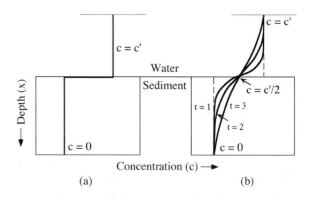

FIG. 5.3. (a) At the beginning of worked problem 5.2, the concentration of the mobile species is equal to zero at all depths in the sediment, and is equal to c' throughout the overlying water. (b) Diffusion across the water-sediment interface produces profiles described by equation 5.5. Notice that the concentration at the interface remains equal to $c'/2$ at all times after $t = 0$.

and

for $x < 0$ at $t = 0$, $c = c'$ (in the lake).

To address this problem, we need something more complicated than the thin-film solution, but we can use the thin-film solution as a guide. We imagine a column of lake water with unit cross-sectional area, divided into a set of n horizontal slices, each of thickness Δa. Each slice acts as a thin-film source for solute molecules that eventually reach a depth x_p in the sediment. We can therefore calculate the total concentration at x_p after a time t has passed by taking the sum of all the contributing thin-film solutions.

From worked problem 5.1, we know these solutions to have the form:

$$c(x, t) = \frac{c'\Delta a}{2\sqrt{\pi Dt}} \exp\left(\frac{a_i^2}{4Dt}\right),$$

where a_i is the distance from depth x to the center of the ith slice. The term α, which we used in worked problem 5.1 to equal the total amount of Cu^{2+} in the system, has been replaced by $c'\Delta a$, where Δa is the depth of the lake. If each slice is infinitesimally thin and we treat the infinite sum of all slices as an integral,

$$c(x, t) = \frac{c'}{2\sqrt{\pi Dt}} \int_x^\infty \exp\left(-\frac{z^2}{4Dt}\right)dz.$$

It is customary to rewrite this analytical result in a form that is easy to evaluate numerically with a computer or even a pocket calculator. To do this, we substitute the variable $\eta = z/2\sqrt{Dt}$, so that:

$$\begin{aligned}
c(x, t) &= \frac{c'}{\sqrt{\pi}} \int_{x/(2\sqrt{Dt})}^\infty \exp(-\eta^2)d\eta \\
&= \frac{c'}{\sqrt{\pi}} \left[1 - \int_0^{x/(2\sqrt{Dt})} \exp(-\eta^2)d\eta\right] \quad (5.5) \\
&= \frac{c'}{2} \left[1 - \mathrm{erf}\left(\frac{x}{2\sqrt{Dt}}\right)\right].
\end{aligned}$$

This may not look like much of a simplification, but it is. The error function, $\mathrm{erf}(x)$, is a standard mathematical function that can be approximated algebraically (see appendix A).

Figure 5.3b illustrates the concentration profile generated by this equation at several times after the onset of diffusion. Notice that the profile is symmetrical around the sediment-water interface, where concentration at all times is equal to $c'/2$. Unless the water column is unusually stagnant, as in some swamps, we would not expect to see a measurable gradient like this above the interface. Instead, the lake water should be uniformly stirred by winds or thermal convection, with the result that the concentration for all values of $x \le 0$ is c' at all times. We won't use the space to prove it here, but it can be shown that $c(x, t)$ in the sediment column with this additional boundary condition differs only by a factor of two from our first answer:

$$c(x, t) = \left[1 - \mathrm{erf}\left(\frac{x}{2\sqrt{Dt}}\right)\right].$$

Suppose, then, that we add enough Cu^{2+} to the lake water to raise its concentration to 500 ppm. If the diffusion coefficient for Cu^{2+} in water, as before, is 5×10^{-6} cm^2 sec^{-1}, what should be the copper concentration in pore waters 10 cm below the water-sediment interface after a year? Applying these values in our final equation for $c(x, t)$, we get:

$$c(10 \text{ cm}, 3.1 \times 10^7 \text{ sec}) = 500 \text{ ppm } [1 - \\ \mathrm{erf}(10 \text{ cm } /2\sqrt{[5 \times 10^{-6} \text{ cm}^2 \text{ sec}^{-1}][3.1 \times 10^7 \text{ sec}])})]$$

$$= 285 \text{ ppm}.$$

In the discussion so far, we have considered only the diffusion of dissolved material through water. This is appropriate, because diffusion through any solid medium is almost negligibly slow compared with diffusion in water. The bulk diffusive flux within a sediment column therefore depends on how much volume is *not* occupied by solid particles (that is, it depends on the *porosity* (φ)). Diffusion may also be slowed significantly because of the geometrical arrangement of particles, because the effective path length for diffusion must increase as dissolved material detours around sediment particles. This second factor is known as the *tortuosity* (θ), defined by:

$$\theta = dl/dx,$$

where dl is the actual path length a diffusing species must traverse in a given distance dx. Tortuosity cannot be measured directly, but is usually determined by comparing values of D determined in open water with those measured in pore water (Li and Gregory 1975), or by comparing the electrical conductivities of sediments and pore waters (Manheim and Waterman 1974). Taking both porosity and tortuosity into account, the diffusion coefficient in the sediment column, D_s, is related to the value of D in pure water by:

$$D_s = \varphi D/\theta^2. \quad (5.6)$$

Worked Problem 5.3

Measurements of Mg^{2+} concentration are made in the pore waters of a marine red clay, and it is found that Mg^{2+} decreases linearly from 1300 ppm at the sea floor to 1057 ppm at a depth of 1 m. Assuming that the profile is entirely due to diffusion, how rapidly is Mg^{2+} diffusing into the sediment?

Geochemists Yuan-Hui Li and Sandra Gregory (1975) have reported that the diffusion coefficient for Mg^{2+} in pure water close to 0°C is 3.56×10^{-6} cm^2 sec^{-1} and the average porosity and tortuosity in marine red clay are 0.8 and 1.34, respectively. We calculate from equation 5.6 that the effective diffusion coefficient, D_s, is therefore:

$$D_s = (0.8)(3.56 \times 10^{-6} \text{ cm}^2 \text{ sec}^{-1})/(1.34)^2$$
$$= 1.6 \times 10^{-6} \text{ cm}^2 \text{ sec}^{-1}.$$

Because $\partial c/\partial x$ does not vary with depth, we conclude that the profile represents a steady-state flux of Mg^{2+} into the sediment. Thus, we can easily calculate J at the interface from Fick's First Law. The concentration gradient is 243 ppm per meter, or 1.0×10^{-2} mol l^{-1} m^{-1}, or 0.1 μmol cm^{-4}. Therefore,

$$J = -D_s \partial c/\partial x$$
$$= (1.6 \times 10^{-6} \text{ cm}^2 \text{ sec}^{-1})(0.1 \text{ μmol cm}^{-4})$$
$$= 1.6 \times 10^{-6} \text{ μmol cm}^{-2} \text{ sec}^{-1}$$
$$= 4.96 \text{ μmol cm}^{-2} \text{ yr}^{-1}$$
$$= 0.12 \text{ mg cm}^{-2} \text{ yr}^{-1}.$$

In addition to porosity and tortuosity, other chemical factors affect diffusion in a diagenetic environment. These arise because most material transported in pore fluids is in the form of charged particles. A flux of cations into a sediment column must be accompanied by an equal flux of anions or must be balanced by a flux of cations of some other species in the opposite direction. We may expect that electrostatic attractions or repulsions between ions will affect their mobilities. The greater the charge (z) on an ion, the more likely it is that it will be slowed by interaction with other ions. Figure 5.4 illustrates the magnitude of this effect for most simple aqueous ions by plotting the value of D against the ionic potential ($|z|/r$) for a variety of common ionic species. Geochemist Antonio Lasaga (1979) has shown that ionic *cross-coupling* can change the value of D for some species by as much as a factor of ten.

Charged species not only interact with each other but also interact with sediment. Clay particles, in particular, develop a negatively charged surface when placed in water, as interlayer cations are lost into solution. Several important properties of clays arise from this behavior. Some cations from solution are adsorbed onto the clay surface to form a *fixed layer*, which may be held purely by electrostatic forces or may form complexes. Other cations form a *diffuse layer* farther from the clay-water interface, in which ions constitute a weakly bonded

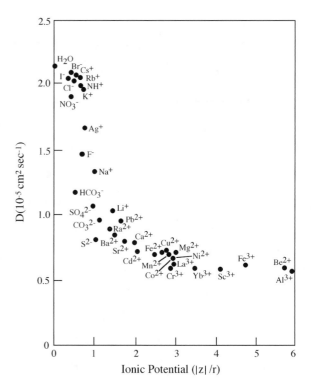

FIG. 5.4. Tracer diffusion coefficients D for common ionic species as a function of ionic potential. (Modified from Li and Gregory 1975.)

cloud. This double layer of positively charged ions repels similar structures around adjacent clay particles. In open water, these electrostatic effects support small particles in stable suspensions known as *colloids* and prevent them from settling to the bottom. For example, we observe that sediment particles are carried as colloids in river water but are commonly deposited as they enter the ocean, where their diffuse layer is overwhelmed by the abundance of free ions in salt water. In general, this tendency to maintain a colloid decreases with increasing ionic strength. In a sediment column, the double layer promotes a very open structure, like a house of cards, and maintains a high porosity.

The way in which surface properties of clays affect ionic diffusion can be shown by considering the adsorption of a mobile ion A onto a clay particle as a chemical reaction:

$$A_{\text{free}} \rightleftarrows A_{\text{bound}}.$$

Ignoring any effects of nonideality, the equilibrium constant for this reaction is given by:

$$K = C/c,$$

where c is the concentration of A in solution and C is its concentration in bound form, in units of moles per liter of interstitial solution. Because the total amount of A is constant, the rate of change in c must be equal and opposite to the rate of change in C. For a rapid exchange between solution and solids, then, we can write:

$$K(\partial c/\partial t) = -(\partial C/\partial t).$$

Equation 5.2, with this modification, becomes:

$$\partial c/\partial t = D\partial^2 c/\partial x^2 + \partial C/\partial t$$

$$= D\partial^2 c/\partial x^2 - K\partial c/\partial t,$$

or

$$\partial c/\partial t = [D/(1 + K)]\partial^2 c/\partial x^2.$$

The rate of diffusion, therefore, becomes much slower as the role of ion exchange or adsorption by clay surfaces increases. If we take chemical interactions with solids into account, the effective diffusion coefficient defined in equation 5.6 now becomes:

$$D_s = \frac{\varphi D}{\theta^2(1 + K)}. \qquad (5.7)$$

Advection

Water can flow through pores in a sediment as the result of compaction or a regional hydraulic gradient. The first of these is most likely to dominate in a marine or lacustrine environment and is therefore most commonly associated with early diagenesis. Regional flow, however, can be important when dissolved matter is transported and interacts with sediments in an aquifer or tectonic pressure induces fluid flow in deep sediments.

As sediment accumulates, we can examine the progress of diagenesis in a given bed in either of two reference frames. In one frame, we locate the bed by referring to its depth below the sediment-water interface. With continued deposition, the depth to the bed increases. Viewed from this perspective, both the sediment and any interstitial fluid seem to move with time. Unless the sediment is compacted as it is buried, however, the sediment particles are not actually moving relative to each other.

Because there is also no movement of fluid relative to the sediment if it, too, is simply buried, we call its apparent flow *pseudoadvection*. To avoid being misled by pseudo-advection, we can choose an alternate reference frame centered on the bed itself instead of one that is centered on the sediment-water interface.

True fluid advection in marine or lacustrine sediments is most commonly driven by the gradual reduction in porosity that occurs as sediments are compacted with depth. If we choose a bed-centered reference frame, we usually anticipate a gentle upward flow, as water is squeezed out of sediments in deeper layers. This effect should be most prominent in clay-rich sediments, which initially have high porosities because of electrostatic repulsion between particles. Within a few tens of meters of the surface, these weak repulsive forces are overwhelmed by the weight of accumulating sediment, and porosity drops from 70–80% to <40%, gradually approaching an asymptotic limit within a few tens of meters of the surface. Porosity decreases very rapidly within a few tens of meters of the sediment surface, and then only very slowly for the next few thousand meters, as shown in the schematic profile in figure 5.5. For our purposes, it is reasonable to treat the porosity below a depth of ~30–50 m as a constant.

Even in high-porosity clays, however, the gentle upward flow of fluid is too slow to measure, because clays typically accumulate and compact very slowly. In sandy sediments, it is even slower. Particularly in those sands with poor sorting, porosity is initially low and does not

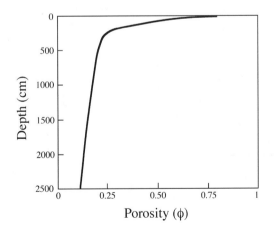

FIG. 5.5. Porosity in clastic sediments decreases to a nearly asymptotic limit within a few tens of meters of the surface. This approximation makes it possible to estimate advective flow rates caused by compaction.

change much with depth. The angularity of coarse clastic particles makes reorientation and compaction difficult. Regardless of particle size and burial rate, then, we cannot directly measure fluid advection due to compaction during marine or lacustrine diagenesis. We can only estimate its rate by constructing theoretical models of the compaction process. The simplest model we can devise (and the only one we consider in this book) assumes that the sedimentation rate is constant and $\partial\varphi/\partial t = 0$ at any depth in the sediment column. This is very nearly true within a few meters of the sediment-water interface.

Let us adopt the symbols ω and u to indicate the rates at which sediment and pore waters are buried, respectively. If there were no compaction, ω would be equal to u, and both would simply be equal to the rate of sediment accumulation. Under steady-state compaction, however, we find that as porosity decreases toward its asymptotic limit (φ_d), ω and u each decrease as well, gradually approaching a limit where $\omega_d = u_d$. As long as the change in porosity with depth is fairly small (as it is below about 10 meters in the sediment), we conclude that:

$$u\varphi = \varphi_d u_d = \varphi_d \omega_d.$$

Assuming steady-state compaction, therefore, it is possible to calculate the rate of fluid burial at any depth if we have a porosity-depth profile and can estimate ω.

Clearly, however, at any depth less than the one where φ, u, and ω all approach their asymptotic limits, fluids are buried less rapidly than sediment. Expelled waters, therefore, flow upward at a rate (U) equal to:

$$U = u - \omega.$$

If we multiply both sides of this expression by φ and combine it with the previous equation, we see finally that it is possible to calculate the rate of upward flow from:

$$U = (\varphi_d \omega_d / \varphi) - \omega. \tag{5.8}$$

Worked Problem 5.4

A team of researchers has measured porosity as a function of depth in a sediment column at the bottom of mythical Puzzle Pond. They find that they can describe the data adequately with the mathematical function:

$$\varphi(x) = 0.5e^{(-0.05x)} + 0.2,$$

where x is depth in meters. How can we use this information to estimate the rate of compaction-driven fluid flow in the sediments?

Yale University geochemist Robert Berner (1980) showed that if we assume steady-state compaction, $\partial[(1 - \varphi)\omega]/\partial x = 0$. That is, $[1 - \varphi(x)]\omega(x)$ has a fixed value. To find what that value is, note that the sediment burial rate at the sediment interface, ω_0, is equal to the sediment deposition rate. We can evaluate $\varphi(x)$ at $x = 0$ to see that $\varphi(0) = 0.5e^{(0.0)} + 0.2 = 0.3$. Therefore,

$$[1 - \varphi(x)]\omega(x) = 0.3\omega_0.$$

The burial rate of the sediment in this column, therefore, is given by:

$$\omega(x) = 0.3\omega_0 / [1 - (0.5e^{(-0.05x)} + 0.2)]$$
$$= 3\omega_0 / [8 - 5e^{(-0.05x)}].$$

In Puzzle Pond, the sedimentation rate is 0.2 cm yr^{-1}. Applying our expressions for $\varphi(x)$ and $\omega(x)$, we calculate that φ and ω approach nearly constant values ($\varphi_d = 0.22$ and $\omega_d = 0.077$ cm yr^{-1}) ~60 meters down. These, then, can be inserted in equation 5.7 to yield:

$$U(x) = [0.22(0.077 \text{ cm yr}^{-1}) / \varphi(x)] - \omega(x).$$

The fluid advection rate at selected depths can be calculated by substituting appropriate values for $\varphi(x)$ and $\omega(x)$:

x (m)	$\varphi(x)$	$\omega(x)$	U(cm yr^{-1})
5	0.59	0.15	−0.117
10	0.50	0.12	−0.087
15	0.44	0.11	−0.068
20	0.38	0.097	−0.053
30	0.31	0.087	−0.033
40	0.27	0.082	−0.019
50	0.24	0.079	−0.009
60	0.22	0.077	−0.002

Notice that the advection rates are all negative, indicating upward flow.

Because these rates of fluid advection depend so strongly on grain size, sorting, and sedimentation rate, it is hard to devise reliable models that are more detailed than those we have described above. Given a semi-empirical model like this, however, it is easy to compare rates of diffusion and advection in a sediment and see which is the major mechanism for transport during diagenesis.

Worked Problem 5.5

Which is the dominant process for interstitial transport of Mg^{2+} in a typical marine mud, diffusion or advection? In worked problem 5.3, we calculated that the effective diffusion coefficient (D_s) for Mg^{2+} under these circumstances is ~1.6 × 10^{-6} cm^2 sec^{-1}. To answer the question, then, we compare that value of D_s against the advection rates we calculate with equation 5.8, using the same parameter values we used in worked problem 5.4. To express D_s and U in compatible units, we will first need to adjust U for porosity, multiplying it by φ to produce an effective advection rate, U_s. To convert its dimensions of cm yr^{-1} into cm sec^{-1}, we then divide U_s by 3.1×10^7 sec yr^{-1}.

The ratio D_s/U_s has the dimensions of distance (L), in centimeters. To evaluate the balance between diffusion and advection, then, we can divide D_s/U_s by L to create the *Peclet number*, D_s/LU_s, which is dimensionless. If the Peclet number for an environment is <<1, then advection dominates; if D_s/LU_s >> 1, then diffusion is the more effective transport process. The critical distance L_c at which the Peclet number equals 1.0, therefore, defines the physical boundary between diffusion-dominated and flow-dominated regimes in a sediment. Because U_s is a function of depth, L_c also varies with depth. In this problem, we find:

x (m)	L_c (m)
5	7.2
10	11.3
15	16.8
20	24.3
30	48.6
40	99.4
50	234
60	1058

From these results, we conclude that because the actual distance Mg^{2+} has to travel to escape from the sediment column into the overlying water is always significantly less than L_c, diffusion is the major transport process during diagenesis of marine muds. This result would not be significantly different if we had chosen to follow a different ion (see fig. 5.4).

To illustrate how much this inference depends on the value of U, consider the transport of Mg^{2+} in a sandy aquifer, where the rate of flow is imposed by some regional pressure gradient. To calculate such a rate, we might use a form of Darcy's law, the flow equation familiar to hydrologists:

$$U_s = k \, (1/h),$$

where k is a measure of permeability and $1/h$ is the local hydraulic gradient. It would not be unusual to find flow rates in excess of 100 cm yr^{-1}. Substituting in the expression above, we find that the Peclet number is equal to 1.0 when $L_c = 0.5$ cm. Not surprisingly, then, we verify that flow, not diffusion, is the major transport mechanism in an aquifer.

KINETICS OF MINERAL DISSOLUTION AND PRECIPITATION

In our discussion so far, we have considered changes in the sediment column from a mechanical point of view, in which diffusing or advecting ions are carried as inert particles. Variations in fluid composition during diagenesis, however, are not due to diffusion or advection alone. In addition, dissolved material can be added to solution (or lost from it) in a variety of chemical reactions. Reactions between primary minerals in the sediment may produce or consume dissolved species, as can the growth of new (authigenic) minerals.

Biological processes associated with decay or respiration can act in a number of ways to change the abundance of organic compounds in the sediment column and modify the composition and redox state of pore waters. We look carefully at some of these processes in chapter 6. We consider dissolution-precipitation reactions from a thermodynamic perspective in chapters 7 and 8. The thermodynamics of redox reactions, which may affect the concentration of transition metal ions like Mn^{2+} or Fe^{2+} in solution, is also the subject of chapter 7. Although we devote a lot of attention to thermodynamics in these chapters, it is important to recognize that chemical reactions at low temperature are commonly arrested in midflight. That is, the phase assemblages and compositions we observe in a sediment column frequently represent some intermediate step along the way toward equilibrium rather than equilibrium itself. Because this is so, thermodynamics is often inadequate to interpret the chemical variability observed in sediments. Instead, geochemists find that they can apply principles of kinetics to minerals and fluids in the sediment and thus deduce the reaction pathways that dominate during diagenesis, as well as the physical or transport pathways. We demonstrate two very different approaches in this final section of the chapter, first considering kinetics at the molecular level and later at a macroscopic level.

Antonio Lasaga (1998) makes a useful distinction between *elementary* chemical reactions, which occur as written at the molecular level, and *overall* reactions, which describe a net change that involves several intermediate steps and take place along several competing, parallel pathways. We have also encountered this distinction briefly in chapter 1, where thermodynamic and kinetic approaches to geochemistry were contrasted. As we show here, this distinction is also reminiscent of the contrast between atomistic and empirical perspectives on chemical change, with which we opened this chapter. Let us consider a few short examples to see how this distinction affects the way we treat reactions during diagenesis.

The chemical reactions that occur during diagenesis include some processes like carbonate precipitation, for which we can write an elementary reaction:

$$Ca^{2+} + CO_3^{2-} \rightarrow CaCO_3.$$

The rate at which molecules of $CaCO_3$ are produced is directly proportional to the probability that a Ca^{2+} ion will collide with a CO_3^{2-} ion in solution. Therefore, the appropriate rate law in this case is:

$$dn_{CaCO_3}/dt = kn_{Ca^{2+}}n_{CO_3^{2-}},$$

in which $n_{Ca^{2+}}$, $n_{CO_3^{2-}}$, and n_{CaCO_3} are the concentrations of each species and k is a linear rate constant. In fact, for any elementary reaction, the rate is simply proportional to the abundance of reactant species.

To put this into a more theoretical framework, remember that in chapter 4, we introduced notation for writing a generic chemical reaction in the form:

$$v_{j+1}A_{j+1} + \ldots + v_{i-1}A_{i-1} + v_iA_i \rightleftarrows v_1A_1 + v_2A_2 + \ldots + v_jA_j.$$

We introduced the notion of flux in chapter 1 in equation 1.3:

$$dA_i^{out}/dt = \sum k_{ij}A_i.$$

Applying this equation, it can be shown that the rate of a generic forward reaction (v_f) is represented by:

$$v_f = dA_{forward}/dt = k_f(a_{j+1}^{v_{j+1}} a_{j+2}^{v_{j+2}} \ldots a_i^{v_i}),$$

and the rate of the reverse reaction (v_r) by:

$$v_r = dA_{reverse}/dt = k_r(a_1^{v_1} a_2^{v_2} \ldots a_j^{v_j}),$$

in which the quantities k_f and k_r are first-order rate constants. Thus, as in the party example of chapter 1, v_f and v_r are each proportional to the abundance of reactant (or product) species.

At equilibrium, the forward and reverse reaction rates must be equal if there is to be no net change in the system. Therefore, there is a direct relationship between the rate constants and the equilibrium constant for the reaction:

$$K_{eq} = k_f/k_r.$$

Furthermore, the degree of disequilibrium can be estimated from the ratio v_f/v_r and the activities of species in the system, because:

$$k_f/k_r = (v_f/v_r)(a_{j+1}^{v_{j+1}} a_{j+2}^{v_{j+2}} \ldots a_i^{v_i})/(a_1^{v_1} a_2^{v_2} \ldots a_j^{v_j}).$$

This expression can be written, for greater clarity, in the form:

$$k_f/k_r = K_{eq} = (v_f/v_r)Q. \qquad (5.9)$$

This expression degenerates to $K_{eq} = Q$ (equation 4.12) for the equilibrium case, where $v_f/v_r = 1$, but it tells us more than that. If $K_{eq}/Q > 1$, then the rate of the forward reaction exceeds the rate of the reverse reaction. If $K_{eq}/Q < 1$, the opposite is true. Furthermore, we can see that a reaction proceeds most rapidly (in either a forward or a backward direction) when it is farthest from equilibrium; that is, when K_{eq}/Q is very much greater or less than 1. For this reason, the direction and rate of reaction are often good indicators of how far a system may be from equilibrium.

Worked Problem 5.6

Reactions involving carbon dioxide and water are important in many geologic environments, including those undergoing diagenesis. One important reaction in this system involves the equilibrium between dissolved CO_2 gas and the bicarbonate ion:

$$CO_2(aq) + H_2O \rightleftarrows HCO_3^- + H^+.$$

This is actually an overall reaction that proceeds first by the slow step of hydration:

$$CO_2(aq) + H_2O \rightleftarrows H_2CO_3,$$

an elementary reaction, and then much more quickly by dissociation:

$$H_2CO_3 \rightleftarrows HCO_3^- + H^+,$$

another elementary reaction. Hydration is such a slow step, comparatively, that $CO_2(aq)/H_2CO_3$ is ~600 at 25°C. The overall reaction, therefore, behaves as if it were a much simpler elementary reaction and follows approximately a first-order rate law. For the sake of this worked problem, then, let's pretend that it really is an elementary reaction. Can we use its kinetic behavior to describe the approach to equilibrium and estimate the value of K_{eq}?

The combined rate constants for the two-step reaction (Kern 1960) are $k_f = 3 \times 10^{-2}$ sec^{-1} and $k_r = 7.0 \times 10^4$ sec^{-1}. Figure 5.6 shows the progress of reaction in a system presumed to consist initially of 1×10^{-5} mol liter^{-1} $CO_2(aq)$ at pH = 7(a_{H^+} = 1×10^{-7}). Hypothetically, the concentrations of $CO_2(aq)$ and HCO_3^- could be measured as functions of time, but this is a nearly impossible task in the CO_2-H_2O system. To construct figure 5.6, therefore, the concentrations were calculated algebraically from the experimentally determined rate constants. For details of the method, see Stumm and Morgan (1995).

Our interest here is in the rates of reaction as the overall reaction approaches equilibrium. These can be calculated at any given time from:

$$v_f = k_f a_{CO_2}(aq),$$

and

$$v_r = k_r a_{HCO_3^-} a_{H^+}.$$

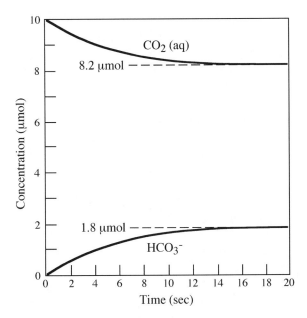

FIG. 5.6. Computed concentrations of $CO_2(aq)$ and HCO_3^- as a function of time for the reaction $CO_2(aq) + H_2O \rightleftarrows HCO_3^- + H^+$ at 25°C in a closed aqueous system. The total concentration of carbon species remains constant at 1×10^{-5} moles. (Modified from Stumm and Morgan 1995.)

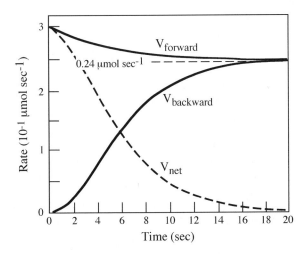

FIG. 5.7. Computed reaction rates for the forward and reverse directions of the approximate first-order reaction illustrated in figure 5.6 as functions of time. At equilibrium, both reaction rates approach 2.4×10^{-4} mole sec^{-1}. (Modified from Stumm and Morgan 1995.)

The results are shown in figure 5.7. Notice that the rate profiles for the forward and reverse reactions are not mirror images of each other. During the approach to equilibrium, $v_f/v_r > 1$, so $K_{eq}/Q > 1$. Both ratios decrease with time, however, and within roughly 20 seconds, v_f and v_r approach the same value (2.4×10^{-4} mole sec^{-1}) to <1% error. The dashed line in figure 5.7 indicates the progress of the net reaction rate, $v_f + v_r$, which approaches zero. The value of K_{eq}, calculated from the rate constants by equation 5.8, is 4.3×10^{-7}. This is virtually the same as the value (4.4×10^{-7}) calculated from the free energies of formation of dissolved species.

Unlike the elementary reactions in this simple example, unfortunately, the rates at which overall reactions occur are commonly nonlinear functions of concentration. They may be influenced by processes that occur at the surface of a growing or dissolving particle, by the production of transient, metastable species, or by transport of dissolved material toward or away from the interface. As a result, rate laws may be quite complex, reflecting a variety of inhibiting or enhancing processes. Robin Keir (1980), for example, has determined by experiment that the rate of calcite dissolution, R, in seawater follows the nonlinear relationship:

$$R = k[1 - (a_{Ca^{2+}}a_{CO_3^{2-}}/K_{cal})]^{4.5},$$

in which K_{cal} is the K_{eq} for the reaction:

$$CaCO_3(\text{calcite}) \rightleftarrows Ca^{2+} + CO_3^{2-},$$

and k is an empirical constant that varies widely among samples and is most strongly a function of grain size (fig. 5.8). The large exponential factor has been hypothesized to arise from the absorption of phosphate ions on crystal surfaces, where they serve as dissolution inhibitors. Similarly, Robert Cody and Amy Hull (1980) have demonstrated in the laboratory that some organic acids are preferentially adsorbed on nuclei of gypsum, inhibiting further growth and leading instead to low-temperature formation of anhydrite from saturated calcium sulfate solutions. In both of these cases, we cannot predict the rate law without detailed knowledge of the reaction mechanism or, as is more often the case, an exhaustive set of empirical observations.

For many overall reactions that affect diagenesis, we have little theoretical knowledge and few relevant observations, and we cannot therefore adopt an atomistic approach to studying them. The formation of sulfide in sediments, for example, is associated with bacterial

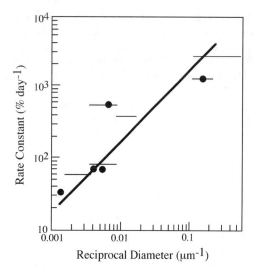

FIG. 5.8. The rate of dissolution of calcite (expressed as the rate constant) increases rapidly with decreasing grain size because of the increased surface area in fine-grained materials. (Modified from Keir 1980.)

reduction of sulfate in pore waters. We can write a reasonable reaction to express this process,

$$7CH_2O + 4SO_4^{2-} \rightleftharpoons$$
$$2S^{2-} + 7CO_2 + 4OH^- + 5H_2O,$$

but it is valuable only as a guide to system mass balance. The species labeled "CH_2O" is a generalized compound with the average stoichiometry of organic matter in sediments. The rate of sulfide formation during diagenesis depends on a large number of factors that are not indicated in the schematic form of the overall reaction. These may include the identity of specific organic molecules in the fresh sediment and any that form after burial, surface reactions on growing sulfides, and absorption by clay surfaces. We examine these factors in greater detail in chapter 6. Our understanding of reaction kinetics in such cases is primitive.

Fortunately, it is often possible to model diagenetic environments that are dominated by mineral growth without dealing with overall reactions at an atomistic or molecular level. Before we finish this section, let us consider another specific diagenetic problem and demonstrate an empirical approach to handling the kinetics of overall reactions.

Bottom waters throughout most of the ocean are oxygenated. Within the sediment column below the ocean floor, however, oxygen is rapidly consumed by bacteria that feed on organic debris. As mineral dissolution reactions take place in the sediment, then, any iron and manganese become mobilized as reduced Fe^{2+} and Mn^{2+} rather than remaining oxidized and relatively immobile. These ions migrate, primarily by diffusion, toward the sediment surface, where they become reoxidized to Fe^{3+} and Mn^{3+} or Mn^{4+} and precipitate to form a stable layer of Fe-Mn-oxyhydroxides. Manganese nodules are formed on the ocean floor by a closely related process. Examined in detail, this redistribution of metal ions to form new minerals involves a complicated network of organic and inorganic reactions, many of which are nonlinear. Burdige and Gieskes (1983), however, developed an analytical solution that does not make explicit reference to the overall reactions that are occurring in the sediment column. Figure 5.9a describes Mn^{2+} variations in pore water with depth in a sediment core from the east equatorial Atlantic Ocean. The solid curve is Burdige and Gieskes's solution to a differential equation that includes measured values of porosity, sedimentation rate, and other key parameters in the sample area. Rather than lay out the details of their rather complicated model, let's follow the example of Robert Berner (1980) by using a simpler model to illustrate the same principles.

Worked Problem 5.7

How does the distribution of manganese in pelagic sediments change with time? Berner's model, illustrated in figure 5.9b, assumes that the zone of mineral precipitation grows upward from the sediment-water interface ($x = 0$) and consists only of oxyhydroxide minerals and fluid-filled pore spaces. This is possible only if both the sedimentation rate and bioturbation are insignificant. Berner further assumes that there are no porosity gradients, and that Mn^{2+} is not adsorbed on clay surfaces, so that advective flow is negligible and the rates of dissolution reactions that initially mobilize Mn^{2+} are rapid enough to not limit the eventual precipitation process. He has therefore reduced the complex set of elementary reactions in this overall system to a single precipitation reaction. He describes this reaction implicitly by monitoring how rapidly the zone of mineral precipitation grows instead of including the reaction rate itself in the model. Finally, because the concentration gradient in the transition zone between the precipitation layer and the top of the source region ($x = L$) is nearly linear, he makes the approximation:

$$\partial C/\partial x = (C_L - C_0)/L,$$

where C_L and C_0 are the concentrations of Mn^{2+} at $x = L$ and $x = 0$, respectively.

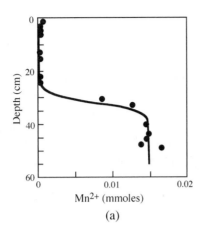
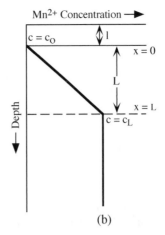

(a)   (b)

FIG. 5.9. (a) Plot of the manganese content of pore water as a function of depth in a sediment core from the equatorial Atlantic Ocean. Solid curve is the model profile calculated by Burdige and Gieskes (1983). (b) Concentration profile in the simplified redox model for marine Mn redistribution discussed by Berner (1980).

The concentration profile in this model consists, then, of an uninterestingly constant segment below depth L, and a linear segment above it, the slope of which changes with time. It bears a crude resemblance to the measured curve in figure 5.9a, offset vertically so that the zone of mineral deposition lies entirely above the sediment. Rather than using this model simply to calculate the concentration profile, however, Berner shows how it is possible to estimate how long it takes to grow a monomineralic surface layer of any given thickness.

The flux of dissolved Mn at $x = 0$ at any given time can be calculated from Fick's First Law, adjusted for sediment porosity and tortuosity:

$$J_0 = -D_x \partial C/\partial x = -D_x(C_L - C_0)L.$$

The thickness of the transition zone (that is, the value of L) varies as a function of time that is directly proportional to J_0:

$$dL/dt = V_d D_s(C_L - C_0)/F_d L.$$

Here, V_d = the volume of dissolving mineral matter that contains one mole of Mn^{2+} and F_d = the volume fraction of dissolving matter in the bulk sediment.

In a similar way, the thickness of the oxyhydroxide layer (l) grows according to:

$$dl/dt = V_p D_s(C_L - C_0)/F_p L,$$

in which V_p and F_p are similar to V_d and F_d, but refer to the mineral matter in the precipitation zone.

The thickness of the transition zone can be calculated by integrating dL/dt for the boundary conditions $L = 0$ when $t = 0$ and $L = L$ when $t = t$:

$$L = [(2V_d D_s/F_d)(C_L - C_0)t]^{1/2}.$$

If we then insert this result in the expression for dl/dt and integrate for the boundary conditions $l = 0$ when $t = 0$ and $l = l$ when $t = t$, we have derived an expression for t:

$$t = l^2 F_p^2 V_d/(2V_p^2 F_d D_s[C_L - C_0]).$$

We can now determine how long it takes to form an oxidized surface layer of an arbitrary thickness (1 cm) by placing reasonable values for each of the variable parameters in the last two equations:

$$D_s = 2.4 \times 10^{-6}\ cm^2\ sec^{-1},$$

$$V_p = V_d = 20\ cm^3\ mol^{-1},$$

$$F_p = 0.2,$$

and

$$C_0 = 0.0\ M.$$

For several pairs of F_d and C_L values, we calculate t and L:

F_d	C_L (mol cm^{-3})	L (cm)	t (yr)
0.001	1×10^{-7}	200	133,500
0.002	1×10^{-7}	100	66,770
0.005	1×10^{-7}	40	26,700
0.010	1×10^{-7}	20	13,350
0.010	1×10^{-6}	20	1335

The geometry of this model is similar in many ways to the more complex example presented by Burdige and Gieskes, and the numerical results may be generally comparable as well. Notice, however, that the predicted time scale for growth of the authigenic layer is long enough to stress some of the initial assumptions of the model. In nature, it is unlikely that a shallow sediment column would remain undisturbed by bioturbation and chemical perturbations for 10^3–10^5 years.

The primary appeal of this approach is that it avoids the necessity of describing directly each of the many chemical reactions in the overall system. By dealing instead with the observed transport of chemical species and focusing on the few elementary reactions whose rates limit the overall system, this approach makes an otherwise impossibly difficult problem easy. Depending on the complexity of the dissolution/precipitation problem, then, geochemists can adopt the atomistic or molecular-level perspective that we have identified with Lasaga and with the CO_2–H_2O example in worked problem 5.6, or they can choose a more empirical perspective, such as the one demonstrated in worked problem 5.7.

THE DIAGENETIC EQUATION

With this brief survey, we have introduced each of the important processes that affect the mass balance in a sediment during diagenesis. Our final task is to combine them, in what Robert Berner (1980) has called a *general diagenetic equation*. The rate at which concentration of some interesting chemical species in a particular volume of sediment changes can be expressed by:

$$\partial C/\partial t = -\partial J/\partial x + \sum R.$$

Here, the first term on the right side describes the total flux of the species through the volume per unit time, and the second term includes the rates, R, of all relevant chemical reactions proceeding in the volume. The distance variable, x, is taken relative to the sediment-water interface. From discussions in previous sections, we know that J must include both the diffusive and the advective flux:

$$J = -D\partial C/\partial x + UC.$$

The chemical reaction rates, R, may be explicit rates of elementary reactions or may be generalized in a more empirical fashion, as we have just shown. The general form of the diagenetic equation, then, becomes:

$$\frac{\partial C}{\partial t} = \frac{\partial \left(D \frac{\partial C}{\partial x} \right)}{\partial x} - \frac{\partial (UC)}{\partial x} + \sum R. \qquad (5.10)$$

This general equation implicitly includes each of the factors we have discussed in previous sections, as well as many we have not, such as the effect of bioturbation.

SUMMARY

We have spent this chapter discussing the kinetics of transport and reaction, a departure from our emphasis on thermodynamics in earlier chapters. This is appropriate, because so many of the changes we associate with diagenesis are incomplete—limited by the rates at which reactants move from one place to another or by how slowly they combine. We have introduced and contrasted diffusion and advection as transport mechanisms, and have experimented with mathematical methods for evaluating both. We use these methods in later chapters as well, particularly in the very different context of melt crystallization, in chapter 11. We have tried to show how these tools can be applied in the form of a general diagenetic equation to problems of practical interest in sedimentary geochemistry. In this survey of diagenesis, we have deliberately postponed most references to the fate of organic matter, which is a large enough topic to deserve much of our attention in chapter 6.

SUGGESTED READINGS

Few books have been written on the geochemical side of diagenesis, although there is a great deal of interest among geologists. Most of the informative writing is disseminated in journal articles. For general background, you may wish to consult the following sources.

Berner, R. A. 1980. *Early Diagenesis: A Theoretical Approach.* Princeton: Princeton University Press. (This is the best single reference in the field. Much of what is written today about the chemistry of diagenesis, including what we have said in this chapter, was first considered by Berner. The first five chapters are particularly useful; they establish the fundamental principles we have introduced here.)

Lasaga, A. C. 1998. *Kinetic Theory in the Earth Sciences.* Princeton: Princeton University Press. (This is a comprehensive volume, presenting what geochemists know about kinetics over a wide range of environments. It is heavy on theory, but very clearly written.)

Scholle, P. A., and P. R. Schluger, eds. 1979. *Aspects of Diagenesis.* Special Publication 26. Tulsa: Society of Economic Paleontologists and Mineralogists. (This volume contains widely quoted papers from two symposia on clastic diagenesis.)

Stumm, W., and J. J. Morgan. 1995. *Aquatic Chemistry: Chemical Equilibria and Rates in Natural Waters,* 3rd ed. New York: Wiley. (A detailed yet highly readable text with an emphasis on the chemical behavior of natural waters. Chapter 6, dealing with precipitation and dissolution, is particularly well written.)

The following sources were referenced in this chapter. The reader may wish to examine them for further details.

Burdige, D. J., and J. M. Gieskes. 1983. A pore water/solid phase diagenetic model for manganese in marine sediments. *American Journal of Science* 283:29–47.

Cody, R. D., and A. Hull. 1980. Experimental growth of primary anhydrite at low temperatures and water salinities. *Geology* 8:505–509.

Crank, J. 1975. *The Mathematics of Diffusion.* London: Oxford University Press.

Keir, R. S. 1980. The dissolution of biogenic calcium carbonate in seawater. *Geochimica et Cosmochimica Acta* 44:241–254.

Kern, D. B. 1960. The hydration of carbon dioxide. *Journal of Chemical Education* 37:14–23.

Lasaga, A. C. 1979. The treatment of multi-component diffusion and ion pairs in diagenetic fluxes. *American Journal of Science* 279:324–346.

Li, Y.-H., and S. Gregory. 1975. Diffusion of ions in sea water and in deep-sea sediments. *Geochimica et Cosmochimica Acta* 38:703–714.

Manheim, F. T., and L. S. Waterman. 1974. Diffusimitry (diffusion coefficient estimation) on sediment cores by resistivity probe. *Initial Reports of the Deep Sea Drilling Project* 22: 663–670.

PROBLEMS

(5.1) During a catastrophic flood, a barrel of liquid nuclear waste containing radium is carried away by a major stream and dropped on the floodplain, where it bursts and is spread in a thin layer. Assume that overbank deposits cover the material rapidly, and that no process affects the distribution of radium other than diffusion. Using the diffusion coefficient for Ra^{2+} from figure 5.4, calculate the effective vertical distance over which radium would be distributed in six months.

(5.2) A zinc smelter is required to dispose of slurry from its pickling operation, containing 50 ppm Cd^{2+}, into a "safe" environment. Management wants to pump it into a clay pit abandoned by a neighboring brick factory. Use equation 5.5 to calculate the one-dimensional distribution of cadmium in the clay after the pit has been used for 3 years. Use a value of D from figure 5.4 and assume that porosity is 0.1 and tortuosity is 1.25. Assume that no process other than diffusion takes place.

(5.3) It is determined that porosity in sediments at the bottom of a lake follows the relationship:

$$\varphi = 0.8e^{(-0.07x)} + 0.08.$$

Assume that sediment accumulates at the rate of 0.15 cm yr^{-1} and compaction is steady state. Calculate and graph the porosity, rate of sediment burial, ω, and vertical fluid advection rate as functions of depth.

CHAPTER SIX

ORGANIC MATTER AND BIOMARKERS

A Different Perspective

OVERVIEW

Organic geochemistry is built on the application of organic chemical principles and analytical techniques to sedimentary geology. Carbon, nitrogen, oxygen, sulfur, hydrogen, calcium, and iron are the primary elements that living organisms utilize in their structural tissues, for energy harvesting, and for replication. Accumulation of organic matter in recent and ancient sediments is the most important link between the biosphere and geology. Not only are the organic materials in sedimentary rocks an economically important resource (coal and petroleum) but, as we shall see, they also provide a molecular record of life. The sedimentary burial of organic matter is also important to the global cycles of carbon, sulfur, and oxygen over geologic time.

In this chapter, we focus on understanding the production, degradation, and preservation of organic matter in sedimentary environments. We introduce the role of the carbon cycle in organic matter production, diagenesis, and preservation. Our focus will be on biochemically important compounds and the relationship of specific biomarkers to their precursor biota. We discuss mechanisms and factors that influence the accumulation of organic-rich sediments and evaluate biomarker distribution in recent sediments and ultimately in the sedimentary record. We close this chapter with a few examples illus-

trating how biomarkers are used to reconstruct paleoclimatic and paleoenvironmental conditions.

ORGANIC MATTER IN THE GLOBAL CARBON CYCLE

Organic geochemistry emerged in 1936, when Alfred Treibs discovered and described porphyrin pigments isolated from shale, oil, and coal. Treibs demonstrated that porphyrins originated from the degradation of chlorophyll, thereby linking the biochemicals in living organic matter to the ancient sedimentary record.

To understand the fate of organic matter in the biosphere and in sediments, it is useful to have an appreciation of how carbon is distributed among different geochemical environments and how it is transferred from one of them to another. We examine this topic at greater length in chapter 8, where we consider various ways to characterize environments and transfer processes for the purpose of creating analytical models of ocean chemistry. As you will see, there is no "best" way to model the global carbon cycle. Rather than begin that discussion now, let's just consider the highly simplified model in figure 6.1, which is reminiscent of the party illustration we used in chapter 1. The inputs and outputs from the various reservoirs are roughly in balance, resulting in what can be considered a steady state system.

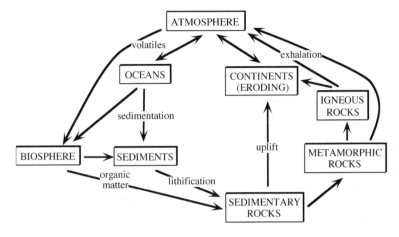

FIG. 6.1. Simplified version of the global carbon cycle indicating the principal reservoirs (boxes; see table 6.1 for amounts of carbon in most boxes) and pathways (arrows).

The Earth contains ~10^{23} g of carbon, disseminated between sedimentary materials and active surficial reservoirs. Most of this carbon is sequestered in carbonate rocks (6.5×10^{22} g C) and organic materials, such as kerogen or coal (1.56×10^{22} g C) (Schlesinger 1997). The inventories show that in shales and other sedimentary rocks, essentially one out of every five carbon atoms is organic (table 6.1) and ~90% of this preserved organic matter is amorphous kerogen, with the remaining 10% bitumen. *Kerogen* is insoluble, high molecular weight organic matter derived from algae and woody plant material that can yield petroleum products when heated. *Bitumen* is soluble in organic solvents and is formed from kerogen during petroleum generation.

Although a large fraction of the Earth's organic and inorganic carbon is sequestered, only ~0.1% (40×10^{18} g) of the carbon in the Earth's upper crust is cycled throughout active surface reservoirs (table 6.1). The largest active reservoir is dissolved inorganic carbon (DIC) contained in the global ocean. Other inorganic reservoirs, such as soil carbonates and atmospheric carbon, and the organic reservoirs (soil humus, land plant tissue, dissolved organic carbon [DOC] in seawater, and carbon preserved in marine sediments) are one to two orders of magnitude smaller than the DIC reservoir.

Atmospheric carbon exists mainly as carbon dioxide, which is used by plants and other photosynthetic organisms, thereby linking the atmosphere with the biosphere

TABLE 6.1. Major Reservoirs of Inorganic and Organic Carbon

Reservoir Type	Amount (10^{18} g C)	Reference
Sedimentary rocks		
Inorganic		
Carbonates	60,000	Berner (1989)
Organic		
Kerogen, coal, etc.	15,000	Berner (1989)
Active (surficial) reservoirs		
Inorganic		
Marine DIC	38	Olson et al. (1985)
Soil carbonate	1.1	Olson et al. (1985)
Atmospheric CO_2	0.66	Olson et al. (1985)
Organic		
Soil humus	1.6	Olson et al. (1985)
Land plant tissue	0.95	Olson et al. (1985)
Seawater DOC	0.60	Williams and Druffel (1987)
Marine surface sediments	0.15	Emerson and Hedges (1988)

After Hedges and Keil (1995).

and oceans (fig. 6.1). The ocean DIC reservoir moderates changes in atmospheric CO_2 concentration and, at equilibrium, the oceans contain ~56 times the carbon in the atmosphere.

The organic reservoirs constitute only ~8% of the total active surficial reservoirs and are typically composed of a complex mixture of heavily degraded substances (kerogen and bitumen), in which there are generally few readily identifiable biochemicals. The soil reservoir contains the largest concentration of organic and inorganic carbon in the terrestrial environment. Organic matter is almost totally recycled in terrestrial soils; therefore, terrestrial organic-rich deposits are typically confined to peat formations in bogs and low-lying swamps. Although the modern peat reservoir is small compared with organic-rich marine sediments, the presence of large coal deposits in the sedimentary record suggests that peat accumulation may have been an important process in the geologic past.

The amount of carbon in land plant tissue is similar to that contained in ocean waters and sediments. Together, these reservoirs contain approximately half as much carbon as is sequestered in soils. We have more to say about organic matter in soils in chapter 7, and we return to the topic later in this chapter, when we focus on diagenesis. For now, let's turn our attention to the areas of the globe where most of the organic biomass is produced—the oceans.

ORGANIC MATTER PRODUCTION AND CYCLING IN THE OCEANS

Organic matter is derived from the tissues of living organisms. Photosynthetic organisms (land and marine plants) capture sunlight energy, store it in organic compounds, and release the O_2 that sustains other organisms. The organic matter produced by these organisms is almost completely utilized in the biosphere, but a small portion is preserved. Phytoplankton are the primary producers in the oceans, much as vegetation is in the

terrestrial realm. Compared with the massive forests, marine phytoplankton are easily overlooked, owing to their small size and ephemeral nature. However, these organisms occupy almost 362×10^{12} m^2 (Schlesinger 1997) of the Earth's surface and consequently their presence in the open ocean accounts for almost half of the Earth's photosynthesis (table 6.2). A simple version of the carbon cycle within the oceans is shown in figure 6.2.

Phytoplankton production takes place in the upper 100 m of the ocean, where enough sunlight and nutrients delivered by rivers are available for phytoplankton to reduce carbon dioxide to form carbohydrate while splitting a mole of water and releasing a mole of oxygen:

$$6CO_2 + 6H_2O \xrightarrow[\text{chlorophyll}]{\text{sunlight}} H_2O + O_2. \qquad (6.1)$$

The most productive regions of the oceans are upwelling areas (table 6.2), where cold nutrient-rich bottom waters are brought to the surface and warm surface waters are pushed offshore by onshore winds. Coastal zones are also highly productive (table 6.2). Although the open ocean has the lowest mean production, its vast area is one to two orders of magnitude greater than the coastal and upwelling areas. Consequently, the open ocean constitutes 42% of the total global production, whereas coastal and upwelling zones make up only 9% and <1% of the global total, respectively. Oceanographers refer to the total mass of photosynthetic organisms at the ocean's surface as *primary production.*

FATE OF PRIMARY PRODUCTION: DEGRADATION AND DIAGENESIS

Regardless of where the organic matter is produced—in large terrestrial forests, the soil reservoir, or lacustrine or marine environments—it is subject to degradation during deposition and to chemical changes after deposition (diagenesis), as we discussed in chapter 5. In the ocean water column, most marine primary production is consumed by animal plankton (zooplankton) and free-

TABLE 6.2. Estimates of Marine Primary Production

Province	Percentage of Ocean	Area (10^{12} m^2)	Mean Production (g C m^{-2} yr^{-1})	Total Global Production (10^{15} g C m^{-2} yr^{-1})
Open ocean	90	326	130	42
Coastal zone	9.9	36	250	9.0
Upwelling area	<0.1	0.36	420	0.15

After Hedges and Keil (1995).

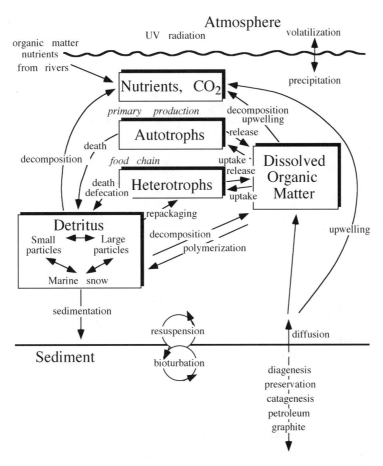

FIG. 6.2. Schematic depicting the fate of organic matter in the oceans. Boxes represent organic reservoirs and arrows indicate pathways and processes.

floating bacteria. The bacteria also decompose a large fraction of dissolved organic carbon and organic colloids produced by phytoplankton. Cole and coworkers (1988) found that net bacterial production is about twice that of zooplankton and accounts for the disappearance of 30% of primary production from the photic zone and, in some areas, as much as 70% of primary production may be degraded. Bacterial communities are important to the biogeochemical cycling of nutrients and organic matter in the oceans. The organic matter produced by phytoplankton may be consumed by zooplankton and eventually make its way up the food chain to higher trophic level organisms, such as fish. Alternatively, phytoplankton may be consumed by bacteria, which, in turn, may be consumed by bacteriovores. This process ultimately mineralizes nutrients and releases CO_2 back to the surface waters. If the bacterial abundance

is high, then a large fraction of the carbon fixed during photosynthesis is not passed to higher trophic levels. In chapter 8, we confirm that between 80 and 90% of the primary production is degraded to inorganic compounds (CO_2, NO_3, PO_2, and the like) in surface waters and the remainder sinks below the *euphotic zone* (the depth to which light penetrates, where most primary production occurs) and into the deep ocean as particulate organic matter.

Most of the organic matter sequestered in marine sediments is ultimately derived from organic matter synthesized by marine organisms inhabiting the surface waters of the oceans and transported to the seafloor (fig. 6.2). However, only a small fraction of the sinking organic matter survives transport to the seafloor to be preserved in the sediments. Extensive alteration of the organic matter in the water column can yield sedimentary organic matter having a chemical composition markedly different from that of the original material. Despite this alteration, a small fraction of particulate and dissolved organic material eventually reaches the sediment-water interface. There, the organic matter either undergoes further degradation by microbial communities or is preserved as kerogen and possibly petroleum if conditions are favorable.

FACTORS CONTROLLING ACCUMULATION AND PRESERVATION

In general, the preservation of organic materials depends on a complex interaction between the oxygen content of a system and the type of organic matter deposited within it (fig. 6.3). In the presence of oxygen, organic matter can be remineralized or converted to CO_2. However, if this material is protected in some manner or is deposited in a suboxic or anoxic environment, it is likely to be preserved (fig. 6.3). Preservation is also enhanced in sediments underlying highly productive surface waters. In these environments, production of organic matter is much greater than its oxic degradation, so that much more organic matter reaches the sediments and is preserved. Although we have identified the key variables that influence organic matter preservation, you should realize that the mechanisms governing this process remain unclear.

In this section, we address some of the mechanisms proposed to explain how organic matter is preserved in the sedimentary record.

Preservation by Sorption

Most organic matter in marine environments is concentrated in deltaic and continental shelf/slope sediments, whereas lower organic carbon (OC) concentrations are associated with high productivity areas and anoxic basins (table 6.3). Despite differences in OC concentration and key variables that influence organic matter preservation, a common mechanism appears to control organic matter preservation in these environments. More than 90% of the sedimentary OC from marine environments cannot be physically removed from its mineral matrix (Hedges and Keil 1995). Consequently, organic matter must be sorbed to mineral surfaces. The concentration of sorbed organic matter covaries with mineral surface area, such that more OC is associated with sediments that have highly irregular surfaces with small pores (Mayer 1994). Sorption of organic matter forms protective coatings that typically approach a single molecule-thick covering (often expressed as a *monolayer equivalent*, ~0.5–1.0 mg OC m^{-2}; Hedges and Keil 1995). These coatings contain both *refractory* (not easily degraded) and *labile* (easily degraded) organic matter that is pre-

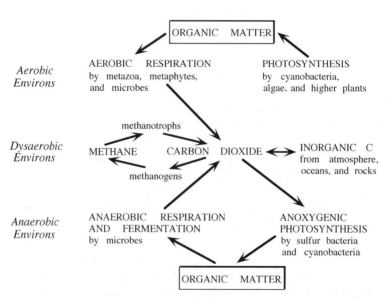

FIG. 6.3. The fate of organic matter in different redox environments. Arrows show pathways along which organic matter is produced or altered to various byproducts by different processes.

TABLE 6.3. Organic Carbon Preservation
in Various Marine Environments

Sediment Type	Organic Content $(10^{12}$ g C yr$^{-1})$
Deltaic sediments	70
Shelves and upper slopes	68
High-productivity slope and pelagic zone	10
Shallow-water shelf carbonates	6
Low-productivity pelagic sediments	5
Anoxic basins (e.g., Black Sea)	1

From Hedges and Keil (1995).

served in the underlying anoxic sediments, because it is protected from mineralization during transport through the water column and oxygenated surface sediments. Formation of monolayer equivalents of organic matter on marine sediments implies that, at one time, the organic matter was in the dissolved phase, because it is unlikely that particulate organic matter would spread uniformly over mineral surfaces.

Surface area is the primary control on organic preservation within continental shelf and slope environments, despite differences between these settings in primary production, bottom water oxygen concentration, sediment accumulation rate, and water depth (Keil et al. 1994; Mayer 1994). This control plays an important role in global organic matter preservation, because continental shelf and slope environments account for ~45% of the Earth's carbon burial (table 6.3).

Organic matter sorption is also the key mechanism in deltaic sediments, which account for another 45% of the total preserved OC, but the coatings on deltaic sediments are typically less than monolayer equivalents. Some of the sorbed OC to river sediments is removed when the sediments are deposited in a marine environment. The exact mechanism that causes this loss of coatings is unclear, but it may be related to desorption by seawater or direct oxidation (discussed later) on the mineral surface during sediment transport to the marine environment.

Sediments underlying highly productive O_2-poor waters are typically enriched in organic matter, but these marine environments are rare and constitute only ~5% of the global organic preservation. Monolayer equivalent sorption cannot explain the high organic matter concentrations (>5%) observed in these sediments. Less than 15% of the OC in these sediments can be separated from the mineral matrix, so these mineral surfaces must have coatings equivalent to several monolayers.

Many factors may lead to the formation of these thicker coatings. Among these are elevated DOC in sediment porewaters, condensation reactions, the presence of sulfides, or very brief O_2 exposure times. Elevated DOC concentrations may lead to increased organic sorption to the mineral surfaces. DOC may also form organic-bearing sulfides that are resistant to degradation. Degraded biomolecules may combine to form complex high-molecular-weight substances (condensation reactions). The coating that these substances produce is most likely resistant to microbial degradation and strongly bound to the mineral surface. In addition, the coating may promote further condensation reactions, thereby protecting the organic matter sorbed to the mineral surface. Rapid sedimentation rates in highly productive areas may enhance preservation, as the O_2 exposure time may be limited. This limited exposure may permit the preservation of labile organic matter and enhance the formation of resistant substances through condensation.

Degradation in Oxic Environments

Organic matter preservation in deep-sea sediments is typically poor (table 6.3), which suggests that some degradation mechanism may overcome surface area protection. Oceanic sediments have coatings with less mass than monolayer-equivalents, potentially formed from direct oxidation during transport. Slow accumulation rates and oxygenated waters in deep-sea environments give rise to long O_2 exposure times and may inhibit organic preservation.

The most direct evidence of this oxygen degradation is seen in "oxidation fronts" in deep-sea turbidites, where slumping exposes sedimentary organic matter to O_2-rich bottom waters for long periods of time. Molecular O_2 reacts with OC and reduced minerals along sharp redox fronts, which can reduce the organic matter content by up to 75%. Oxidation continues until the deposit is covered by either another turbidite flow or by gradual pelagic sedimentation (Hedges and Keil 1995).

Diagenetic Alteration

We have seen how organic matter is produced and where it is likely to accumulate and be preserved. Now we examine the transformations that organic matter undergo during diagenesis, as described in chapter 5.

Microbial communities and other biologic agents primarily control diagenetic transformations, although some chemical transformations catalyzed by mineral surfaces do occur.

During diagenesis, sediments undergo compaction and consolidation, with a simultaneous decrease in water content and increase in temperature. Biologic alteration of OC eventually ceases as temperatures increase (>50°C) during the later stages of burial, a process called *catagenesis*. The boundary between diagenesis and catagenesis is not well defined but it essentially coincides with the onset of oil formation.

As we have seen, degradation of OC begins in the water column and continues after sedimentation (fig. 6.2). Different compounds in the organic matter degrade at different rates and only some compounds survive in a recognizable form. We examine this aspect of degradation more closely in the next section. During diagenesis, residual organic matter from microbial degradation undergoes condensation to yield macromolecules of insoluble brown organic material. The result of diagenesis is a condensed organic residue, or *geopolymer*, which contains varying amounts of largely unaltered refractory organic material. In soil environments, diagenesis yields humin, brown coal (or lignite) in coal-forming swamps, and kerogen in marine and deep lacustrine sediments.

Humic substances are found in soils, terrestrial and marine sediments, coal deposits, and all aquatic environments. Humics can account for almost all the organic carbon in freshwater environments, giving waters a brown color or possibly a green tint. Humin in soils is derived from terrestrial plants that decompose, forming OC that condenses during diagenesis to form the insoluble humic residue.

Coals may also form during diagenesis of organic matter, and the type of coal formed typically reflects its precursor plant material. For example, humic coals are formed mainly from the woody tissues of vascular plant remains. These coals are typically stratified, have a lustrous, black/dark brown appearance, and go through a peat-forming stage. Sapropelic coals, on the other hand, are not stratified and are dull in appearance. They form in quiet, anoxic shallow waters from fine-grained, organic-rich sediments that contain varying amounts of algal remains and degraded peat swamp plants. Unlike humic coals, formation of sapropelic coals does not typically involve a peat-forming stage. Instead, these coals follow a pathway similar to kerogens, which we discuss next.

Coals typically form as a result of two main processes: *peatification*, dominated largely by biologic activity, followed by *coalification*, in which heat and pressure are the most important agents of change. Peatification and early coalification are equivalent to diagenesis. Late-stage coalification, which is essentially catagenesis, parallels the onset of metamorphism. The boundary between the early and late stages of coalification is not sharp, because the actions of the biologic and physicochemical agents may overlap. The principal stages of coal formation begin with peat formation followed by the formation brown coal, bituminous coal, and eventually anthracite.

Kerogen is the highly complex organic material from which hydrocarbons are produced as the material is subject to increasing burial and heating. It is by far the most abundant form of organic carbon in the Earth's crust and it occurs primarily in sedimentary rocks as finely disseminated organic *macerals* (chunks of organic matter, analogous to minerals in rocks). Until recently, kerogen was thought to form during the late stages of diagenesis from the alteration of huminlike material. This material was supposedly derived from the condensation of insoluble geopolymers, which formed from microbial degradation of organic matter. Kerogen is now thought to form during the early stages of diagenesis from mixtures of partially altered, refractory biomolecules (Tregelaar et al. 1989).

Kerogen is modified by temperature ultimately to yield petroleum hydrocarbons. The potential yield of petroleum products typically depends on the type of kerogen. There are four types of kerogens distinguished by their maceral groups: lignite, exinite, vitrinite, and inertinite. Type I kerogens (*lignite*), which are derived from algal or bacterial remains, are relatively rare but have a high oil potential. These materials formed in fine-grained, organic-rich muds deposited under anoxic conditions in quiet, shallow water environments, such as lagoons and lakes. Type II kerogens (*exinite*) are the most common and are usually formed in marine environments from mixtures of phytoplankton, zooplankton, and microbial organic matter under reducing conditions, but they can also be formed from higher plant debris. Type II kerogen has a lower yield of hydrocarbons than Type I but it has still produced oil shales of commercial value and sourced a large number of oil and gas fields (Killops and Killops

A CRASH COURSE IN ORGANIC NOMENCLATURE

You need some knowledge of organic chemistry to understand the principal aims, methods, and results of organic geochemical investigations, so let's briefly examine some geochemically important structures and outline their nomenclature.

Organic compounds consist principally of carbon atoms linked to nitrogen, oxygen, sulfur, and other carbon atoms. The basic carbon skeleton can be arranged in simple straight chains, branched chains, one or more rings, or combinations of these structures. The simplest organic compounds are *hydrocarbons*, which consist solely of carbon and hydrogen atoms. Carbon atoms have four valence electrons that must be satisfied by either single covalent bonds (see chapter 2) to four separate atoms or some combination of single and double bonds. A double bond is one that involves two valence electrons. Hydrocarbons that have only single bonds are *saturated*, whereas those that contain double bonds are referred to as *unsaturated*.

The carbon atoms can be linked together to form either straight-chain (*aliphatic*) or simple cyclical (*alicyclic*) structures. Saturated aliphatic hydrocarbons are *alkanes*, where the suffix *-ane* refers to the lack of any double bonds in the structure. An unsaturated aliphatic hydrocarbon is called an *alkene*, where the suffix *-ene* denotes a double bond. Alkanes (and alkenes) are named according to the number of carbon atoms in the chain: methane (1), ethane (2), propane (3), butane (4), pentane (5), hexane (6), heptane (7), octane (8), and so on. These compounds are typically represented by two-dimensional ("sawtooth") drawings, in which the kink at each line segment represents a carbon atom. Leftover valence electrons (that is, ones not involved in bonding carbon atoms to each other) attach hydrogen atoms to the structure. For example, figure 6.4a shows dodecane, a straight chain, aliphatic hydrocarbon with 12 carbon atoms and no double bonds. It is understood that two hydrogen atoms attach to the carbon atom at each kink in the chain, and three attach at each end of the chain. The compound shown in figure 6.4b, dodecene, is essentially the same compound shown in figure 6.4a with

one double bond. The carbon atoms at each end of the double bond would have only one attached hydrogen atom.

Stable configurations of carbon atoms also occur where double (C=C) bonds alternate with single (C—C) bonds to form a pattern like ~C—C=C—C=C—C=C~. This configuration, referred to as *conjugated*, is common in polyunsaturated compounds. If conjugation occurs in a ring structure with three double and three single bonds (fig. 6.4c), the configuration is *aromatic*. Aromatic compounds are unsaturated and are typically very stable.

The groups of elements bonded to the carbon backbone (whether chain or ring) are called *functional groups*, because they are usually reactive and thus influence the chemical behavior of the compound. Functional groups involve combinations of hydrogen atoms with one or more atoms of another element, most commonly oxygen, nitrogen, or sulfur. Some of these, like hydroxyl (–OH), are familiar from inorganic chemistry. *Alcohols* and *phenols* are based

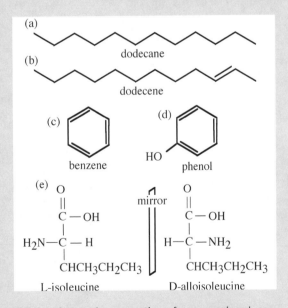

FIG. 6.4. Structural representations of some organic molecules. (a) Dodecane; (b) dodecene; (c) benzene; (d) phenol; (e) L- and D-enantiomers of isoleucine.

TABLE 6.4. Some Functional Groups in Organic Molecules

Symbol[1]	Function Group	Compounds
R—OH	Hydroxyl	Alcohol (R = aliphatic group) Phenol (R = aromatic group)
$-\overset{\displaystyle \|}{\underset{\displaystyle R}{C}}=O$	Carbonyl	Aldehyde (R = H) Ketone (R = aliphatic group) Alkanone (R = aromatic group)
$-\overset{\displaystyle \|}{\underset{\displaystyle OH}{C}}=O$	Carboxyl	Acid
—NH$_2$	Amino	Amine
pyrrole ring structure	Pyrryl	Pyrrole

[1]R denotes groups of atoms that may be joined to the function group.

on combinations of −OH with aliphatic or aromatic groups, respectively (see the phenol in fig. 6.4d). *Carbohydrates,* such as *sugars* and *starches,* are a major class of aliphatic compounds with a large number of hydroxyl groups in place of hydrogen atoms. Organic compounds are often grouped under the name of the functional group that they contain. For example, carboxylic acids are compounds that bear a carboxyl group. Some geochemically important functional groups are listed in table 6.4.

Although we typically represent organic compounds with two-dimensional drawings, these compounds actually have three-dimensional structures. Different spatial configurations of atoms attached to a central carbon atom can exist. This phenomenon is called *stereoisomerism.* Several compound classes exhibit this property, in which two structurally identical compounds exist as mirror images that cannot be superimposed. Your hands provide a straightforward illustration of this property: each hand has the same structure of four fingers and a thumb, but you cannot superimpose them (palms facing up). Similarly, certain amino acids and carbohydrates have identical structures but differ by the placement of their functional groups on one side of the central carbon, called the *chiral carbon.* These mirror images, called *enantiomers,* have similar physical properties but rotate light in different directions. Compounds that rotate light clockwise are called *dextrorotatory* and are denoted by the prefix D-, whereas counterclockwise rotation is referred to as *levorotatory* and denoted by the prefix L-. For example, the amino acid isoleucine can exist in either L-or D-forms (fig. 6.4e). The only difference between these two compounds is the placement of an amino group, which is on opposite sides of the central carbon atoms. (As you can see, isoleucine has two chiral carbon atoms, which is a special case. When the L-form of this compound changes to its D-form, this process is called *epimerization* and the D-form is referred to as *alloisoleucine.* You will encounter an example of this in worked problem 6.3.)

1993). Type III kerogen (*vitrinite*) is derived from vascular plants and may contain identifiable woody plant debris. Vitrinite is usually not extensively altered by microbial degradation, which may be a result of the material's rapid sedimentation and burial. Vitrinite generally occurs as coals or coaly shales; therefore, it is similar to coal in terms of its composition and behavior with increasing burial. Type IV kerogen (*inertinite*) is composed of primarily black opaque debris potentially formed from the deposition of highly oxidized, higher plant material. It has no hydrocarbon generating potential and is sometimes not considered a true kerogen.

CHEMICAL COMPOSITION OF BIOLOGIC PRECURSORS

Despite all the potential changes that may alter organic matter along various pathways from production to preservation, some organic compounds still reflect their original source materials, whether those are higher plants, marine or lacustrine algae, or bacteria. It is important to understand the chemical composition of these sources so that we can relate these compounds to those preserved in the sedimentary record.

All organisms are composed of compounds derived from carbohydrates, proteins, and lipids. In addition, significant amounts of lignin (structural tissues) are present in higher plants. Let's now review the composition of the main chemical classes of compounds and their biochemical functions in living organisms. The structural representations and important functional groups for these compounds are discussed in an accompanying box.

Carbohydrates

The carbohydrates are aliphatic compounds containing only carbon, hydrogen, and oxygen, with a ratio of hydrogen to oxygen similar to water. Their general formula is represented by $(C_n(H_2O)_n)$. Organisms use carbohydrates as food reserves and structural materials. Polysaccharides (many saccharide units) are major components of cell membranes in plants, bacteria, and fungi.

Cellulose, the main structural unit in higher plants, is the most common carbohydrate. Cellulose is composed of roughly 10,000 glucose (C_6 monosaccharide) units that are, in essence, energy reserves. This energy is stored as starch, which is a component of the cellulose. Energy is derived from this reserve by breaking down the polysaccharides into individual glucose units.

Hemicellulose compounds are the next most abundant group of carbohydrates. These compounds are complex mixtures of polysaccharides containing between 50 and 2000 monosaccharide units that form a matrix around the cellulose fibers in plant cell walls. These compounds are present in nonwoody tissues and fruits but are only a minor component in woody tissues.

Whereas cellulose is the dominant structural material in plants, the polysaccharide *chitin* is the principal structural compounds in most fungi, some algae, mollusks, and arthropods (insects and crustaceans). This compound is composed of chains of glucose units with an attached amino (NH_2) functional group. Eubacterial cell walls contain *murein*, another polysaccharide. It is possible to distinguish some eubacteria based on the compounds that are contained on the exterior of their cell walls.

Similarly, other carbohydrates (specifically enantiomers; see accompanying box) may be characteristic of, but not necessarily exclusive to, different organisms. Fungal polysaccharides are predominantly D-glucose, D-galactose, and D-mannose, whereas marine algae such as Chlorophyta contain a large proportion of L-rhamanose, and Phaeophyta are rich in D-ribose. Freshwater algae and higher aquatic plants, however, contain significant amounts of L-arabinose and D-xylose (Killops and Killops 1993).

Proteins

Proteins, which consist of linked amino acids, are the principal storage sites for nitrogen in organisms. The amino acids contain an amino group (NH_2) and a carboxyl group (COOH) that can bond through the elimination of water to form peptide linkages. Structurally, all amino acids are chiral molecules with the exception of glycine, which has two hydrogen atoms bonded to the chiral carbon atom. There are only 20 amino acids involved in protein synthesis and the L-configuration of these compounds is utilized exclusively, owing to the stereospecificity of enzymes in all organisms except bacteria. The predominant use of D-amino acids by bacteria may help identify the source of these compounds in complex mixtures, provided that diagenesis has not occurred.

Plants are able to synthesize all the amino acids necessary for biomass production, and most amino acids are derived from glutamic acid. Animals, on the other hand, are not able to synthesize all the amino acids required for protein synthesis; therefore, their essential amino acids are supplied by eating plants.

Proteins typically constitute a substantial portion of the bulk OC in an organism. Owing to their ability to form fibers, proteins provide supportive tissues such as skin and bone (collagen) or hooves and claws (keratin) in animals. The function of these protein-derived structural materials is analogous to that of cellulose and lignin materials in plants.

Lipids

Lipids are those compounds that are insoluble in water but are extractable in solvents known to dissolve fats (such as chloroform, hexane, toluene, and acetone). Most organic geochemists use this broad definition of lipids, because it encompasses all the compound classes of geologic importance. The most abundant and geochemically important lipids (such as those that are principal components of oil source rocks) include the following groups.

Glycerides are esters of the alcohol glycerol that can bond with one, two, or three carboxylic acids (-COOH) to form mono-, di-, and triglycerides, respectively. Among the triglycerides, fats are important compounds.

Fats are composed of fatty acids, which are straight-chain carboxylic acids (RCOOH) that typically contain between 12 and 36 carbon atoms. Saturated fatty acids are found in animals, whereas unsaturated and poly-unsaturated fatty acids are common in plants. Unsaturated acids have a lower melting point for a given chain length, which is why some unsaturated plant-derived fats are oils whereas saturated animal fats are solids. The C_{16} and C_{18} saturated fatty acids are most abundant in animals, whereas the C_{18} mono-, di-, and triunsaturated fatty acids are the dominant forms in plant tissues. Polyunsaturated fatty acids are more common in algae than in higher plants. Fatty acids are predominantly even-numbered carbon compounds, because their biosynthetic pathway incorporates two carbon atoms (acetyl units) that are derived from glucose. Similar to carbohydrates, fats are used as an energy reserve by animals and plants, but the principal difference is that fats provide twice the energy as carbohydrates during oxidation. This is particularly useful during processes that require a substantial amount of energy, such as seed and fruit generation.

Waxes serve mostly to protect organism membranes, as in, for example, the waxes that coat plant leaves. These biochemicals are mixtures of fatty acid esters and fatty alcohols, whose carbon chains range from C_{24} to C_{28}. Waxes are mostly even-carbon–numbered owing to their synthesis from fatty acids (also even-carbon numbered). Hydrocarbons, principally long-chain *n*-alkanes, are also a major constituent of waxes, but these compounds contain predominantly odd numbers of carbon atoms in the range of C_{23} to C_{33} with the majority of these being C_{27}, C_{29} and C_{31}. The odd-number preference arises because *n*-alkanes are formed from an aliphatic acid

($CH_3(CH_2)_nCH_2COOH$) by the elimination of the carboxyl group (COOH), which contains one carbon atom. Higher plants and fungi contain a similar distribution of *n*-alkanes, whereas these compounds are almost absent in most bacteria.

Terpenoids are a class of lipids whose structures and functions vary markedly from the basic building block of chlorophyll and gums of higher plants to volatile sex pheromones. All the different terpenoids, however, have a similar structural component made of five carbon atoms called an *isoprene unit*. Terpenoids are classified by the number of constituent isoprene units.

Monoterpenoids consist of two isoprene units (hence, they are C_{10} compounds) and some are highly volatile. They are usually found in algae and constitute the essential oils of higher plants. These oils, like citronella or menthol in peppermint, give the plant a specific odor. Due to their volatility, some of these compounds, such as insect pheromones, act as attractants. Others, such as chrysanthemic acid in pyrethrum flower heads, are natural insecticides.

Sesquiterpenoids (C_{15} compounds), formed from three isoprene units, function not only as the essential oils of higher plants but also as fungal antibiotics. Farnesol is a common acyclic compound found in many plants and in the chlorophyll of some bacteria.

Diterpenoids (C_{20} compounds) contain four isoprene units and form resins, which seal damaged tissues with protective coatings and inhibit insect and animal attack. Other compounds in this class cause some bitter taste in plants, which can also act as a protective mechanism.

Triterpenoids (C_{30}) contain six isoprene units and are believed to develop from squalene, which is a ubiquitous 30-carbon compound in many organisms. Most of the triterpenoids are either tetracyclic (four rings) or pentacyclic (five rings). The pentacyclic triterpenoids typically constitute the resins of higher plants, but another group of these compounds (hopanoids) is found in bacteria. In addition, other triterpenoids have been identified as the precursors to certain petroleum hydrocarbons. Although you may be not familiar with the majority of these compounds, you probably recognize the tetracyclic triterpenoids known as *steroids* (or *sterols*). These compounds contain four rings in their structure, which result from the oxidation of squalene followed by cyclization (ring formation). This reaction produces two different compounds that are the precursors to all plant (cyclo-artenol) and animal (lanosterol) sterols. Oxidation of lanosterol produces cholesterol (C_{27} compound), which is the precursor to all other animal steroids. The majority of the sterols that are found in the geologic record range between C_{27} and C_{30}. Sterols are rare in bacteria, but hopanoids are abundant, which may provide a means to distinguish the source of organic matter in sedimentary environments.

Tetraterpenoids (C_{40}) consist of eight isoprene units. The most important group of these compounds is the carotenoid pigments. Carotenoids are essential for the production of vitamin A and are found most organisms. Carotenoid pigments are responsible for the red coloration in "red tide" blooms of dinoflagellates. Carotenes and xanthophylls (from the Greek for *contain oxygen*) are highly unsaturated and can absorb relatively short wavelengths. These compounds are major photosynthetic pigments that aid in the capture of sunlight energy and transfer this energy to chlorophyll. Certain carotenoids are characteristic of different photosynthetic organisms.

Tetrapyrrole pigments consist of four pyrrole units (a five-member ring structure, in which four of the atoms are carbon and the fifth member is a NH group linked by a double-bonded CH group). These structures can either be a ring or large open chain. The pigments are involved in photosynthesis as either primary or accessory pigments. Chlorophylls are members of this group that have a large ring structure called a *porphyrin*. Chlorophyll is found in all algae, higher plants, and cyanobacteria.

Lignin

Lignin, characterized by phenolic structures (aromatic six-member carbon rings bonded to OH groups, illustrated in fig. 6.4d), is a major, unique component of the structural tissues of higher plants. These compounds are derived from monosaccharides. Lignin forms around the fibrous woody core (xylem) of terrestrial plants and acts as a support structure as the plant grows. Cellulose constitutes up to 40–60% of wood and lignin essentially makes up the remainder. This high-molecular-weight material has a highly complex structure consisting of many different types of alcohols that undergo condensation reactions to ultimately form lignin.

BIOMARKERS

Now that we are familiar with some of the biochemically important compounds, we can use these as biomarkers to gain insight into the types of organisms that contribute to the total organic matter in sedimentary environments. Biomarkers survive diagenesis almost intact or their diagenetic pathways can be traced, whereas other organic compounds are altered and may not necessarily reflect their original source. Here we investigate the role of biomarkers as source indicators in sedimentary environments, using lignin and certain lipids as examples.

There are many differences in the biochemical compositions of the principal groups of organisms. For example, the presence of lignin in sediments may indicate a higher plant contribution to the total organic matter of a sedimentary environment, because, as we have seen, lignin is present only in higher plants. We have also seen that higher plants contain protective waxes, in which even-carbon-number fatty acids predominate, on their leaves. The predominance of even- over odd-carbon-number fatty acids in organic residue, then, is a further biomarker for higher plants. Hydrocarbons (*n*-alkanes) are also a major constituent of the waxes, but these compounds contain predominantly odd numbers of carbon atoms, so an odd-over-even predominance of *n*-alkanes also suggests input from higher plants.

Compounds derived from the steroid class can also be indicative of certain groups of organisms. Phytoplankton primarily consist of abundant C_{28} sterols, whereas zooplankton typically contain C_{27} sterols, particularly cholesterol. In contrast, major plants are dominated by the C_{29} sterols, and C_{27}–C_{29} sterols are typically associated with fungi.

Lignin and its associated phenolic units can be used to differentiate among the types of vascular plants that contribute organic matter in complex sedimentary environments. To do this in the laboratory, a geochemist liberates the phenolic units from the lignin structure through an oxidation process. Oxidation produces three groups of structurally related compound groups, vanillyl (V), syringyl (S), and cinnamyl (C), which are diagnostic of lignin. These compound groups are absent from nonvascular plants and vary according to vascular plant type (table 6.5), which makes source determination possible. For example, nonwoody angiosperms can be distinguished from the other three groups of vascular plants based on their S/V ratios, and woody angiosperms have a distinct C/V ratio. Another approach is to plot the S/V ratios versus C/V ratios, which can also provide a means to differentiate these sources. Let's see how this works with a real example.

TABLE 6.5. Distribution of Lignin Oxidation Products in Various Plant Tissues

Plant Type	Syringyl (mg/g)	Cinnamyl (mg/g)	Vanillyl (mg/g)	S/V	C/V
Nonvascular plants	0	0	0	—[1]	—
Nonwoody angiosperms	1–3	0.4–3.1	0.6–3.0	0.5–2.5	0.4–1.1
Woody angiosperms	7–18	0	2.7–8.0	0	1.1–5.2
Nonwoody gymnosperms	0	0.8–1.2	1.9–2.1	0	0.4–0.6
Woody gymnosperms	0	0	4–13	0	0

[1]—, not applicable.

Worked Problem 6.1

The syringyl, cinnamyl, and vanillyl concentrations have been measured at different depths within a lacustrine sediment core. The calculated S/V and C/V ratios are shown below. How can we determine the source of organic matter that has accumulated in this lake basin over time?

Sample Depth (m)	Age (kyr)	S/V	C/V
0	1.0	0.44	0.22
2		0.65	0.20
4	7.2	1.00	0.08
6		0.97	0.10
8	10.0	0.96	1.00
10		0.95	0.90
12	13.8	0.05	0.38

We can compare the S/V and C/V values with table 6.5 to determine planet sources for each core sample and investigate any vegetation changes. The basal portion of the core has low S/V and C/V ratios, which suggests input from nonwoody gymnosperms, such as ferns. Between 8 and 10 m depth, both the S/V and C/V ratios are high, reflecting a change to nonwoody angiosperms (marsh or grassland plants) and possibly a tundra landscape. High S/V and low C/V ratios between 4 and 6 m indicate input from woody angiosperms, such as deciduous trees, making another vegetation change to a boreal biome. A decrease in S/V ratios from 4 m to the top of the core with relatively constant C/V ratios suggests a progressive enrichment in woody gymnosperms (conifers), mostly likely at the expense of the woody angiosperms (deciduous trees). Changes in the lignin parameters (S/V and C/V) thus indicate that climate changed from a fern-dominated landscape 13 kyr ago to tundra by 10 kyr. Climate change forced the tundra to give way to boreal forest by 7 kyr, which eventually gave way to a pine forest.

APPLICATION OF BIOMARKERS TO PALEOENVIRONMENTAL RECONSTRUCTIONS

Biomarkers can also be used to quantify specific environmental conditions. For example, the distributions of some biomarkers are sensitive to temperature (because of the metabolic pathways involved in their synthesis) and time (because of the kinetics of their reactions).

Alkenone Temperature Records

One family of temperature-sensitive lipids is the (C_{37}–C_{39}) unsaturated ketones or alkenones. These compounds have a structure similar to long chain alkenes (remember, the -*ene* suffix denotes double bonds) and can have between two and four double bonds. The number of double bonds is a function of temperature. A higher proportion of unsaturated compounds is associated with colder temperatures. For $C_{37:3}$ and $C_{37:4}$, 37 indicates the number of carbon atoms and 3 or 4 is the number of double bonds, respectively. These unsaturated alkenones are produced by the coccolithophore algae *Emiliani huxleyii*, which first appeared in the geologic record during the late Pleistocene approximately 250,000 years ago (Killops and Killops 1993). By culturing these organisms at various temperatures, Simon Brassell and coworkers (1986) established a linear relationship between the degree of unsaturation and growth temperature. This relationship is known as the U_{37}^K index:

$$U_{37}^K = \frac{[C_{37:2}] - [C_{37:4}]}{[C_{37:2}] + [C_{37:3}] + [C_{37:4}]},$$

where the square brackets are the concentrations of the di-, tri-, and tetra- unsaturated ketones. The di- and tri- unsaturated compounds are dominant in most sediments, so this index can be simplified to:

$$U_{37}^{K'} = \frac{[C_{37:2}]}{[C_{37:2} + C_{37:3}]}, \tag{6.2}$$

Values of U_{37}^K can be related to the growth temperature and hence sea surface temperature (SST) in the case of marine sediments through the following relationship (Prahl et al. 1988):

$$U_{37}^{K'} = 0.034(\text{SST}) + 0.039. \qquad (6.3)$$

Equation 6.3 holds well for most oceans, but there are some, such as the Southern Ocean, for which a slightly different temperature calibration has been developed. Regardless of the calibration, the U_{37}^{K} index appears to provide reasonable temperatures in the modern oceans and consequently has been used to determine SST on glacial/interglacial time scales.

Worked Problem 6.2

How can we determine changes in SST in the Pacific Ocean over glacial/interglacial time scales at midlatitudes and in the tropics? Oceanographers have recovered ocean sediment cores that span the past 50,000 years, and we have determined the U_{37}^{K} values from the concentrations of the di- and tri-unsaturated C_{37} alkenones for several radiocarbon dated samples. The U_{37}^{K} values for the tropical core for the last glacial maximum and the Holocene are 0.925 and 0.99, respectively. In the midlatitudes, the last glacial maximum and Holocene U_{37}^{K} values are significantly smaller at 0.53 and 0.65, respectively.

Using equation 6.3, the SST during the last glacial maximum in the tropical Pacific Ocean was:

$$(U_{37}^{K'} - 0.039)/0.034 = (\text{SST}) = (0.925 - 0.039)/0.034$$
$$= 26.2°C.$$

and the Holocene temperature was:

$$(\text{SST}) = (0.99 - 0.039)/0.034 = 28.0°C.$$

This corresponds to a 1.8°C temperature increase from the last glacial maximum, which is consistent with other ways of estimating SST. If we examine the temperature changes in the midlatitudes, we see that during the last glacial period,

$$(\text{SST}) = (0.53 - 0.039)/0.034 = 14.4°C,$$

and Holocene temperatures were on the order of:

$$(\text{SST}) = (0.65 - 0.039)/0.034 = 18.0°C,$$

which is ~4°C warmer. This is also consistent with other proxy records that suggest that midlatitude glacial oceans were much colder than tropical oceans and that the glacial/Holocene temperature rise was more pronounced in the midlatitudes.

Amino Acid Racemization

We noted earlier that all living organisms (except bacteria) manufacture L-amino acids, rather than D-amino acids. The chiral (L- versus D-) configuration of amino acids is sensitive to both time and temperature. L-amino acids convert or *racemize* (for one chiral carbon atom and *epimerize* for two chiral carbon atoms) to their D-amino acid equivalents over time after the proteins are formed and effectively protected from the biologic processes of the organism. The change from L- to D-amino acids continues until an equal mixture of the two forms exists.

Bacteria preferentially produce D-amino acids. These compounds are nearly ubiquitous in the biosphere and can come into contact with a potential sample of preserved nonbacterial material, thereby introducing amino acids not indigenous to the sample. To reduce the risk of contamination, most organic geochemists prefer to sample materials that are least likely to have been accessible to bacteria. They sample shells of marine organisms (mollusks and foraminifera), land snails, and ostrich eggs, because the amino acids in these materials are protected from the surrounding environment owing to a layer of calcium carbonate deposited over protein layers in the shell.

Reliable ages and temperatures can be calculated from the *D/L* values of a sample provided that a reasonable kinetic model is established to quantify the relationships among time, *D/L* ratios, and temperature. This is typically accomplished through artificial heating experiments of age-dated fossil shells. The relationship between the age of a sample and measured *D/L* value is expressed as (McCoy 1987):

$$t = \ln([1 + D/L]/[1 - K'D/L]) - \ln([1 + D_0/L_0]/[1 - K'D_0/L_0])/(1+K')Ae^{-Ea/RT}, \qquad (6.4)$$

where t is the age of the sample, *D/L* is the ratio of D to L enantiomers, K' is $1/K_{eq}$ (equilibrium constant ~0.77 for alloisoleucine/isoleucine [A/I]), D_0/L_0 is the *D/L* ratio of the sample at $t = 0$ (~0.01, owing to a small amount of racemization during sample preparation), E_a (the activation energy in J mol^{-1}), and A (the entropy factor) are the Arrhenius parameters of the racemization reaction, which are determined experimentally. R is the gas constant (1.9872 cal K^{-1} mol^{-1}), and T is the effective diagenetic temperature in K.

Application of this equation to generate numerical age estimates is limited by the temperature dependence of the racemization reaction and the variable temperature history of many sample sites. This variability leads to imprecise and potentially inaccurate age dates. However, if samples are collected from different environments with similar postdepositional temperature histories, then differences in *D/L* ratios can be used to identify relative age differences between samples. This approach is known as *aminostratigraphy*.

Age estimates may also be made from an empirical relationship between the measured D/L ratios and the sample age. The D/L ratios of independently dated samples can be used to calibrate ages of other samples on the basis of their D/L ratios, provided that all the samples have had similar postdepositional temperature histories.

By solving equation 6.4 for temperature (T), we can estimate the effective diagenetic temperature of a sample terms of its age, measured D/L values, and the Arrhenius parameters (McCoy 1987):

$$T = -Ea/R \ln(\ln\{((1 + D/L]/[1 - K'(D/L]) \\ - ([1 + D/L]/[1 - K'D/L])\}/At[1 + k']). \quad (6.5)$$

This equation can be simplified for an interval of time bracketed by two independently dated samples to:

$$T_{(t_2-t_1)} = E_a/R \ln(A/k), \quad (6.6)$$

where

$$k = (k_2 t_2 - k_1 t_1)/(t_2 - t_1) \\ = \ln(c/b)/[(1 - K')(t_2 - t_1)], \quad (6.7)$$

in which $b = (1 + D_1/L_1)/(1 - K'D_1/L_1)$ and $c = (1 + D_2/L_2)/(1 - K'D_2/L_2)$. D_1/L_1 and D_2/L_2 refer to the D/L ratios for the younger and older samples, respectively. The numerical ages of the younger and older samples are represented by t_1 and t_2, respectively.

Worked Problem 6.3

How can we determine the effective diagenetic temperature of fossil shells deposited at different times following the last glacial maximum? Eric Oches and coworkers (1996) collected fossil gastropod shells from silt in the Mississippi Valley and measured their D/L ratios. They also determined the Arrhenius parameters ($A = 1.658 \times 10^{17}$, K' for alle/ile, and $E_a = 29,235$ cal mol^{-1}), measured the equilibrium constant for isoleucine epimerization ($K' = 0.77$), and obtained radiocarbon dates for each sample.

Locality	Temp. (°C)	D_1/L_1 (alle/ile)	Age (yr)	D_2/L_2 (alle/ile)	Age (yr)
Finley, Tenn.	15.7	0.11	26,040	0.13	32,230
Vicksburg, Miss.	18.4	0.11	20,820	0.14	25,110
Vicksburg, Miss.	18.4	0.12	20,820	0.13	25,110
Vicksburg, Miss.	18.4	0.11	20,820	0.14	27,930
Pond, Miss.	18.8	0.14	18,320	0.16	25,610

The values of b and c in equation 6.7, for the Finley, Tenn., samples, are therefore:

$b = (1 + 0.11)/(1 - [0.77 \times 0.11]) = 1.21,$

$c = (1 + 0.13)/(1 - [0.77 \times 0.13]) = 1.25.$

Substituting these values into equation 6.7, we can determine the value of k:

$k = \ln(1.25/1.21)/[(1 + 0.77)(32,320 - 26,040)]$
$= 3.18 \times 10^{-6}.$

Using this value in equation 6.6, we can find T:

$T = (29,235$ cal mol$^{-1})/(1.9872$ cal K^{-1} mol$^{-1})\ln(1.658$
$\times 10^{17}/3.18 \times 10^{-6}) = 279.7$ K $= 6.5$°C.

Determining the temperatures for all sample sites using this procedure (see below), we find that temperatures between 20,000 and 30,000 years ago were much cooler than the mean annual temperature (MAT) for all locations. These results agree with the trend in midlatitude sea-surface temperatures that we calculated in worked problem 6.2. These temperatures were much cooler during the last glacial maximum than at present.

Locality	MAT (°C)	Effective Diagenetic Temp. (°C)
Finley, Tenn.	15.7	6.5
Vicksburg, Miss.	18.4	12.3
Vicksburg, Miss.	18.4	6.4
Vicksburg, Miss.	18.4	9.5
Pond, Miss.	18.8	7.2

SUMMARY

Although the field of organic geochemistry has its origins in the study of petroleum, here we have concentrated on the application of organic geochemistry to paleoenvironmental and paleoclimatic reconstructions. We have examined the origin of organic matter and how it is cycled within the geosphere. The organic carbon reservoir is only a small fraction of the total global carbon content and represents only 8% of the active surface reservoirs. Despite its relatively small size, this reservoir is intricately linked to the biosphere and geosphere.

We showed that the oceans are the primary source of organic carbon in the sedimentary record and discussed how organic matter is transformed and altered during deposition and diagenesis. Most of the organic matter produced in surface waters is completely recycled within the water column and only a very small fraction gets preserved in sediments. Further alteration of this material occurs during diagenesis, owing to microbial degradation of labile organic matter. Degradation is typically followed by condensation to form kerogen, which is ultimately preserved in the sedimentary record.

We examined several mechanisms that may influence the accumulation and preservation of organic matter in marine environments. Preservation of organic matter is controlled by sorption to mineral surfaces. Organic matter preservation may actually be a competition between sorptive preservation and oxic degradation. We highlighted the geochemically relevant biologic compounds to give you an idea of which compounds can be used to identify inputs of organic matter from specific organisms. The lipid and lignin compounds are potentially the most useful biomarkers. Using alkenones, we were able to reconstruct sea surface temperatures from midlatitude and tropical regions of the Pacific over the past 35,000 years. We also examined how amino acid racemization can be used as a paleoenvironmental tool.

There are many more questions that organic geochemists try to answer, not only in the petroleum and paleoclimate fields but also from the modern environmental/contamination point of view. To gain insight into these fields, you may consider investigating some of the sources listed in the suggested readings for this chapter.

SUGGESTED READINGS

Only a few books dedicated to organic geochemistry have been written, and these are already dated. Most findings, especially on the biomarker front, are disseminated in articles in the following technical journals: *Organic Geochemistry, Geochimica et Cosmochimica Acta, Marine Geochemistry,* and *Paleoceanography.* For general background, you may wish to consult the following texts.

Eglinton, G., and M. T. J. Murphy. 1969. *Organic Geochemistry: Methods and Results.* New York: Springer-Verlag. (This book was the first definitive text in organic geochemistry and still provides background information on the occurrence and evolution of organic matter in sediments.)

Engel, M. H., and S. A. Macko, eds. 1993. *Organic Geochemistry: Principles and Applications.* New York: Plenum. (This text nicely shows applications of organic geochemistry to the formation of petroleum, reconstructing paleoenvironments, and understanding the fate of organic compounds.)

Killops, S. D., and V. J. Killops. 1993. *An Introduction to Organic Geochemistry.* New York: Wiley. (Top-of-the-line text that introduces basic concepts of organic geochemistry in an easily digestible manner.)

The following articles were referenced in this chapter.

Berner, R. A. 1989. Biogeochemical cycles of carbon and sulfur and their effect on atmospheric oxygen over Phanerozoic time. *Palaeogeography, Palaeoclimatology, Palaeocology* 73: 97–122.

Brassell, S. C., G. Eglinton, I. T. Marlowe, U. Pflaumann, and M. Sarnthein. 1986. Molecular stratigraphy: A new tool for climatic assessment. *Nature* 320:129–133.

Cole, J. J., S. Findlay, and M. L. Pace. 1988. Bacterial production in fresh and saltwater ecosystems: A cross-system overview. *Marine Ecology-Progress Series* 43:1–10.

Emerson, S., and J. Hedges. 1988. Processes controlling the organic carbon content of open ocean sediments. *Paleoceanography* 3:621–634.

Hedges, J. I., and R. G. Keil. 1995. Sedimentary organic-matter preservation—an assessment and speculative synthesis. *Marine Chemistry* 492–493:81–115.

Keil, R. G., E. Tsamakis, C. B. Fuh, C. Giddings, and J. I. Hedges. 1994. Mineralogical and textural controls on the organic composition of coastal marine sediments—hydrodynamic separation using SPLITT-fractionation. *Geochimica et Cosmochimica Acta* 582:879–893.

Mayer, L. M. 1994. Surface area control of organic carbon accumulation in continental shelf sediments. *Geochimica et Cosmochimica Acta* 114:347–363.

McCoy, W. D. 1987. The precision of amino acid geochronology and paleothermometry. *Quantitative Science Review* 6: 43–54.

Oches, E. A., W. D. McCoy, and P. U. Clark. 1996. Amino acid estimates of latitudinal temperature gradients and geochronology of loess deposition during the last glaciation, Mississippi Valley, United States. *Geological Society of America Bulletin* 1087:892–903.

Olson, J. S., R. M. Garrels, et al. 1985. *The Natural Carbon Cycle. Atmospheric Carbon Dioxide and the Global Carbon Cycle.* J. R. Trabalka, ed. Washington, D.C.: U.S. Department of Energy, pp. 175–213.

Prahl, F. G., L. A. Muehlhausen, and D. L. Zahnle. 1988. Further evaluation of long-chain alkenones as indicators of paleoceanographic conditions. *Geochemica et Cosmochemica Acta* 52:2303–2310.

Schlesinger, W. H. 1997. *Biochemistry: An Analysis of Global Change.* San Diego: Academic.

Tregelaar, E. W., J. de Leeuw, S. Derenne, and C. Largeau. 1989. A reappraisal of kerogen formation. *Geochimica et Cosmochimica Acta* 53:3103–3106.

Williams, P. M., and E. R. M. Druffel. 1987. Radiocarbon in dissolved organic matter in the central North Pacific Ocean. *Nature* 330(6145):246–248.

PROBLEMS

(6.1) Briefly discuss how the chemical compositions of higher plants differ from those of bacteria and determine which biochemical compounds you would expect to survive diagenesis and why.

(6.2) You recover a sediment core from a Miocene age lake deposit. Knowing that Miocene woody forests were replaced by grasslands, discuss: (a) which organic compounds you would analyze to determine the organic matter sources to sinkhole sediments; and (b) what you would expect to find if indeed forests were replaced by grasslands.

(6.3) Identify three areas of the oceans where monolayer equivalents of various masses would potentially form, and discuss the key factors that control their formation in each environment.

(6.4) Draw the structures of the following organic compounds: octane, pentene, ethanol (ethyl alcohol), a phenol, and an alkenone with 37 carbon atoms and 3 double bonds (an unsaturated straight-chain ketone).

(6.5) Using the relationship $U_{37}^K = 0.040(SST) - 0.110$, recalculate the values in worked problem 6.2. What difference does this calibration make in the resulting values, and is this significant?

(6.6) Using the information in worked problem 6.3, calculate the age of fossil snail shells that have alle/ile ratios of: 0.09, 0.13, 0.19, 0.23, 0.35. Is this a linear relationship?

CHAPTER SEVEN

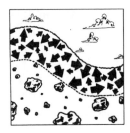

CHEMICAL WEATHERING

Dissolution and Redox Processes

OVERVIEW

The geochemical processes that most profoundly affect the surface of continents—the environment with which we have the closest personal familiarity—can be classed together as chemical weathering processes. These include a variety of reactions involving water and either acids or oxygen in the decomposition of rocks. In chapter 5, we referred to these implicitly as the source of sediments and ionic species for diagenetic processes. In this chapter, we examine them explicitly. Our focus is on the weathering of silicates, which constitute the bulk of continental rocks.

We begin the chapter by considering simple reactions that dominate the low-temperature SiO_2-MgO-Al_2O_3-(Na_2O, K_2O)-water system, to show how these control the stability of key minerals such as quartz, feldspars, and clays on the Earth's surface. We then look more closely at the roles of organic compounds and CO_2 in the weathering process, and finish by examining oxidation-reduction reactions. Throughout the chapter, we emphasize the graphical representation of equilibria, using activity-activity diagrams.

The results of this theoretical treatment can be applied to two closely related geochemical processes: the formation of residual minerals and the chemical evolution of surficial and groundwaters. We introduce both of these

by example in this chapter and end the chapter by showing how our knowledge can also be applied to problems of economic importance.

FUNDAMENTAL SOLUBILITY EQUILIBRIA

Silica Solubility

Equilibria between water and silica are important in two contexts. First, many sedimentary rocks and most igneous and metamorphic rocks contain quartz. Chemical weathering of these involves the removal of SiO_2 from a discrete phase. In the second, broader context, because virtually all common crustal minerals contain SiO_2, our appreciation of the simple silica-water system is important to understanding more general reactions with silicates.

For any silica mineral, the primary reaction with water is:

$$SiO_2 + 2H_2O \rightleftarrows H_4SiO_4. \qquad (7.1)$$

If the silica phase is quartz, the equilibrium constant K_q at 25°C is 1×10^{-4}. If it is opaline or amorphous quartz, the equilibrium constant K_{qa} at 25°C is 2×10^{-3}. Assuming that both water and the silica phase have unit activity, K_q (or K_{qa}) is also numerically equal to the activity of H_4SiO_4 that enters the solution as a result of reaction 7.1.

H_4SiO_4 is a weak acid, by which we mean that it does not break down readily to release hydrogen ions in aqueous solution. It dissociates first to form $H_3SiO_4^-$ by the reaction:

$$H_4SiO_4 \rightleftarrows H^+ + H_3SiO_4^-, \qquad (7.2)$$

for which:

$$K_1 = a_{H^+}a_{H_3SiO_4^-}/a_{H_4SiO_4} = 1.26 \times 10^{-10} \text{ at } 25°C,$$

and then to form $H_2SiO_4^{2-}$ by the reaction:

$$H_3SiO_4^- \rightleftarrows H^+ + H_2SiO_4^{2-}, \qquad (7.3)$$

for which:

$$K_2 = a_{H^+}a_{H_2SiO_4^{2-}}/a_{H_3SiO_4^-} = 5.01 \times 10^{-11} \text{ at } 25°C.$$

No further dissociation takes place. With the exception of some polymerization reactions that occur only in very alkaline solutions and have a negligible effect on total solubility, the concentration of all silica species is given by:

$$\sum Si = m_{H_4SiO_4} + m_{H_3SiO_4^-} + m_{H_2SiO_4^{2-}}. \qquad (7.4)$$

We can now combine equations 7.2 through 7.4 to write:

$$\sum Si = m_{H_4SiO_4}(1 + K_1/a_{H^+} + K_1K_2/a_{H^+}^2), \qquad (7.5)$$

in which all activity coefficients ($\gamma_{H_3SiO_4^-}$, $\gamma_{H_2SiO_4^{2-}}$, etc.) have been omitted for simplicity. This equation is shown graphically in figure 7.1, in which we follow standard chemical convention by using the quantity pH ($= -\log_{10}a_{H^+}$) as our measure of acidity. At pH values below ~9, only H_4SiO_4 contributes significantly to $\sum Si$. At pH 9.9, however, the first and second terms in equation 7.5 are numerically equal, so $m_{H_4SiO_4} = m_{H_3SiO_4^-}$. In increasingly alkaline solutions, $m_{H_4SiO_4}$ becomes even less important. At pH 11.7, according to equation 7.5, $m_{H_3SiO_4^-} = m_{H_2SiO_4^{2-}}$. The silica-water system, then, has a pH *buffering capacity* in natural waters. By this, we mean that pH is resistant to change when we add an acid or base to a solution near either 9.9 or 11.7. At pH 9.9, any added hydrogen ions are likely to be involved in forcing reaction 7.2 to the left, producing H_4SiO_4 rather than remaining as free H^+ in solution. Similarly, at pH 11.7, "extra" H^+ forces reaction 7.3 to produce $H_3SiO_4^-$. Because pH values as high as 9 are uncommon in nature, SiO_2-H_2O buffering has only a small influence in natural waters. Buffering is an important concept, how-

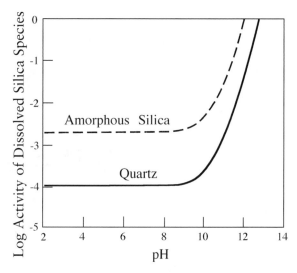

FIG. 7.1. Activities of aqueous silica species at 25°C in equilibrium with quartz and with amorphous silica as a function of pH. The lines are $\sum Si$. (Modified from Drever 1997.)

ever, so we will consider it in greater depth in chapter 8, where we study the influence of CO_2-H_2O on seawater chemistry.

Solubility of Magnesian Silicates

Just as for silica solubility, it is convenient to write reactions between magnesian silicates and water in such a way that they eventually lead to solubility expressions in terms of dissolved silica and some function of a_{H^+}. For now, we consider only *congruent* reactions (that is, those that produce only dissolved species) rather than those that produce an intermediate hydrated phase. We disregard, for example, the reaction:

$$3Mg_2SiO_4 + 2H_2O + H_4SiO_4 \rightleftarrows 2Mg_3Si_2O_5(OH)_4,$$
forsterite   serpentine

and consider only:

$$Mg_2SiO_4 + 4H^+ \rightleftarrows 2Mg^{2+} + H_4SiO_4,$$

for which:

$$K_{fo} = a_{H_4SiO_4}a_{Mg^{2+}}^2/a_{H^+}^4 = a_{H_4SiO_4}(a_{Mg^{2+}}/a_{H^+}^2)^2,$$

and

$$\log a_{H_4SiO_4} = \log K_{fo} - 2\log(a_{Mg^{2+}}/a_{H^+}^2).$$

One justification for ignoring incongruent reactions is that reactions between magnesian silicates and water in nature strongly favor total dissolution, so that little magnesium is retained in solid phases after weathering. Petit et al. (1987) have verified this experimentally by means of a highly sensitive resonant nuclear imaging technique. They found that dissolution of diopside ($CaMgSi_2O_6$) takes place as water diffuses slowly into the silicate to produce a hydrated zone ~1000 Å thick. This zone is very porous near the grain surface, but shows no evidence of secondary precipitates. This observation strongly suggests that diopside dissolves congruently, and lends support to the notion that other magnesian silicates may do so as well.

We can write many such congruent reactions. For example, we have:

$$Mg_3Si_2O_5(OH)_4 + 6H^+ \rightleftarrows 3Mg^{2+} + 2H_4SiO_4 + H_2O,$$
$$\text{serpentine}$$

for which (assuming that $a_{H_2O} = 1$):

$$K_{serp} = a^2_{H_4SiO_4}a^3_{Mg^{2+}}/a^6_{H^+} = a^2_{H_4SiO_4}(a_{Mg^{2+}}/a^2_{H^+})^3,$$

and

$$\log a_{H_4SiO_4} = \tfrac{1}{2}\log K_{serp} - \tfrac{3}{2}\log(a_{Mg^{2+}}/a^2_{H^+}).$$

The similar reaction for talc,

$$Mg_3Si_4O_{10}(OH)_2 + 6H^+ + 4H_2O \rightleftarrows$$
$$\text{talc}$$
$$3Mg^{2+} + 4H_4SiO_4,$$

yields:

$$K_{tc} = a^4_{H_4SiO_4}a^3_{Mg^{2+}}/a^6_{H^+} = a^4_{H_4SiO_4}(a_{Mg^{2+}}/a^2_{H^+})^3$$

and

$$\log a_{H_4SiO_4} = \tfrac{1}{4}\log K_{tc} - \log(a_{Mg^{2+}}/a^2_{H^+}).$$

For sepiolite, the reaction:

$$Mg_4Si_6O_{15}(OH)_2 \cdot 6H_2O + 8H^+ + H_2O \rightleftarrows$$
$$\text{sepiolite}$$
$$4Mg^{2+} + 6H_4SiO_4$$

leads to the equations:

$$K_{sep} = a^4_{Mg^{2+}}a^6_{H_4SiO_4}/a^8_{H^+} = a^6_{H_4SiO_4}(a_{Mg^{2+}}/a^2_{H^+})^4,$$

and

$$\log a_{H_4SiO_4} = \tfrac{1}{6}\log K_{sep} - \tfrac{2}{3}\log(a_{Mg^{2+}}/a^2_{H^+}).$$

Finally, we consider the solubility of brucite, although it is not a silicate, and find that:

$$Mg(OH)_2 + 2H^+ \rightleftarrows Mg^{2+} + 2H_2O$$
$$\text{brucite}$$

yields:

$$K_{br} = a_{Mg^{2+}}/a^2_{H^+},$$

so that $\log a_{H_4SiO_4}$ is independent of $\log(a_{Mg^{2+}}/a^2_{H^+})$.

Notice how cleverly we have described each of these equilibria as linear, first-order equations in $\log a_{H_4SiO_4}$ and $\log(a_{Mg^{2+}}/a^2_{H^+})$, so that they can be shown easily in graphical form, as in figure 7.2. The solubility of quartz, which is independent of $\log a_{H_4SiO_4}$ is also shown. We have already shown that H_4SiO_4 dissociates appreciably above pH 9, so this diagram is not reliable for highly alkaline environments. Each of the magnesium silicate equilibria plots as a straight line whose y intercept (at infinitesimal H_4SiO_4 activity) is equal to a reaction K, calculated from standard free energies of formation. Their slopes in each case depend on the stoichiometry of the

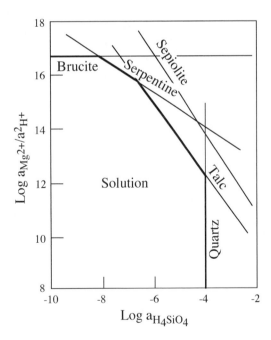

FIG. 7.2. Solubility relationships among magnesian silicates in equilibrium with water at 25°C. Lines represent reactions discussed in the text. The dissolution of forsterite lies too far to the right to appear on this diagram (see problem 7.7 at the end of the chapter). (Modified from Drever 1997.)

solubility reaction (that is, on the Mg/Si ratio in the silicate mineral). The serpentine reaction, for example, has a slope of $-\frac{3}{2}$, and talc that of $-\frac{3}{4}$.

Notice that the dissolved species are stable on the lower left side of each equilibrium line, so that each line divides the diagram into a region in which fluid is supersaturated with respect to a silicate mineral (to the upper right) and one in which it is undersaturated (to the lower left). Therefore, the bold boundary of the solution field shown in figure 7.2, combining segments of the brucite, serpentine, talc, and quartz equilibria, represents the only set of log $a_{H_4SiO_4}$ and log$(a_{Mg^{2+}}/a_{H^+}^2)$ values for which fluid and a mineral are stable. Sepiolite and forsterite, whose reaction lines are entirely to the right of the boundary are unstable (or metastable) in contact with water at 25°C.

Solubility of Gibbsite

Leaving the relatively simple realm of magnesium silicate equilibria, let's now try our luck with aluminosilicates. Dissolved aluminum in acid solutions exists as the free ion Al^{3+}. With increasing alkalinity, however, Al^{3+} reacts with water to form ion pairs, which dominate at pH >5. When we considered the solubility of magnesian silicates, there was no mention of complex species, because they have little effect in that system. The ion pair $Mg(OH)^+$, for example, is abundant only for pH >11, outside the range of acidity in most natural waters. By contrast, the solubility relations of $Al(OH)_3$ and aluminosilicate minerals are greatly influenced by the balance among complex species, and therefore are more complicated than those of $Mg(OH)_2$ and magnesium silicates.

Beginning in highly acid solutions, dissolution of gibbsite is described by:

$$Al(OH)_3 + 3H^+ \rightleftarrows Al^{3+} + 3H_2O,$$

for which:

$$K_1 = a_{Al^{3+}}/a_{H^+}^3.$$

As pH increases beyond 5, $Al(OH)_2^+$ dominates instead. The dissolution equilibrium is described by:

$$Al(OH)_3 + H+ \rightleftarrows Al(OH)_2^+ + H_2O,$$

and the constant:

$$K_2 = a_{Al(OH)^{2+}}/a_{H^+}.$$

Finally, above pH 7, the major species in solution is $Al(OH)_4^-$, which forms by:

$$Al(OH)_3 + H_2O \rightleftarrows Al(OH)_4^- + H^+,$$

and for which:

$$K_3 = a_{Al(OH)_4^-} a_{H^+}.$$

Thus, to calculate the total concentration of dissolved aluminum (ΣAl) in waters in equilibrium with gibbsite, we write:

$$\sum Al = a_{Al^{3+}} + a_{Al(OH)^{2+}} + a_{Al(OH)_4^-},$$

and substitute appropriately from the equations for K_1, K_2, and K_3 to get:

$$\sum Al = K_1 a_{H^+}^3 + K_2 a_{H^+} + K_3/a_{H^+}. \tag{7.6}$$

Equation 7.6 differs from equation 7.5 (for ΣSi) in that it predicts high solubility in both very alkaline and very acid solutions, but low solubility throughout most of the range of pH observed in streams and most groundwaters. To see this quantitatively, let's do a worked problem.

Worked Problem 7.1

Decomposition of organic matter in the uppermost centimeters of a soil may decrease the pH of percolating water to 4 or lower. As weathering reactions consume H^+, we expect the dominant form of dissolved aluminum and the total solubility, ΣAl, to change. Ignoring any possible reactions involving the organic compounds themselves, by how much should ΣAl vary between, say, pH 4 and pH 6?

The three equilibrium constants in equation 7.6 can be calculated from tabulated values of $\Delta \bar{G}_f^\circ$. We use molar free energies for water from Helgeson and coworkers (1978) and for aluminum species at 25°C from Couturier and associates (1984), except for $Al(OH)_2^+$, which is from Reesman and coworkers (1969):

Species	$\Delta \bar{G}_f^\circ$ (kcal mol^{-1})
$Al(OH)_3$	−276.02
Al^{3+}	−117.0
$Al(OH)_2^+$	−216.1
$Al(OH)_4^-$	−313.4
H_2O	−56.69

We calculate K_1 from:

$$\Delta \bar{G}_1 = \Delta \bar{G}_f^\circ(Al^{3+}) + 3\Delta \bar{G}_f^\circ(H_2O) - \Delta \bar{G}_f^\circ(Al(OH)_3)$$

$$= -117.0 + 3(-56.69) - (-276.02) = -11.05,$$

$$K_1 = e^{-\Delta \bar{G}_1/RT} = 1.27 \times 10^8,$$

K_2 from:

$$\Delta \bar{G}_2 = \Delta \bar{G}_f^\circ(\text{Al(OH)}_2^+) + \Delta \bar{G}_f^\circ(\text{H}_2\text{O}) - \Delta \bar{G}_f^\circ(\text{Al(OH)}_3)$$

$$= -216.1 + (-56.69) - (-276.02) = 3.23$$

$$K_2 = e^{-\Delta \bar{G}_2/RT} = 4.28 \times 10^{-3},$$

and K_3 from:

$$\Delta \bar{G}_3 = \Delta \bar{G}_f^\circ(\text{Al(OH)}_4^-) - \Delta \bar{G}_f^\circ(\text{Al(OH)}_3) - \Delta \bar{G}_f^\circ(\text{H}_2\text{O})$$

$$= 311.4 - (-276.02) - (-56.69) = 21.31,$$

$$K_3 = e^{\Delta \bar{G}_3/RT} = 2.34 \times 10^{-16}.$$

At pH 4, $a_{\text{H}^+} = 1 \times 10^{-4}$, so we calculate from equation 7.6 that:

$$\sum \text{Al} = 1.27 \times 10^{-4} + 4.28 \times 10^{-7} + 2.34 \times 10^{-12}$$

$$= 1.27 \times 10^{-4},$$

and conclude that $a_{\text{Al}^{3+}}/a_{\text{Al(OH)}_2^+} = 297$. At pH 6 ($a_{\text{H}^+} = 1 \times 10^{-6}$),

$$\sum \text{Al} = 1.27 \times 10^{-10} + 4.28 \times 10^{-9} + 2.34 \times 10^{-10}$$

$$= 4.64 \times 10^{-9},$$

so that:

$$a_{\text{Al}^{3+}}/a_{\text{Al(OH)}_2^+} = 2.97 \times 10^{-2}.$$

With a shift of two pH units, total aluminum concentration has dropped by four orders of magnitude and, as expected, Al(OH)_2^+ has eclipsed free Al^{3+} as the major dissolved species. Figure 7.3 shows the solubility of gibbsite, calculated from the data in this worked problem over the range pH 3–12.

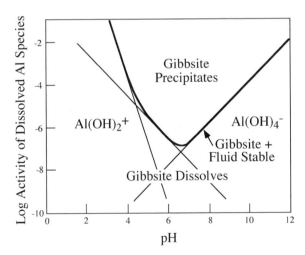

FIG. 7.3. Activities of aqueous aluminum species in equilibrium with gibbsite at 25°C. The bold curve, marking the boundary of the fluid field, is $\sum$Al. (Modified from Drever 1997.)

Solubility of Aluminosilicate Minerals

In the same way that we constructed figure 7.2 to illustrate the solubility of magnesium silicates, it would be convenient to diagram the solubility behavior of aluminum silicates. Now that we have studied Al(OH)_3 solubility, however, we see that this cannot be done easily. For example, consider the dissolution of kaolinite. Over the range of natural pH values, each of the following three reactions contributes to $\sum$Al:

$$\tfrac{1}{2}\text{Al}_2\text{Si}_2\text{O}_5(\text{OH})_4 + 3\text{H}^+ \rightleftarrows \text{Al}^{3+} + \text{H}_4\text{SiO}_4 + \tfrac{1}{2}\text{H}_2\text{O},$$
kaolinite

$$\tfrac{1}{2}\text{Al}_2\text{Si}_2\text{O}_5(\text{OH})_4 + \text{H}^+ + \tfrac{3}{2}\text{H}_2\text{O} \rightleftarrows$$
$$\text{Al(OH)}_2^+ + \text{H}_4\text{SiO}_4,$$

$$\tfrac{1}{2}\text{Al}_2\text{Si}_2\text{O}_5(\text{OH})_4 + \tfrac{7}{2}\text{H}_2\text{O} \rightleftarrows$$
$$\text{Al(OH)}_4^- + \text{H}^+ + \text{H}_4\text{SiO}_4.$$

Equation 7.6 now takes the form:

$$\sum \text{Al} = (K_1 a_{\text{H}^+}^3 + K_2 a_{\text{H}^+} + K_3/a_{\text{H}^+})/a_{\text{H}_4\text{SiO}_4}.$$

This result is shown graphically in figure 7.4. The solubility surface for kaolinite has cross sections perpendicular to the $a_{\text{H}_4\text{SiO}_4}$ axis that are qualitatively similar to figure 7.3. With increasing silica activity, however, kaolinite solubility at all pH values decreases. A three-dimensional diagram such as figure 7.4 or contours drawn at constant $a_{\text{H}_4\text{SiO}_4}$ on diagrams as in figure 7.3 can help us visualize the stability fields for such minerals as kaolinite or pyrophyllite. More complicated minerals like feldspars and micas, however, cannot be considered in this way. Furthermore, because the aluminum concentration in most natural waters is too low to measure reliably with ease, diagrams such as figures 7.3 and 7.4 are less practical than figure 7.2 was for magnesian silicates. For this reason, geochemists commonly write dissolution reactions for aluminum silicates that produce secondary solid aluminous phases instead of dissolved aluminum. Such reactions are said to be *incongruent*.

One example of such a reaction is:

$$\tfrac{1}{2}\text{Al}_2\text{Si}_2\text{O}_5(\text{OH})_4 + \tfrac{5}{2}\text{H}_2\text{O} \rightleftarrows \text{Al(OH)}_3 + \text{H}_4\text{SiO}_4,$$

which can be viewed loosely as an intermediate to any of the kaolinite reactions just considered. Notice that the relative stabilities of kaolinite and gibbsite, written this way, depend only on the activity of H_4SiO_4 (that is,

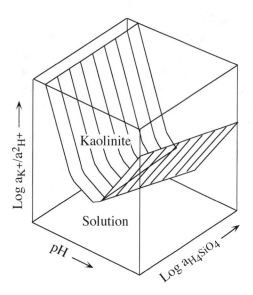

FIG. 7.4. Schematic diagram illustrating the kaolinite solubility surface as a function of $\log(a_{K^+}/a_{H^+}^2)$, pH, and $\log a_{H_4SiO_4}$. Notice the similarity in form between crosssections at constant $a_{H_4SiO_4}$ and the gibbsite solubility curve in figure 7.3. (Modified from Garrels and Christ 1965.)

$K_{kao-gib} = a_{H_4SiO_4}$). Any information about dissolved aluminum species is "hidden," because this reaction focuses entirely on the solid phase $Al(OH)_3$. *We have no information about the stability of either mineral with respect to a fluid unless we also know* $\sum Al$. We emphasize this point before proceeding to develop equations and diagrams commonly used in aluminosilicate systems because, unfortunately, it is easy to overlook and can lead to misinterpretation of those equations and diagrams.

With that warning, let's now consider the broader set of common minerals that includes not only silica and alumina, but other cations as well. In the system K_2O-Al_2O_3-SiO_2-H_2O, for example, we can write the following set of reactions between minerals:

$$KAlSi_3O_8 + H^+ + \tfrac{9}{2}H_2O \rightleftarrows$$
K-feldspar
$$\tfrac{1}{2}Al_2Si_2O_5(OH)_4 + K^+ + 2H_4SiO_4,$$
$$\text{kaolinite}$$

for which:

$$K_{ksp-kao} = (a_{K^+}/a_{H^+})a_{H_4SiO_4}^2;$$
$$KAlSi_3O_8 + H^+ + 2H_2O \rightleftarrows$$
K-feldspar
$$\tfrac{1}{2}Al_2Si_4O_{10}(OH)_2 + K^+ + H_4SiO_4,$$
$$\text{pyrophyllite}$$

for which:

$$K_{ksp-pyp} = (a_{K^+}/a_{H^+})a_{H_4SiO_4};$$
$$KAlSi_3O_8 + \tfrac{2}{3}H^+ + 4H_2O \rightleftarrows$$
K-feldspar
$$\tfrac{1}{3}KAl_3Si_3O_{10}(OH)_2 + \tfrac{2}{3}K^+ + 2H_4SiO_4,$$
$$\text{muscovite}$$

for which:

$$K_{ksp-mus} = (a_{K^+}/a_{H^+})a_{H_4SiO_4}^3;$$
$$KAl_3Si_3O_{10}(OH)_2 + H^+ + \tfrac{3}{2}H_2O \rightleftarrows$$
muscovite
$$\tfrac{3}{2}Al_2Si_2O_5(OH)_4 + K^+,$$
$$\text{kaolinite}$$

for which:

$$K_{mus-kao} = (a_{K^+}/a_{H^+});$$

and

$$\tfrac{1}{2}Al_2Si_4O_{10}(OH)_2 + \tfrac{5}{2}H_2O \rightleftarrows$$
pyrophyllite
$$\tfrac{1}{2}Al_2Si_2O_5(OH)_4 + H_4SiO_4,$$
$$\text{kaolinite}$$

for which:

$$K_{pyp-kao} = a_{H_4SiO_4}.$$

Each of these equations plots as a straight line on a graph of $\log a_{H_4SiO_4}$ versus $\log(a_{K^+}/a_{H^+})$, as shown in figure 7.5. This diagram looks superficially like the one we constructed for magnesium silicates. Because figure 7.5 has been constructed on the basis of incongruent reactions, there is one major difference between the two diagrams: instead of marking the conditions under which a mineral is in equilibrium with a fluid, each line segment in figure 7.5 indicates that two minerals coexist stably. Consequently, four reactions (kao-gib, mus-kao, ksp-kao, and pyp-kao) bound a region inside which kaolinite is stable relative to any of the other minerals we have considered. Each of the other areas bounded by reaction lines similarly represents a region in which one aluminosilicate mineral is stable.

Several times in earlier chapters we warned that thermodynamic calculations are valid only for the system as it has been defined, and we raise the same caution flag again. Just as figure 7.2 showed both forsterite and sepiolite to be unstable in the presence of water, you would find that reactions between muscovite and pyrophyllite, K-feldspar and gibbsite, or pyrophyllite and gibbsite are

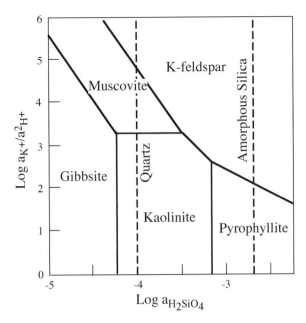

FIG. 7.5. Stability relationships among common aluminosilicate minerals at 25°C. Precise locations of field boundaries may vary because of uncertainties in thermodynamic data from which the equilibrium constants (K) are calculated.

all metastable with respect to the reactions we have plotted in figure 7.5. You can verify this for yourself (or simply assume) that we have already written expressions for all potential equilibrium constants in the system and inserted free energy data to determine which reactions are favored. It's not an easy task. Particularly as other components like CaO and Na$_2$O are added and graphical representation once again becomes difficult, it is no simple matter to identify each of the reactions that bounds a mineral stability field. Figure 7.6 shows how unwieldy graphical methods become with the addition of even one more compositional variable (a_{Na^+}/a_{H^+}). Today, computer programs do the job by brute force, but accurately, so that no significant reaction is inadvertently overlooked. The most likely sources of error, then, are those that arise because the thermodynamic data we input to the program have experimental uncertainties or we accidentally omit phases from consideration.

Both of these sources of error are more likely than you might think. The crystal chemistry of phyllosilicates is complex: cationic and anionic substitutions are common, as are mixed-layer architectures in which two or more mineral structures alternate randomly. Among

(Ca, Na)-aluminosilicates, this is particularly true. Natural smectites, for example, include both dioctahedral (pyrophyllite-like) clays like montmorillonite and trioctahedral (talc-like) clays such as saponite. All show high ion-exchange capacity and readily accept Ca^{2+}, Na$^+$, K$^+$, Fe^{2+}, Mg^{2+}, and a host of minor cations in structural sites or interlayer positions. Their variable degrees of crystallinity and generally large surface areas also contribute to large uncertainties in their thermodynamic properties. As a result, thermodynamic data for many aluminosilicates are gathered on phases that bear only a loose resemblance to minerals we encounter in the world outside the laboratory. The exact placement of field boundaries in diagrams such as figures 7.5 and 7.6 is subject to considerable debate. Complicating the picture are a number of common zeolites, like analcime and phillipsite, which are stable at room temperature and should therefore appear on diagrams of this sort, but are remarkably easy to overlook.

These comments are not intended to be discouraging, but should point out the necessity of choosing data carefully when you try to model a natural aluminosilicate system rather than a hypothetical one. One moderately successful approach to verifying diagrams like figure 7.5 involves the assumption that subsurface and stream

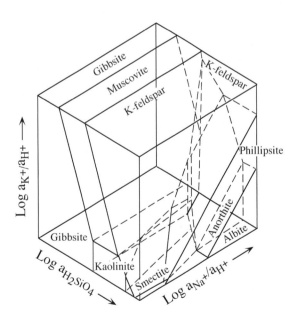

FIG. 7.6. Stability relationships among common sodium and potassium aluminosilicate minerals at 25°C. (Modified from Drever 1997.)

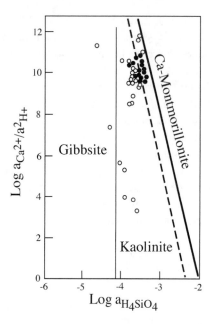

FIG. 7.7. Activity diagram illustrating the compositions of water samples from the Rio Tanama, Puerto Rico. Solid circles indicate that sediment samples associated with the water contain both kaolinite and calcium montmorillonite. Open circles identify samples containing only kaolinite. The dashed boundary line, based on these observations, compares well with the solid bold boundary line, which was calculated from thermodynamic data. Rio Tanama water lies to the right of the gibbsite stability field. (Modified from Norton 1974.)

waters are in chemical equilibrium with the weathered rocks over which they flow. If this is so, it should be possible to place field boundaries by analyzing water from wells and streams for pH, dissolved silica, and major cations and plotting the results as representative of the coexisting minerals identified in the rocks at the sampling site. Norton (1974) followed this method in his analysis of springs and rivers in the Rio Tanama basin, Puerto Rico. As indicated in figure 7.7, the theoretical boundary between stability fields for kaolinite and Ca-montmorillonite corresponds closely to the measured composition of natural waters in equilibrium with both minerals. Drever (1997) has questioned the assumption of equilibrium in studies of this type, but the agreement often shown between theory and nature in such studies justifies hesitant optimism.

So much for theory. Let's see how to use what we have learned.

Worked Problem 7.2

Assuming that reliable stability diagrams can be drawn to illustrate solubility relationships among aluminosilicate minerals, how can we use this information to predict the course of chemical weathering? As an example, consider the reactions that take place as microcline reacts with acidified groundwater in a closed system.

In principle, we can predict a pathway by treating the problem as a titration of acid by microcline, just as you can predict how solution pH varies as you add Na_2CO_3 to a dilute acid in chemistry lab. We consider exactly that problem in chapter 8. The difference here is that incongruent mineral reactions occur along the reaction path, so that species other than aqueous ions have to be taken into account. This is a cumbersome exercise, commonly performed today by computer. The results shown in figure 7.8 and discussed here were produced by the program EQ6 (Wolery 1979).

As microcline dissolves in a fixed volume of water, $a_{H_4SiO_4}$ and the ratio a_{K^+}/a_{H^+} are both initially very low. The fluid composition, then, begins beyond the lower left corner of the diagram and gradually creeps upward and to the right. If pH is below ~4, the dominant reaction at this early stage is:

$$KAlSi_3O_8 + 4H_2O + 4H^+ \rightleftarrows K^+ + Al^{3+} + 3H_4SiO_4.$$

But remember our earlier warning. No secondary phase can appear until $\sum Al$ is high enough to saturate the fluid, even though the water composition lies in the field marked

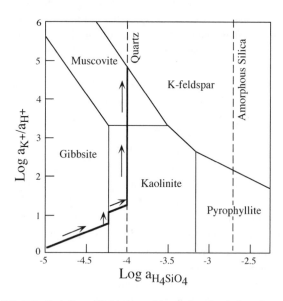

FIG. 7.8. Evolution of fluid composition during dissolution of microcline in a closed system at 25°C, calculated by the program EQ6 (Wolery 1979). Reactions along each segment of the path indicated by arrows are discussed in worked problem 7.2.

"Gibbsite." Once enough aluminum is in solution, gibbsite forms while microcline continues to dissolve. The net reaction releases potassium ions, OH^-, and silica to the solution, causing the fluid composition to move toward the upper right as both $a_{H_4SiO_4}$ and a_{K^+}/a_{H^+} increase:

$$KAlSi_3O_8 + 8H_2O \rightleftarrows Al(OH)_3 + K^+ + OH^- + 3H_4SiO_4.$$

Eventually, the fluid composition reaches the gibbsite-kaolinite boundary. At this point, microcline dissolution continues to release H_4SiO_4. Instead of continuing to accumulate, however, that H_4SiO_4 is now consumed in the production of kaolinite at the expense of gibbsite:

$$2KAlSi_3O_8 + 4Al(OH)_3 + H_2O \rightleftarrows$$
$$3Al_2Si_2O_5(OH)_4 + 2K^+ + 2OH^-.$$

The fluid composition cannot move farther to the right as long as some gibbsite remains. Potassium ions continue to accumulate in solution, however, so the fluid composition follows the gibbsite-kaolinite boundary toward higher a_{K^+}/a_{H^+}.

When all of the gibbsite is finally converted to kaolinite, the fluid composition is no longer fixed along the field boundary, and both $a_{H_4SiO_4}$ and a_{K^+}/a_{H^+} can increase as kaolinite precipitates:

$$2KAlSi_3O_8 + 11H_2O \rightleftarrows$$
$$Al_2Si_2O_5(OH)_4 + 2K^+ + 2OH^- + 4H_4SiO_4.$$

Once the activity of H_4SiO_4 reaches 1×10^{-4} (the value of K_q) it cannot increase further, because the fluid is saturated with respect to quartz, which begins to precipitate as microcline dissolution continues. This reaction does not change the identity of the stable aluminosilicate. Kaolinite remains in equilibrium with the fluid as a_{K^+}/a_{H^+} rises.

At the muscovite-kaolinite boundary, the situation is similar to that at the gibbsite-kaolinite boundary reached earlier, except that neither $a_{H_4SiO_4}$ nor a_{K^+}/a_{H^+} can change until all kaolinite is converted to muscovite by the reaction:

$$KAlSi_3O_8 + Al_2Si_2O_5(OH)_4 + H_2O \rightleftarrows$$
$$KAl_3Si_3O_{10}(OH)_2 + SiO_2.$$

Still prevented from becoming more silica-rich by the saturation limit for quartz, the fluid continues to follow a vertical pathway across the muscovite field:

$$3KAlSi_3O_8 + 2H_2O \rightleftarrows$$
$$KAl_3Si_3O_{10}(OH)_2 + 2K^+ + 2OH^- + 6SiO_2,$$

until it finally reaches the edge of the microcline stability field. No further dissolution takes place, because the fluid is now saturated with respect to microcline.

It is important, once again, to emphasize that the preferred pathway in nature may be distinctly different from what we have calculated in this worked problem. For example, quartz generally nucleates less readily than amorphous silica, so that H_4SiO_4 might continue to accumulate when $a_{H_4SiO_4} = 10^{-4}$. In that case, the reaction path would continue diagonally across the kaolinite field, eventually reaching the kaolinite-pyrophyllite boundary instead of moving vertically through the kaolinite and muscovite fields. The fluid path would then move vertically to the K-feldspar field without $a_{H_4SiO_4}$ ever reaching 2×10^{-3}, where amorphous silica could precipitate.

Although we devised this problem to illustrate weathering in a closed system, chemical weathering more commonly takes place in an open system. If water flows through a parent material instead of stagnating in it, then fluid composition not only evolves with time but also varies spatially. As a result, residual minerals accumulate in a layered weathering sequence that can provide direct information about the fluid reaction path. A feldspathic boulder, for example, might be surrounded by concentric zones of muscovite-quartz, kaolinite-quartz, and gibbsite, suggesting that open-system weathering had followed the fluid path indicated in worked problem 7.2.

RIVERS AS WEATHERING INDICATORS

As we have shown in several contexts so far, the type and intensity of weathering reactions is reflected in an assemblage of residual minerals and the composition of associated waters. Sometimes it is appropriate to examine the pathways of chemical weathering by comparing soil profiles with mineral sequences that we derive from model calculations, as in worked problem 7.2. In other cases, it is more reasonable to study the waters draining a region undergoing weathering. This might be true if soils were poorly developed or mechanically disturbed, or if we were interested in alteration in a subsurface environment from which residual weathering products are not easily sampled. It might also be true if we were concerned about weathering on a regional scale and wanted to integrate the effects of several types of parent rock, or if we were interested in seasonal variations in weathering.

In preceding sections, we focused on the SiO_2-Al_2O_3 framework of silicate minerals and its stability in a weathering environment. We have seen, however, that mineral stabilities during chemical weathering are functions of activity ratios such as a_{K^+}/a_{H^+} or $a_{Mg^{2+}}/a_{H^+}^2$, so this has also been a study of alkali and alkaline earth element behavior. What happens, however, when several cations are released along a reaction pathway? How are they partitioned, relative to each other, among the clay products and the fluid?

Nesbitt and coworkers (1980) analyzed a progressively weathered sequence of samples from the Toorongo

Granodiorite in Australia, which provides an interesting perspective. The parent rock consists largely of plagioclase, quartz, K-feldspar, and biotite, with some minor hornblende and traces of ilmenite. During initial stages of weathering, sodium, calcium, and strontium in the rock decrease rapidly in comparison with potassium. The primary reason for this decline is the early dissolution of plagioclase, which is much less stable than K-feldspar or biotite, the major potassic minerals. The clay mineral that appears in the least altered samples is kaolinite.

As weathering progresses, vermiculite and then illite join the clay mineral assemblage. These further influence the loss or retention of cations. As the initially low pH of migrating fluids begins to increase, charge balance on surfaces and in interlayer sites in clays is satisfied by large cations (K^+, Rb^+, Cs^+, and Ba^{2+}) rather than by H^+. The smaller cations (Ca^{2+}, Na^+, and Sr^{2+}) are less readily accepted in interlayer sites and so are quantitatively removed in groundwater and surface runoff. The sole exception is Mg^{2+}, which is largely retained in the soil, as biotite is replaced by vermiculite, which has a small structural site to accommodate it. As a result, progressive weathering of the Toorongo Granodiorite produces a residual that is greatly depleted in calcium and sodium and only slightly depleted in potassium and magnesium.

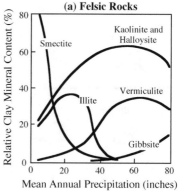

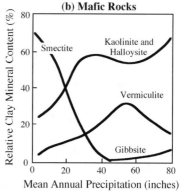

FIG. 7.10. Clay minerals in residual soils in California formed on (a) felsic and (b) mafic igneous rocks. (After Barshad 1966.)

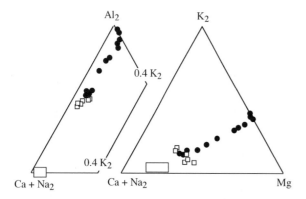

FIG. 7.9. Chemical changes during weathering of the Toorongo Granodiorite. Measured compositions of rock samples are indicated by solid circles. Average unweathered rock compositions are indicated by small open boxes, and average stream water composition by the larger boxes. With progressive weathering, rock compositions move away from the Ca + Na$_2$ corner on both diagrams, becoming more aluminous and relatively enriched in potassium and magnesium. Water compositions show complementary trends. (Modified from Nesbitt et al. 1980.)

Figure 7.9 illustrates this trend, and shows that the composition of local stream waters is roughly complementary (that is, they are relatively rich in Na^+ and Ca^{2+} and poor in K^+ and Mg^{2+}). Here, then, is an illustration of how analyses of natural fluids can confirm what geochemists learn about weathering by analyzing residual minerals.

These observations are generally consistent with the thermodynamic predictions we made in the last section and with observed clay mineral abundances on a global scale (see, for example, fig. 7.10). Gibbsite only appears in weathered sections that are intensively leached (that is, ones in which the associated waters are extremely dilute). Kaolinite dominates where waters are less dilute, and smectites where waters have higher concentrations of dissolved cations. Vermiculite and illite are common as intermediate alteration products where soil waters have concentrations between these extremes, as is true in most regions of temperate climate and moderate annual rainfall.

AGENTS OF WEATHERING

Carbon Dioxide

All surface environments are exposed to atmospheric CO_2 and H_2O. The chemical combination of the two is a powerful and ubiquitous weathering agent. Data compiled by Heinrich Holland (1978) suggest that water in wells and springs draining limestone terrains contains between 50 and 90 mg Ca^{2+} kg^{-1}, due almost entirely to contact with CO_2-charged waters. When we study carbonate equilibria in greater detail in chapter 8, however, we show that this is much more dissolved calcium than we ought to expect from weathering by atmospheric CO_2 alone. The measured calcium concentrations reported by Holland suggest that the partial pressure of CO_2 in equilibrium with limestones in nature is 10–100 times higher than atmospheric P_{CO_2}. These surprisingly high values are actually common in air held in the pore spaces of soil. Measurements made in arable midlatitude soils indicate that P_{CO_2} is between 1.5×10^{-1} and 6.5×10^{-3} atm—roughly 5–20 times that in the atmosphere. Analyses of a tropical soil reported in Russell (1961) indicate P_{CO_2} as high as 1×10^{-1} atm—330 times the atmospheric value!

These findings indicate that if we want to study the effectiveness of CO_2 as a chemical weathering agent, we must look primarily to those processes that control its abundance in soils. A small amount of CO_2 is generated when ancient organic compounds in sediments are oxidized during weathering. This amount, however, is almost negligible. Most of the CO_2 is produced by respiration from plant roots and by microbial degradation of buried plant matter (see, for example, Wood and Petratis 1984; Witkamp and Frank 1969). Cawley and coworkers (1969) provided dramatic confirmation of this CO_2 source by sampling streams in Iceland. They found that streams draining vegetated regions had bicarbonate concentrations two to three times higher than streams in areas with the same volcanic rock but without plant cover.

CO_2 generation follows an annual cycle in response to variations in both temperature and soil moisture. Values rise rapidly at the onset of a growing season and fall again as lower air and soil temperatures mark the end of the growing season. The rate of chemical weathering, therefore, should be greatest in most regions during the summer. Even in midwinter, however, soil P_{CO_2} may sometimes be several times the atmospheric value. Solomon and Cerling (1987) found that snow cover can trap CO_2 in some intermontane soils, so that P_{CO_2} in winter can actually exceed the value in summer at shallow depths (see fig. 7.11). Because carbon dioxide is more soluble in cold than in warm water, this seasonal pattern may lead to weathering rates that are significantly enhanced during the winter.

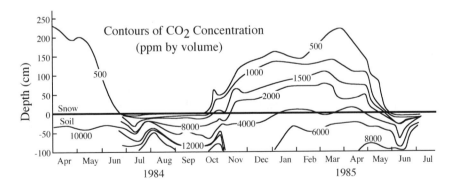

FIG. 7.11. Soil CO_2 as a function of depth and time from April 1984 to July 1985 at Brighton, Utah. In general, CO_2 levels are highest during the warm months, when biological activity in the soil is highest. Also, P_{CO_2} decreases toward the surface due to diffusive loss to the atmosphere. Notice, however, that P_{CO_2} in shallow soils is higher during the winter months than during the summer, even though biological CO_2 production is lower. The snow may act as a cap to prevent CO_2 from escaping to the atmosphere. (After Solomon and Cerling 1987.)

Organic Acids

Despite considerable study, the role of organic matter in chemical weathering is poorly understood. We suggested in the previous section that some of the elevated CO_2 concentrations in soil are the result of biodegradation, so that organic compounds affect weathering rates indirectly by serving as a source for CO_2. Laboratory and field studies, however, indicate that they have a much stronger direct influence.

Upon decomposition, the organic remains of plants and animals yield a complex mixture of acids. A number of simple compounds, such as acetic and oxalic acid, are commonly found around active root systems and fungi. These and other acids (such as malic and formic acids) serve the same role as HCl or H_2SO_4 and act as H^+ donors. Depending on the acidity of groundwaters, any of these acids can increase mineral solubility.

Low-molecular-weight organic acids may play a role in mineral dissolution by forming complexes with polyvalent cations on mineral surfaces. Organic ligands such as oxalate, malonate, succinate, salicylate, phthalate, and benzoate have been shown to accelerate dissolution rates of simple oxides (Al_2O_3 and iron [III] oxides) compared with rates determined for inorganic acids alone. This mechanism appears to have a limited effect on quartz solubility, although Bennett (1991) observed direct evidence of increased quartz dissolution in a hydrocarbon-contaminated aquifer and inferred surface complexation. Additional laboratory studies show that quartz dissolution at 25°C increases in the presence of citric and oxalic acids (0.02 M) with the greatest effect between pH 5.5 and 7. In contrast, organic complexes on the surfaces of feldspars appear to greatly enhance solubility. The Si/Al ratio of the mineral influences the extent to which organic ligands enhance feldspar dissolution, with the greatest solubility corresponding to the lowest ratio. This inverse relation suggests that the organic ligand enhances dissolution by forming a complex with available Al^{3+} ions. Amrhein and Suarez (1988) found that at pH 4 and above, 2.5 mM oxalate enhanced anorthite (Si/Al = 1) dissolution by a factor of two, and Stillings and others (1996) showed that in 1 mM oxalate solutions, feldspar dissolution rates were enhanced by a factor of 2–5 at pH 3 and by factors of 2–15 at pH 5–6.

The role of high-molecular-weight acids is less well understood. Humic and fulvic acids are composed of a bewildering array of aliphatic and phenolic structural units, with a high proportion of –COOH and phenolic OH radicals, as suggested schematically in figure 7.12. At least some of these acids are soluble or dispersible in water, where limited dissociation of these radicals can occur, particularly if soil pH is low. The effect is twofold. First, dissociation releases hydrogen ions into solution, where they influence solubility equilibria in the same way that inorganic acids do. Second, the organic acid molecule, following dissociation, develops a surface charge that helps to repel other similarly charged particles. It therefore adopts a dispersed or colloidal character, which increases its mobility in soil waters.

This second role is particularly significant, because several studies indicate that organic colloids may be important in controlling both solubility and transport of mineral matter (Schnitzer 1986). Provided that free cations are not so abundant that they completely neutralize the surface charge, humic and fulvic acids can attach significant amounts of Fe^{3+} and Al^{3+}, and lesser amounts of alkali and alkaline earth cations, without losing their mobility as colloids. In this way, the stability of colloidal organic matter may, for example, be a controlling factor in the downward transport of iron and aluminum in acidic forest soils (spodosols and alfisols) (DeConick 1980).

Finally, organic acids in soils commonly act as chelating agents. Chelates of fulvic acids and of oxalic, tartaric, and citric acid are thought to be responsible for most of the dissolved cations in forest soils (Schnitzer 1986). As we first saw in chapter 4, dissolved cations can contribute significantly to the mobility of metals. Fulvic and humic acids also both form mixed ligand complexes with phosphate and other chemically active groups.

Although each of these factors can contribute to mineral dissolution, their net effect is slight at best. In fact, high-molecular-weight organic molecules are more refractory than light molecules (that is, they are not as easily metabolized by extant microbial communities), and they appear to decrease the feldspar dissolution rates rather than increase them, except at low pH. Ochs and co-workers (1993) investigated the pH dependence of α-Al_2O_3 dissolution in the presence of humic acids and found that at pH 3, there was a slight enhancement, whereas at pH 4 and 4.5, the acids inhibited dissolution. They concluded that under strongly acidic conditions, most of the functional groups on the humic molecule are protonated and can therefore detach Al^{3+} ions by forming mononuclear complexes. At higher pH, the surface

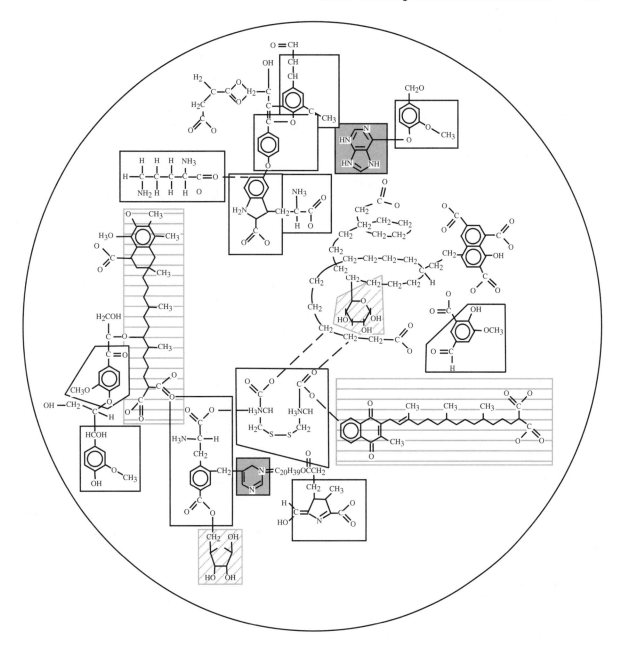

FIG. 7.12. Chemical-structural model of a humic acid with a molecular weight of ~5000. Boxes outline various biological precursor units incorporated into the humic acid. Light shaded boxes are lignin degradation products. Dark shaded boxes are nucleic acids, boxes with horizontal lines are vitamins, boxes with diagonal lines are sugars, and white boxes are amino acids or other compounds. (Modified from Leventhal 1986.)

of the mineral is coated with poorly protonated humic molecules that block the release of cations and decrease the dissolution rate.

Solubility in some lab settings can be striking. For example, Baker (1986) performed a series of experi-
ments in which various minerals were held in a 500-ppm solution of humic acids extracted from bracken fern. Because of the possibility that CO_2 derived from organic acids (rather than the acids themselves) affects mineral solubility in nature, Baker also performed dissolution

TABLE 7.1. Dissolution of Selected Minerals
after 24 Hours at 25°C

Mineral	Ion	Atomspheric H_2O/CO_2	Humic Acid	Humic Acid (H_2O/CO_2)
Calcite	Ca	0.65	17.40	2.75
Dolomite	Ca	0.25	7.50	0.45
Dolomite	Mg	0.14	5.65	0.33
Hematite	Fe	0.06	0.58	
Pyrite	Fe	0.04	0.24	
Galena	Pb	0.06	2.62	0.24
Gold	Au	10 ng/mL	190 ng/mL	

All solution concentrations are in mg L^{-1} except as noted.
Based on data selected from Baker (1986).

experiments in water charged with enough CO_2 to simulate the effect of converting all of the organic matter in the humic acid solution to carbon dioxide. Some of his results are shown in table 7.1. The data indicate that organic acids can greatly enhance the dissolution process, especially when reactions take place in an open, flow-through system. Unfortunately, stability constants for most relevant organic complexes are not known, so measurements of this sort cannot generally be used to constrain thermodynamic models of chemical weathering. Also, in general, silicate dissolution rates derived from experiments are much faster than those observed in natural settings. We conclude that the quantitative effect of organic acids is highly variable, but generally nowhere near as great as the effect of other weathering agents.

OXIDATION-REDUCTION PROCESSES

Thermodynamic Conventions for Redox Systems

In previous sections and chapters, we confined our discussion to reactions that involve atoms with only one common oxidation state. Solubility equilibria in these systems are characterized by either hydrolysis or acid-base reactions. Most natural systems, however, also contain transition metal elements, which may form ions in two or more different oxidation states.

Reactions in which the oxidation states of component atoms change are called *redox reactions*. These are not restricted to reactions involving transition metals, although it is most often those systems that are of interest.

In many cases, redox reactions involve the actual exchange of oxygen between phases. For example:

$$2Fe_3O_4 + \tfrac{1}{2}O_2 \rightleftarrows 3Fe_2O_3.$$

In a great many cases, however, oxygen is not a reactant in redox systems. In fact, it is misleading to emphasize the role of oxygen, because it is the change in the transition metal ion that characterizes these systems. It is more practical in many cases, therefore, to write redox reactions to emphasize the exchange of *electrons* between reactant and product assemblages. The example above can be rewritten as:

$$2Fe_3O_4 + H_2O \rightarrow 3Fe_2O_3 + 2H^+ + 2e^-,$$

in which the fundamental change involves the oxidation of ferrous iron:

$$Fe^{2+} \rightarrow Fe^{3+} + e^-.$$

Geochemists may write reactions according to either convention. At low temperatures or when a fluid phase is present, it is usually convenient to write redox reactions in terms of electron transfer. As we shall see, reaction potentials in aqueous systems are easily measured as voltages. At high temperatures, where these measurements are much more difficult to perform, it is generally easier to determine oxygen fugacity and write redox equilibria in terms of oxygen exchange. In this chapter, we devote most of our attention to aqueous systems.

A reaction written to emphasize the production of free electrons, although conceptually useful, cannot take place in isolation in nature, because an electron cannot exist as a free species. (To emphasize this important distinction, we are using a single arrow [$\rightarrow$] to indicate reactions of this type, instead of the double arrows [$\rightleftarrows$] used for all other reactions in this book.) Such a reaction should be viewed in tandem with another similar reaction that consumes electrons. Together, the pair of *half-reactions* represents the exchange of electrons between atoms in one phase, which become oxidized by loss of electrons, and those in another, which gain electrons and are therefore reduced. A companion to the half-reaction above, for example, might be:

$$\tfrac{1}{2}H_2 \rightarrow H^+ + e^-.$$

A natural reaction involving half-reactions but resulting in no free electrons is:

$$2Fe_3O_4 + H_2O \rightarrow 3Fe_2O_3 + H_2.$$

By writing these illustrative reactions, we introduce two standard conventions. First, geochemists usually write half-reactions with free electrons on the right side;

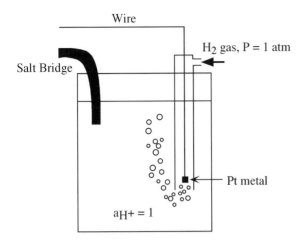

FIG. 7.13. Schematic diagram of the standard hydrogen electrode (SHE). The half-cell reaction is $\frac{1}{2}H_2 \rightarrow H^+ + e^-$. Platinum metal serves as a catalyst on the electrode. The salt bridge is a tube filled with a KCl solution and closed with a porous plug.

that is, they are written as *oxidation* reactions. It should be clear that the progress of a pair of half-reactions depends on their relative free energies of reaction, so that there is no special significance in this convention. Fe_3O_4 may actually be oxidized to form Fe_2O_3 if it is coupled with a half-reaction with a higher free energy of reaction, but may be reversed if its companion half-reaction has a lower $\Delta \bar{G}_r$.

The second standard convention involves the oxidation of H_2 to form H^+, which we used above. It is arbitrarily chosen to serve as an analytical reference for all other half-reactions. To explain this standard and its value in solution chemistry, we have drawn a schematic illustration (fig. 7.13) of a lab apparatus known as the *standard hydrogen electrode* or *SHE*. It consists of a piece of platinum metal immersed in a 25°C solution with pH = 1, through which H_2 gas is bubbled at 1 atm pressure. The platinum serves only as a reaction catalyst and a means of making electrical contact with the world outside the apparatus. If the SHE had a means of exchanging electrons with some external system, it would be the physical embodiment of the half-reaction:

$$\frac{1}{2}H_2 \rightarrow H^+ + e^-.$$

By agreement, chemists declare that the $\Delta \bar{G}_f$ of H^+ and of e^- is zero. Hence, the SHE has its free energy of reaction defined arbitrarily as zero. Because $a_{H^+} = 1$ in

the SHE, $a_{e^-} = 1$ as well. The "activity of electrons" is a slippery concept, because it is a measure of a quantity that has no physical existence. In this sense, a_{e^-} is not equivalent to ion activities we have considered before. Instead, it is best to treat it as a measure of the solution's tendency to release or attract electrons from a companion half-reaction cell. By agreeing that there is no electrical potential between the platinum electrode and the solution, we have created a laboratory device against which the reaction potential of any other half-reaction cell can be measured.

Let's see how this works. Figure 7.14 illustrates an experiment in which the SHE is paired with another system containing both Fe^{2+} and Fe^{3+} in solution. The platinum electrode of the SHE is connected by a wire with a similar electrode in the $Fe^{2+}|Fe^{3+}$ cell, and electrical contact between the solutions is maintained through a KCl solution in a U-tube between them. When the circuit is completed, a current will flow, as electrons are transferred from the solution with the higher activity of electrons to the one with the lower activity. If a voltage meter is placed in the line, it will measure the difference in electrical potential (E) between the SHE and the $Fe^{2+}|Fe^{3+}$ cell.

In this example and most others of interest to aqueous geochemists, the voltage measurement (E) is given the special symbol E_h, where the h informs us that the reported value is relative to the standard hydrogen electrode. The implication of the last few paragraphs is that chemical potential and electrical potential are interrelated, so that it is fair to speak either of the E_h of a half-cell or of its free energy. To be explicit, recall from chapter 3 that the decrease in free energy of a system undergoing a reversible change at constant temperature and pressure is equal to the nonmechanical (i.e., non-pressure–volume) work that can be done by the system. That is, $dG = -dw_{rev}$. The reversible work done by an electrochemical half-cell is equal to the product of its electrochemical potential (E_h) and the current (Q) flowing from it. Q, in turn, is simply equal to the product of Faraday's constant ($F = 23.062$ kcal volt^{-1} eq^{-1}) and the number of moles of electrons released (n):

$$Q = -nF.$$

The negative sign appears because the charge on an electron is negative. This gives us, then:

$$\Delta G = -\Delta w = -E_hQ = nFE_h. \tag{7.7}$$

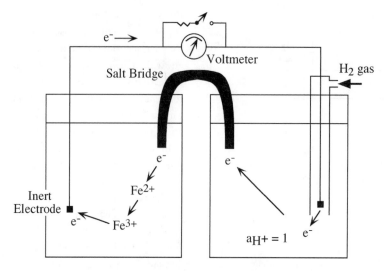

FIG. 7.14. A pair of half-cells described by the SHE and the reaction $Fe^{2+} \rightarrow Fe^{3+} + e^-$. When the switch is closed, electrons can be transferred through the salt bridge and wires between the two cells. When it is opened, the voltage meter measures the potential difference between them. The direction of current flow depends on the Fe^{2+}/Fe^{3+} ratio in the left hand cell.

For the two half-cells in this example, we can write equilibrium constants:

$$K_{eq} = a_{Fe^{3+}} a_{e^-}/a_{Fe^{2+}},$$

and

$$K_{SHE} = a_{H^+} a_{e^-}/P_{H_2}.$$

We calculate the value of K_{eq} in now familiar fashion from tabulated $\Delta \bar{G}_f$ values for Fe^{2+} and Fe^{3+}, and we see that a_{e^-} is proportional to the ratio $a_{Fe^{2+}}/a_{Fe^{3+}}$. For the net reaction,

$$Fe^{2+} + H^+ \rightleftarrows Fe^{3+} + \tfrac{1}{2}H_2,$$

the free energy of reaction is given by:

$$\Delta \bar{G}_r = \Delta \bar{G}_r^0 + RT \ln(K_{eq}/K_{SHE}),$$

which is simply:

$$\Delta \bar{G}_r = \Delta \bar{G}_r^0 + RT \ln(a_{Fe^{3+}}/a_{Fe^{2+}}).$$

From equation 7.7, we see that this result is equivalent to:

$$E_h = E^0 + \left(\frac{RT}{nF}\right) \ln(a_{Fe^{3+}}/a_{Fe^{2+}}),$$

or

$$E_h = E^0 + \left(\frac{2.303RT}{nF}\right) \log_{10}(a_{Fe^{3+}}/a_{Fe^{2+}}).$$

The quantity E^0, called the *standard electrode potential*, is analogous to the standard free energy of reaction. It is the E_h that the system would have if all of the chemical species in both half-cells were in their standard states (that is, if each had unit activity). The logarithmic term is simply the ratio of the activity product of oxidized species (here, $a_{Fe^{3+}}$) to the activity product of reduced species (here, $a_{Fe^{2+}}$). This equation, in its general form, is called the *Nernst equation*. (We have used the notation $\log_{10}$ for base 10 logarithms sparingly elsewhere in this book. We use it for the rest of this chapter, however, to call attention to the numerical conversion factor between base 10 and natural logarithms in most equations.)

From this general discussion, it should now be clear that the thermodynamic treatment of redox reactions in terms of E_h is fully compatible with the procedure we have observed beginning with chapter 4. In the same way that the direction of a net reaction depends on whether the products or the reactants have the lower free energy, a half-cell reaction may be viewed as "oxidizing" or "reducing" relative to the SHE, depending on the arithmetic sign on E_h. The advantage of using E_h rather than $\Delta \tilde{G}_r$ is that electrochemical measurements are relatively easier to perform than is calorimetry.

Worked Problem 7.3

Iron and manganese are commonly carried together in river waters, but are separated quite efficiently upon entering the ocean. Ferric oxides and hydroxides precipitate in near-shore environments or estuaries, but manganese remains as soluble Mn^{2+} even in the open ocean, except where it is finally bound in nodules on the abyssal plain. Some of this behavior is due to colloid formation. Can we also describe it in terms of electrochemistry?

Two half-reactions are of interest in this problem. Iron may be converted from fairly soluble Fe^{2+} to relatively insoluble Fe^{3+} by the half-reaction:

$$Fe^{2+} \rightarrow Fe^{3+} + e^-,$$

for which $E^0 = +0.77$ volt. The manganese half-reaction is:

$$Mn^{2+} + 2H_2O \rightarrow MnO_2 + 4H^+ + 2e^-,$$

for which $E^0 = +1.23$ volt. We construct the net reaction by doubling the stoichiometric coefficients in the $Fe^{2+}|Fe^{3+}$ reaction and subtracting it from the manganese reaction to eliminate free electrons:

$$MnO_2 + 4H^+ + 2Fe^{2+} \rightleftarrows Mn^{2+} + 2H_2O + 2Fe^{3+}.$$

We do not calculate the standard electrode potential for the net reaction in the way you might expect. Unlike enthalpies and free energies, voltages are not multiplied by the coefficients we used to generate the net reaction. The reason is apparent from equation 7.7, from which we see indirectly that:

$$E^0 = \frac{\Delta G_r^0}{nF}.$$

As the stoichiometric coefficients of the $Fe^{2+}|Fe^{3+}$ reaction are doubled, the value of $\Delta \tilde{G}_r^0$ and the number of electrons in the reaction, n, also double. The half-cell E^0, therefore, is unaffected. The standard potential for the net reaction is $+0.77 - (+1.23) = -0.46$ volt. This corresponds to a net $\Delta \tilde{G}_r^0$ of -21.2 kcal mol^{-1}. E_h for this problem is calculated from:

$$E_h = -0.46 + \frac{2.303R(298)}{2F} \log_{10} \frac{a_{Mn^{2+}} a_{Fe^{3+}}^2}{a_{H^+}^4 a_{Fe^{2+}}^2}$$

$$= -0.46 + \frac{0.059}{2} \log_{10} \frac{a_{Mn^{2+}} a_{Fe^{3+}}^2}{a_{H^+}^4 a_{Fe^{2+}}^2}.$$

Sundby and coworkers (1981) found that waters entering the St. Lawrence estuary have a dissolved manganese content of

10 μg/L^{-1}, or approximately 1.8×10^{-7} mol L^{-1}. On average, the ratio of total iron to manganese in the world's rivers is about 50:1, approximately the ratio in average crustal rocks. We estimate, therefore, that $a_{Fe^{2+}} + a_{Fe^{3+}} = 9 \times 10^{-6}$. Unfortunately, river concentrations of Fe^{3+} are rarely reported, so we must choose a value for the activity ratio of Fe^{3+} to Fe^{2+} arbitrarily. Divalent iron is almost certainly the more abundant species but, for reasons that soon become apparent, $a_{Fe^{3+}}$ cannot be much below 1×10^{-6}. For the purposes of this problem, we assume that $a_{Fe^{3+}}/a_{Fe^{2+}}$ is 1:8, so that:

$$a_{Fe^{3+}} = 1.0 \times 10^{-6},$$

and

$$a_{Fe^{2+}} = 8.0 \times 10^{-6}.$$

We also assume that the pH of water entering the St. Lawrence system is 6.0, which is typical of streams draining forest-covered terrain in temperate climates.

By making these assumptions and substituting appropriate values for R and F in our earlier expression for E_h, we calculate that:

$$E_h = -0.46 + \frac{0.059}{2} \log_{10} \frac{1.8 \times 10^{-7} (1.0 \times 10^{-6})^2}{a_{H^+}^4 (8.0 \times 10^{-6})^2}.$$

At pH = 6, the value of E_h predicted by this expression is very close to zero. This indicates, therefore, that the net reaction we have written is very nearly in equilibrium. However, ferric ions cannot accumulate by much before the solubility product for a number of ferric oxides and hydroxides is exceeded. As ferric iron is lost to sediments, the bulk reaction favors gradual oxidation of Fe^{2+} and thus a reduction in the total amount of dissolved iron in the estuary. The same reaction favors the production of Mn^{2+} ions, which remain in solution and are washed out to sea.

E_h-pH Diagrams

As we have shown in the previous section, many redox reactions are functions of both E_h and pH. The net reaction in worked problem 7.3, for example, consumes four hydrogen ions and involves the transfer of two electrons. It can be easily shown that reactions in many systems of interest to aqueous geochemists can be illustrated in diagrams of the type we have already seen in figures 7.5 and 7.8, but using E_h and pH as variables instead of the activities of H_4SiO_4 and aqueous cations. We show this by discussing figure 7.15, which represents a number of equilibria among iron oxides and water.

The uppermost and lowest lines in figure 7.15 establish the stability limits of water in an atmosphere with a

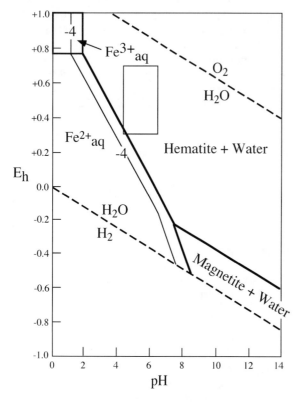

FIG. 7.15. Stability fields for hematite, magnetite, and aqueous ions in water at 25°C and 1 atm total pressure. Boundaries involving aqueous species are calculated by assuming that the total activity of dissolved species is 10^{-6}. Lighter-weight lines indicate the field boundaries when the sum of dissolved species activities is 10^{-4}. The boxed area near pH 4–6 indicates the range of rainwater compositions in many parts of the world.

total pressure of 1 atm (that is, at the surface of the Earth). These limits are derived from the net reaction:

$$2H_2O \rightleftarrows 2H_2 + O_2,$$

which is the sum of the half-reaction:

$$2H_2O \rightarrow O_2 + 4H^+ + 4e^-,$$

and the reaction that defines the SHE. E_h for this reaction at 25°C, therefore, is given by:

$$E_h = E^0 + \frac{2.303RT}{4F} \log_{10} \frac{K_q}{K_{SHE}}$$

$$= E^0 + \frac{0.059}{4} \log_{10} \frac{P_{O_2}^4 a_{H^+}}{a_{H_2O}}.$$

The activity of water, as a pure phase, is unity. The partial pressure of O_2 in the atmosphere at sea level is 0.2 atm. We make only a small difference in the final result if we choose instead to let $P_{O_2} = 1$ atm to simplify the expression for E_h:

$$E_h = E^0 + (0.059/4)\log_{10} a_{H^+}^4$$

$$= E^0 - 0.059 \text{ pH}.$$

This is the equation for a straight line with slope -0.059 and y intercept (E^0) equal to +1.23 volts, which is a graphical representation of the half-reaction producing O_2 from H_2O. As shown in figure 7.15, it divides the E_h-pH plane into a region in which water is stable and one in which O_2 is.

The lower stability limit for water is derived from the SHE half-reaction itself, because it is that half of the net reaction that defines the oxidation of H_2. The appropriate expression is:

$$E_h = E^0 + (0.059/2)\log_{10} a_{H^+}^2.$$

Because E^0 for the SHE is defined to be zero, and P_{H_2} cannot be higher than the total atmospheric pressure of 1 atm, this expression simplifies to:

$$E_h = -0.059 \text{ pH}.$$

Again, this is a line in the E_h-pH plane (fig. 7. 15) representing the SHE graphically; in this case, it defines a region in which H_2 is stable and another in which H^+ (and therefore H_2O) is stable. The net reaction, therefore, is represented in the E_h-pH plane as a region between the half-reactions, in which H_2O is stable. As a point of reference within this large region, it may be convenient to focus on the shaded box around $E_h = +0.5$ and pH = 5.5, which is typical of rainwater in many parts of the world and thus a reasonable starting point for many weathering reactions.

The stability fields of iron oxides are also bounded by half-reactions. The reaction forming hematite by oxidation of magnetite, for example, can be written:

$$2Fe_3O_4 + \tfrac{1}{2}O_2 \rightleftarrows 3Fe_2O_3,$$

or can be written in terms of electron exchange by combining this equation with the half-reaction:

$$2H_2O \rightarrow O_2 + 4H^+ + 4e^-.$$

We saw the result earlier as:

$$2Fe_3O_4 + H_2O \rightarrow 3Fe_2O_3 + 2H^+ + 2e^-.$$

Following the procedure we just used for the stability of water, we find that:

$$E_h = E^0 + (0.059/2) - \log_{10}(a^3_{Fe_2O_3}a^2_{H^+}/[a^2_{Fe_3O_4}a_{H_2O}])$$

$$= E^0 + (0.059/2)\log_{10}a^2_{H^+}$$

$$= E^0 - 0.059 \text{ pH}.$$

The value of E^0, calculated from standard free energies, is +0.221 volt.

The remaining half-reactions shown in figure 7.15 do not have a slope of -0.059, because they all involve water as a reactant or product. Consider, for example, the half-reaction forming hematite from dissolved Fe^{2+}:

$$2Fe^{2+} + 3H_2O \rightarrow Fe_2O_3 + 6H^+ + 2e^-.$$

Here the equilibrium constant involves not only hydrogen ions and electrons, but ferrous ions as well, so that:

$$E_h = E^0 + (0.059/2)\log_{10}(a^6_{H^+}/a^2_{Fe^{2+}})$$

$$= E^0 - 0.059\log_{10}a_{Fe^{2+}} - 0.177 \text{ pH},$$

where $E^0 = +0.728$ volt. To draw a line to represent this reaction in the E_h-pH plane, therefore, we need to specify an activity of ferrous ions. In drawing the Fe^{2+}-hematite boundary in figure 7.15, we have assumed that $a_{Fe^{2+}} = 10^{-6}$, and have indicated by use of a lighter line the way that the boundary would shift if $a_{Fe^{2+}}$ were 10^{-4}. The boundary between Fe^{2+} and magnetite can be calculated in the same way.

The line between the aqueous Fe^{3+} and Fe^{2+} fields indicates the conditions under which activities of the two species are identical. The redox reaction is:

$$Fe^{2+} \rightarrow Fe^{3+} + e^-,$$

which involves no hydrogen ions and is therefore independent of pH. We see easily that:

$$E_h = E^0 + 0.059\log_{10}(a_{Fe^{3+}}/a_{Fe^{2+}}).$$

Because $a_{Fe^{2+}} = a_{Fe^{3+}}$, however, this simply reduces to $E_h = E^0$, which in this case is +0.771 volt.

Finally, the boundary between aqueous Fe^{3+} and hematite is defined by the net reaction

$$Fe_2O_3 + 6H^+ \rightleftarrows 2Fe^{3+} + 3H_2O.$$

We could write this as the sum of two half-reactions, but we can save ourselves a lot of trouble by noticing that this net reaction does not involve a change of oxidation state for iron. The field boundary, therefore, must be a

horizontal line, independent of E_h. The standard molar free energy of reaction, $\Delta\bar{G}^0_r$, is equal to 2 kcal mol^{-1}, so that:

$$\log_{10}K_{eq} = -\Delta\bar{G}^0_r/(2.303RT) = \log_{10}(a^2_{Fe^{3+}}/a^6_{H^+})$$

$$= -1.45.$$

Rearrange this result to solve for pH, and we conclude that:

$$\text{pH} = -0.24 - \tfrac{1}{3}\log_{10}a_{Fe^{3+}},$$

which plots as a vertical line with a position that depends, again, on the activity of dissolved iron.

Redox Systems Containing Carbon Dioxide

Redox reactions in weathering environments always occur in an atmosphere whose P_{CO_2} is at least as high as the atmospheric value and, as we saw earlier in this chapter, may be several hundred times greater. Under some conditions, therefore, we might expect to find siderite ($FeCO_3$) stable in the weathered assemblage. Including carbonate equilibria in the redox calculations we have just considered is not difficult. It merely requires us to write half-reactions to include the new phases. For the equilibrium between siderite and magnetite, we write:

$$3FeCO_3 + H_2O \rightarrow Fe_3O_4 + 3CO_2 + 2H^+ + 2e^-,$$

and

$$E_h = 0.319 + 0.089\log_{10}P_{CO_2} - 0.059 \text{ pH}.$$

For the equilibrium between siderite and hematite:

$$2FeCO_3 + H_2O \rightarrow Fe_2O_3 + 2CO_2 + 2H^+ + 2e^-,$$

and

$$E_h = 0.286 + 0.059\log_{10}P_{CO_2} - 0.059 \text{ pH}.$$

Worked Problem 7.4

What values of E_h, pH, and P_{CO_2} define the stability limits for siderite formed in weathering environments? The answer is shown in graphical form by the shaded area in figure 7.17. To see how it is derived, take a look at the two equations for E_h above. Notice that each of these would plot as a diagonal line with slope of -0.059 on an E_h-pH diagram if we were to choose a fixed value of P_{CO_2}. For successively higher values of P_{CO_2}, the boundary lines would be drawn through higher values of E_h at any selected pH.

HOW DO WE MEASURE pH?

Electrodes are commercially available in a variety of physical configurations, and can be designed for sensitivity to a particular ion or group of ions, to the exclusion of others. Regardless of the details of design, however, all rely on ion exchange to do their job. The transfer of ions between the electrode and a test solution results in a current, which can then be registered in a metering circuit.

One standard design is known as the *glass-membrane electrode;* the basic operating principles of this design are illustrated in figure 7.16. The device consists of a thin, hollow bulb of glass, filled with a solution with a known activity of a cation to be measured. The glass of the bulb contains the same cation, introduced

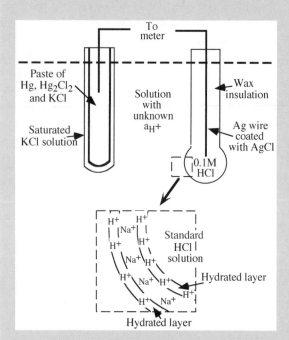

FIG. 7.16. A standard glass-membrane pH electrode (on the right) consists of a Na-glass bulb filled with 0.1 M HCl, at the tip of a Ag-AgCl reference electrode. Inset shows the ion exchange path between the electrode and a solution with unknown pH. This electrode must be paired with a second one, which serves as a reference half-cell to complete the electrical circuit. The one shown on the left in this figure is a calomel electrode, and is based on the redox pair Hg⁰|Hg⁺.

during manufacture. A wire immersed in the filling solution completes the electrode.

In operation, the bulb is placed in a solution to be analyzed. Because the solution inside the bulb and the solution outside generally have different activities of the cation to be measured, a chemical potential gradient is immediately established between them. Ions on the more concentrated side minimize the difference by attaching themselves to the glass, whereas those on the other side leave the glass to join the less concentrated solution. The glass thus serves as a semipermeable membrane. Because charged particles are passing from one side to the other, an electrical imbalance is created. By connecting the wire inside the bulb to an inert or reference electrode immersed in the test solution, we allow electrons to flow and neutralize the charge difference. A voltage measured in this system is the electrode potential.

A standard pH electrode, found in almost any chemistry lab, is a glass electrode of this type, filled with an HCl solution of known a_{H^+}. The outer bulb is made of a Na-Ca-silicate glass, from whose surfaces sodium ions are readily leached and replaced by H^+. When it is placed in a test solution, the glass gains extra hydrogen ions on one side and yields them on the other. Instead of a simple wire, its internal electrical connection is actually a second electrode, which consists of a silver wire coated with silver chloride. This serves as an internal reference electrode. The charge imbalance that develops within the filling solution in the glass bulb is resolved by the silver wire in the inner electrode, which reacts with the filling solution to release or gain an electron in the half-reaction:

$$Ag + Cl^- \rightarrow AgCl + e^-.$$

The charge imbalance in the test solution is ultimately resolved by transferring electrons between the silver wire and a reference electrode (such as the SHE or a more portable equivalent) that is also immersed in the test solution. In this rather roundabout way, the a_{H^+} difference between the test solution and the filling solution is expressed as a measurable voltage.

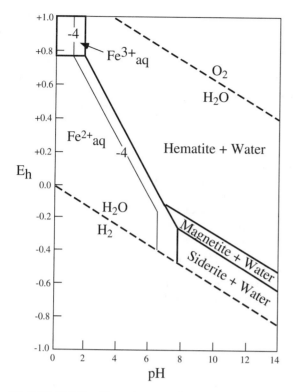

FIG. 7.17. Stability of hematite, magnetite, siderite, and aqueous species at 25°C and 1 atm total pressure. The limits of the siderite field (shaded) are drawn to show its probable maximum stability in surficial weathering environments. Boundaries involving aqueous species have been drawn by assuming that the total activity of dissolved species is 10^{-6}.

To form siderite in a weathering environment by either of these reactions, water must be present. Consequently, siderite's lower stability limit is imposed by the lower stability limit of water, which we found to be:

$$E_h = 0.0 - 0.059 \text{ pH}.$$

The siderite-hematite line coincides with this limiting condition when both lines have the same y intercept; that is, when $0.0 = 0.286 + 0.059 \log_{10} P_{CO_2}$, which occurs when $P_{CO_2} = 1.4 \times 10^{-1}$ atm. The siderite-magnetite line, however, coincides with the breakdown of water when $0.0 = 0.319 + 0.089 \log_{10} P_{CO_2}$, which is true when P_{CO_2} is almost 20 times higher, at 2.6×10^{-4} atm. We conclude, therefore, that if the partial pressure of CO_2 in a weathering environment $< 2.6 \times 10^{-4}$ atm, no siderite is formed. The stable phase, instead, would be magnetite.

The upper stability limit of siderite produced during weathering is set by the natural availability of CO_2. An absolute upper limit on P_{CO_2} is 1 atm, which we also adopted earlier as the maximum possible value for either P_{O_2} or P_{H_2}. A more realistic limit, perhaps, is 10^{-1} atm, measured by Russell (1961) in a tropical soil. At this P_{CO_2} value, the siderite-hematite boundary is given by:

$$E_h = 0.227 - 0.059 \text{ pH},$$

which is the upper edge of the shaded siderite field in figure 7.17.

In very alkaline solutions, siderite is stable over the range of E_h and P_{CO_2} values we have just found. In more acidic solutions, however, siderite breaks down according to the reaction:

$$FeCO_3 + 2H^+ \rightleftarrows Fe^{2+} + CO_2 + H_2O.$$

No free electrons are involved, so this boundary is a vertical line. For the maximum P_{CO_2} we chose above, assuming that $a_{Fe^{2+}} = 10^{-6}$, this line lies at pH = 6.79. This condition puts the siderite-Fe^{2+} boundary at almost the same place as the magnetite-Fe^{2+} boundary shown in figure 7.16.

We conclude from these various constraining equations that siderite does not form easily on exposed rock and soil, where P_{CO_2} is at the atmospheric value of 3.2×10^{-4} atm. In more CO_2-rich environments such as tropical soils, however, siderite may form under roughly the same low E_h conditions in which magnetite is normally stable. These might exist, for example, where decay of organic matter creates a local reducing environment. If we drew figure 7.17 at P_{CO_2} values between 2.4×10^{-4} atm and 1×10^{-1} atm and assembled the drawings in an animated sequence, we would see the siderite field grow at the expense of the magnetite field, squeezing it totally out of existence when P_{CO_2} reached $\sim 8 \times 10^{-2}$ atm.

Activity-Activity Relationships: The Broader View

Instead of thinking of figure 7.17 as a single frame in an animated sequence, as suggested in worked problem 17.4, you might think of it as a slice out of a three-dimensional diagram that recognizes P_{CO_2} as a variable in addition to E_h and pH. Similar diagrams for systems containing sulfur species or phosphate or nitrate are also common in the geochemical literature. Two-dimensional sections through any of these can be drawn and interpreted by analogy with the E_h-pH diagram for the system iron-water. Depending on how you slice the multivariable figure, your finished product might be a P_{CO_2}-pH diagram, a P_{CO_2}-P_{S_2} diagram, an E_h-P_{S_2} diagram, or any number of other possibilities. It is worth noting, in fact, that all of the redox equations we have discussed are similar in form to the solubility expressions we considered in the opening sections of this chapter, and therefore lead to similar graphical representations. You may have been struck with the similar appearance of E_h-pH diagrams and earlier figures such as figure 7.5. In each case, the stabilities of minerals or fluids have been expressed in terms of relative activities of key species. Generically, these are called *activity-activity* diagrams. We will introduce a number of such diagrams in later chapters.

We want to emphasize that activity-activity diagrams are not the exclusive property of geochemists who study chemical weathering, even though we have presented them in that context. In fact, figure 7.5 first gained broad use among geochemists through studies of wall-rock alteration by hydrothermal fluids (Hemley and Jones 1964; Meyer and Hemley 1967), and is in wide use by economic geologists. E_h-pH diagrams also are applied to a variety of problems in mining and mineral extraction. Consider the following problem as an example.

Worked Problem 7.5

Much of the world's mineable uranium occurs as UO_2 (uraninite) in sandstone-hosted deposits, in association with both vanadium and copper minerals. What redox conditions are necessary to form such a deposit, and how can that information be used as a guide to mining uranium?

Before we address the question directly, let's review a few features of solution chemistry relevant to the geologic occurrence of uranium. Uranium can exist in several oxidation states, of which U^{4+} and U^{6+} are most important in nature. In highly acidic solutions, U^{4+} can occur as a free ion under reducing con-

ditions or can take part in fluoride complexes. More commonly, however, tetravalent uranium combines with water to form soluble hydroxide complexes. Under reducing conditions, uranium may also be present in solution in its 5+ state as UO_2^+. The tendency for uranium to combine with oxygen is even more pronounced when it is in the hexavalent state. Throughout most of the natural range of pH, U^{6+} forms strong complexes with oxygen-bearing ions such as CO_3^{2-}, HCO_3^-, SO_4^{2-}, PO_4^{3-}, and AsO_4^{3-}, which are present in most oxidized stream and subsurface waters. At 250°C and with a typical groundwater P_{CO_2} of ~10^{-2} atm, the most abundant of these are the uranyl carbonate species, which are stable down to a pH of ~5. Langmuir (1978), however, has shown that phosphate and fluoride complexes can be at least as abundant as the uranyl carbonate species in some localities (see fig. 7.18). Below pH = 5, U^{6+} is generally in the form of UO_2^{2+}. In addition to these inorganic species, a large number of poorly understood organic chelates also contribute to uranium solubility.

In most near-surface environments, therefore, uranium is easily transported in natural waters. Very little uranium can be deposited in the 6+ state. In some localities where P_{CO_2} is probably close to the atmospheric value, uranyl complexes combine with vanadate to precipitate $K_2(UO_2)_2(VO_4)_2$ (carnotite), or $Ca(UO_2)_2(VO_4)_2$ (tyuyamunite), or with phosphate or silica to form autunite or uranophane. These constitute only a small fraction of mineable uranium, however. Most uranium is deposited as U^{4+} in the major ore minerals uraninite, UO_2 and coffinite, $USiO_4$.

Field observations that form the basis for this brief overview are consistent with figure 7.19, in which we have calculated the stability fields of abundant species in terms of E_h and pH from experimentally determined stability constants. This diagram,

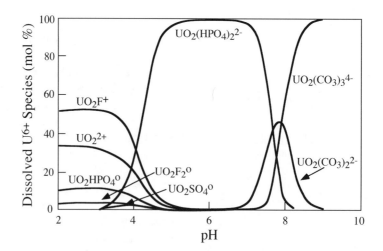

FIG. 7.18. Distribution of uranyl complexes versus pH for some typical ligand concentrations in groundwaters of the Wind River Formation, Wyoming, at 25°C. $P_{CO_2} = 10^{-2.5}$ atm, $\Sigma F = 0.3$ ppm, $\Sigma Cl = 10$ ppm, $\Sigma SO_4^{2-} = 100$ ppm, $\Sigma PO_4^{3-} = 0.1$ ppm, $\Sigma Si = 30$ ppm. (After Langmuir 1978.)

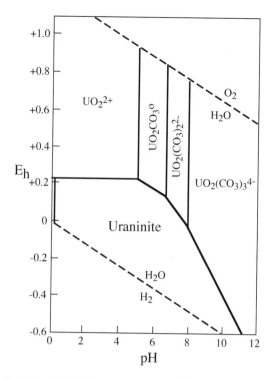

FIG. 7.19. Stability of uraninite and aqueous uranium complexes at 25°C for $P_{CO_2} = 10^{-2}$ atm. Boundaries involving aqueous species are drawn by assuming the total activity of dissolved species is 10^{-6}. (Modified from Langmuir 1978.)

$$4[UO_2(CO_3)_3^{4-}] + HS^- + 15H^+ \rightleftharpoons$$
$$4UO_2 + SO_4^{2-} + 12CO_2 + 8H_2O,$$

or Fe^{2+} may be oxidized to form an insoluble hydroxide mineral by:

$$UO_2(CO_3)_3^{4-} + 2Fe^{2+} + 3H_2O \rightleftharpoons$$
$$UO_2 + 2Fe(OH)_3 + 3CO_2.$$

In either case, a uranyl complex is reduced to yield insoluble UO_2. Another commonly hypothesized reduction mechanism involves the adsorption of uranyl complexes on hydrocarbons, leading to gradual oxidation of the organic matter and precipitation of uraninite.

One mining technique used at several prospects in the southwestern United States involves reversing these precipitation pathways by artificially increasing E_h within an ore body. Typically, a system of wells is drilled into a uranium-bearing sandstone unit. Through some of these, miners inject fluorine-rich acidic solutions with a moderately high E_h. Fluids are then withdrawn by pumping from the remaining wells. The extracted solution contains hexavalent uranium as a uranyl fluoride complex or as UO_2^{2+}. In this way, uranium can be mined efficiently without a significant amount of excavation.

SUMMARY

In this chapter, we explored ways to apply our knowledge of solubility equilibria to problems in chemical weathering. We have shown that, although silica minerals and magnesian silicates dissolve congruently, most major rock-forming minerals decompose to produce secondary minerals. Prominent among these secondary products are clay minerals and various oxides and hydroxides.

We can examine the chemical pathways along which these changes occur by studying either the residual phases or associated waters. To appreciate the driving forces for weathering, however, we need to understand processes that control the abundance of major weathering agents: CO_2, organic acids, and oxygen. We have seen that it is often desirable to express the last of these in redox reactions, written to show the exchange of electrons between oxidized and reduced species.

Finally, we have shown that the concepts and methods used in this chapter can be used for problems other than those in chemical weathering. Economic geologists, in particular, are fond of using E_h-pH diagrams and other activity-activity diagrams to interpret the environments of ore transport and deposition. We explore other types of activity-activity diagrams and their uses in later chapters.

although quantitative, has been constructed to show only those species in the U-O$_2$-CO$_2$-H$_2$O system that are stable at 25°C with $P_{CO_2} = 10^{-2}$ atm, ignoring the role of other species, such as those shown in figure 7.18. Even with this simplification, however, we can deduce qualitative fluid paths that may characterize the formation of sandstone-hosted uranium deposits.

The primary source of sedimentary uranium is probably granitic rocks, which may have as much as 2–15 ppm uranium. Surface waters, as we have found, have pH values that typically range from ~5 to 6.5. E_h values of these waters are usually >0.4 volt. Therefore, during weathering of the igneous source rocks, uranium is readily carried in solution as neutral $UO_2CO_3^0$ or as $UO_2(CO_3)_2^{2-}$. In earlier discussions (for example, in worked problem 7.2), we showed that progressive weathering of aluminosilicate minerals causes an increase in pH. When this happens, $UO_2(CO_3)_3^{4-}$ may also play an important role in solution. Any of these mobile uranyl complexes, which occupy a wide range of moderate to high E_h and pH values in figure 7.19, may be reduced to precipitate uraninite. These often involve the simultaneous oxidation of iron, carbon, or sulfur to create a low E_h environment. For example, Langmuir (1978) suggests that HS^- generated by bacterial processes may be oxidized at pH 8 to SO_4^{2-} by the reaction:

SUGGESTED READINGS

The literature on chemical weathering is voluminous. Many good references are available in the applied fields of soil science, water chemistry, and economic geology. The following list is intended to be representative, not exhaustive.

Barnes, H. L., ed. 1997. *Geochemistry of Hydrothermal Ore Deposits,* 3rd ed. New York: Wiley Interscience. (This massive volume contains basic papers on a variety of geochemical topics of interest to the economic geologist.)

Bowers, T. S., K. J. Jackson, and H. C. Helgeson. 1984. *Equilibrium Activity Diagrams.* New York: Springer-Verlag. (This book contains only eight pages of text, but is the most extensive compilation of activity-activity diagrams available.)

Brookins, D. G. 1988. E_h-*pH Diagrams for Geochemistry.* New York: Springer-Verlag. (This is an excellent, slim book, crammed with E_h-pH diagrams for geologically relevant systems.)

Colman, S. P., and D. P. Dethier, eds. 1986. *Rates of Chemical Weathering of Rocks and Minerals.* New York: Academic. (This is an excellent compendium of papers at the introductory and advanced level on mechanisms and kinetics of chemical weathering.)

Drever, J. I. 1997. *The Geochemistry of Natural Waters,* 3rd ed. New York: Simon and Schuster. (This text provides a good introduction to the geochemistry of natural waters. Chapters 7, 8, and 12 are particularly relevant to topics in this chapter.)

Garrels, R. M., and C. L. Christ. 1965. *Solutions, Minerals, and Equilibria.* New York: Harper and Row. (This book, now out of print, was the standard text in aqueous geochemistry for a generation of geologists. It is still a model of clarity.)

Langmuir, D. 1996. *Aqueous Environmental Chemistry.* Upper Saddle Brook: Prentice-Hall. (A good introductory level text with examples from biochemistry and environmental chemistry as well as geology.)

The following papers were referenced in this chapter. You may wish to consult them for further information.

Amrhein, C., and D. L. Suarez. 1988. The use of a surface complexation model to describe the kinetics of ligand-promoted dissolution of anorthite. *Geochimica et Cosmochimica Acta* 51:2785–2793.

Baker, W. E. 1986. Humic substances and their role in the solubilization and transport of metals. In D. Carlisle, W. L. Berry, I. R. Kaplan, and J. R. Watterson, eds. *Mineral Exploration: Biological Systems and Organic Matter.* Englewood Cliffs: Prentice-Hall, pp. 377–407.

Barshad, I. 1966. The effect of variation in precipitation on the nature of clay mineral formation in soils from acid and basic igneous rocks. *Proceedings of the International Clay Conference (Jerusalem)* 1:167–173.

Bennett, P. C. 1991. Quartz dissolution in organic-rich aqueous systems. *Geochimica et Cosmochimica Acta* 55:1781–1797.

Cawley, J. L., R. C. Burrus, and H. D. Holland. 1969. Chemical weathering in central Iceland: An analog of pre-Silurian weathering. *Science* 165:391–392.

Couturier, Y., G. Michard, and G. Sarazin. 1984. Constantes de formation des complexes hydroxydés de l'aluminum en solution aqueuse de 20 à 70°C. *Geochimica et Cosmochimica Acta* 48:649–660.

DeConick, F. 1980. Major mechanisms in formation of spodic horizons. *Geoderma* 24:101–128.

Helgeson, H. C., J. M. Delany, H. W. Nesbitt, and D. K. Bird. 1978. Summary and critique of the thermodynamic properties of rock-forming minerals. *American Journal of Science* 278A:1–229.

Hemley, J. J., and W. R. Jones. 1964. Chemical aspects of hydrothermal alteration with emphasis on hydrogen metasomatism. *Economic Geology* 59:538–569.

Holland, H. D. 1978. *The Chemistry of the Atmosphere and Oceans.* New York: Wiley Interscience.

Langmuir, D. 1978. Uranium solution-mineral equilibria at low temperatures with applications to sedimentary ore deposits. *Geochimica et Cosmochimica Acta* 42:547–569.

Leventhal, J. S. 1986. Roles of organic matter in ore deposits. In W. E. Dean, ed. *Organics and Ore Deposits.* Wheat Ridge, Colorado: Denver Region Exploration Geologists Society, pp. 7–20.

Li, Y.-H. 2000. *A Compendium of Geochemistry.* Princeton: Princeton University Press.

Meyer, C., and J. J. Hemley. 1967. Wall rock alteration. In H. L. Barnes, ed. *Geochemistry of Hydrothermal Ore Deposits,* 1st ed. New York: Holt, Rinehart, and Winston, pp. 166–235.

Ochs, M., I. Brunner, W. Stumm, and B. Cosovic. 1993. Effects of root exudates and humic substances on weathering kinetics. *Water, Air and Soil Pollution* 681–682:213–229.

Nesbitt, H. W., G. Markovics, and R. C. Price. 1980. Chemical processes affecting alkalis and alkaline earths during continental weathering. *Geochimica et Cosmochimica Acta* 44:1659–1666.

Norton, D. 1974. Chemical mass transfer in the Rio Tanama system, west-central Puerto Rico. *Geochimica et Cosmochimica Acta* 38:267–277.

Ohmoto, H., and R. O. Rye. 1979. Isotopes of sulfur and carbon. In H. L. Barnes, ed. *Geochemistry of Hydrothermal Ore Deposits,* 2nd ed. New York: Holt, Rinehart, and Winston, pp. 509–567.

Petit, J.-C., G. DellaMea, J.-C. Dran, J. Schott, and R. A. Berner. 1987. Mechanism of diopside dissolution from hydrogen depth profiling. *Nature* 325:705–707.

Reesman, A. L., E. E. Pickett, and W. D. Keller. 1969. Aluminum ions in aqueous solutions. *American Journal of Science* 267:99–113.

Russell, E. W. 1961. *Soil Conditions and Plant Growth.* New York: Wiley.

Schnitzer, M. 1986. Reactions of humic substances with metals and minerals. In D. Carlisle, W. L. Berry, I. R. Kaplan, and J. R. Watterson, eds. *Mineral Exploration: Biological Systems and Organic Matter.* Englewood Cliffs: Prentice-Hall, pp. 408–427.

Solomon, D. K., and T. E. Cerling. 1987. The annual carbon dioxide cycle in a montane soil: Observations, modeling, and implications for weathering. *Water Resources Research* 23:2257–2265.

Stillings, L. L., J. I. Drever, S. L. Brantley, Y. T. Sun, and R. Oxburgh. 1996. Rates of feldspar dissolution at pH 3–7 with 0–8 mM oxalic acid. *Chemical Geology* 1321(4): 79–89.

Sundby, B., N. Silverberg, and R. Chesselet. 1981. Pathways of manganese in an open estuarine system. *Geochimica et Cosmochimica Acta* 45:293–307.

Witkamp, M., and M. L. Frank. 1969. Evolution of CO_2 from litter, humus, and subsoil of a pine stand. *Pedobiologia* 9: 358–365.

Wolery, T. J. 1979. *Calculation of Chemical Equilibrium between Aqueous Solution and Minerals: The EQ3/6 Software Package.* NTIS Document UCRL-52658. Livermore: Lawrence Livermore Laboratory.

Wood, W. W., and M. J. Petratis. 1984. Origin and distribution of carbon dioxide in the unsaturated zone of the southern high plains of Texas. *Water Resources Research* 20:1193–1208.

PROBLEMS

(7.1) Using appropriate data from this chapter or literature sources, calculate the value of $a_{H_4SiO_4}$ for which gibbsite and kaolinite are in equilibrium with an aqueous fluid at 25°C.

(7.2) Following the approach used in worked problem 7.2, give a qualitative description of the fluid reaction path for pure water during dissolution of potassium feldspar, assuming that amorphous silica rather than quartz is precipitated. Write all relevant reactions.

(7.3) How much total silica (ΣSi in ppm) is in equilibrium with quartz at pH 9? At pH 10? At pH 11?

(7.4) A typical bottled beer has a H^+ activity of $\sim 2 \times 10^{-5}$. Assuming that it remains in the bottle long enough to reach equilibrium with the glass and that the effect of organic complexes is negligible, what concentrations of H_4SiO_4, $H_3SiO_4^-$, and $H_2SiO_4^{2-}$ should we expect to find in the beer?

(7.5) Assume that two species, A and B, can each exist in either an oxidized or a reduced form. For the redox reaction $A_{ox} + B_{red} \rightleftarrows A_{red} + B_{ox}$, in which only one electron is transferred, what must be the value of E^0 at 25°C if the equilibrium ratios A_{red}/A_{ox} and B_{ox}/B_{red} are found to be 1000:1?

(7.6) What is the oxidation potential (E_h) for water in equilibrium with atmospheric oxygen ($P_{O_2} = 0.21$ atm) if its pH is 6?

(7.7) Using data from Helgeson (1969) or Li (2000), calculate the value of $\log 10_{K_{fo}}$ and plot a line representing the congruent dissolution of forsterite on figure 7.2. Explain why forsterite is metastable with respect to the other magnesian silicates shown.

(7.8) Using data from worked problem 7.1, estimate the total dissolved aluminum content (ΣAl in ppm) of water near the mouth of the Amazon River. Assume that runoff is in equilibrium with kaolinite, ΣSi is 11 ppm, and the pH is 6.3.

(7.9) By writing reactions in the system $Fe-O_2-H_2O$ in terms of transfer of oxygen rather than transfer of electrons, redraw figure 7.15 as an f_{O_2}-pH diagram.

(7.10) E^0 for the reaction $Ce^{3+} \rightarrow Ce^{4+} + e^-$ is -1.61 volt. The ratio Ce^{3+}/Ce^{4+} in surface waters in the ocean has been estimated to be 10^{17}. If the redox half-reaction for cerium is in equilibrium with the half-reaction $Fe^{2+} \rightarrow Fe^{3+} + e^-$ at 25°C, what should be the Fe^{2+}/Fe^{3+} ratio in these waters? How does this value compare with the ratio you would calculate from the reaction $4Fe^{3+} + 2H_2O \rightleftarrows 4Fe^{2+} + 4H^+ + O_2$, assuming that the pH of seawater is 8.2?

(7.11) Write the half-reaction for the reference electrode shown on the left side of figure 7.17. What is its standard potential at 25°C? What is the net reaction for this pH measurement system?

CHAPTER EIGHT

THE OCEANS AND ATMOSPHERE AS A GEOCHEMICAL SYSTEM

OVERVIEW

In this chapter, we introduce three broad problems that have occupied much of the attention of geochemists who deal with the ocean-atmosphere system. The first of these concerns the composition of the oceans and the development of what has been called a "chemical model" for seawater. This is a comprehensive description of dissolved species, constrained by mass and charge balance and the principles of electrolyte solution behavior we discussed in chapter 4.

Closely linked to this problem are questions about the dynamics of chemical cycling between the ocean-atmosphere and the solid Earth. Aside from the geologically rapid processes that control the electrolyte species in seawater, there are large-scale processes that maintain its bulk chemistry by balancing inputs from weathering sources against a variety of sinks. We discussed many of these inputs and outputs in chapters 5 and 7. We consider their global effect here, and add hydrothermal reactions at midocean ridges and the formation of evaporites as examples of this class of problem.

Finally, to assure you that oceanic geochemistry has room for visionaries, we discuss several ideas relating to the history of seawater and air. This third area of study draws on our understanding of electrolyte chemistry and our understanding of cycling processes to help interpret ancient marine environments and events.

COMPOSITION OF THE OCEANS

A Classification of Dissolved Constituents

The oceans provide geochemists with a very large workshop in electrolyte solution chemistry. In one sense, this workshop is boringly uniform: its composition is dominated by six major elements (Na, Mg, Ca, K, Cl, and S). As early as 1819 Alexander Marcet, professor of Chemistry in Geneva, determined that the relative abundances of these six elements are very nearly the same all over the world. This is true even though the total dissolved salt content (the salinity) of the open ocean varies from 33‰ to 38‰. (It is common practice to report salinity measurements in parts per thousand by weight, or *per mil* [‰]. Thus, seawater that has a total dissolved salt content of 3.45 wt % has a salinity of 34.5‰.) Of these six elements, only calcium has been shown to vary in its ratios to the other five by a small but measurable amount. Constituents whose relative concentrations remain constant in seawater are identified as *conservative*. Variations in their absolute abundance can be attributed solely to the addition or subtraction of pure water to the oceans.

137

Most often, geochemists focus their attention on the remaining elements, the minor constituents of seawater. Some of them (such as B, Li, Sb, U, and Br) are also conservative, but many vary dramatically relative to local salinity. The latter dissolved species are, therefore, *nonconservative*. Most of these variations are associated with the oceans' major role as a home for living organisms. Plants live only in near-surface waters, where they can receive enough sunlight to permit photosynthesis. They and the animals that feed on them deplete those waters in nutrient species such as phosphate, nitrate, dissolved silica (ΣSi), and dissolved inorganic carbon ($\Sigma CO_2 = CO_2 + HCO_3^- + CO_3^{2-}$), which are converted to organic and skeletal matter. These materials settle toward the ocean floor in dead organisms or in fecal matter. As they settle, hard body parts made of carbonate redissolve and tissues are rapidly consumed by bacteria, causing a relative enrichment of nutrients in bottom water. Of the elements associated with ocean life, the only significant exception to this pattern is oxygen, which is released by photosynthesis at the surface and consumed by respiration at depth. Since the mid-1970s, geochemists have gradually discovered that a number of trace elements, such as Sr, Cd, Ba, and Cr, also follow nutrient-correlated abundance patterns, although it is not clear in many cases how these elements are involved in the oceans' biochemical mechanisms. From a geochemical viewpoint, these measurable gradients are essential to our ability to trace the pathways and processes of biologic, sedimentary, and circulatory behavior in the oceans.

We have indicated the abundance and behavior of a number of elements in the ocean in table 8.1, dividing them into conservative and nonconservative groups. Dissolved gases (except for those like O_2, CO_2, H_2S, and to some degree N_2 and N_2O, which are involved in metabolic processes) follow abundance patterns that are controlled largely by their solubility. To a good first approximation, concentration in surface waters is fixed by the temperature of the water and the atmospheric partial pressure of the gas. Because atmospheric gases are also trapped in bubbles during stirring by waves, however, their measured abundances are not controlled solely by solubility, but vary by an unpredictable (but usually small) amount. Human activity contributes significantly to the abundance or distribution of a few elements. Lead, injected into the ocean-atmosphere system by automobile emissions, is a good example.

It is apparent from the last column of table 8.1 that organisms play a central role in determining the chemistry of the oceans, in addition to the six major conservative elements already mentioned. Of the elements listed as nutrients or nutrient-related, five are particularly noteworthy. Phosphate, nitrate, silica, Zn, and Cd are very nearly depleted in surface waters, and are often referred to as *biolimiting* constituents of seawater, because they are scavenged so efficiently by microorganisms and, once depleted, place a limit on the biomass of surface waters. Only one of these five is universally biolimiting, however. Phosphorus (as phosphate) alone is required in the metabolic cycles of all organisms, and therefore is the ultimate biolimiting element. The supply of silica limits only the population of organisms (diatoms and radiolarians) that form hard body parts from SiO_2, but not the vast population of those that build skeletons with calcium carbonate. Nitrate is a limiting nutrient for a great many species, but not for the large family of blue-green algae, which can fix nitrogen directly from dissolved N_2. Cadmium and zinc are apparently depleted drastically in surface waters only because plants mistakenly incorporate these elements in place of phosphorus.

One element that is conspicuously absent from the list of biolimiting constituents is carbon, about which we have much more to say as this chapter unfolds. Notice that, although surface waters have a lower concentration of total dissolved carbon species than deep waters, organisms deplete surface waters by only ~10%—not a startlingly large depletion. They are therefore leaving ~90% of the dissolved carbon around them untouched.

One way to appreciate the nonlimiting nature of carbon more fully is to compare its abundance with that of elements that are biolimiting. The atomic ratio of phosphorus to nitrogen to carbon in organic tissue formed in surface waters is very nearly constant at 1:15:105, a proportion known as the *Redfield ratio*, after Alfred C. Redfield, the oceanic geochemist who first determined it. As organic matter settles into the deep oceans, it is almost entirely consumed and returned to solution as dissolved species. It is not surprising, then, that the atomic ratio P:N in deep waters is 1:15. The deep water ratio of P:C, however, is roughly 1:1000. This implies that ~90% of the dissolved carbon in deep water cannot have been carried down in organic matter. Some was carried down as carbonate, but this, too, is a rather small fraction. For every four atoms of carbon fixed in organic matter, ap-

TABLE 8.1. Abundance and Behavior of Selected Elements in Seawater

Element	Surface Concentration	Concentration at Depth	Behavior
Li	178 µg kg^{-1}		Conservative[1]
B	4.4 mg kg^{-1}		Conservative[1]
C	2039 µmol kg^{-1}	2264 µmol kg^{-1} (714 m)	Nutrient
N (N$_2$)		560 µmol kg^{-1} (995 m)	Nonnutrient gas
(NO$_3^-$)	<3×10^{-7} mol kg^{-1}	41 µmol kg^{-1} (891 m)	Nutrient
O (O$_2$)	205 µmol kg^{-1}	47 µmol kg^{-1} (891 m)	Nutrient-related
F	1.3 mg kg^{-1}		Conservative
Na	10.781 g kg^{-1}		Conservative
Mg	1.284 g kg^{-1}		Conservative
Al	1.0 µg kg^{-1}	0.6 µg kg^{-1} (1000 m)	?
Si	2.0 µg kg^{-1}	100.8 µmol kg^{-1} (891 m)	Nutrient
P	0.08 µg kg^{-1}	2.84 µmol kg^{-1} (891 m)	Nutrient
S	2.712 g kg^{-1}		Conservative
Cl	19.353 g kg^{-1}		Conservative
K	399 mg kg^{-1}		Conservative
Ca	417.6 mg kg^{-1}	413.6 mg kg^{-1} (1000 m)	Conservative[2]
Cr	268 ng kg^{-1}	296 ng kg^{-1} (1000 m)	Nutrient-related
Mn	34 ng kg^{-1}	38 ng kg^{-1} (985 m)	Nutrient-related
Fe	8 ng kg^{-1}	45 ng kg^{-1}	Nutrient-related
Ni	146 ng kg^{-1}	566 ng kg^{-1} (985 m)	Nutrient-related
Cu	34 ng kg^{-1}	130 ng kg^{-1} (985 m)	Nutrient-related
Zn	6–7 ng kg^{-1}	438 ng kg^{-1} (985 m)	Nutrient-related
As	1.1 µg kg^{-1}	1.8 µg kg^{-1} (1000 m)	Nutrient-related
Br	67 mg kg^{-1}		Conservative
Rb	124 µg kg^{-1}		Conservative
Sr	7.404 mg kg^{-1}	7.720 mg kg^{-1} (700 m)	Nutrient-related
Cd	0.3 ng kg^{-1}	117 ng kg^{-1} (985 m)	Nutrient-related
Sn	1.0 ng kg^{-1}		Anthropogenic
I	48 µg kg^{-1}	60 µg kg^{-1} (1036 m)	Nutrient-related
Cs	0.3 µg kg^{-1}		Conservative
Ba	4.8 µg kg^{-1}	13.3 µg kg^{-1} (997 m)	Nutrient-related
Hg	3 ng kg^{-1}	4 ng kg^{-1} (1000 m)	Nutrient-related
Pb	13.6 ng kg^{-1}	4.5 ng kg^{-1} (1000 m)	Anthropogenic
U	3.2 µg kg^{-1}		Conservative

Modified from Quimby-Hunt and Turekian (1983).

[1]These values are for illustration only. They are not to be taken as average values, which in most cases are poorly known.

[2]Although these data indicate that typical surface and deep waters have similar calcium contents, calcium is commonly removed from surface waters by biological precipitation of carbonates. The net loss of Ca, however, is a small fraction of the total amount present and is not easily discerned from global values. Therefore, Ca is nearly conservative.

proximately one atom is combined with a calcium ion and precipitated biochemically as carbonate. (This process accounts, qualitatively, for the slightly nonconservative behavior of calcium.) We must conclude that, per phosphorus atom in the ocean, there are roughly 870 more carbon atoms than are necessary to satisfy the growth demands of the biomass. If you were skeptical before, it should now be clear that carbon is by no means biolimiting, despite its central role in the architecture of living matter.

Chemical Variations with Depth

It should be clear that the major compositional gradients in the oceans are vertical, driven by the biologic cycling process we have just outlined. As a way of building a greater appreciation of these gradients, we presented in figure 8.1 several sets of graphical data from the northern Pacific Ocean gathered by the Geochemical Ocean Sections program (GEOSECS) during the 1970s.

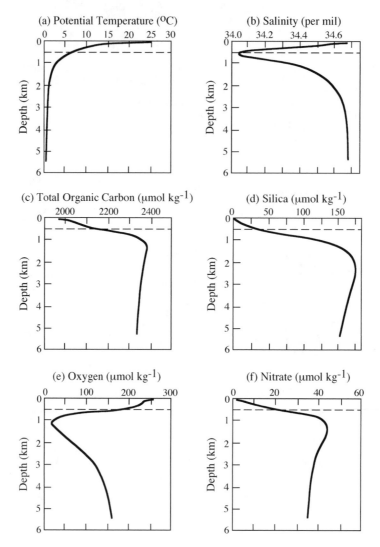

FIG. 8.1. Variations in chemical and physical properties of the ocean with depth at GEOSECS station 214 in the North Pacific Ocean. (a) Potential temperature, (b) salinity, (c) total organic carbon, (d) silica, (e) oxygen, and (f) nitrate. (Modified from Broecker and Peng 1982.)

The first two graphs in the figure, showing vertical gradients in temperature and salinity, are important for understanding oceanic environments, but are controlled by physical rather than geochemical processes. The thermal gradient shown in figure 8.1a is the result of solar heating at the surface. Water is so opaque to visible and infrared radiation that sunlight penetrates only a few tens of meters before it is absorbed almost completely. This opacity and thermal exchange with the atmosphere result in significant warming of the top of the ocean. Mixing by winds and currents causes limited downward

heat transport, so that the water temperature throughout most of the deep ocean is nearly constant at 2°C. (Seawater experiences a minor temperature increase with depth [~0.1°C/km] due to compression, but this is an adiabatic effect. No net internal energy change would result if the water were slowly brought to the surface again. For that reason, water temperatures at depth are generally reported as *potential temperatures*, from which adiabatic heating has been subtracted.) It is convenient, therefore, to describe the ocean in thermal terms as if it consisted of two zones: a warm surface zone, mixed by winds to an average depth of 70 m, and a deep water zone of constant cold temperatures that begins at ~1 km below the surface. The transitional region between these zones is called the *thermocline*. Depending on distance from the equator and the time of year, the thermocline may be very thin and close to the surface or (as in fig. 8.1a) much thicker and deeper. Because warm water is, in general, less dense than cold water, the shape of the temperature profile suggests that the oceans should be stable against thermal convection.

Salinity profiles are more complex than temperature profiles. At most latitudes, solar heating causes enough evaporation from the surface to cause an increase in salinity in waters above the thermocline. Despite their higher salt content, these waters are warm and thus less dense than the less saline waters below. Near the poles, however, evaporation is minimized and the production of sea ice tends to form cold, dense saline water, which descends to the ocean floor. Global circulation draws large amounts of this cold bottom water (largely from the North Atlantic Ocean and the Weddell Sea off of Antarctica) into more temperate latitudes, where it gradually displaces local bottom water and causes a gentle upward circulation. Over large areas of the world, this pattern is further complicated by the lateral transport of cold, low-salinity melt waters (as in the North Pacific, for which the profile in fig. 8.1b was constructed) or warm, high-salinity waters from shallow seas (as in the mid-Atlantic, where Mediterranean waters enter through the Straits of Gibraltar). As a result, the oceans have multiple layers of high and low salinity, created as global circulation injects waters of varying density into the water column.

Although figure 8.1a,b is important to our appreciation of the physical behavior of the oceans, the remaining profiles (fig. 8.1c–f) are more interesting geochemically, because they help to illustrate the effects of the biologic "pumping" mechanism. Total inorganic carbon and silica show depth profiles that are roughly mirror images of the thermal profile. Together, these delineate the zone of photosynthesis and surficial biologic activity. The only difference between them is that silica, being biolimiting, begins at nearly zero concentration at the surface, whereas carbon begins at ~90% of its value in bottom water, as we discussed earlier. Barium is shown in table 8.1 as an example of a nutrient-related, but not biolimiting, constituent. It is not yet clear how most elements like barium are incorporated into organic matter or skeleta, but their association with such constituents as silica (in hard body parts) and carbon (largely in tissues) is quite apparent from their similar depth profiles.

Oxygen and nitrate deserve particular attention, because their profiles demonstrate the influence of both biologic mechanisms and global circulation. Near the surface, the oxygen content of seawater is constrained by its solubility in thermodynamic equilibrium with the atmosphere. In fact, surface waters generally have excess O_2 as the result of local photosynthesis. Just below the thermocline, however, the oxygen level drops precipitously as O_2 is consumed by the oxidation of organic matter raining down from above. Nitrate is released during this process, so its profile is almost a mirror image of that for oxygen. Dissolved oxygen content has not been found to reach zero at any place in the modern oceans, but there are inland seas such as the Black Sea that are anoxic below the thermocline. The sedimentary record indicates that anoxia was more common in pre-Cenozoic oceans.

In polar waters, the story is somewhat different. Because of the low temperatures, organisms do not consume all of the nutrient nitrate and phosphate at the surface. The rain of organic matter toward to bottom is therefore less intense. As a result, the oxygen minimum below the thermocline is less pronounced and the nutrient content of polar bottom water is slightly lower than is observed at the base of the thermocline in warmer latitudes. When this water is transported by the deep global currents discussed earlier, the result is that average bottom water is much more oxygen-rich and somewhat less nutrient-rich than average seawater. This is illustrated in figure 8.1e,f. A profile of phosphate would be qualitatively similar to that for nitrate.

COMPOSITION OF THE ATMOSPHERE

The three major reactive components of the atmosphere—oxygen, water, and carbon dioxide—are exchanged readily between the air and the oceans. Equilibrium or kinetic factors that influence their distribution, therefore, have an important effect on marine chemistry. In chapter 7, we considered how the atmosphere engenders weathering on land and thus provides further controls on the bulk composition of streams and the ocean. We look at some of these controls again shortly.

Although some photochemical reactions that take place in the upper atmosphere may affect geochemical processes on the Earth, we can largely disregard the composition above ~12 km. The lower portion of the atmosphere, known as the *troposphere,* contains the bulk of its mass and is vigorously stirred by convection due to solar heating at the planetary surface. As a result, its composition, presented in table 8.2, is nearly uniform. Water vapor, of course, is a major exception. Because atmospheric P_{H_2O} is, on average, very close to the saturation vapor pressure, relatively small local changes in temperature can make the difference between evaporation and precipitation. The conversion of liquid water to vapor and back again is a major means for energy transport at the Earth's surface.

In addition to the species listed in table 8.2, there are many gases with concentrations <1 ppm. Variations in some of these, such as ozone (O_3), can have a significant effect on organisms, but rarely have a measurable role in controlling geochemical environments.

The noble (also called rare) gases are chemically inert, so they have no tendency to form compounds during chemical weathering or rock-forming processes. To a first approximation, then, they are gradually accumulating in the atmosphere with time. The picture is much more complicated than this, however. Helium, for example, is light enough to escape from the top of the atmosphere in a geologically short time, argon is produced continually in the crust by radioactive decay, and all of the rare gases have been shown to bond weakly to clay surfaces and undergo limited reburial during sedimentation. With allowances for these complexities, geochemists have used rare gas abundances to interpret the origin and evolution of the atmosphere and of the Earth itself, precisely because they are nonreactive. This is a topic of discussion in chapters 14 and 15, and we preview it briefly at the end of this chapter.

Nitrogen is also nearly inert, although a limited amount of N_2 is fixed directly by microorganisms and incorporated by growing plants. Harvard geochemist Heinrich Holland (1978) estimated that of the 4700×10^{12} g of nitrogen used in marine photosynthesis each year, only 85×10^{12} g (<2%) is converted from atmospheric N_2. Most nitrogen in the nutrient cycles of the ocean or continents, therefore, is recycled within the biosphere. The rates of nitrate burial in sediments and nitrogen release by weathering and volcanism are very small and nearly balanced, although estimates are difficult to verify. Additional values from Holland indicate that less than one part of the atmospheric mass of N_2 in 2×10^8 is lost to sedimentation each year. As a result, it is a very good approximation to call the atmospheric mass of nitrogen inert.

The two remaining constituents of interest in this quick survey are oxygen and carbon dioxide, both of which owe their abundance in the atmosphere to the processes of the biosphere. Each year, ~83×10^{15} g of carbon (12% of the atmospheric reservoir) is used in photosynthesis. The process liberates 2.2×10^{17} g of O_2. Estimating the total mass of atmospheric O_2 to be 1.2×10^{21} g, it is easy to calculate that the biosphere could produce that amount of oxygen once every 5500 years. The shorter time in which all CO_2 could theoretically be consumed by photosynthesis (~9 yr) is largely a function of the relative amounts of both gases in the atmosphere. In fact, the biogeochemical cycles for O_2 and CO_2 are very nearly balanced. The yearly mass of CO_2 fixed in organic matter is almost quantitatively oxidized by the yearly mass of "new" O_2 and returned to the atmosphere. Neither gas, in any case, can be regarded nonreactive, in comparison with nitrogen and the noble gases.

Despite the dominance of biologic processes in controlling the fates of oxygen and CO_2, both gases also play a major role in the inorganic cycle of weathering and sedimentation. In a year, ~3.6×10^{14} g of carbon (as CO_2) are consumed in weathering continental rocks. This amounts to <0.5% of the mass of carbon fixed annually in organic matter. The O_2 consumed in all weathering reactions is ~4×10^{14} g yr^{-1}—again, a very small mass compared with the annual amount linked to photosynthesis. These small amounts, however, dominate the weathering process.

TABLE 8.2. The Composition of the Atmosphere at Ground Level

Constituent	Concentration (ppm by volume)
Nitrogen (N_2)	78.084 ($\pm$0.004) $\times 10^4$
Oxygen (O_2)	20.946 ($\pm$0.002) $\times 10^4$
Argon	9340 ($\pm$10)
Water	40 to 4×10^4
Carbon dioxide	320
Neon	18.18 ($\pm$0.04)
Helium	5.24 ($\pm$0.04)
Methane (CH_4)	1.4
Krypton	1.14 ($\pm$0.01)

Modified from Holland (1978).

It is almost true that the biologic cycle for oxygen and carbon dioxide is independent of the weathering-sedimentation cycle. Less than 0.1% of the carbon fixed yearly in organic matter is trapped in sediments rather than immediately reoxidized, and an equivalent amount is weathered out of continental rocks. The constituents of both cycles meet in the oceans, however, and are a major source of its chemical stability. In the next section, we concentrate on the chemistry of marine carbonate species.

CARBONATE AND THE GREAT MARINE BALANCING ACT

Some First Principles

A good way to begin is by considering a simplified "ocean" that consists only of pure water in equilibrium with carbon dioxide. We can then write four major reactions relating species in the system to one another:

$$CO_2 \text{ (gas)} \rightleftarrows CO_2 \text{ (aq)},$$

$$CO_2 \text{ (aq)} + H_2O \rightleftarrows H_2CO_3,$$

$$H_2CO_3 \rightleftarrows HCO_3^- + H^+, \tag{8.1}$$

$$HCO_3^- \rightleftarrows CO_3^{2-} + H^+. \tag{8.2}$$

As we demonstrated in chapter 7, the first two of these reactions can easily be combined into one. Because the hydration of CO_2 is more than two orders of magnitude faster than the formation of H_2CO_3 from CO_2(aq), the combined reaction follows nearly a first-order kinetic expression. It is customary, therefore, to use the expression:

$$CO_2 \text{ (gas)} + H_2O \rightleftarrows H_2CO_3 \tag{8.3}$$

to describe the hydration of CO_2 and to speak of all CO_2(aq) as H_2CO_3. No thermodynamic validity is lost by this simplification, because the equilibrium constant for the overall reaction is simply the product of the constants for the two elementary reactions. The set of thermodynamic equations to describe this system, therefore, consists of a solubility expression for CO_2 and two dissociation reactions. At 25°C, the equilibrium constants for these reactions have the values:

$$K_1 = a_{HCO_3^-} a_{H^+} / a_{H_2CO_3} = 10^{-6.35}$$

(corresponding to equation 8.1),

$$K_2 = a_{CO_3^{2-}} a_{H^+} / a_{HCO_3^-} = 10^{-10.33}$$

(corresponding to equation 8.2), and

$$K_{CO_2} = a_{HCO_3^-} / P_{CO_2} = 10^{-1.47}$$

(corresponding to equation 8.3).

We have now written three equations to relate five unknowns (P_{CO_2} and the activities of H_2CO_3, HCO_3^-, CO_3^{2-}, and H^+). If we are given any two other pieces of nonredundant information about the system, therefore, it should be possible to determine the activities of all five species. The other two pieces of information vary from one problem to another, because they are conditions of the particular environment we wish to study.

Worked Problem 8.1

Suppose that we do not know the partial pressure of CO_2 in the atmosphere. During this century, after all, atmospheric P_{CO_2} has increased significantly, so a published value is likely to be out of date (see fig. 8.2). How can we determine it?

One way is to collect a beaker of pure rainwater and measure its pH, on the reasonable assumption that rainwater is in equilibrium with atmospheric CO_2. (For simplicity, we ignore other gases commonly dissolved in "pure" rainwater, even though some of them do affect pH.) The activity of H^+, therefore, is one piece of information we did not have before. One of our two remaining equations, then, is simply:

$$a_{H^+} = 10^{-pH},$$

where pH is the measured value.

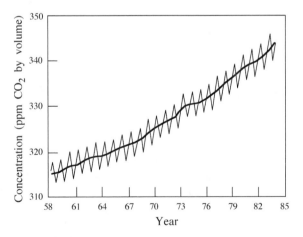

FIG. 8.2. Atmospheric CO_2 concentrations recorded at the Mauna Loa Observatory, Hawaii. Annual oscillations around the increasingly CO_2-rich average (solid line) follow the growing cycle in the Northern Hemisphere. (After Sundquist 1985.)

Because it is necessary to maintain electrical neutrality, we can often write a charge balance equation that provides another constraint. In this case, that equation (our fifth and final one) is:

$$m_{H^+} = m_{HCO_3^-} + 2m_{CO_3^{2-}}.$$

To calculate P_{CO_2}, we solve the five equations simultaneously. Rainwater is a very dilute solution, so we simplify the calculation by assuming that the activities of all aqueous species are equal to their concentrations and the activity of water is equal to 1:

1. From reaction 8.2, we know that:

    $$m_{CO_3^{2-}} = K_2 m_{HCO_3^-}/m_{H^+}.$$

    By combining this with the charge balance expression and performing a little algebra, we find that:

    $$(m_{H^+})^2 = 2K_2 m_{HCO_3^-}/(1 - m_{HCO_3^-}).$$

2. Reaction 8.1 gives us a way to express $m_{HCO_3^-}$ as a function of $m_{H_2CO_3}$ and m_{H^+}. If we combine it with the expression for K_{CO_2}, the result is:

    $$m_{HCO_3^-} = P_{CO_2} K_{CO_2} K_1/m_{H^+}.$$

3. Finally, we can substitute this expression into the result of step one. A little more algebraic manipulation yields the answer:

    $$P_{CO_2} = (10^{-pH})^3/(K_{CO_2} K_1 [10^{-pH} + 2K_2]).$$

To calculate the atmospheric partial pressure of CO_2, then, we insert the measured pH. If we had measured a pH of 5.65, for example, we would conclude that $P_{CO_2} = 331$ ppm.

While on this topic, let's consider the inverse of this problem. Suppose that we had measured atmospheric P_{CO_2} and we wanted to know what rainwater pH to expect. The final equation produced above cannot be rewritten to get a_{H^+} by itself on one side, so we are forced to extract pH by solving a cubic equation. There are several ways to do this. One way is to place all of the terms on one side, so that the equation is in the form:

$$f(a_{H^+}) = 0,$$

write an expression for its first derivative:

$$df/da_{H^+} = 3(a_{H^+})^2 - (K_{CO_2} K_1 P_{CO_2}),$$

and use the Newton-Raphson method described in appendix A. We recommend this as a simple programming exercise.

A much more common method, however, involves using our knowledge of the H_2O-CO_2 system to simplify the problem. When carbon dioxide is dissolved in water at room temperature, the resulting solution always has a pH <7. As we show shortly, this implies that $a_{CO_3^{2-}}$ is negligibly small. (In fact, it is ~2 × 10^{-5} times $a_{HCO_3^-}$ under the conditions we are using.) If this is the case, the charge balance equation is very nearly:

$$a_{H^+} = a_{HCO_3^-}.$$

The expression for K_1, therefore, may be written as:

$$(a_{H^+})^2 = K_1 a_{H_2CO_3}.$$

By combining this with the expression for K_{CO_2}, we get:

$$(a_{H^+})^2 = K_1 K_{CO_2} P_{CO_2},$$

or:

$$pH = -0.5 \log_{10}(K_1 K_{CO_2} P_{CO_2}).$$

Using a little chemical sophistication, therefore, we have derived an analytical solution that is easy to use with paper, pencil, and a pocket calculator.

How do we know when it is safe to make simplifying approximations like the one we used in worked problem 8.1? Figure 8.3a is a plot illustrating the relative importance of H_2CO_3, HCO_3^-, and CO_3^{2-} in solutions of varying pH but a fixed total activity of dissolved carbon species. This sort of diagram was introduced in 1914 by the Swedish chemist Niels Bjerrum and gained wide use in chemical oceanography through the many studies of physical chemist Lars Sillén in the 1960s. From it, we see that there are two pH values at which the activities of a pair of carbon species are equal. We can derive these easily. The expression for K_1 can be written to give a ratio of:

$$a_{H_2CO_3}/a_{HCO_3^-} = a_{H^+}/K_1.$$

This ratio will be equal to 1.0 if a_{H^+} is equal to K_1. Similarly, the expression for K_2 can be written in the form:

$$a_{HCO_3^-}/a_{CO_3^{2-}} = a_{H^+}/K_2,$$

from which we see that the activities of bicarbonate and carbonate will be equal if a_{H^+} equals K_2. At 25°C, the magic pH values are 6.35 and 10.33. Because these crossover points on the Bjerrum plot are several pH units apart, we are justified in neglecting $a_{CO_3^{2-}}$ at pH values much below 8 or $a_{H_2CO_3}$ at pH values much above 9.

Because the activities of two dissolved carbon species are roughly the same in the vicinity of the crossover points, a CO_2-H_2O solution is said to have a high *buffering capacity* in these regions. By this, we mean that the solution pH is extremely resistant to change when we add an acid or a base to it. We can start to explain this feature of solution chemistry qualitatively by recognizing that hydrogen ions added around pH 6.35 or pH 10.33 are likely to be involved in reactions 8.1 or 8.2,

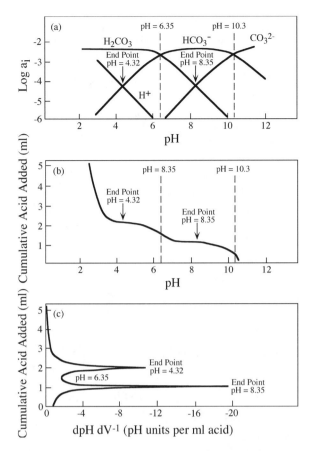

FIG. 8.3. (a) Bjerrum plot for a solution with $\Sigma CO_2 = 1 \times 10^{-3}$ m. (b) Titration curve, showing the effect on pH of adding measured volumes of 1.0 m HCl to the solution in (a). (c) d(pH)/dV versus volume of added acid for the titration in (b). Notice the changes in each curve at the titration end points.

rather than remaining as free ions in solution. Charge balance is maintained in the solution, even as hydrogen ions (or OH⁻) are added, simply by adjusting the ratio of H_2CO_3 to HCO_3^- or the ratio of HCO_3^- to CO_3^{2-}. This qualitative explanation should not satisfy you, however, because reactions 8.1 and 8.2 must consume or release hydrogen ions in *any* pH region. To see why buffering is strongest around pH 6.35 and 10.33, we consider the following titration problem, which was alluded to in chapter 7 in the discussion of the weathering of carbonate rocks.

Worked Problem 8.2

Imagine a closed container filled with one liter of a 10^{-3} m Na_2CO_3 solution. How does the pH in the container change as we add successive small amounts of 1.0 m HCl? To make the calculation easier, assume that there is never any gas phase in the container, and pretend that the volume of solution always remains 1.0 L. Furthermore, let's assume that all species in this system behave ideally. If we are ever concerned about this last assumption, we can use the procedure of worked problem 4.7 to calculate the necessary activity coefficients.

Before we tackle the general problem, let's address a warm-up question: what is the pH before we add any acid? There are four unknown quantities in this particular problem: $m_{CO_3^{2-}}$, $m_{HCO_3^-}$, $m_{H_2CO_3}$, and pH. We therefore need to find four equations to define the system. Two are the defining equations for K_1 and K_2 (we cannot use K_{CO_2}, because there is no gas phase present). Charge balance must be maintained, so the third equation is:

$$m_{Na^+} + m_{H^+} = m_{HCO_3^-} + 2m_{CO_3^{2-}} + m_{OH^-}.$$

We can simplify this equation by drawing on observations in the laboratory, where we know from experience that a Na_2CO_3 solution is quite alkaline. Before we begin adding acid, therefore, m_{H^+} should be negligibly small, so the charge balance expression is very nearly:

$$m_{Na^+} = m_{HCO_3^-} + 2m_{CO_3^{2-}} + m_{OH^-}.$$

(Don't worry about the sudden appearance of OH⁻ in the problem. We have, of course, added another unknown species, but we have also implicitly gained another equation to describe the dissociation of water:

$$K_w = a_{H^+}a_{OH^-}/a_{H_2O},$$

or, in this problem:

$$K_w = m_{H^+}m_{OH^-} = 10^{-14}.$$

We therefore include the activity of OH⁻ and K_w explicitly as a fifth unknown and a fifth constraint without really making the problem more difficult.)

The final equation could be a statement fixing the value of $m_{H_2CO_3}$, $m_{HCO_3^-}$, or $m_{CO_3^{2-}}$ at some measured value, as we did in worked problem 8.1. In this case, however, we would find it difficult to measure any of these three concentrations independently. Instead, we write a statement of *mass balance*. Because the system is closed, the only source of dissolved carbon species is the Na_2CO_3. We know, therefore, that:

$$\Sigma CO_2 = m_{H_2CO_3} + m_{HCO_3^-} + m_{CO_3^{2-}}.$$

This, too, can be simplified, because we know that $m_{H_2CO_3}$ must be negligible in alkaline solutions. Therefore:

$$\Sigma CO_2 = m_{HCO_3^-} + m_{CO_3^{2-}}.$$

We know both ΣCO_2 and m_{Na^+} from the stoichiometry of Na_2CO_3 and the initial condition of the problem we are solving:

$$\Sigma CO_2 = \tfrac{1}{2} m_{Na^+} = 10^{-3} \text{ m}.$$

The equations for K_1, K_2, K_w, charge balance, and mass balance, solved simultaneously, yield a quadratic expression in m_{H^+}:

$$m_{Na^+}(m_{H^+})^2 - 2K_w(m_{H^+} + K_2) = 0.$$

By inserting numerical values for m_{Na^+}, K_2, and K_w, we can solve for m_{H^+}. The initial pH turns out to be 10.56. This is clearly high enough, in retrospect, to justify the simplifications we made in writing the charge balance and mass balance equations.

Now, back to the original question: what happens when we start to add 1.0 m HCl? To begin, we need to rewrite the charge balance equation to accommodate the addition of Cl^-. In doing this, we should no longer ignore m_{H^+}, which becomes increasingly significant as more acid is added. Therefore,

$$m_{Na^+} + m_{H^+} = m_{HCO_3^-} + 2m_{CO_3^{2-}} + m_{OH^-} + m_{Cl^-}.$$

We use the symbol V to represent the volume of acid added (in milliliters) by the end of any stage in our experiment. If V is small compared to the 1.0 L volume we started with, then m_{Cl^-} in the container at any time is defined by:

$$m_{Cl^-}(\text{mol L}^{-1}) = 1.0\,(\text{mol L}^{-1}) \times V(\text{ml})/[10^3 + V]\,(\text{ml})$$
$$\approx 10^{-3}V.$$

The charge balance equation, therefore, can be recast in terms of V instead of m_{Cl^-}. This is a convenient substitution, because the volume of added acid is the control variable we want to remain at the end of this calculation. The charge balance expression now becomes:

$$V = (m_{Na^+} + m_{H^+} - m_{HCO_3^-} - 2m_{CO_3^{2-}} - m_{OH^-}) \times 10^3.$$

We now combine this equation with the expressions for K_1, K_2, and K_w to reduce the number of compositional variables, producing the intermediate result:

$$V = (m_{Na^+} + m_{H^+} - [K_1 m_{H_2CO_3}/m_{H^+}] - [2K_2 m_{HCO_3^-}/m_{H^+}]$$
$$- [K_w/m_{H^+}]) \times 10^3.$$

We're almost finished, but we need to rewrite the mass balance among dissolved carbon species to include $m_{H_2CO_3}$, because it is increasingly important as we add acid to the system. Therefore, we manipulate the equations for K_1 and K_2 once more to derive expressions for $m_{HCO_3^-}$ and $m_{CO_3^{2-}}$ in terms of $m_{H_2CO_3}$, m_{H^+}, and the equilibrium constants. We substitute the results in the original mass balance equation and get:

$$\Sigma CO_2 = m_{H_2CO_3} + (K_1 m_{H_2CO_3}/m_{H^+})$$
$$+ (K_1 K_2 m_{H_2CO_3}/[m_{H^+}]^2).$$

The last step is to eliminate $m_{H_2CO_3}$ from this equation by combining it with the intermediate result to get:

$$V = \left(m_{Na^+} - \frac{\Sigma CO_2 K_1(m_{H^+} + 2K_2)}{(m_{H^+})^2 + K_1 m_{H^+} + K_1 K_2} - K_w/m_{H^+} + m_{H^+} \right)$$
$$\times 10^3.$$

This equation describes a *titration curve* for the Na_2CO_3 solution. We can examine it graphically by inserting the values for m_{Na^+} and ΣCO_2 and solving for V at a number of selected values of pH. The result is shown in figure 8.3b. Below it, in figure 8.3c, is a plot of $d(\text{pH})/dV$.

While the experience of deriving and plotting the titration curve is still fresh, we can ask again how buffering in the carbonate system works. We see now that in the immediate vicinity of pH 6.35, m_{H^+} is very nearly the same as K_1. This simplifies the titration equation greatly:

$$V \approx \left(m_{Na^+} - \frac{\Sigma CO_2(m_{H^+} + 2K_2)}{2m_{H^+} + K_2} - K_w/m_{H^+} + m_{H^+} \right)$$
$$\times 10^3,$$

or, since $K_2 \ll m_{H^+}$ and $K_w \ll m_{H^+}$ under these circumstances,

$$V \approx (m_{Na^+} - \Sigma CO_2/2 + m_{H^+}) \times 10^3.$$

Around pH 6.35, therefore, m_{H^+} is almost directly proportional to V. As we move away from pH 6.35 in either direction, however, this approximation breaks down and the quadratic behavior of m_{H^+} in the denominator of the second term of the titration equation gradually speeds up the rate of change of pH with V. From a mathematical perspective, then, the buffering region around pH 6.35 arises because the titration curve becomes steeper and more nearly linear when:

$$m_{H_2CO_3}/m_{HCO_3^-} = m_{H^+}/K_1 = 1.$$

A similar situation arises when pH is ~10.3, for which value the solution is buffered by nearly equal amounts of carbonate and bicarbonate, so that $m_{H^+} \approx K_2$. Beyond pH 11, the largest changes in the titration equation are due to the K_w/m_{H^+} term; below about pH 3, the last term is most significant. In both regions, pH is again relatively less affected by the second term and, therefore, changes slowly with the addition of acid.

Notice, however, that there are two regions in which the pH of the solution changes very rapidly with only a small addition of acid—that is, where $d(\text{pH})/dV$ is maximized. These regions are centered on the *end points* of the titration at pH 8.35 (where HCO_3^- is the dominant species) and 4.32 (where H_2CO_3 is dominant). At these points, the $(m_{H^+})^2$ in the denominator of the second term in the titration equation makes the titration curve very sensitive to small added volumes of acid. It is commonly said that in these regions, all of the dissolved carbon species have been titrated to form either H_2CO_3 or HCO_3^-. In fact, this is not true, because we can always find all three species at any pH. However, it is true that one dissolved carbon species is significantly more abundant than the other two.

Earlier, we used the terms *conservative* and *nonconservative* to separate elements in seawater on the basis of whether their relative abundances remain constant from place to place in the ocean. Now we see that the same concept can be applied to ions as well, and that it has a special significance when we consider the ocean's response to processes that generate or consume hydrogen ions. Most ions in seawater, such as Na^+, Ca^{2+}, K^+, Mg^{2+}, SO_4^{2-}, NO_3^-, and Cl^-, are unaffected by changes in acidity, because they are not associated with species (like bicarbonate) that contain one or more hydrogen atoms. These are conservative ions. A handful of others, including not only HCO_3^- and CO_3^{2-} but also $B(OH)_4^-$, $H_3SiO_4^-$, and a variety of organic ions, are nonconservative. Their relative abundances can therefore vary (sometimes considerably) from one part of the ocean to another.

To understand further how the carbonate system helps maintain charge balance in the oceans, let's sum the charge equivalents in seawater due to dominant conservative cations and compare them with the summed charge equivalents due to conservative anions:

Species	Concentration (mol/kg)	Charge (eq/kg)
Cations		
Na$^+$	0.470	0.470
Mg^{2+}	0.053	0.106
K$^+$	0.010	0.010
Ca^{2+}	0.010	0.020
Sum of cation charges		0.606
Anions		
Cl$^-$	0.547	0.547
SO$_4^{2-}$	0.028	0.056
Br$^-$	0.001	0.001
Sum of anion charges		0.604

We commit only a small error by ignoring the remaining conservative ions, which are much less abundant.

The difference between our totals, therefore, is 0.002 eq kg^{-1}. This charge deficit, referred to as the *alkalinity* of the seawater, must be balanced by some combination of nonconservative anions. To a good first approximation, this means that:

$$\text{alkalinity} = m_{HCO_3^-} + 2m_{CO_3^{2-}},$$

because these two ions make up the bulk of the remaining dissolved species in seawater. More correctly, this quantity is known as the *carbonate alkalinity*. Unless we are making very careful measurements, the only difference between it and the true alkalinity is the omission of $m_{B(OH)_4^-}$, which is a small but nonnegligible quantity in the oceans.

With this perspective, it should now be apparent why the titration we described at length in worked problem 8.2 is often called an *alkalinity titration*. The amount of acid necessary to bring a solution's pH to its first end point (8.35) is a measure of $m_{CO_3^{2-}}$, because that is the pH at which "all" CO_3^{2-} has been converted to bicarbonate. Titration to the second end point (pH 4.32) gives us an indirect measure of ΣCO_2, because at this point, "all" of the original CO_3^{2-} and HCO_3^- have been converted to H_2CO_3. Because $\Sigma CO_2 = m_{CO_3^{2-}} + m_{HCO_3^-}$, we can use the last equation above to calculate the alkalinity of the solution.

Why should we expect $m_{HCO_3^-}$ and $m_{CO_3^{2-}}$ in seawater to be different from one part of the ocean to another? Early in this chapter, we discussed how carbon is carried from surface to deep waters in the organic and inorganic remains of planktonic organisms. Two sorts of chemical changes take place in this process. On the one hand, photosynthesis allows plants to consume carbon from seawater and fix it in organic matter. As a result, ΣCO_2 is lower in the surface ocean than it is in the deep ocean. A very slight alkalinity increase results from the uptake of dissolved NO_3^- by tissues. Except for that minor contribution, however, the alkalinity of surface water is unaffected by the formation of organic matter, because the balance of other conservative ions is left intact. In contrast, as carbonate shells form, both ΣCO_2 and the alkalinity of surface seawater change. Each mole of carbon that leaves the surface ocean as a constituent of calcite takes a mole of calcium ions with it, carrying two moles of positive charge. Formation of calcite, therefore, causes the alkalinity (expressed in milliequivalents per liter) to change by exactly twice the amount that ΣCO_2 changes (in mmoles per liter).

To appreciate the combined effects of tissue and shell formation, recall that four out of every five carbon atoms transferred from surface to deep ocean waters are carried in organic matter. Therefore, if we momentarily ignore the alkalinity change due to nitrate, we find that every five millimoles of carbon leaving the surface ocean are accompanied by one mole of Ca^{2+} ions, thus reducing the alkalinity by two milliequivalents. Even correcting for the slight charge increase due to nitrate reduction, we conclude that alkalinity decreases only 28% as fast as ΣCO_2 during growth of marine organisms. In deep ocean waters, the reverse takes place. As respiration, fermentation, and dissolution return the constituents of dead organisms to seawater, alkalinity increases at 28% of the rate of total carbon increase.

These relative changes can also be seen as changes in the ratio of carbonate to bicarbonate with depth. With a little algebra, we can express $m_{CO_3^{2-}}$ and $m_{HCO_3^-}$ in terms of ΣCO_2 and alkalinity:

$$m_{CO_3^{2-}} = \text{Alkalinity} - \sum CO_2, \tag{8.4}$$

and

$$m_{HCO_3^-} = 2\sum CO_2 - \text{Alkalinity}. \tag{8.5}$$

This little algebraic shuffling shows us, therefore, that the ratio of $m_{CO_3^{2-}}$ to $m_{HCO_3^-}$ increases as surface organisms consume carbon, nitrogen, and calcium ions. In deep waters, $m_{CO_3^{2-}}/m_{HCO_3^-}$ decreases by the same factor. Relative variations in dissolved carbon species with depth, then, are a direct result of the life-death cycle in the oceans. These variations occur laterally as well, as a result of differences in biologic activity from place to place on the globe, as illustrated in figure 8.4.

Calcium Carbonate Solubility

By introducing carbon species from a source other than atmospheric CO_2, worked problem 8.2 brought us closer to understanding carbonate equilibria in the oceans, where the abundance of dissolved carbonate species depends not only on equilibrium with the atmosphere, but also on reactions involving continental silicates and marine carbonate sediments. Chemical weathering provides a continuous source of dissolved ions, including bicarbonate, to river water by processes already discussed

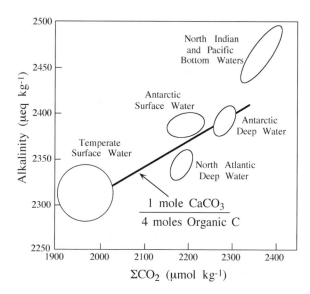

FIG. 8.4. Plot of alkalinity versus ΣCO_2 for major surface and deep waters in the world's oceans. The slope of the solid line confirms our deduction that particulate carbon is transferred from the surface ocean to the bottom primarily in the form of organic debris. The ratio of $CaCO_3$ to organic carbon transported vertically is 1:4. (Modified from Broecker and Peng 1982.)

in chapter 7. These accumulate in seawater until they reach steady-state limits imposed by solubility or adsorption equilibria. In addition to the reactions that give us K_1, K_2, and K_{CO_2}, then, two solubility equilibria are important. These are:

$$CaCO_3 \rightleftarrows Ca^{2+} + CO_3^{2-},$$
calcite

and

$$CaCO_3 \rightleftarrows Ca^{2+} + CO_3^{2-}.$$
aragonite

Carbonates of Na, Mg, and K, the other major cations in seawater, are significantly more soluble than either calcite or aragonite, so they have little effect on alkalinity or ΣCO_2. For kinetic reasons, dolomite does not precipitate in the open ocean, so we can disregard it, too. Calcite and aragonite, therefore, dominate carbonate equilibria in most ocean chemistry problems. Because most $CaCO_3$ is precipitated from surface waters by plankton, the relative abundances of calcite and aragonite follow variations in the population of microbiota. Their solubility equations, then, are each important in different marine settings. At 25°C, the solubility product constants for these reactions are:

$$K_{cal} = 10^{-8.34},$$

and

$$K_{arag} = 10^{-8.16}.$$

With the addition of $CaCO_3$, problems in oceanic carbon chemistry become only slightly more complicated. There are now six potentially variable quantities: P_{CO_2} and the activities of H_2CO_3, HCO_3^-, CO_3^{2-}, H^+, and Ca^{2+}. Their values can be determined, therefore, by solving any six independent equations that relate them. Four of these are the equations for K_1, K_2, K_{CO_2}, and either K_{cal} or K_{arag} (depending on which $CaCO_3$ polymorph is present). As in previous examples, the remaining two may include explicit values for one or two of the activities, or for some combination of two or more activities (ΣCO_2, for example), or a charge balance expression.

Worked Problem 8.3

A sample of limestone consisting of pure calcite is placed in a beaker of water and allowed to reach equilibrium with the water and a CO_2 atmosphere above it. Suppose that we can control the partial pressure of CO_2 experimentally, and that $m_{HCO_3^-}$ can be measured. How does the solubility of limestone vary as a function of $m_{HCO_3^-}$ and P_{CO_2}?

We know that:

$$K_1 = a_{H^+}a_{HCO_3^-}/a_{H_2CO_3},$$

and

$$K_2 = a_{H^+}a_{CO_3^{2-}}/a_{HCO_3^-}.$$

If we divide one of these by the other to eliminate a_{H^+} and then rearrange the result to get an expression for the activity of CO_3^{2-}, we find that:

$$a_{CO_3^{2-}} = \frac{K_2 a_{HCO_3^-}^2}{K_1 a_{H_2CO_3}}.$$

Because we also know that:

$$K_{cal} = a_{Ca^{2+}}a_{CO_3^{2-}},$$

if the limestone is pure calcite, we can once again combine expressions to eliminate $a_{CO_3^{2-}}$:

$$a_{Ca^{2+}} = \frac{K_{cal}K_1 a_{H_2CO_3}}{K_2 a_{HCO_3^-}^2}.$$

In addition, P_{CO_2} is related to $a_{H_2CO_3}$ by:

$$K_{CO_2} = a_{H_2CO_3}/P_{CO_2},$$

so we can write:

$$a_{Ca^{2+}} = \frac{K_{cal}K_1 K_{CO_2}P_{CO_2}}{K_2 a_{HCO_3^-}^2}.$$

Finally, we want an answer written in terms of the measurable quantities $m_{Ca^{2+}}$ and $m_{HCO_3^-}$ instead of activities, so we must include the appropriate activity coefficients:

$$m_{Ca^{2+}} = \frac{K_{cal}K_1 K_{CO_2}P_{CO_2}}{K_2 a_{HCO_3^-}^2}\frac{1}{\gamma_{Ca^{2+}}\gamma_{HCO_3^-}^2}.$$

To obtain the most reliable numerical result from the final expression in worked problem 8.3, it is best to perform the calculation iteratively, as in worked problem 4.7, to refine $\gamma_{Ca^{2+}}$ and $\gamma_{HCO_3^-}$. To see how this is done, we consider another worked problem.

Worked Problem 8.4

If we make the simplifying assumption that the acidity of surface ocean waters is controlled by equilibrium between calcite, seawater, and atmospheric CO_2 and we use a value of 340 ppm (3.4×10^{-4} atm) for P_{CO_2}, what is the equilibrium value of pH?

Once again, our challenge is to solve several equations simultaneously to reduce the problem to a single equation that will let us calculate a dependent variable (pH) as a function of an independently measurable quantity (P_{CO_2}). By combining the expressions for K_1 and K_{CO_2}, we find that:

$$a_{HCO_3^-} = K_{CO_2}P_{CO_2}K_1/a_{H^+}.$$

Because we also know that:

$$K_2 = a_{H^+}a_{CO_3^{2-}}/a_{HCO_3^-},$$

we can eliminate $a_{HCO_3^-}$ by substitution to get:

$$a_{CO_3^{2-}} = K_1 K_2 K_{CO_2}P_{CO_2}/a_{H^+}^2.$$

We can now eliminate $a_{CO_3^{2-}}$ as a variable by combining this expression with the solubility product constant for calcite:

$$a_{H^+}^2 = a_{Ca^{2+}}K_1 K_2 K_{CO_2}P_{CO_2}/K_{cal}.$$

The remaining uncontrolled variable is $a_{Ca^{2+}}$. The best method for isolating it involves the charge balance equation:

$$2m_{Ca^{2+}} + m_{H^+} = 2m_{CO_3^{2-}} + m_{HCO_3^-} + m_{OH^-}.$$

We defined a simplified "seawater" for the purposes of this problem. You should be a little worried now, because you know that the bulk composition of natural seawater is dominated by many species that are missing from this charge balance expression. Stay worried; we address this shortcoming in a moment. Meanwhile, we anticipate that pH will be somewhere between 7 and 9, so it is fair to simplify the charge balance expression by disregarding m_{H^+}, m_{OH^-}, and $m_{CO_3^{2-}}$, each of which should be much smaller than either $m_{Ca^{2+}}$ or $m_{HCO_3^-}$. Therefore,

$$2m_{Ca^{2+}} \approx m_{HCO_3^-},$$

and

$$a_{Ca^{2+}} \approx \frac{a_{HCO_3^-}\gamma_{Ca^{2+}}}{2\gamma_{HCO_3^-}}.$$

As before, we eliminate $a_{HCO_3^-}$ from this equation by replacing it with $K_{CO_2}P_{CO_2}K_1/a_{H^+}$. This yields an expression for $a_{Ca^{2+}}$ that, when inserted into the intermediate result above, finally yields:

$$a_{H^+}^3 = \frac{P_{CO_2}^2 K_1^2 K_{CO_2}^2 K_2\gamma_{Ca^{2+}}}{2K_{cal}\gamma_{HCO_3^-}},$$

from which we can extract pH. We're done, almost.

Let's get back to that worry. It is tempting to suggest that we should iterate toward "best" values for $\gamma_{Ca^{2+}}$ and $\gamma_{HCO_3^-}$, as we did in chapter 4. Because we cannot ignore all of the other species in seawater, however, the exercise should not be taken lightly. Each cycle of iteration requires us to calculate improved values not only for $m_{Ca^{2+}}$ and the concentrations of carbonate species, but also for those of the other major ions and complexes in seawater, plus each of their activity coefficients. The product of this computationally intense exercise has been called a *chemical model for seawater*. Several such models have been produced and refined by geochemists, beginning with Robert Garrels and Mary Thompson (1962), whose model we describe briefly in a later section. Rather than expend the effort to construct and justify such a model here, we complete this worked problem by using activity coefficients derived from experiments with synthetic seawater (for example, those in Pytkowicz [1983], table 5.89). Using $\gamma_{Ca^{2+}} = 0.261$ and $\gamma_{HCO_3^-} = 0.532$, we calculate that surface ocean water should have an equilibrium pH of 8.39. The presence of calcite in the ocean, therefore, should maintain a pH very near to the HCO_3^-/CO_3^{2-} titration end-point (pH 8.32) we discussed in worked problem 8.2. In fact, pH measurements in surface seawater (between 8.1 and 8.4) cluster very close to this value.

The solubility of calcite varies as a function of depth in the oceans in ways that are directly related to the steady downward rain of dead organisms. To see how this works, consider our earlier discussion of alkalinity and ΣCO_2 variations with depth, in which we concluded that the ratio $m_{CO_3^{2-}}/m_{HCO_3^-}$ decreases from the surface downward. In fact, although the total concentration of dissolved carbon is 10% higher in bottom waters, $m_{CO_3^{2-}}$ is lower by almost two-thirds. Figure 8.5 illustrates a vertical profile of $m_{CO_3^{2-}}$ at a station in the South Atlantic Ocean, in which this decrease is apparent. Notice that the most dramatic change occurs within the wave-mixed

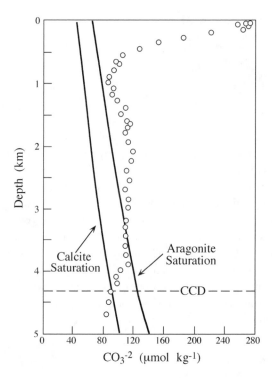

FIG. 8.5. Carbonate ion variations with depth at a station in the South Atlantic Ocean. The depth at which measured concentrations are the same as the theoretical saturation limit for calcite is the carbonate compensation depth (CCD). (After Broecker and Peng 1982.)

surface zone, and that rather gradual changes continue below ~500 m.

The two solid lines on figure 8.5 indicate the theoretical values of $m_{CO_3^{2-}}$ in equilibrium with calcite and aragonite. The slight increase in each with depth is due primarily to pressure, but is also a function of decreasing temperature. To the right of either line, seawater is supersaturated with respect to a carbonate mineral; to the left, it is undersaturated. Notice that the measured profile, in this example, lies above both solid lines at depths less than ~3300 m. Therefore, a carbonate particle falling through the upper portion of the ocean should act as a nucleus for further growth. Below 3300 m, though, aragonite should dissolve in seawater. Below about 4300 m, calcite also is unstable. The lower of these two saturation levels has been called the *carbonate compensation depth* (CCD), representing the level below which no carbonate mineral is stable.

In practice, the CCD is generally defined in kinetic terms as the level at which the rates of carbonate growth

SEARCHING FOR THE CCD

At first glance, it might seem that the easiest way to locate the saturation depth in the oceans would be to look for dissolution features in pelagic sediment. The abyssal plains, however, have remarkably low relief. Except on the flanks of seamounts, much of the ocean floor is well below the CCD or the lysocline, and is therefore virtually carbonate-free.

Several people have devised clever ways to measure the CCD in situ experimentally. An early, elegant method was introduced by Melvin Peterson of Scripps Oceanographic Institute (1966), who hung preweighed spheres of calcite at various depths in the Pacific Ocean, suspending them on cables from a current-monitoring buoy. After 250 days, he retrieved the spheres and reweighed them to assess their net gain or loss.

Several scientific teams repeated and improved upon Peterson's original experiment. One of the most elegant was performed by Susumo Honjo and J. Erez (1978), who reasoned that there could be a significant uncertainty in Peterson's measurements if the carbonate samples had seen different current flow rates. Their solution was to use a small pump at each sample depth to pull water at a constant rate

through a container with a preweighed mass of calcite or aragonite.

Direct measurements have also been made by an apparatus that oceanographer Peter Weyl (1961) has called a saturometer. In this apparatus, seawater is pumped at a high flow rate through a tube filled with calcite crystals and then stopped abruptly. The pH of water trapped in the tube is monitored by an embedded glass electrode. If the water is initially undersaturated, then the pH rises as its acidity is titrated by dissolving calcite. If it is initially supersaturated, then calcite precipitation causes a drop in pH. With time, the pH value in the tube approaches a saturation value asymptotically, and the initial CO_3^{2-} concentration can be calculated from:

$$m_{CO_3^{2-}}(\text{initial}) = m_{CO_3^{2-}}(\text{final}) \\ \times m_{H^+}(\text{final})/m_{H^+}(\text{initial}).$$

This apparatus has been modified since the mid-1970s, so that it is now possible to make in situ measurements by lowering a saturometer from a ship into the deep waters (see, for example, Ben-Yaakov et al. 1974). Figure 8.5 was constructed from measurements of this type.

and dissolution are balanced. Rates of reaction at 2°C are slow enough that some carbonate persists in pelagic sediments even below the CCD. Therefore, many studies refer instead to the *lysocline*, the depth at which dissolution features are first observed in carbonate sediments. The position of the lysocline varies from place to place in the oceans, depending on the rate of carbonate supply and burial, its composition, and the local degree of undersaturation.

Chemical Modeling of Seawater: A Summary

Seawater is the most widely available natural electrolyte solution and serves both to control and to record the effects of many global geochemical processes. As we have seen, biologic processes cause significant variations in seawater chemistry from place to place, particularly in vertical profiles. It is not surprising, then, that geo-

chemists have invested a considerable amount of work in understanding how ionic species are distributed in the oceans.

To this point in the chapter, we have emphasized the role of carbonate equilibria in controlling the chemistry of many natural waters, including seawater, through the mechanism of charge balance. As we have seen in worked problem 8.4, however, the other major ions in seawater (Mg^{2+}, Na^+, K^+, Cl^-, and SO_4^{2-}) may modify the simple equilibria we have described. In addition, the high ionic strength of the oceans (~0.7) favors the formation of ion pairs, which may influence solution chemistry significantly.

Garrels and Thompson (1962) recognized ten complexes whose association constants are large enough that they contribute significantly to the ionic strength of seawater: $MgHCO_3^+$, $MgCO_3^0$, $MgSO_4^0$, $CaHCO_3^+$, $CaCO_3^0$, $CaSO_4^0$, $NaHCO_3^0$, $NaCO_3^-$, $NaSO_4^-$, and KSO_4^-. They estimated the activity coefficients for these and for the six major ions by using the mean salt method (see chapter 4) or, in the case of uncharged species, they assumed that the coefficients were equal to unity. Then, by extension of the methods we have just discussed, they wrote a series of mass and charge balance equations. These, together with appropriate expressions for dissociation and association constants, were solved simultaneously to calculate the equilibrium abundances of all major dissolved species.

Many geochemists have applied Garrels and Thompson's method to other brine systems, or have developed more sophisticated means of estimating activity coefficients in seawater. Rudolph Pytkowicz and coworkers have made some of the most impressive advances in the laboratory and in theoretical modeling of activity-concentration relationships. It is now apparent that other complex species (particularly those involving chloride) are also present in significant concentrations. Instead of the mean salt method, investigators now generally employ a variety of comprehensive theoretical equations (in the spirit of the Debye-Hückel equation, although different in form) to calculate activity coefficients. A good review of seawater models can be found in chapter 5 of Pytkowicz (1983).

Despite the sophistication of these studies, a variety of problems are not easily interpreted by any thermodynamic model of seawater. The relative magnitudes of K_{cal} and K_{arag} for example, suggest that aragonite should always be metastable relative to calcite. We know, however, that both carbonates are common in marine sediments and aragonite is particularly abundant in tropical waters. To a certain degree, this observation can be explained by recognizing that Mg^{2+} enters most readily into the structure of calcite, where it increases its solubility slightly. Aragonite accepts magnesium ions only sparingly, and is therefore only marginally less stable than "impure" calcite. That explanation is only part of the answer, however. Virtually all carbonate precipitated from the sea is nucleated biochemically, and aragonite depends on the peculiarities of ocean life for its tenuous stability. Some organisms, particularly among the scleractinian corals, pelecypods, and gastropods, build shells and skeletons exclusively with aragonite. Others precipitate only calcite, and some build with both polymorphs interchangeably. Once formed, aragonite persists metastably until it is gradually replaced in ancient carbonate rocks

by calcite. This, then, is a kinetic problem outside the scope of thermodynamic models of seawater.

Geochemists also observe that the concentration of CO_3^{2-} in the surface ocean is roughly five times greater than should be expected for seawater in equilibrium with calcite (see fig. 4.5). Carbonate precipitation by organisms, as we found earlier in this chapter, is limited by the availability of phosphate. Why, then, doesn't excess carbonate precipitate inorganically? Once again, the apparent answer lies with kinetics and biology, rather than with thermodynamics. Plankton are such voracious and nondiscriminating eaters that they ingest not only nutrients, but also a large fraction of the suspended submicroscopic calcite particles that might otherwise act as seeds for inorganic crystal growth. When these are later excreted by the plankton, they carry a thin coating of organic matter that is sufficient to retard further growth. Carbonate ions, therefore, accumulate in surface seawater well past the concentration at which K_{cal} is exceeded.

In summary, the distribution of major chemical species in the ocean can be understood with an impressive degree of confidence if we recognize the central role of carbonate equilibria and manipulate the proper thermodynamic equations. Few other geochemical systems have been as well described. As the two brief examples given above show, however, there are still problems in seawater chemistry that defy solution by standard thermodynamic modeling.

GLOBAL MASS BALANCE AND STEADY STATE IN THE OCEANS

Examining the Steady State

To this point, this chapter has emphasized the equilibrium behavior of nonconservative species in the oceans. Another class of geochemical problems is equally challenging. Put simply, these problems compel us to ask what kinetic factors control the abundance of conservative or nonconservative species. Because dissolved matter is constantly being supplied to oceans in river water, the influx of each conservative species must be exactly balanced by removal processes. Without such a balance, a species abundance cannot be in steady state: it must either accumulate or vanish with the passage of time. In a similar way, nonconservative species such as HCO_3^- and CO_3^{2-} may vary in concentration from one place to another in the ocean, but there are kinetic factors that operate to keep the HCO_3^-/CO_3^{2-} ratio nearly constant at each location. A steady state for nonconservative species is therefore maintained among various smaller reservoirs within the ocean. Thus, it is a matter of great importance to identify each of the sources and sinks for dissolved material in seawater and estimate the magnitudes of these sources and sinks, so that we can evaluate the chemical pathways for change in the oceans.

Worked Problem 8.5

What does the oceanic budget for sodium look like? At first glance, this appears to be an easy problem. Sodium is the most abundant cation in seawater (see table 8.1). Its concentration, on average, is 10.78 g kg⁻¹. The total mass of sodium in the oceans, then, is:

$$A_{Na^+} = 10.78 \text{ g kg}^{-1} \times 1.4 \times 10^{21} \text{ kg} = 15.1 \times 10^{21} \text{ g}.$$

The average concentration of Na^+ in river water is 6.9 mg kg⁻¹. Rivers supply ~4.6 × 10¹⁶ kg of water to the oceans each year, so the rate of sodium supply is:

$$dA_{Na^+}/dt = 6.9 \text{ mg kg}^{-1} \times 4.6 \times 10^{16} \text{ kg yr}^{-1}$$
$$\times 0.001 \text{ g mg}^{-1} = 3.2 \times 10^{14} \text{ g yr}^{-1}.$$

At that rate, it would take the world's rivers only 4.7 × 10⁷ years to deliver the ocean's entire present mass of sodium. Using earlier and somewhat less accurate numbers, the nineteenth-century Irish geologist James Joly calculated a value of 9.0 × 10⁷ years, which he then concluded was the age of the Earth. This was a clever calculation, and the results were remarkably good for the period. His interpretation was flawed, however, because he failed to recognize the existence of sinks as well as sources for sodium.

We now know the age of the Earth to be much greater than 47 or even 90 million years. Furthermore, by studying the compositions of formation brines and the stratigraphic record of evaporate formation, Heinrich Holland (1984) concluded that the sodium content of the oceans probably has not varied by >30% during the Phanerozoic. This implies that the river input of sodium is very nearly balanced by processes that remove sodium from the oceans; that is, the oceanic concentration of Na^+ has been at or very close to a steady state value during the past 600 million years. Holland has presented further observations that suggest that the concentration of NaCl in seawater has varied by less than a factor of ±2 in the past 3.5 × 10⁹ years. Instead of yielding the age of the oceans, as Joly supposed, the balanced source and loss rates give us a measure of the *mean residence time* for sodium in the ocean. The average sodium ion, in other words, spends 4.7 × 10⁷ years in the ocean before it is removed by some loss process.

The best available data (Holland 1978) suggest that a little more than 40% of the sodium in river water is derived from the solution of halite in evaporates, ~35% is derived from weathering of all other rocks, and the remaining 25% comes from sea spray carried inland by winds. The major removal processes from seawater include precipitation in evaporates, burial in pore waters in marine sediments, reaction with silicates in submarine geothermal systems, and atmospheric transport of sea spray. On the basis of experimental results reported from several labs, Holland estimated that the oceans annually lose between 1.0 and 1.4 × 10¹⁴ g Na^+ as seawater circulates through midocean ridge systems. This corresponds roughly to 30 or 40% of the annual river flux. Burial of sodium in interstitial waters in the sediment column may account for another 15% of the annual river flux. Estimates of sea salt transport by winds are highly variable, but it is fairly clear that sodium derived from sea spray is carried onshore at roughly the same annual rate that rivers return it to the oceans. We reported this value above to be ~25% of the river flux of sodium per year. The remainder, ~40% of the annual input of sodium, must be lost by formation of evaporates.

We present these estimates here and in figure 8.6 as an abbreviated example of the sort of mass balance exercise that is needed to account for the provenance of each dissolved constituent in seawater and to evaluate the geochemical pathways they follow. You should not be misled, however, by the apparent confidence with which we have reported magnitudes of the loss processes. The rates of these processes for most dissolved species are poorly known. By some estimates, for example, formation of evaporates may account for as much as 60% of the annual river flux of sodium. Consequently, this implies that the importance of other loss processes has been exaggerated.

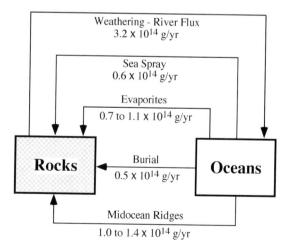

FIG. 8.6. A simple box model illustrating major reservoirs, pathways, and fluxes for sodium.

Even the oceanic budget for sodium, then, is not a simple matter to calculate.

On occasion, ignorance about some fundamental pathways has led geochemists to overestimate greatly the importance of other pathways. The mass balance cycle for magnesium is a good example. Rivers annually carry 1.4×10^{14} g of Mg^{2+} per year to the oceans (Meybeck 1979), the result of weathering both silicates and dolomitic carbonates on the continents. If the magnesium budget of the oceans is in a steady state, then some combination of processes must remove an equal amount of magnesium each year. Until the early 1970s, geochemists believed that the most important loss process was the formation of authigenic clays (glauconite) in marine sediments. The only other known loss process was dolomitization, widely recognized as negligible in today's oceans.

As more analyses of marine clays became available, however, it quickly became apparent that authigenic clay formation is insufficient to balance the river flux. The "magnesium problem" became most acute when James Drever's (1974) research could not authenticate any clearly authigenic magnesian marine clays. Laboratory experiments and the observations gathered during the Deep Sea Drilling Program of the 1970s and 1980s have shown that reactions between seawater and hot basalt, previously thought to be inconsequential, remove virtually all of the magnesium from seawater as it cycles through mid-ocean ridges. Estimates of the cycling rate itself depend on the inferred distribution of heat and its rate of loss around ridges. Because these estimates involve significant uncertainties, calculations of the total amount of magnesium removed from the oceans by this process per year are still imprecise. It is now clear, however, that a major portion of the annual river flux may be balanced this way. Some observations suggest that a substantial amount of magnesium is also lost during alteration of basalts at bottom water temperatures (~0°C) far from ridge systems. Minor amounts are also deposited in magnesian calcite or are involved in exchange reactions with marine clays, and some, in fact, may be lost during authigenic mineralization. Although uncertainties persist, in any event, the "magnesium problem" seems to have vanished.

Each of these two examples—the sodium and magnesium cycles—illustrates the difficulty of describing a steady state in a system of global scale. In most cases, we must accept estimates of reservoir contents and inter-reservoir fluxes that have large uncertainties, and we must be prepared to revise them frequently.

How Does the Steady State Evolve?

Geochemists have an interest in kinetic modeling that extends beyond understanding the steady state. Clearly, geologic conditions change through time. Paleogeographic studies indicate that the rates of seafloor spreading and volcanism have not remained constant, and that climatic fluctuations have induced changes in rates of weathering, erosion, and evaporation. If the ocean has managed to remain nearly in steady state through all of these fluctuations, it is logical to ask what keeps the system under control. What sorts of feedback mechanisms allow seawater to recover from external perturbations, and how rapidly do these mechanisms operate?

As an example of a familiar elemental cycle whose evolving steady state has been debated quite intensely, let's consider the fate of carbon. In a sense, this is an arbitrary choice, because our purpose here is to demonstrate how global-scale cycles stay in balance, not necessarily to study carbon. However, as we have already seen in earlier parts of this chapter, the carbon cycle is of pivotal importance to the chemistry of the oceans and the ocean biosphere. Thus, by looking closely at the carbon cycle, we accomplish two goals at once.

Since the late 1960s, ocean chemists have been interested in models that predict how the global carbon balance is likely to adjust to the increase in atmospheric P_{CO_2} illustrated in figure 8.2. These calculations often also attempt to estimate the climatic changes that may occur as more of Earth's outgoing infrared radiation is absorbed by CO_2 and retained by the atmosphere (the so-called *greenhouse effect*). Such predictions have great value for agricultural, economic, and political planners. A large number of these models have been produced, each differing from the others in either the number of reservoirs and pathways it considers or the numerical approach it follows in evaluating them. We briefly consider a few of these to demonstrate that no model is "best," but that each is appropriate for addressing a different problem.

Box Models

The models most commonly used by geochemists describe the oceans and adjoining environments as a finite set of reservoirs or "boxes," each of which is internally homogeneous, trading mass along discrete kinetic pathways. This is the approach we used for the party example in chapter 1. As in that example, oceanic box models assume that the mass inputs and outputs from each reservoir normally balance each other, so that the system as a whole is in steady state. A geochemist can then examine variations in ocean chemistry with time by introducing a perturbation in one or more of the reservoirs or kinetic pathways. In studying twentieth-century anthropogenic CO_2, for example, geochemist Eric Sundquist (1985) developed a series of models in which the world ocean is divided first into four geographically distinct regions and then conceptually into a dozen or more reservoirs. Figure 8.7 is a schematic illustration of one of

these models. Reservoirs within each geographic region (the Atlantic, Pacific, Arctic, and Antarctic Oceans) represent vertical differences in water chemistry and the intensity and style of biologic activity. These oceanic reservoirs are supplemented by an atmospheric box and several boxes on the continents, each characterized by a distinct biologic community with a different metabolic role in the carbon cycle. Sundquist then wrote a set of differential equations to describe the transfer of carbon among these reservoirs. The equations describe geochemical and biochemical processes (such as weathering, photosynthesis, decay) as well as physical transfer by currents, particle settling, and advection.

Box models like Sundquist's, describing short-term chemical variations, can be complex. In addition to the

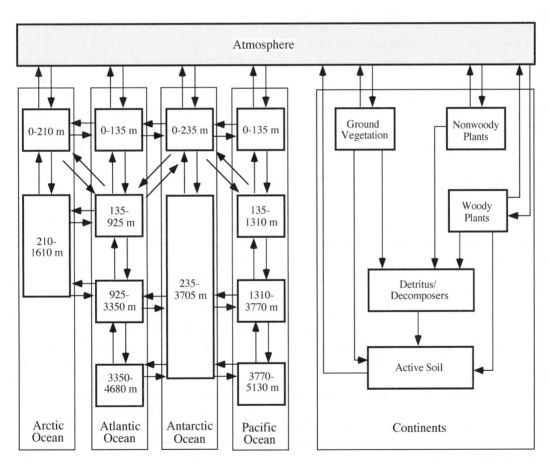

FIG. 8.7. A schematic box model for the short-term carbon cycle described by Sundquist (1985). The world ocean is divided into four separate reservoirs, each of which is further divided vertically. The indicated pathways between oceanic reservoirs describe the physical movement of water and dissolved carbon by currents and advection. The five continental reservoirs are all in the biosphere, and the pathways between them are biochemical. In this model, continental and oceanic reservoirs communicate only through the atmosphere. (Modified from Sundquist 1985.)

reservoirs shown in figure 8.7, for example, some short-term models recognize geologic and biologic differences among the continents. Others emphasize processes along the continental margins or in marine sediments. Changes in seawater chemistry that take place in less than a few thousand years (the time scale over which global stirring occurs) do not affect the entire ocean at once. The same also is true on the continents. Therefore, the shorter our attention span is, the more we must acknowledge that the ocean behaves as a series of semi-independent reservoirs, and the more boxes we build into the model.

Oceanic box models that focus on changes over a few years to a few hundred years also have to be sensitive to biologic processes with rapid response times. The surface ocean boxes in Sundquist's models adjust to changes in atmospheric CO_2 within two to seven years, in part because populations of living organisms respond quickly to the environmental pressures of food supply and climate. In the twentieth century, we saw rapid environmental changes caused by human activity. By comparison, most geologic processes, particularly those that cause slow changes in the positions of continents or the average height of mountains, have operated nearly at a constant rate during our lifetimes. Although rates of tectonic processes change, they change so slowly that biologic communities do not notice. This means that box models designed to study the oceans over short time spans can ignore many geologically distinct boxes and pathways that are crucial in long-term models.

As the time scale of interest becomes longer, biologic factors have less effect than geologic factors on system kinetics. To develop a midrange box model, Sundquist simplified the model in figure 8.7 by lumping the atmosphere together with all surface ocean reservoirs, land plants, and detritus to make a single super-reservoir box. The mean residence times of these reservoirs are all so short that, from the perspective of several hundred years or more, they all appear to be in equilibrium. With increasing time scale, however, Sundquist found it necessary to add reservoirs for marine and terrestrial sediments, which interact with the ocean-atmosphere system along weathering and diagenetic pathways. Response times for the modified geochemical cycles are appropriately longer as a result.

For a more detailed account of what happens when we expand our time frame beyond a few hundred years, let's take a closer look at another midrange box model. In 1982, Wallace Broecker at Columbia University hy-pothesized that the late Pleistocene ice sheets may have fluctuated in response to changes in the distribution of carbon and nutrients in the oceans. He reasoned that because organic carbon burial takes place almost exclusively on continental shelves, it should be least effective during a glacial epoch, when the sea level is lowest. During deglaciation, however, the sea level rises, the proportion of shelf area increases, and more organic carbon should be buried. As submarine shelf area fluctuated, therefore, the balance of carbon species in shallow and deep waters should also have shifted, as should the amount of CO_2 in the atmosphere. Because global temperatures follow P_{CO_2}, this constitutes a climatic feedback system. Broecker observed that the carbon contents of pelagic and shelf sediments and of atmospheric gases trapped in ancient polar ice follow qualitative trends consistent with this scheme.

Robin Keir and W. H. Berger (1983, 1985) tested Broecker's hypothesis with the box model shown in figure 8.8, which consists of seven reservoirs rather than the eighteen shown in figure 8.7. Four of these are peripheral to the ocean/atmosphere system, and are large enough that changes in their carbon contents over 10^5 years are almost certainly negligible: (1) *continental rocks* provide a continuous flux (H_{riv}) of dissolved HCO_3^- to the oceans, (2) the *shelves* act as a sink for $CaCO_3$ (E_{carb}) and (3) organic matter (E_{org}), and (4) *deep-sea sediments* exchange both organically and inorganically derived carbon with seawater in the deep oceans. These last interactions involve inorganic (I_{carb}) and organic (I_{org}) carbon deposition in the ocean floor, and carbonate dissolution (ξ) from pelagic sediments. Keir and Berger characterized the transfer of carbon among reservoirs with a set of mathematical functions and then wrote four differential equations to examine the evolution of ΣCO_2 and alkalinity (A) in the remaining two boxes (surface and deep ocean waters). Rather than describe their approach in detail, we invite you to recall the definitions of alkalinity and ΣCO_2 and compare their equations with figure 8.8. The atmosphere and surface waters follow the relations:

$$V_s d\left(\Sigma CO_{2(tot)}\right)/dt = U\left(\Sigma CO_{2(d)} - \Sigma CO_{2(s)}\right) - I_{org} \\ - I_{carb} - E_{org} - E_{carb} + H_{riv},$$

and

$$V_s d(A_s)/dt = U(A_d - A_s) - 2I_{carb} - 2E_{carb} \\ + \nu_o I_{org} + 2H_{riv}.$$

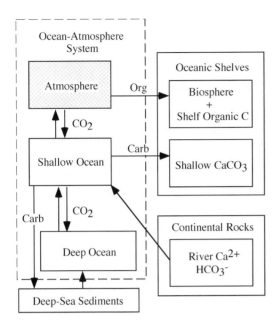

FIG. 8.8. A box model for the intermediate-term carbon cycle described by Keir and Berger (1983) to test Broecker's (1982) glacial hypothesis. Inorganic solids follow the paths identified as "Carb"; organic solids follow the "Org" path. Other pathways are dominated by dissolved or gaseous carbon species.

In the deep ocean,

$$V_d\, d\!\left(\sum CO_{2(d)}\right)\!/dt = U\!\left(\sum CO_{2(d)} - \sum CO_{2(s)}\right) + I_{org} + \xi,$$

and

$$V_d\, d(A_d)/dt = -U(A_d - A_s) - v_o I_{org} + 2\xi.$$

By inspection, you can see that each equation consists of source terms (with positive signs) and loss terms (with negative signs), which may combine to produce a net increase or a decrease during any time interval of interest. The total surface $\sum CO_2$ $\left(\sum CO_{2(tot)}\right)$ is partitioned into an atmospheric fraction and a dissolved fraction $\left(\sum CO_{2(s)}\right)$ on the basis of solubility equilibria, and the rate of dissolution (ξ) is controlled by the solubility of calcite in deep waters and the rate of pelagic sedimentation. Other parameters appearing in these equations are the volumes of water in surface and deep ocean (V_s and V_d), the rate of physical overturning of deep and surface waters (U), and the ratio of alkalinity loss to CO_2 production due to oxidation of organic matter (v_o). This last factor takes into account any changes in alkalinity due to conversions between NO_3^- and organic N.

To test the model, Keir and Berger introduced perturbations in E_{org} and I_{carb}, as inferred from the geologic record of Pleistocene sea levels. Given these functions and reasonable initial choices for each of the model parameters, they then solved the four differential equations repeatedly at closely spaced time steps to watch the evolution of atmospheric P_{CO_2} and the preservation rate for pelagic carbonate. Their numerical results are consistent with the carbonate record in deep-sea drill cores and with the CO_2 trapped in air bubbles in polar ice caps, supporting Broecker's hypothesized link between Pleistocene glacial fluctuations and the global carbon cycle.

Notice that Keir and Berger's model ocean has only a vertical dimension. Because global currents tend to homogenize waters within the surface and deep layers, this simplification is characteristic of midrange models. Midrange box models also characteristically recognize that the ocean's slow rate of overturn (the parameter U) means that chemical perturbations within the surface reservoir are not readily communicated to the deep ocean, and vice versa. Consequently, short-term biologic perturbations within surface and deep reservoirs, such as those addressed by Sundquist, constitute a background "noise" that is insignificant over the time scale that Keir and Berger consider. Their model not only has far fewer boxes, therefore, but it is also largely geologic.

Again, the longer the time span to be modeled, the less important biologic factors are. Processes that change only with geologic slowness act as natural feedback mechanisms to keep short-term fluctuations from getting out of hand. For example, a planktonic community that overeats may grow quickly, but it grows only until it reaches the limits of its food supply. The river supply of phosphate or nitrate is a function of the global weathering rate, which remains nearly constant over periods of hundreds or thousands of years. Boxes dominated by biologic activity can be quite noisy, therefore, but their effects on the ocean-atmosphere system as a whole are severely limited by the slower kinetics of geologic processes.

If we consider even longer time scales, biology and structure in the ocean-atmosphere system can be ignored altogether. For example, figure 8.6, which we used in discussing the long-term geochemical cycle for sodium, includes only two boxes: one for the ocean and one for the continents. In a similar vein, Yale geochemist Robert Berner and his colleagues (1983, 1985) have shown that variations in the seafloor spreading rate can be related to long-term fluctuations in atmospheric P_{CO_2} during the past 100 million years. Their model, in its simplest form, includes an undifferentiated ocean reservoir with no hint of biologic control. Mass transfer between the ocean and various solid-earth reservoirs takes place only by the slow geologic processes of weathering, metamorphism, volcanism, and seawater cycling at ridge crests. The response of seawater to a perturbation, in this model, is ultimately governed by the mean residence time for bicarbonate (10^5 yr), calcium (10^7 yr), and magnesium (10^7 yr) in the oceans. Even the interglacial fluctuations considered by Keir and Berger generate only negligible "noise" at this time scale, and they are therefore ignored.

As in Keir and Berger's model, the atmosphere follows changes in ocean chemistry very closely. Berner and associates, however, have incorporated an empirical equation in their long-term model that describes the rate of chemical weathering as a nonlinear function of P_{CO_2}. As the amount of carbon dioxide in the atmosphere increases, both temperature and rainfall increase as well, and the rates of chemical weathering for continental rocks rise rapidly. This natural mechanism provides a negative feedback that prevents P_{CO_2} from varying by more than a factor of two or three. Calculated values of P_{CO_2} since the late Mesozoic era, from Berner et al. (1983), are shown in figure 8.9. Kasting and Richardson (1986) have used this model's prediction of a rise in P_{CO_2} ~45 million years ago to account for the 2–3°C global warming that is generally inferred to have taken place during the Eocene. Considered over the appropriate time scale, therefore, the kinetic behavior of the oceans can be exploited to show a causal link between plate tectonics and paleoclimate.

Continuum Models

If the general rule is that oceanic box models are most complex when they are designed to examine processes within short time frames, it should follow that the ultimate short-term model must identify an infinite number of boxes. Before the advent of high-speed computers, such an idea would have been impractical. A growing class of continuum models, however, does essentially that by describing chemical variations as continuous functions that can be sampled at any random point in the ocean. Rather than describing box-to-box mass transfer, differential equations in a continuum model describe the physical, biologic, and chemical processes that occur in any part of the ocean. The steady state is described by a global-scale database of measured temperatures, salinities, ΣCO_2 values, and so forth that anchor the differential equations at specific sites. Assuming that the various parameters vary smoothly from one site to another, a computer program solves numerical forms of the differential equations to determine the variables at intermediate locations. A chemical oceanographer can then test the response of the model system to perturbations by altering the parameters at one or more sites and monitoring the propagation of the calculated changes to other locations. Models of this type are similar to the global atmospheric circulation models that have steadily improved the accuracy of weather forecasts over the past 25 years. Although they are of growing interest to those who study the short-term sensitivity of the ocean to natural and human events, continuum models are highly complex mathematically and are beyond the scope of this book. Curious readers may want to consult issues of the journals *Chemical Oceanography* or *Ocean Modelling*.

A Summary of Ocean-Atmosphere Models

We have had two goals in this discussion of models. One is simply to introduce you to concepts that geochemists consider in modeling the ocean-atmosphere system. No single model is appropriate for studying the full range of kinetic behavior in the system. Depending on the processes that interest us, time scales can range from less than a year to many tens of millions, and reservoirs can be grouped and regrouped in many ways. The ultimate tests of success are internal consistency and the ability to elucidate some facet of the system's behavior in a way that is compatible with the geologic record. Our

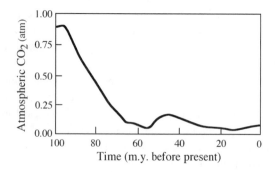

FIG. 8.9. Variations in P_{CO_2} during the Cenozoic, as calculated by the model of Berner et al. (1983, 1985).

other goal is to demonstrate some dynamic features of the chemistry of the ocean-atmosphere system and, in particular, of the carbon cycle. The chemistry of the oceans is resilient. If we examine the dynamic redistribution of an element such as carbon, we find that its abundance in seawater is rarely at a true steady-state value, but that it is also rarely far from such a state. Short-term perturbations have a limited effect on the ocean-atmosphere system, because the response times of reservoirs dominated by biology are very short and long-term geologic processes place stringent limits on the growth of biologic communities. The geologic processes that tend to disturb the ocean's chemistry are themselves gently cyclical and have periods that are long compared with the mean residence times for many dissolved species in the oceans. Therefore, their abundances and rates of movement into and out of neighboring reservoirs follow gradual changes in global tectonics from one geologic period to another without deviating markedly from the evolving steady state.

GRADUAL CHANGE: THE HISTORY OF SEAWATER AND AIR

Beyond the time scales we have just considered lie broader geochemical questions about the evolution of our planet. What did the earliest ocean and atmosphere look like? What pathways did they follow in arriving at their present compositions? How has evolution of the ocean-atmosphere system been linked with evolution of the crust and the history of life on the Earth? These are highly speculative realms, but by careful interpretation of the rock record and the application of thermodynamic constraints, geochemists have been able to peer through the fog of time and glimpse the outlines of the early state of the ocean-atmosphere system.

Early Outgassing and the Primitive Atmosphere

Among the most useful bits of information we have to work with is the Earth's *excess volatile inventory*. This tabulation, first suggested in a landmark paper by William W. Rubey (1951), is an estimate of the total amount of carbon, nitrogen, sulfur, and water that have been released from the solid Earth since the time of its formation. A certain fraction of this inventory is now in the ocean-atmosphere system, but much of it has been recombined with crustal rocks during chemical weathering and now resides primarily in the sedimentary column as a constituent of carbonates, clay minerals, and fossil organic matter. In particular, roughly a third of the outgassed water and nitrogen now reside in the crust, along with almost all of the carbon and sulfur. These are regularly recycled through the surface reservoirs of the Earth as a result of volcanism, metamorphism, and erosion.

Geochemists' understanding of the history of volatiles has been dominated by two schools of thought. According to one, the volatile inventory has accumulated gradually through geologic time as the result of volcanic emanations. The other holds that degassing of the planet was an early, catastrophic event associated with accretion and differentiation. Clearly, some outgassing has continued throughout the Earth's history, so the answer lies somewhere between these two extremes. The best evidence we have comes not from the major volatiles listed in table 8.2, but from the noble gases that are presumed to have followed similar outgassing histories. The advantage of working with noble gases is that they are chemically nonreactive and therefore accumulate in the atmosphere through time rather than being partially recycled into the crust. Abundance ratios of their stable and radiogenic isotopes can, in many cases, yield valuable clues to the history of outgassing.

When we discuss radiogenic isotopes in chapter 14, we show how geochemists can read these clues. For now, we present a summary diagram (fig. 8.10) derived from Holland (1984) to illustrate outgassing curves that result from several different assumptions regarding the present volatile content and homogeneity of the mantle. A key consideration is the degree to which primordial gases (those that were part of the Earth at its time of formation) have escaped from the mantle. We show in chapter 14 that, although the mantle has lost some of its primordial gas content, it has by no means lost all of it. Figure 8.10 indicates that unless <30% of those gases remain in the mantle today, most of the planet's volatile inventory must have been outgassed catastrophically during the earliest Archean era. In other words, although outgassing has continued throughout the Earth's history, most of the volatiles now in the atmosphere and oceans and trapped in crustal rocks were released quite early. This result is compatible with our modern conception of planetary formation, which is discussed in chapter 15. We should keep in mind, though, that it depends on many glaring uncertainties and may change dramatically as we continue to gather data.

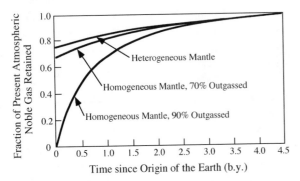

FIG. 8.10. Accumulation of noble gases in the atmosphere. The upper curve was calculated by assuming a heterogeneous mantle. The others were calculated by assuming that the mantle is homogeneous. Notice that the two models of mantle structure have indistinguishable effects unless the noble gas inventory in a homogeneous mantle is at least 70% outgassed. (Based on calculations by Holland 1984.)

In addition to determining the composition of the Earth's excess volatile inventory and the rate at which it was released, a significant effort has been made to deduce the species abundances in the ocean-atmosphere system through geologic time. Although some attempts have been made, geochemical cycling models such as those for the carbon cycle that we discussed earlier have not proven useful in pre-Phanerozoic studies. Our knowledge of initial reservoir contents and rate constants is too scanty. Also, we understand so little about indirect controls, such as the configuration of continents and the average depth of seas, that it is very difficult to construct a reliable kinetic model. Instead, the chemical history of the Precambrian oceans and atmosphere has been reconstructed by studying a succession of presumed equilibrium states through which they passed.

During planetary accretion, it seems likely that primordial gases were released from the solid Earth in equilibrium with silicate magmas. The species abundances in the gas mixture, therefore, can be estimated theoretically by assuming that their bulk composition was that of the present excess volatile inventory and then using the tools of thermodynamics to find the mixture whose free energy, in association with the magma, is lowest. The major controlling parameter is oxygen fugacity, which is determined in most magmatic systems by the oxidation state of iron.

No samples of volcanic rock from the first half billion years of the Earth's existence have survived, so we have no direct information about their composition. It is almost certainly true, however, that the oxidation state of volcanic gases during the early Archean was no higher than it is today. Modern volcanic gases behave as if their oxidation state is controlled by the reaction:

$$3SiO_2 + 2Fe_3O_4 \rightleftarrows 3Fe_2SiO_4 + O_2,$$
quartz magnetite   fayalite

commonly referred to as the QFM oxygen fugacity buffer. This places an upper limit on the oxidation state of the primordial atmosphere. The lower limit is probably best defined by the reaction:

$$SiO_2 + 2Fe + O_2 \rightleftarrows Fe_2SiO_4,$$
quartz   iron   fayalite

which may have been dominant before metallic iron was segregated to form the Earth's core. The oxygen fugacity controlled by this assemblage, known as QFI, is about four orders of magnitude lower than QFM at any temperature of interest in a magmatic system (see fig. 8.11). As we shall see in chapter 15, prevailing ideas about the differentiation of the primitive Earth favor conditions near this lower limit.

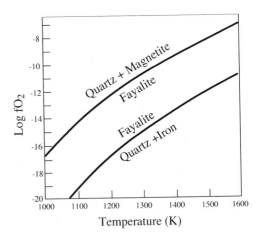

FIG. 8.11. Calculated curves indicating the oxidation state of magmatic systems buffered by quartz + fayalite + magnetite (QFM) or quartz + fayalite + iron (QFI) as a function of temperature. The two curves are approximately four orders of magnitude apart over the temperature range for most magmas.

Worked Problem 8.6

Consider a closed system containing a gas phase and the solid QFM oxygen buffer at 1100°C. The total pressure on the gas is 1.0 atm, and it consists of a mixture of C-H-O species. What are the partial pressures of each of those species if the system is at equilibrium and the bulk atomic ratio C:H is 0.05?

As phrased, this is a very difficult problem to do on paper, because it involves a great number of potentially stable species. To make it easier, let's assume that there are only four gases in the mixture, other than O_2: H_2, H_2O, CO, and CO_2. The partial pressures of these five, then, constitute the unknown values we must calculate.

To solve for the five unknowns, we must write five equations. Two of these follow directly from the bulk chemistry of the system. Because total pressure is 1.0 atm, we know that:

$$P_{H_2} + P_{H_2O} + P_{CO} + P_{CO_2} + P_{O_2} = 1.0.$$

Because the atomic ratio C:H is 0.05, we also know that:

$$(P_{CO} + P_{CO_2})/(P_{H_2} + P_{H_2O}) = 0.1.$$

The partial pressure of O_2 is defined by the QFM buffer at the temperature specified. An empirical equation commonly used for this purpose is:

$$\log f_{O_2} = 9.00 - (25738/T),$$

where T is the temperature in kelvins (Eugster and Wones 1962). Our third equation, then, is

$$f_{O_2} = 1.8 \times 10^{-10}.$$

The final two equations are based on the stoichiometry of the gas species. From the potential reaction:

$$2CO_2 \rightleftarrows 2CO + O_2,$$

we calculate that:

$$f_{CO_2}/f_{CO} = (f_{O_2}/K_c)^{1/2},$$

in which K_c is the equilibrium constant for this reaction between carbon species. It has a value, under these conditions, of 3.79×10^{-13}.

The only other independent chemical reaction that can be written is:

$$2H_2O \rightleftarrows 2H_2 + O_2,$$

from which:

$$f_{H_2O}/f_{H_2} = (f_{O_2}/K_h)^{1/2}.$$

The equilibrium constant, K_h, has the value of 8.82×10^{-14} at 1373 K.

Assuming ideality, each of the fugacities in these last three equations is numerically equivalent to a partial pressure in at-mospheres. The five equations can therefore be solved simultaneously to yield:

$$P_{H_2} = 1.97 \times 10^{-2} \text{ atm,}$$

$$P_{H_2O} = 0.89 \text{ atm,}$$

$$P_{CO} = 4.00 \times 10^{-3} \text{ atm,}$$

$$P_{CO_2} = 8.70 \times 10^{-2} \text{ atm.}$$

This, of course, is a rough answer, because we limited the selection of gas species and then made the simplifying assumption of ideality. To do the problem more carefully, we would use the tabulated $\Delta \bar{G}_f^0$ values and fugacity coefficients for a large number of possible C-O-H gas species to compute equilibrium mole fractions of each for the specified bulk composition and f_{O_2}.

The method used in worked problem 8.6 becomes cumbersome if the number of potential gas species is large, so a computer is a more appropriate tool than pencil and paper. To investigate model atmospheres for the very early Archean, we have calculated the mole fractions of 84 gas species in the system C-O-H-S, with its bulk composition derived from the volatile inventory in table 8.3. The mixture, therefore, has atomic ratios C:H = 0.032, C:N = 16.5, and C:S = 42.0. We have assumed a total atmospheric pressure of 1.0 bars and a magma temperature of 1200°C, but have found that the results of the model calculation are very insensitive to changes in either parameter. The results are applicable, therefore, throughout the probable range of outgassing pressures and temperatures. Values were computed at both the QFM and QFI oxidation state limits. Here are the concentrations of the major species and some important abundance ratios:

Species	QFM	QFI
H_2O	0.91	0.30
H_2	0.02	0.64
CO_2	0.057	0.0097
CO	0.0031	0.051
SO_2	0.0013	2.0×10^{-7}
H_2S	1.3×10^{-1}	0.0014
N_2	0.0018	0.0018
O_2	3.4×10^{-9}	4.0×10^{-13}
CH_4	4.7×10^{-14}	2.7×10^{-7}
CO_2/CO	18.4	0.19
SO_2/H_2S	100.0	1.4×10^{-4}
H_2O/H_2	45.5	0.46

Today's volcanic gases are similar to the calculated mixture in the QFM column. They are dominated by

TABLE 8.3. Distribution of Carbon, Nitrogen, Sulfur, and Water in Near-Surface Geochemical Reservoirs

	Carbon ($\times 10^{18}$ g)	Nitrogen ($\times 10^{18}$ g)	Sulfur ($\times 10^{18}$ g)	Water ($\times 10^{21}$ g)
Atmosphere	0.69	3950	<0.001	≤0.2
Biosphere	1.1	0.02	<0.1	1.6
Hydrosphere	40	22	1300	1400
Crust	90,000	2000	15,000	600
Inventory	90,041.79	5972.02	16,300.1	2001.8

After Holland (1984), table 3.5.

H_2O. Among the carbon and sulfur species, CO_2 and SO_2 are most abundant, and N_2 is the major nitrogen species. The values in the QFI column, in contrast, indicate that if metallic iron were present in very early Archean magmas, H_2, a major reduced gas species, should have been roughly twice as abundant as water vapor in erupting gases. Carbon monoxide should have been five times as abundant as CO_2, and H_2S should have been the dominant sulfur species. We are only partway to our goal, however. Partial pressures of atmospheric gas species should differ from those in volcanic gases because of reactions between gases during cooling or as the result of various loss processes. Therefore, to the extent that thermodynamic rather than kinetic factors determined the state of the primitive atmosphere, we can estimate the relative abundances of gas species during the earliest Archean (before ~4.0×10^9 years ago) by allowing the mixture in the QFI column to reequilibrate at low temperature.

Carbon monoxide, for example, is not stable at low temperatures in a hydrogen-rich atmosphere, and should react to form CH_4 plus water or, as P_{H_2} drops below 10^{-4} atm, to form graphite plus water. Therefore, the dominant carbon species in the earliest Archean atmosphere should have been methane. Similarly, as long as P_{H_2} exceeded ~10^{-3} atm, the dominant nitrogen species at $25°C$ should have been NH_3 rather than N_2. Figure 8.12 illustrates the way in which each of these equilibria depends on the partial pressure of H_2. The actual equilibrium partial pressures of CH_4 and NH_3, of course, would also have depended on the amount of outgassing, as suggested by the pair of curves in figure 8.12b.

Hydrogen is lost from the upper atmosphere to space at a rate that is roughly proportional to its partial pressure and is rapid enough that the mean residence time for atmospheric H_2 today is four to seven years. To maintain the high initial partial pressure suggested by our calcula-

tions, the rate of volcanic supply of H_2 had to meet or exceed its rate of loss by chemical reactions and escape to space. The change from QFI to QFM buffering was probably complete within the first 5×10^8 years of the Earth's history, when segregation of free iron from the Earth's mantle to the core had ended. For the next 10^9 years, P_{H_2} was probably sustained at its early high value by photodissociation of water in the upper atmosphere and other photochemical reactions. Kasting (1986) suggested that the dominant reactions were the photochemical reaction producing hydrogen peroxide from water vapor,

$$2H_2O \rightleftarrows H_2O_2 + H_2,$$

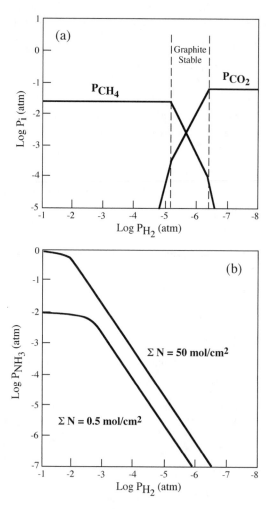

FIG. 8.12. Equilibrium partial pressures of major carbon- and nitrogen-bearing gases as functions of P_{H_2} at $25°C$. (a) P_{CO_2} and P_{CH_4} calculated for a total atmospheric carbon concentration of 1.5 moles cm^{-3}. (b) P_{NH_3} calculated for two values of total nitrogen abundance. (After Holland 1984.)

and the photooxidation of ferrous iron in seawater,

$$Fe^{2+} + H^+ \rightleftarrows Fe^{3+} + \tfrac{1}{2}H_2.$$

Several lines of evidence suggest that the delicate balance between H_2 production and loss was maintained until 3.7×10^9 years ago, when P_{H_2} began to decrease.

Studies of ancient basalts indicate that the oxidation state of the mantle has remained essentially unchanged since at least 3.7×10^9 years ago and possibly 5×10^8 years earlier. The average composition of volcanic gases derived from the mantle, then, has probably also stayed roughly the same throughout much of geologic history. NASA scientists David Catling, Kevin Zahnle, and Christopher McKay (2001), therefore, have ascribed the atmospheric P_{H_2} decrease during the mid- to late Archean not to a change in the composition of outgassed volatiles, but to a gradual increase in the rate at which H_2 was lost to space. Noting that significant amounts of organic carbon began to appear in sediments ~3.7×10^9 years ago, they hypothesized that anaerobic decomposition would have released increasing amounts of methane into the atmosphere, even though new volcanic gases buffered by QFM should have contained a vanishingly low volume fraction of CH_4. For >10^9 years, until ~2.5×10^9 years ago, P_{CH4} may have been as high as 10^2–10^3 ppm by volume, compared with 1.7 ppmv today. Such a CH_4-rich atmosphere, they estimate, would have lost H_2 to space orders of magnitude faster than today's atmosphere does. This, coupled with a decrease in the total amount of mantle outgassing as the crust became progressively thicker during the late Archean, would have led to gradually lower atmospheric P_{H_2}.

Hydrogen sulfide, easily destroyed by solar ultraviolet radiation or consumed in redox reactions with the crust, should also have become less abundant during this period. Also, as P_{H_2} continued to decrease, P_{CO_2} should have increased as a result of the reactions:

$$CH_4 + 2H_2O \rightleftarrows CO_2 + 4H_2,$$

followed by:

$$C(graphite) + 2H_2O \rightleftarrows CO_2 + 2H_2,$$

as shown in figure 8.12a. As we discuss below, however, the continual burial of CO_2 as organic carbon probably delayed the rise of atmospheric P_{CO_2} until almost 2.7×10^9 years ago. Only as the Archean period drew to a close, then, did the atmosphere gradually become a mixture of CO_2 and N_2.

It is important to reemphasize that this interpretation is based on thermodynamic arguments and may not bear any resemblance to the actual history of Archean atmosphere-ocean chemistry. Today's atmosphere, in fact, does not contain the mixture of NO_x gases that would be predicted by equilibrium thermodynamics. It is likely that the early atmosphere's composition was also controlled by reaction kinetics rather than thermodynamics. The true course of events may never be known. Regardless of the pathways, however, it is fairly certain that the atmosphere had become predominantly a CH_4-CO_2-N_2 mixture by the time our stratigraphic record began 3.7×10^9 years ago. The earliest surviving sedimentary rocks already contain mature chemical weathering products, which are formed most readily by the interaction of CO_2-charged rainwater with crustal rocks.

The greenhouse effect may have been important in stabilizing surface temperatures. Of the likely atmospheric gases in the C-O-H-S system, only water, CO_2, and CH_4 are effective in this role. If any of these species is present, outgoing radiation is partially trapped, and the surface temperature of the planet is raised. The degree of warming is a nonlinear function of the abundance of greenhouse gases and the amount of incoming solar radiation. Using the estimated Pleistocene global surface temperatures as a guide, James Kasting (1986) calculated that a CO_2 partial pressure of ~0.05 bars, or roughly 100 times the present atmospheric level, would have been necessary to generate the same surface temperature by the greenhouse effect during the Huronian glacial event. Archean paleosols, however, show no evidence of the intense acidic weathering that should have occurred in a high-P_{CO_2} atmosphere. Catling and coworkers (2001) suggest that the greenhouse was largely supported instead by methane. It seems likely, then, that the evolution of P_{CO_2} during the Precambrian followed a path closer to the bottom than the center of the envelope hypothesized by Kasting in 1986 (see fig. 8.13).

As we discussed in chapter 7, H_2CO_3 produced in CO_2-charged rainwater is one of the most common participants in reactions like:

$$2KAlSi_3O_8 + 2H_2CO_3 + 9H_2O \rightleftarrows$$
K-feldspar
$$2K^+ + Al_2Si_2O_5(OH)_4 + 4H_4SiO_4.$$
kaolinite

Clay-forming reactions, therefore, would have served as a sink for CO_2 and water from the early Archean and

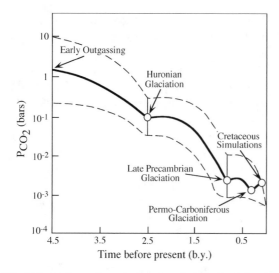

FIG. 8.13. Limits on the evolution of atmospheric CO_2, inferred primarily from the greenhouse-supported temperatures necessary to avoid runaway glaciation. (Modified from Kasting 1986.)

would have increased in efficiency as P_{CO_2} increased. In the same way, oxidation of volcanic gases would have served as a sink for any free molecular oxygen, creating a primitive "acid rain" that, in turn, helped to sequester both oxygen and sulfur in sediments during weathering. Therefore, as chemical weathering progressed, a sizable portion of the excess volatile inventory was transferred to residual minerals and buried in sedimentary rocks. The mean residence times for most dissolved species in the oceans are short enough that these weathering reactions would probably also have led to seawater of nearly modern composition within a few hundred million years. As we remarked earlier, the earliest sedimentary record contains carbonates, evaporates, and shales that appear to have been deposited from an ocean similar to today's.

The Rise of Oxygen

The final story we have to tell in this chapter concerns the evolution of oxygen, which has been parallel to the evolution of life. Prebiotic O_2 levels were very low, partly because there were no volumetrically significant oxygen sources prior to photosynthesis and partly because crustal and atmospheric sinks were plentiful. The dissociation of H_2O by solar ultraviolet radiation, followed by the escape of hydrogen into space, would have provided a small but continuous supply of O_2 in the upper atmosphere. Molecular oxygen is easily consumed by back reaction with H_2, however, so much of the

oxygen would have been consumed before it reached the Earth's surface. As we noted earlier, atmospheric P_{H_2} remained high even after hydrogen had become a less abundant component in volcanic emanations. Kasting (1986) has shown that photochemical reactions alone would have guaranteed a P_{H_2} high enough to preclude an O_2 abundance greater than about one part in 10^{12} at the Earth's surface during the prebiotic Archean.

Stromatolites resembling modern algal mats, which are the product of cyanobacteria (blue-green algae), have been identified in rocks $\sim 3 \times 10^9$ years old. The earliest of these, however, may have been formed by photosynthetic bacteria that use reduced compounds like H_2S and thiosulfate rather than H_2O as electron donors. There is a growing consensus that cyanobacteria, and thus photosynthetically generated O_2, first appeared $\sim 2.7 \times 10^9$ years ago.

The oxidation state of various weathering products in Proterozoic and pre-Proterozoic detrital sediments and soils suggests that P_{O_2} rose very slowly, if at all, for the next 400 million years. Several rock units in Precambrian terrain, most notably conglomerates in the Witwatersrand Basin in South Africa and at Blind River and Elliot Lake in Canada, contain grains of uraninite and pyrite that are almost certainly detrital. Over most of the natural range of pH and oxidation conditions today, pyrite is readily oxidized to form insoluble iron hydroxides. Uraninite, at pH less than ~ 5, combines with O_2 and bicarbonate to form the soluble uranyl carbonate complex, $UO_2(CO_3)_2^{2-}$, and is thus easily destroyed. Between pH 4 and 6, reactions with O_2 and H^+ also yield $UO_2(OH)^+$. As a result, except in very arid regions or under conditions of unusually rapid erosion and burial, neither pyrite nor uraninite commonly accumulates in modern sediments. The presence of both in these ancient sediments strongly suggests that the end of the "O_2-free" era did not come until 2.3×10^9 years ago.

Until recently, geologists assumed that P_{O_2} rose only slowly because cyanobacteria struggled to gain a foothold for much of that period. The carbon isotopic record in sediments, however, indicates that the rate at which organic matter is buried has been more or less constant since at least 3.2×10^9 years ago. This suggests that the slow rise in atmospheric P_{O_2} was not due to a biologic limit on O_2 production but instead to highly efficient sinks for O_2.

The primary sink seems to have been thick marine deposits of banded iron formation (BIF), which formed extensively during this period, but not since. Briefly, oxygen

ARCHEAN ATMOSPHERIC GASES AND THE ORIGIN OF LIFE

The composition of the early atmosphere is of interest not only to geochemists but also to biologists, who would like to know what environmental conditions were most likely to have prevailed when the first biologic molecules were formed on the Earth. The Russian biochemist A. I. Oparin opened discussion on the topic in the 1920s by theorizing that lightning or intense solar ultraviolet radiation could have produced amino acids and other organic building blocks from an early atmosphere rich in CH_4, H_2, and NH_3. These amino acids then accumulated in the oceans, forming a primordial "soup" in which more complex hydrocarbons, such as proteins and nucleic acids, finally developed. This notion was further reinforced as it became apparent that the atmospheres of Jupiter and the other outer planets were mixtures of these same reduced C-H-N gases, plus helium. Earth's present atmosphere was assumed by many scientists of the period to have evolved from a gravitationally accreted primordial one like Jupiter's.

The first modern experiments to shed light on this problem were performed by Stanley Miller (1953), who was then a graduate student in chemistry. Miller built a closed apparatus, illustrated in figure 8.14, into which he introduced various mixtures of CH_4, H_2, and NH_3. Water was boiled in the small flask to add water vapor to the gas mixture and encourage circulation through the larger flask, where the gases were subjected to a spark discharge. The resulting mixture was then condensed in a cold trap and fluxed back through the smaller flask. After a week of operation, during which nonvolatile products gradually accumulated in the small flask, samples were withdrawn and analyzed by gas chromatography.

Miller's experiments produced a number of fairly simple organic compounds, many of which are found in living organisms. In all, 15% of the carbon in the system had reacted to form amino acids, aldehydes, or polymeric hydrocarbons. This was a dramatic and somewhat surprising confirmation of Oparin's hypothesis, which seemed to confirm the importance of an early, highly reduced atmosphere.

It is a big leap from producing organic molecules to creating life, but these experiments at least suggested that some organic building blocks for life could have existed in the Archean ocean-atmosphere system. Unfortunately, that's all they tell us. Miller's now classic experiment has been repeated countless times since 1953. Variants using a wide range of starting gas compositions and different energy sources have shown that prebiotic molecules can be formed from almost any nonoxygenated atmosphere. These results have blunted the scientific community's excitement about Miller's initial experiment, because they do not yield unique answers to our questions about the composition of the Earth's early atmosphere or the earliest organic compounds. We have learned that the proportion and absolute abundance of amino acids decreases with an increasing oxidation state of the atmosphere, to the extent that none have been found in experiments on CO_2, N_2, H_2O mixtures. Even in these experiments, however, some biologically significant aldehydes and simple molecules such as HCN have been observed. So far, then, this line of investigation has shed little light on Oparin's original question. They suggest that there was no lack of organic material early in Earth's history, but they bring us no closer to understanding the environmental conditions for early life.

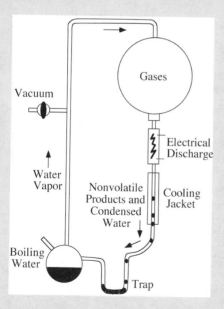

FIG. 8.14. Sketch of the apparatus used by Stanley Miller (1953) to generate prebiotic molecules from mixtures of CH_4, NH_3, H_2, and water vapor.

released by cyanobacteria in surface ocean waters during the Late Precambrian—rather than entering the atmosphere as it does today—was consumed in oxidizing Fe^{2+} to Fe^{3+}. The oxygen, then, was deposited in iron oxides, carbonates, and silicates on the sea floor.

For many years, there has been a debate concerning the primary source of iron in the Precambrian oceans and the ultimate reason why BIF ceased to be an effective oxygen sink. According to one view (for example, Holland 1984), marine Fe^{2+} was derived largely from the global river flux, fed ultimately by continental weathering reactions. As green-plant photosynthesis gained dominance, the river supply decreased, because iron was retained in the soils as Fe^{3+}, and Fe^{2+} remaining in the deep ocean was gradually oxidized. From this perspective, continental soils gradually displaced BIF as an oxygen sink and then were themselves overwhelmed because atmospheric O_2 could only easily affect the upper few meters of soil. This possibility seems to be less in favor today than one articulated in 1983 by Dutch geochemist Jan Veizer. He proposed that iron was introduced into the oceans by seafloor volcanism, presumed to have been more vigorous during the first half of Earth history than it is today. James Kasting (2001) embellished this notion by focusing on the role of seafloor hydrothermal vent activity and the serpentinization of oceanic basalts. Whatever the specific mechanism was, Fe^{2+} was released to the oceans rapidly enough to overwhelm the photosynthetic production of O_2. The rise of P_{O_2} occurred as the Earth's plate tectonic engine gradually slowed down and less iron was being injected into the world ocean. At that point, the Fe^{2+} oxygen sink, BIF, could no longer dominate O_2 production. The rather rapid P_{O_2} increase $\sim 2.3 \times 10^9$ years ago, then, was triggered by a geologic change—gradual cooling of the mantle—rather than a biologic one.

Whether ferrous iron in the oceans was a cause or a side effect of low atmospheric P_{O_2}, it is clear that O_2 was a minor species in the atmosphere as long as the deep oceans were anoxic. Only after the supply of oxygen to the seafloor exhausted the Fe^{2+} sink did it become possible for P_{O_2} to rise to modern levels. The depositional history of banded iron formation—virtually complete by 2.3×10^9 years ago—is therefore a record of the transition to an oxygen-rich atmosphere.

Antonio Lasaga and Hiroshi Ohmoto (2002) have developed an intricate box model of the coupled C-O geochemical cycle, with which they conclude that P_{O_2} has stayed within a narrow range from 0.6 to 2 times its present level since 2.3×10^9 years ago.

SUMMARY

The field of ocean-atmosphere geochemistry is extensive, and we cannot claim to have given it more than a quick glimpse in this chapter. By giving examples, however, we have shown three of the major directions for investigation pursued by geochemists and have suggested other trends in research.

The composition of seawater and its variations with depth and global location are of great interest as indicators of environmental control. In particular, chemical species such as HCO_3^- and CO_3^{2-} play a central role in the ocean's pH and alkalinity regulation and thus affect not only geologic but also biologic processes. The pathways along which change occurs, and the network of kinetic factors that define steady-state conditions, are our gateway to understanding the recent history of the ocean-atmosphere system. Geochemical cycle models can also help to elucidate the connections between geochemistry, biochemistry, and geophysics that force long-term changes in the system. Finally, by examining the chemistry of seawater and air from their earliest days, we gain a perspective on the Earth's history that bears on the evolution of life.

SUGGESTED READINGS

Many books have been written on ocean-atmosphere chemistry. The four listed here were written by some of the undisputed leaders of research in the area and are impressive for their clarity of presentation.

Broecker, W. S., and T.-H.Peng. 1982. *Tracers in the Sea*. New York: Eldigio Press, Columbia University. (A comprehensive book on chemistry of the oceans, written as a textbook, with many worked examples.)

Gray, J. C., and J. C. G. Walker. 1990. *Numerical Adventures with Geochemical Cycles*. London: Oxford University Press. (A good "how to" book for those interested in learning more about modeling geochemical cycles.)

Holland, H. D. 1978. *The Chemistry of the Atmosphere and Oceans*. New York: Wiley Interscience. (An exceptionally readable discussion of processes that relate to the composition of the ocean and atmosphere.)

Holland, H. D. 1984. *The Chemical Evolution of the Atmosphere and Oceans*. Princeton: Princeton University Press. (This is the best single volume of data and ideas regarding the history of the ocean-atmosphere system.)

The following articles were referenced in this chapter. An interested student may wish to explore them further.

Ben-Yaakov, S., E. Ruth, and I. R. Kaplan. 1974. Carbonate compensation depth: Relation to carbonate solubility in ocean waters. *Science* 184:982–984.

Berner, R. A., A. C. Lasaga, and R. M. Garrels. 1983. The carbonate-silicate geochemical cycle and its effect on carbon dioxide over the past 100 million years. *American Journal of Science* 283:641–83.

Berner, R. A., A. C. Lasaga, and R. M. Garrels. 1985. An improved geochemical model of atmospheric CO_2 fluctuations over the past 100 years. In E. T. Sundquist and W. S. Broecker, eds. *The Carbon Cycle and Atmospheric CO_2: Natural Variations Archean to Present.* American Geophysical Union Monograph 32. Washington, D.C.: American Geophysical Union, pp. 397–441.

Broecker, W. S. 1982. Glacial to interglacial changes in ocean chemistry. *Progress in Oceanography* 11:151–197.

Catling, D. C., K. J. Zahnle, and C. P. McKay. 2001. Biogenic methane, hydrogen escape, and the irreversible oxidation of early Earth. *Science* 293:839–842.

Drever, J. I. 1974. The magnesium problem. In E. D. Goldberg, ed. *The Sea*, vol. 5. New York: Wiley Interscience, pp. 337–357.

Eugster, H. P., and D. R. Wones. 1962. Stability relations of the ferruginous biotite, annite. *Journal of Petrology* 3:82–125.

Garrels, R. M., and M. E. Thompson. 1962. A chemical model for sea water at 25°C and one atmosphere total pressure. *American Journal of Science* 260:57–66.

Honjo, S., and J. Erez. 1978. Dissolution rates of calcium carbonate in the deep ocean: An in situ experiment in the North Atlantic. *Earth and Planetary Sciences Letters* 40:226–234.

Kasting, J. F. 2001. The rise of atmospheric oxygen. *Science* 293:819–820.

Kasting, J. F. 1986. Theoretical constraints on oxygen and carbon dioxide concentrations in the Precambrian atmosphere. *Precambrian Research* 34:205–229.

Kasting, J. F., and S. M. Richardson. 1986. Seafloor hydrothermal activity and spreading rates: The Eocene carbon dioxide greenhouse revisited. *Geochimica et Cosmochimica Acta* 49:2541–2544.

Keir, R. S., and W. H. Berger. 1983. Atmospheric CO_2 content in the last 120,000 years: The phosphate extraction model. *Journal of Geophysical Research* 88:6027–6038.

Keir, R. S., and W. H. Berger. 1985. Late Holocene carbonate growth in the equatorial Pacific: Reef growth or neoglaciation? In E. T. Sundquist and W. S. Broecker, eds. *The Carbon Cycle and Atmospheric CO_2: Natural Variations Archean to Present.* American Geophysical Union Monograph 32. Washington, D.C.: American Geophysical Union, pp. 208–220.

Lasaga, A. C., and H. Ohmoto. 2002. The oxygen geochemical cycle: Dynamics and stability. *Geochimica et Cosmochimica Acta* 66:361–381.

Meybeck, M. 1979. Concentration des eaux fluviales en éléments majeurs et apports en solution aux oceans. *Revues de Géologie Dynamique et de Géographie Physique* 21:215–246.

Miller, S. L. 1953. A production of amino acids under possible primitive Earth conditions. *Science* 117:528–529.

Peterson, M.N.A. 1966. Calcite: Rates of dissolution in a vertical profile in the Central Pacific. *Science* 154:1542–1544.

Pytkowicz, R. M. 1983. *Equilibria, Nonequilibria, and Natural Waters*, vol. 1. New York: Wiley Interscience.

Quimby-Hunt, M. S., and K. K. Turekian. 1983. Distribution of elements in sea water. *EOS, Transactions of the American Geophysical Union* 64:130–131.

Rubey, W. W. 1951. Geologic history of seawater, an attempt to state the problem. *Bulletin of the Geological Society of America* 62:1111–1147.

Sundquist, E. T. 1985. Geological perspectives on carbon dioxide and the carbon cycle. In E. T. Sundquist and W. S. Broecker, eds. *The Carbon Cycle and Atmospheric CO_2: Natural Variations Archean to Present.* American Geophysical Union Monograph 32. Washington, D.C.: American Geophysical Union, pp. 5–60.

Veizer, J. 1983. Geologic evolution of the Archean-early Proterozoic Earth. In J. W. Schopf, ed. *The Earth's Earliest Biosphere: Its Origin and Evolution.* Princeton: Princeton University Press, pp. 240–259.

Weyl, P. K. 1961. The carbonate saturometer. *Journal of Geology* 69:32–43.

PROBLEMS

(8.1) A naive geochemist might expect the nitrate concentration in surface waters of the ocean to be maintained at an equilibrium concentration for the reaction:

$$N_2 + \tfrac{5}{2}O_2 + H_2O \rightleftarrows 2H^+ + 2NO_3^-.$$

Use tabulated thermodynamic data and your knowledge of P_{O_2}, P_{N_2}, and seawater pH to estimate the expected oceanic NO_3^- concentration. How does this value compare with the concentration recorded in table 8.1? What factors might account for this difference?

(8.2) What is the pH of a 3×10^{-3} M Na_2CO_3 solution?

(8.3) Using the procedure described in worked problem 8.2, construct titration curves to show the change in solution pH as a 3×10^{-3} m Na_2CO_3 solution is titrated by (a) 0.01 m HCl, (b) 1.0 m HCl, and (c) 10.0 m HCl. Do the end points depend on the molality of the HCl solution?

(8.4) Suppose you were to add Fe^{2+} to the Na_2CO_3 solution in problem 8.2. How high can you raise the Fe^{2+} concentration before $FeCO_3$ begins to precipitate? Assume that the K_{sp} for $FeCO_3$ is 2.0×10^{-11} and that there are no kinetic barriers to precipitation.

(8.5) How is the alkalinity of a beaker of seawater affected by the addition of small amounts of the following substances: (a) $NaHCO_3$, (b) $NaCl$, (c) HCl, and (d) $MgCO_3$?

(8.6) In worked problem 8.1, we made the simplifying assumption that $a_{CO_3^{2-}}$ is negligible in rainwater, and derived an expression to calculate pH, given P_{CO_2}. Use that expression to calculate rainwater pH if $P_{CO_2} = 380$ ppm. Can you estimate the percentage error in your approximate solution?

(8.7) In worked problem 8.6, we calculated the composition of a gas mixture under the moderately oxidizing conditions of the QFM buffer. Redo this calculation, using instead the oxygen fugacity defined by QFI. A useful empirical equation for this buffer is:

$$\log f_{O_2} = 7.51 - 29382/T.$$

CHAPTER NINE

TEMPERATURE AND PRESSURE CHANGES

Thermodynamics Again

OVERVIEW

This chapter expands on the thermodynamic principles introduced in chapter 3 and provides the preparation for the next chapter on phase diagrams. We first give a thermodynamic definition for equilibrium. We also derive the phase rule and explore its use as a test for equilibrium. The effects of changing temperature and pressure on the free energy of a system at equilibrium are analyzed, and a general equation for $d\bar{G}$ at P and T is formulated. P-T and $\bar{G}$-T diagrams for one-component systems are introduced. The utility of the Clapeyron equation in constructing or interpreting phase diagrams is explored. We then introduce concepts that are necessary for understanding systems with more than one component. Raoult's law and Henry's law are defined, and expressions for the chemical potentials of components that obey these laws are derived. The concept of standard states is introduced, and we examine how activity coefficients are defined in mixing models. The usefulness of these concepts is illustrated by discussing formulation of a geothermometer and a geobarometer.

WHAT DOES EQUILIBRIUM REALLY MEAN?

In previous chapters, we have introduced and made use of the concept of thermodynamic equilibrium. By this time,

you should be comfortable with the idea that the observable properties of a system at equilibrium will undergo no change with time, as long as physical conditions imposed on the system remain constant. Although this statement is commonly used to define equilibrium, it is actually a consequence of equilibrium rather than a strict definition of it. A more rigorous definition of the equilibrium state that we have not seen before is: a condition of minimum energy for that portion of the universe under consideration.

We have also learned that Gibbs free energy is a convenient way of describing the energy of a system. The relationship of this thermodynamic quantity to temperature, pressure, and the amounts of the constituent components of a system is given by the following:

$$dG = (\partial G/\partial T)_{P,n_i}dT + (\partial G/\partial P)_{T,n_i}dP$$
$$+ (\partial G/\partial n_i)_{P,T,n_{j\neq i}}dn_i, \qquad (9.1)$$

where n_i represents the number of moles of component i. The quantities $(\partial G/\partial n_i)_{P,T,n_{j\neq i}}$ are chemical potentials, symbolized by μ_i and defined in chapter 3. The goal of this chapter is to learn how to evaluate all of the quantities in this equation.

DETERMINING WHEN A SYSTEM IS IN EQUILIBRIUM

How can we determine whether a geologic system is in a state of equilibrium? Although we can never be completely sure, we can at least argue for equilibrium on the basis of our experience with reactions in the laboratory and the absence of any observational evidence for disequilibrium. Consider, for example, a metamorphic rock consisting of biotite, garnet, plagioclase, sillimanite, and quartz. We know from experiments that these minerals can constitute a stable assemblage under conditions of high temperature and pressure. Let's assume that textural observations of this rock provide no indication of arrested reactions between any of the constituent minerals. Furthermore, grains of all five minerals are observed to be touching each other, and no zoning or other kind of chemical heterogeneity within grains is present. In this case, we might infer that the rock system is in a state of equilibrium, but this indirect evidence may be less than convincing. Is there another way to test for equilibrium?

The Phase Rule

Let's describe the equilibrium configuration of a system in more quantitative terms. The general relationships among phases, components, and physical conditions are given by the *phase rule,* first formulated by J. Willard Gibbs more than a century ago. The phase rule describes the maximum number of phases that can exist in an equilibrium system of any complexity. As we will see, this provides an additional test for equilibrium.

The *variance* (f), also called the *degrees of freedom,* of a system at equilibrium is the number of independent parameters that must be fixed or determined to specify the state of the system. The number of independent parameters equals the number of unknown variables minus the number of known (dependent) relationships between them. Assume that there are p phases and every phase contains one or more of the c chemical components that constitute the system. For each phase p in the system, we can write a Gibbs-Duhem equation:

$$0 = \bar{S}dT - \bar{V}dP + \sum n_c d\mu_c,$$

where μ_c is the chemical potential of each component in the phase. Recall also some of the major conclusions of chapter 3; namely, that at equilibrium, the temperatures and pressures in all phases must be the same, and μ_c must be the same in all phases that contain component c. The total number of potentially unknown variables, therefore, is c (the number of components) plus two (for temperature and pressure, whatever their values). The number of known relationships among them is p (the number of Gibbs-Duhem equations). The variance, therefore, is the difference:

$$f = (c + 2) - p. \tag{9.2}$$

Equation 9.2 is a mathematical statement of the phase rule. To illustrate its use, consider an ordinary glass of water at room temperature and atmospheric pressure. The number of components c necessary to describe the compositions of all the phases in this system is one (H_2O), and the number of phases p is also one (liquid water). Thus, from equation 9.2, the variance of the system is two. This means that two variables (in this case, temperature and pressure) can be changed independently with no resulting effect on the system. Our everyday experience tells us that this is true—it would still be a glass of water if transported from sea level to a mountain top, where the temperature is cooler and atmospheric pressure is less—but there is some finite limit to the permissible variation. For example, the phase in this system (liquid water) would certainly change if we were to lower the temperature drastically by putting the glass of water in a freezer. However, in the case of a frozen glass of water, the variance would still be two, because there is still only one component (H_2O) and one phase (ice in this case). Are there situations in which the variance of this system is other than two?

The *phase diagram* for water, illustrated in figure 9.1, summarizes how this simple system reacts to changing conditions of temperature and pressure. The diagram consists of fields in which only one phase is stable, separated by boundary curves along which combinations of phases coexist in equilibrium. The three boundary lines intersect at a point, called the *triple point,* at which three phases occur. The boundary between the liquid water and water vapor fields terminates in a *critical point,* above which there is no distinction between liquid and vapor. From inspection of this figure, it is obvious that a single phase is stable under most combinations of temperature and pressure. Within these one-phase fields, the variance is two, and the fields are said to be *divariant.*

However, there are certain special combinations of temperature and pressure at which several phases coexist at equilibrium. For example, because two phases coexist along the boundary curves in this diagram, the variance at any point along the boundary curves is one. (Prove this yourself using the phase rule.) These boundary curves are thus *univariant.* Coexistence of three phases

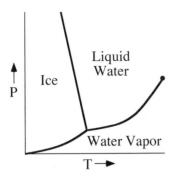

FIG. 9.1. Schematic phase diagram for the system H_2O. Most of the area of the diagram consists of regions in which only one phase is stable at equilibrium. Boundary curves are combinations of pressure and temperature along which two phases are stable, and three phases occur at the triple point. The boundary between liquid water and water vapor terminates in a critical point, above which the distinction between these phases is meaningless.

at the triple point produces a variance of zero, forming an *invariant* point.

For a more complex system, such as a rock, of course, more components are required to describe the system, and there will probably be more phases. In an assemblage of minerals, it is unlikely that every component will occur in every phase. However, the phase rule is a general result and is valid for systems in which some components do not occur in all phases.

Worked Problem 9.1

What is the variance for the metamorphic rock described at the beginning of this section? There are five phases, and the eight components that describe the compositional variation of each phase are listed beside each mineral below. From our discussion of components in chapter 3, you should recognize that the components listed below are not unique, and you might try devising your own set of components to replace the ones given here. Note that it would be necessary to choose additional components if some of these minerals were more complex solid solutions:

biotite: $KMg_3AlSi_3O_{10}(OH)_2$, $KFe_3AlSi_3O_{10}(OH)_2$

garnet: $Fe_3Al_2Si_3O_{12}$, $Ca_3Al_2Si_3O_{12}$

sillimanite: Al_2SiO_5

plagioclase: $NaAlSi_3O_8$, $CaAl_2Si_2O_8$

quartz: SiO_2

As we learned in chapter 3, not all of these components may be necessary to define the rock system, if they are not all independent. For example, we can write a mathematical expres-

sion between the $Ca_3Al_2Si_3O_{12}$ component of garnet, Al_2SiO_5, SiO_2, and the $CaAl_2Si_2O_8$ component of plagioclase (it does not matter that such a reaction might never take place in nature):

$$Ca_3Al_2Si_3O_{12} + 2Al_2SiO_5 + SiO_2 = 3CaAl_2Si_2O_8.$$

Thus we can express anorthite in terms of the other three components. In fact, any one of these four components can be expressed in terms of the other three by rearranging the equation, and so we can delete one of these (take your choice) as a system component. We then are left with five phases and seven components, which give a variance of four. This means that temperature and pressure can be varied independently (within some reasonable limits), as well as the compositions of two phases (for example, biotite and garnet, or garnet and plagioclase), without changing the number of phases in the system at equilibrium.

Open versus Closed Systems

The discussion so far assumes that the system is closed. However, there are examples of geologic systems in which some components are free to migrate into and out of the system. A common example of such a mobile component is the fluid that is present during the metamorphism of some rocks. We may describe this situation using the same Gibbs-Duhem equations that we formulated before, but in this case, we have an additional restriction. In a closed system, the chemical potentials of all components are the same in every phase, but in an open system, the chemical potentials of mobile components are controlled by phases or other parameters outside of the system. Consequently, we have an additional constraining equation for each mobile component. These additional chemical restrictions r are reflected in the phase rule as additional known relationships. The number of known relationships therefore becomes $p + r$. Substitution of this value into equation 9.2 produces the relation:

$$f = (c + 2) - (p + r). \tag{9.3}$$

This open-system modification provides an alternative form of the phase rule for systems in which mobile components can be handled.

Worked Problem 9.2

Consider a rock containing calcite, quartz, and wollastonite, which is undergoing metamorphism in equilibrium with a fluid of H_2O and CO_2. Does the variance of this assemblage change if the system is closed or open to fluid migration?

The system has four phases (H_2O and CO_2 in the fluid are miscible in all proportions under these conditions). The components of each of the phases can be expressed as:

calcite: $CaCO_3$

quartz: SiO_2

wollastonite: $CaSiO_3$

fluid: H_2O, CO_2

But we can reduce the number of system components from the five listed above to four, because of the following relationship:

$$CaCO_3 = CaSiO_3 + CO_2 - SiO_2.$$

Under closed-system conditions, the variance of this system is then given by equation 9.2:

$$f = (4 + 2) - 4 = 2.$$

Now let's suppose that this rock is metamorphosed in an open system with a mobile fluid, whose composition is controlled from outside the rock itself. (If this is hard to visualize, consider a situation in which there is a vast reservoir of fluid and only a thin layer of carbonate rock, so that the reaction in the rock exerts virtually no influence on the composition of the fluid that passes through.) As before, we again have four phases and four system components. However, we also have the additional restriction r that the proportions of CO_2 and H_2O in the fluid are fixed. Solving equation 9.3, we find that the variance is:

$$f = (4 + 2) - (4 + 1) = 1.$$

Therefore, metamorphism of this rock under closed-system conditions is divariant, but the system becomes univariant if the mobile fluid composition is controlled from outside the system.

Now let's return to our discussion of the utility of the phase rule as a test for equilibrium. We still cannot demonstrate equilibrium with certainty, but by examining the variance the system, we can now cite an additional observation that is consistent with equilibrium. In the phase diagram for water (fig. 9.1), it is apparent that most of the area of the diagram is occupied by divariant fields, and those parts of the diagram with lower variance comprise only a tiny fraction of the total area. The same holds true for complex systems such as rocks. Although univariant and invariant mineral assemblages surely are formed under just the right combinations of temperature and pressure, the probability of finding such

assemblages in the field is slight compared with the likelihood of sampling divariant assemblages that are stable over a range of temperatures and pressures. For example, locating univariant mineral reactions precisely in a metamorphic terrain may be difficult, because most outcrops will contain either reactants or products, but not both. Some isograds may mark the locations of univariant reactions, but most are probably the first outcrops to be found on the other side of the reaction. We can therefore conclude that most rocks should have variances of two or higher if they have equilibrium mineral assemblages. Those assemblages that appear to be univariant or invariant probably represent disequilibrium in most cases, unless it can be shown by field mapping that these rocks lie between appropriate divariant assemblages.

CHANGING TEMPERATURE AND PRESSURE

You are probably aware that water boils at different temperatures at the seashore and in the mountains. The boundary curve between liquid water and water vapor in figure 9.1 defines the boiling point of water at any given pressure, and its slope explains this phenomenon. Changing conditions of temperature and pressure obviously influence more complex geological systems as well.

It is most convenient to describe modifications in the state of a geochemical system in terms of changes in free energy (ΔG), because differentials of ΔG with respect to T and P are easily evaluated. Recall that the concept of free energy was developed in chapter 3 for fixed T and P conditions. In the following sections, we see how to evaluate ΔG under different sets of temperature and pressure conditions.

Temperature Changes and Heat Capacity

First we examine the effect of changing only temperature on a system at equilibrium. Differentiating the expression $d\Delta\bar{G} = \Delta\bar{V}dP - \Delta\bar{S}dT$ with respect to temperature at constant pressure gives:

$$(\partial\Delta\bar{G}/\partial T)_P = -\Delta\bar{S}. \tag{9.4}$$

Because it is often not easy to evaluate the entropy change $\Delta\bar{S}$ directly, we must resort to using another more readily measurable quantity, enthalpy ($\Delta\bar{H}$). Since

$\Delta \bar{G} = \Delta \bar{H} - T \Delta \bar{S}$ (equation 3.13), equation 9.4 is equivalent to:

$$(\partial \Delta \bar{G}/\partial T)_P = (\Delta \bar{G} - \Delta \bar{H})/T.$$

Because we are holding pressure is constant in this example, we can rewrite this as:

$$d\Delta \bar{G} = (\Delta \bar{G} - \Delta \bar{H}) dT/T.$$

Dividing both sides of this expression by T and rearranging yields:

$$d\Delta \bar{G}/T - (\Delta \bar{G}) dT/T^2 = -(\Delta \bar{H}) dT/T^2.$$

The left side of this equation is equivalent to $d(\Delta \bar{G}/T)$, so that:

$$d(\Delta \bar{G}/T) = -(\Delta \bar{H}) dT/T^2.$$

If we consider the expression above at standard conditions, then $\Delta \bar{G}^0$ and $\Delta \bar{H}$ become $\Delta \bar{G}^0$ and $\Delta \bar{H}^0$, respectively. For a system at equilibrium, $\Delta \bar{G}^0 = -RT \ln K_{eq}$, so that:

$$d \ln K_{eq}/dT = \Delta \bar{H}^0/RT^2, \qquad (9.5a)$$

or

$$-d \ln K_{eq} = [\Delta \bar{H}^0/R] d(1/T). \qquad (9.5b)$$

Equations 9.5a,b, known in either form as the *van't Hoff equation*, provide a useful way to determine enthalpy changes.

Worked Problem 9.3

Determine $\Delta \bar{H}$ for the breakdown of annite (the Fe end-member component of biotite) by the reaction:

$$KFe_3AlSi_3O_{10}(OH)_2 \rightarrow KAlSi_3O_8 + Fe_3O_4 + H_2.$$
$$\text{annite} \qquad \text{sanidine magnetite}$$

From equation 9.5b, we can see that $\Delta \bar{H}^0$ can be obtained from a plot of $\ln K_{eq}$ versus $1/T$. The slope of this line is equivalent to $-\Delta \bar{H}^0/R$.

Hans Eugster and David Wones (1962) found experimentally that the equilibrium constant for this reaction could be expressed by:

$$\log K_{eq} = -(9215/T) + 10.99.$$

In other words, a plot of $\log K_{eq}$ versus $1/T$ gave a straight line of slope $-9215K$. Therefore,

$$-\left(\frac{\Delta \bar{H}^0}{2.303R}\right) = -9215,$$

and

$$\Delta \bar{H}^0 = 2.303(1.987 \text{ cal K}^{-1}\text{mol}^{-1})(9215K)$$
$$= 42.2 \text{ kcal mol}^{-1}.$$

The van't Hoff equation works well for standard conditions, but how can we determine enthalpy differences under nonstandard conditions? First let us define a new variable, the *heat capacity at constant pressure*, $\bar{C}_P$:

$$\bar{C}_P = (dq/dT)_P = (\partial \bar{H}/\partial T)_P.$$

If a material is heated at constant pressure, then its molar enthalpy increases with temperature, according to the relationship

$$d\Delta \bar{H} = \Delta \bar{C}_P dT, \qquad (9.6)$$

where $\Delta \bar{H}$ is the increase in enthalpy through a temperature interval dT. Tabulations of thermodynamic data for minerals normally give enthalpy at 298 K, and it is necessary to integrate equation 9.6 to obtain enthalpy at any other temperature. Thus, we have:

$$\Delta \bar{H}_T = \Delta \bar{H}^0_{298} \int_{298}^{T} \Delta \bar{C}_P dT. \qquad (9.7)$$

However, heat capacities are not generally independent of temperature, so the integrated form of the expression for $\Delta \bar{C}_P$ is somewhat complex. The experimental data from which integrated $\bar{C}_P$ values are determined are commonly given in the form of a power series:

$$\bar{C}_P = a + bT + c/T^2, \qquad (9.8)$$

where a, b, and c are experimentally determined constants.

Coincident with a change in enthalpy with temperature (equation 9.6), the system will also experience a change in entropy, because

$$d\Delta \bar{H} = T d\Delta \bar{S}$$

at constant pressure. Consequently, entropy can also be expressed in terms of the heat capacity:

$$d\Delta \bar{S} = (\Delta \bar{C}_P dT)/T.$$

We can obtain the entropy at any temperature by integrating this equation:

$$\Delta \bar{S}_T = \Delta \bar{S}^0_{298} = \int_{298}^{T} (\Delta \bar{C}_P/T) dT. \qquad (9.9)$$

We now have all the necessary tools to determine the effect of a change in temperature on the free energy of a system. $\Delta \bar{G}$ at any temperature can be evaluated by substituting enthalpy (equation 9.7) and entropy (equation 9.9) values at the appropriate temperature into the familiar equation $\Delta \bar{G} = \Delta \bar{H} - T\Delta \bar{S}$:

$$\Delta \bar{G}^T = \Delta \bar{H}_{298}^0 + \int_{298}^T \Delta \bar{C}_p dT \\ - T\left(\Delta \bar{S}_{298}^0 + \int_{298}^T (\Delta \bar{C}_p/T)dT\right). \quad (9.10)$$

Worked Problem 9.4

As an example of the use of equation 9.10, determine the free energy change for the following reaction at 800°C and atmospheric pressure:

$MgSiO_3 + CaCO_3 + SiO_2 \rightarrow CaMgSi_2O_6 + CO_2$.
enstatite calcite quartz   diopside
   (En)   (Cc)   (Qz)   (Di)

The thermodynamic data for these phases are given in table 9.1.

After confirming that the reaction is balanced, we calculate Δ values for products ($Di + CO_2$) minus reactants ($En + Cc + Qz$):

$\Delta \bar{H}_{298}^0 = 14,720$ cal mol^{-1}

$\Delta \bar{S}_{298}^0 = 37.01$ cal mol^{-1} deg^{-1}

$\Delta \bar{C}_p = 2.69 - 8.24 \times 10^{-3}T - 2.62 \\ \times 10^5/T^2$ cal deg^{-1} mol^{-1}

Substitution of these values, along with $T = 1073$K, into equation 9.10 gives:

$\Delta \bar{G}_{1073} = 14,720 - 6.38 - 1073(37.01 - .0059) \\ = -23,304$ cal mol^{-1}.

An interesting variation on this problem is to specify the value of P_{CO_2}, say, at 0.1 bar. To solve this problem now, we must first write an expression for the equilibrium constant K_{eq}, which in this case is:

$K_{eq} = P_{CO_2} = 0.1$ bar.

From the relationship:

$\Delta \bar{G} = \Delta \bar{G}^0 + RT \ln K_{eq}$,

we obtain a value of $\Delta \bar{G} = -28,833.65$ cal mol^{-1}.

Pressure Changes and Compressibility

The effect of changing pressure on a system at equilibrium can be determined by differentiating $d\Delta \bar{G} = \Delta \bar{V}dP - \Delta \bar{S}dT$ with respect to pressure at constant temperature:

$$(\partial \Delta \bar{G}/\partial P)_T = \Delta \bar{V}. \quad (9.11)$$

Consequently, the change in free energy of a reaction with respect to pressure alone is equal to the change in the molar volume of products and reactants. Molar volume changes can be determined from simple physical measurements of product and reactant molar volumes. For reactions involving only solid phases under most geologic conditions, the effects of temperature and pressure on $\Delta \bar{V}$ are small and can be ignored. In this case, integration of equation 9.10 gives:

$$\Delta \bar{G} = \Delta \bar{G}_{T,1}^0 + \Delta \bar{V} \int_1^P dP, \quad (9.12)$$

where $\Delta \bar{G}_{T,1}^0$ is the free energy for the reaction at 1 bar and temperature T.

Worked Problem 9.5

Consider the polymorphic transformation of aragonite to calcite. Use equation 9.12 to calculate the pressure at which these two minerals coexist stably at 298 K.

At equilibrium, $\Delta \bar{G} = 0$, so:

$\int_1^P dP = -\Delta \bar{G}_{T,1}^0/\Delta \bar{V}$,

or

$P - 1 = -\Delta \bar{G}_{T,1}^0/\Delta \bar{V}$.

TABLE 9.1. Thermodynamic Data for Worked Problems 9.4 and 9.7

Mineral	$\bar{H}_{298}^0$ (cal mol^{-1})	$\bar{S}_{298,1}^0$ (cal mol^{-1} K^{-1})	$\bar{V}_{298,1}$ (cal mol^{-1} bar^{-1})	$\bar{C}_p$
En	−370,100	16.22	0.752	$24.55 + 4.74 \times 10^{-3}T - 6.28 \times 10^5 T^{-2}$
Cc	−288,420	22.15	0.883	$24.98 + 5.24 \times 10^{-3}T - 6.20 \times 10^5 T^{-2}$
Qz	−217,650	9.88	0.542	$11.22 + 8.20 \times 10^{-3}T - 2.70 \times 10^5 T^{-2}$
Di	−767,400	34.20	1.579	$52.87 + 7.84 \times 10^{-3}T - 15.74 \times 10^5 T^{-2}$
CO$_2$	−94,050	51.06	584.73	$10.57 + 2.10 \times 10^{-3}T - 2.06 \times 10^5 T^{-2}$

From Helgeson et al. (1978).

Appropriate thermodynamic data are: $\Delta \bar{G}^0_{298,1}$ (kJ mol^{-1}) = -1127.793 for aragonite and -1128.842 for calcite; $\bar{V}$ (J bar^{-1}) = 3.415 for aragonite and 3.693 for calcite. For the reaction aragonite → calcite, $\Delta \bar{G}^0 = -1049$ J mol^{-1} and $\Delta \bar{V} = 0.278$ J bar^{-1}. By substituting these values into the equation above, we calculate that $P = 3773$ bars.

Aragonite has the lower molar volume and is thus the phase that is stable on the high-pressure side of this reaction. Therefore, at 298 K, calcite is the stable polymorph to ~3.7 kbar pressure. From this computation, we can readily see why aragonite precipitated by shelled organisms at low pressure is metastable.

Equation 9.12 is an extremely useful relationship for determining the effect of pressure on many equilibria of geologic interest. For any reactions involving fluids (such as many metamorphic reactions), however, or for solid phase reactions at very high pressures and temperatures (such as those under mantle conditions), we cannot assume that $\Delta \bar{V}$ is independent of temperature and pressure. In these cases, we must take into account the expansion or contraction of a fluid or crystal lattice as temperature changes and the compression or relaxation due to changing pressure. For gases, it is permissible to use an equation of state, such as the ideal gas law $\bar{V} = RT/P$, to describe the effects of changing temperature or pressure:

$$(\partial \bar{V}/\partial P)_T = -RT/P,$$

and

$$(\partial \bar{V}/\partial T)_P = R/P.$$

More complex equations of state for fluids have already been introduced in chapter 3.

For solid phases, we can correct $\Delta \bar{V}$ for changing temperature using the thermal expansion, α_P, defined as:

$$\alpha_P = 1/\bar{V}(\Delta \bar{V}/\Delta T)_P,$$

where $\bar{V}$ is the molar volume of the phase at the temperature at which α_P is measured. The temperature dependence of α_P is large near absolute zero, but becomes small at the temperatures at which most geologic processes operate. If we ignore this small variation, this expression can be rearranged to give:

$$\Delta \bar{V} = \bar{V}^0_{298,1}(1 + \alpha_P \Delta T),$$

where $\bar{V}^0$ is the molar volume at 298 K and 1 bar, and $\Delta T = T - 298$.

The effect of changing pressure on $\Delta \bar{V}$ is corrected by using the *compressibility* β_T, defined by:

$$\beta_T = -1/\bar{V}(\Delta \bar{V}/\Delta P)_T,$$

where $\bar{V}$ is the molar volume of the phase at the pressure at which β_T is measured and $\Delta P = P - 1$. The effect of pressure on β_T can as a first approximation be neglected at pressures <2 or 3 GPa, although it is very important in interpreting mantle seismic data. If we regard β_T as constant in the pressure range where most observable geologic processes operate, this expression becomes

$$\Delta \bar{V} = \bar{V}^0_{298,1}(1 - \beta_T \Delta P),$$

where $\bar{V}^0$ is the molar volume at 1 bar and 298°C. By combining the definitions of α_P and β_T in a total differential expression for $\bar{V}$, we obtain

$$d\bar{V} = (\partial \bar{V}/\partial T)_P dT + (\partial \bar{V}/\partial P)_T dP$$
$$= \alpha_P \bar{V} dT - \beta_T \bar{V} dP$$

or

$$\int_{\bar{V}_{298,1}}^{\bar{V}_{T,P}} d\bar{V}/\bar{V} = \int_{298}^{T} \alpha_P dT - \int_{1}^{P} \beta_T dP.$$

This produces the result:

$$\bar{V}_{T,P} = \bar{V}^0_{298,1} \exp(\alpha_P \Delta T - \beta_T \Delta P).$$

Because the term in parentheses is very small, we can use the approximation $\exp x \approx 1 + x$ to yield:

$$\bar{V}_{T,P} = \bar{V}^0_{298,1}(1 + \alpha_P \Delta T - \beta_T \Delta P). \qquad (9.13)$$

This last expression allows us to gauge the effect of changing thermal expansion and compressibility on the volume change for a reaction occurring under high-temperature and high-pressure conditions.

Worked Problem 9.6

What is the change in molar volume of forsterite in going from 25 to 800°C and from 1 bar to 5 kbar? To answer this question, we need the following data for forsterite:

$$\bar{V}^0_{298,1} = 4.379 \text{ cm}^3$$

$$\alpha_P = 41 \times 10^{-6} \text{ K}^{-1}$$

$$\beta_T = 0.8 \times 10^{-6} \text{ bar}^{-1}$$

Inserting these values into equation 9.13 gives the following result:

$$\bar{V}_{800,5000} = 4.379[1 + (41 \times 10^{-6})(1073 - 298)$$
$$- (0.8 \times 10^{-6})(5000 - 1)] = 4.501 \text{ cm}^3.$$

The change in molar volume is thus:

$$\Delta \bar{V} = 4.501 - 4.379 = 0.122 \text{ cm}^3 \text{ mol}^{-1}.$$

Temperature and Pressure Changes Combined

A general equation for the effect of both temperature and pressure on free energy can be obtained by simply combining the equations we have already derived for the effects of temperature alone (equation 9.10) or pressure alone (equation 9.12), as follows:

$$\Delta \bar{G}_{P,T} = \Delta \bar{H}^0_{298} + \int_{298}^{T} \Delta C_P dT - T \left(\Delta \bar{S}^0_{298} + \int_{298}^{T} (\Delta C_P / T) \, dT \right) + \Delta \bar{V} \int_{1}^{P} dP. \quad (9.14)$$

In using this equation, remember that care must be exercised in obtaining the integrated form of ΔC_P, normally done by employing the relationship in equation 9.8. As written, equation 9.14 assumes that thermal expansion and compressibility do not need to be taken into account. However, evaluating $\Delta \bar{V}$ in the last term for reactions at very high temperatures and pressures or when a fluid phase is present may require equation 9.13.

Worked Problem 9.7

Consider again the reaction enstatite + calcite + quartz → diopside + CO_2. In worked problem 9.4, we calculated the free energy of this reaction at 800°C and 1 bar. Now calculate the value of $\Delta \bar{G}$ at 800°C and 2000 bars pressure, assuming that $P_{CO_2} = 0.1$ bar.

To do this, we need to make some assumption about $(\partial \bar{V} / \partial P)_T$. It is clear that $\Delta \bar{V}$ is not constant, because CO_2 is quite compressible. Let's assume instead that all of the compressibility is due to CO_2, and pretend that it is an ideal gas, so that $\Delta \bar{V} = RT / \Delta P$. Which is stable under these conditions, reactants or products?

From the data in table 9.1, we can calculate $\Delta \bar{H}^0_{298,1}$, ΔC_P, and $\Delta \bar{S}^0_{298}$. (In fact, we have already done this in worked problem 9.4). The change in molar volume $\Delta \bar{V}$ of CO_2 between 1 bar and 2000 bar pressure is (1.987 cal K^{-1} mol^{-1})(1073K)/2000 bar = 1.066 cal bar^{-1}. Substituting these values into equation 9.14 gives:

$$\Delta \bar{G}_{800,200} = -30,027.05 \text{ cal mol}^{-1}.$$

Because $\Delta \bar{G}$ is negative, the reaction proceeds as written and diopside + CO_2 is stable under these conditions.

A GRAPHICAL LOOK AT CHANGING CONDITIONS: THE CLAPEYRON EQUATION

A phase diagram of temperature versus pressure summarizes a large quantity of data on the stabilities of various phases and combinations of phases in a system. It is important to recognize, however, that such a diagram gives information about the thermodynamic properties of these phases as well. Let's see how.

The phase diagram for the Al_2SiO_5 system is illustrated in figure 9.2. Like the phase diagram for water, this is a one-component system containing three possible phases: the polymorphs kyanite, andalusite, and sillimanite. Ignore, for the moment, the dashed horizontal lines. At the beginning of this chapter, we noted that equilibrium is defined as a condition of minimum energy for a system. Therefore, the divariant fields in figure 9.2 represent regions of P-T space where the free energy of the system is minimized by the occurrence of only one phase, and the univariant curves and the invariant point define special combinations of pressure and temperature at which two or more coexisting phases provide the lowest free energy.

To illustrate this relationship between free energy and phase diagrams graphically, we construct qualitative $\bar{G}$-T diagrams at constant pressure for the aluminosilicate system. The traces of such isobaric sections are represented by the horizontal dashed lines in figure 9.2. At pressure P_1 and any temperature below T_1, kyanite is the stable phase, so it must have a lower free energy than the

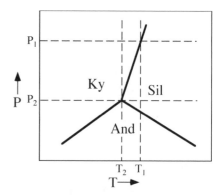

FIG. 9.2. Schematic phase diagram for the system Al_2SiO_5. Dashed lines represent the traces of isobaric sections used to construct $\bar{G}$-T diagrams.

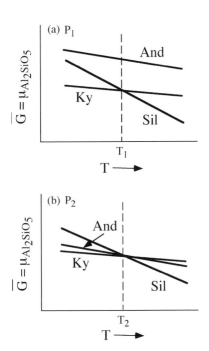

FIG. 9.3. $\bar{G}$-T diagrams for the system Al_2SiO_5 at constant pressure: (a) corresponds to pressure P_1 in figure 9.2, and (b) corresponds to P_2. At any temperature, the stable phase or combination of phases is that with the lowest free energy.

other Al_2SiO_5 polymorphs. This is shown in figure 9.3a. Because $(\partial \bar{G}/\partial T)_P = -\Delta \bar{S}$ and the entropy of any phase must be greater than zero, free energy decreases in all phases as temperature increases. At temperature T_1, the $\bar{G}$-T lines for kyanite and sillimanite intersect, and the co-existence of both phases provides the lowest free energy configuration. At temperatures above T_1, sillimanite has the lowest free energy and is the stable phase. Andalusite never appears at this pressure because its free energy is higher than those of the other polymorphs. The $\bar{G}$-T diagram at pressure P_2, shown in figure 9.3b, is similar to the P_1 case except that the $\bar{G}$-T line for andalusite is depressed so that it intersects those of kyanite and sillimanite at temperature T_2, the aluminosilicate triple point. At temperatures other than T_2, however, the other polymorphs still offer lower free energy configurations.

There are other, even more useful relationships between phase diagrams and thermodynamic quantities. In chapter 3, we derived a Maxwell relationship from $d\bar{H}$ (equation 3.8):

$$(\partial P/\partial T)_{\bar{V}} = (\Delta \bar{S}/\Delta \bar{V})_T,$$

which is equivalent to:

$$dP/dT = \Delta \bar{S}/\Delta \bar{V} = \Delta \bar{H}/ T\Delta \bar{V}. \qquad (9.15)$$

This expression, known as the *Clapeyron equation,* is extremely useful in constructing and interpreting phase diagrams. Notice that the left side of the Clapeyron equation is the slope of any line in a *P-T* phase diagram. Thus, this relationship enables us to calculate phase boundaries from thermodynamic data, provided that one point on the boundary is known. Alternatively, ratios of thermodynamic quantities can be obtained from experimentally determined reactions.

Worked Problem 9.8

To illustrate how the Clapeyron equation is used, calculate the approximate slope of the kyanite-andalusite boundary curve in figure 9.2. The data we need are: $\bar{V}(J\ bar^{-1}) = 4.409$ for kyanite, 5.153 for andalusite, and 4.990 for sillimanite; $\bar{S}_{800}$ (J mol^{-1} K^{-1}) = 242.42 for kyanite, 251.31 for andalusite, and 252.94 for sillimanite. Inspection of these data indicates the relative positions of the phase fields for the three polymorphs. Kyanite, which has the smallest molar volume, should be stable at the highest pressures, and sillimanite should be stable at the highest temperatures because of its high molar entropy. Substituting values of $\bar{S}$ and $\bar{V}$ for kyanite and andalusite into equation 9.15, we obtain:

$$dP/dT = (251.37 - 242.42\ J\ mol^{-1}\ K^{-1})/$$
$$(5.153 - 4.409\ J\ bar^{-1}) = 12.03\ bar\ K^{-1}$$

for the slope of the kyanite-andalusite boundary curve.

To complete the construction of this line, we must fix its position by determining the location of some point along it. The triple point has been located by experiment, although not without controversy (see the accompanying box), and its position can be used to locate this boundary in *P-T* space. There is a considerable amount of potential error in most calculated *P-T* slopes because of uncertainties in thermodynamic data. For this reason, it is preferable to determine phase diagrams experimentally. However, calculations such as these can be valuable in estimating how systems behave under conditions other than those under which experiments have been carried out. Calculations also enable the geochemist to limit experimental runs to the probable location of a line or point, thus saving time and money.

REACTIONS INVOLVING FLUIDS

Many geochemical reactions involve production of a fluid phase; that is, they are dehydration or decarbonation

PHASE EQUILIBRIA IN THE ALUMINOSILICATE SYSTEM

The tortuous history of attempts to determine phase equilibria in the aluminosilicate system illustrates the difficulties that can be encountered in this kind of work. E-An Zen summarized the state of affairs in 1969, after decades of experiments, in the diagram shown as figure 9.4. Because of the constant shifting of the position of the aluminosilicate phase boundaries with each new experimental determination, this situation came to be known popularly as the "flying triple point." Zen discussed a number of causes for this experimental difficulty. One possibility was experimental error, such as incorrect pressure calibration. This might be due to failure to take into account the strength of the material, friction, and other factors in converting the gauge pressure to actual pressure in

the sample. He also observed that most of the experiments to that time were synthesis experiments; that is, they were based only on the first appearance of a phase and either failed to demonstrate reversibility or represented some synthesis reaction that did not occur in metamorphic rocks. For example, in one set of experiments, sillimanite was synthesized from gels or kaolinite. Another potential source of error was that some of the polymorphs were synthesized in the presence of water. If they contained small quantities of H_2O, the polymorphs would not be true one-component phases.

The problem has been evaluated more recently by Daniel Kerrick (1990), who summarized data obtained since Zen's review and critically evaluated potential problems with each experimental determination. Kerrick also considered constraints on the position of the triple point from consistency with thermodynamic data and field calibrations from metamorphic mineral assemblages.

Many metamorphic petrologists now accept the experimental determination of Michael Holdaway and Biswajit Mukhopadhyay (1993) as the best available phase equilibria for this system. The phase diagram of Holdaway and Mukhopadhyay is shown in heavy lines in figure 9.4. Its triple point is located at 504 ± 20°C and 3.75 ± 0.25 kbar. Discrepancies between previous data and Holdaway's experiments may be explained by the difficulty in measuring small amounts of reaction, the occurrence of fibrolite (a rapidly grown, fine-grained sillimanite often intergrown with quartz), and the presence of Fe_2O_3 in sillimanite.

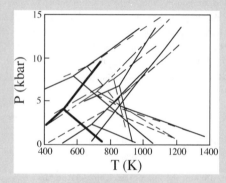

FIG. 9.4. Experimental determinations of phase equilibria in the system Al_2SiO_5. These data, summarized by Zen (1969), illustrate the difficulties in these measurements. The phase relationships determined by Holdaway and Mukhopadhyay (1993), shown in heavy lines, provide the most commonly accepted value for the aluminosilicate triple point.

reactions. At low pressures and high temperatures, H_2O and CO_2 are not very dense, so reactions that liberate these fluids have large, positive $\Delta \bar{V}$. Consequently, the slopes of the reaction lines on P-T diagrams, $dP/dT = \Delta \bar{S}/\Delta \bar{V}$, are small. If the same reactions occur at higher pressures, fluids are more compressed, so that $\Delta \bar{V}$ is smaller and the slopes of the reaction lines are steeper. For this reason, dehydration or decarbonation reactions on P-T diagrams typically are curved upward. Because fluids are more compressible than solids, high pressures

may ultimately so compress the fluid that $\Delta \bar{V}$ for these reactions becomes negative. However, $\Delta \bar{S}$ is still positive, so the reaction curves bend back on themselves with negative slopes. This phenomenon is nicely illustrated by the reaction:

$$NaAlSi_2O_6 \cdot H_2O + SiO_2 \rightarrow NaAlSi_3O_8 + H_2O.$$
$$\text{analcite} \qquad \text{quartz} \qquad \text{albite}$$

For most fluid-release reactions, the slope reversal occurs at very high pressures, but the analcite dehydration

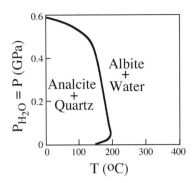

FIG. 9.5. The dehydration reaction for analcite changes slope (*dP/dT*) and finally bends back on itself with increasing pressure, reflecting the large change in volume of the fluid phase.

reaction, common in low-pressure metamorphic rocks, exhibits a bend-back at very modest pressures (fig. 9.5). The reaction has a small slope at very low pressure, which rapidly becomes negative as pressure increases.

RAOULT'S AND HENRY'S LAWS: MIXING OF SEVERAL COMPONENTS

So far in this chapter, we have focused mostly on the effects of changing temperature and pressure on systems with one component. In systems with several components, it is necessary to understand how the various components interact. For example, if we define our system as a grain of olivine, we must determine the behavior of the components Mg_2SiO_4 and Fe_2SiO_4 when mixed together. The mixing characteristics of these two components may be very different from the mixing behavior of some trace olivine component such as Ni_2SiO_4. The interaction of various components in combinations of coexisting minerals, magmatic liquid, or hydrous fluid is in some cases nonideal, resulting in their selective concentration in one phase or another. We now examine the thermodynamic basis for various kinds of mixing behavior.

In an ideal solution, mixing of components with similar volumes and molecular forces occurs without any change in the energy states or total volume of the system. Under these conditions, mixing is neither endothermic nor exothermic, that is, $\Delta \bar{H}_{mixing} = 0$. The activities of components (a_i) mixing on ideal sites are thus equal to their concentrations:

$$a_i = X_i,$$

which is a statement of *Raoult's Law.*

Interactions between molecules or ions, however, may cause the chemical potentials of individual species to increase or decrease on a certain site, resulting in non-ideal mixing. In this case, $\Delta \bar{H}_{mixing}$ is not equal to zero, and the activities will deviate from ideal behavior, as illustrated in figure 9.6. However, as the nonideal component becomes increasingly diluted (as X_i becomes much less than one), the component becomes so dispersed that eventually it is surrounded by a uniform environment of other ions or molecules. Therefore, at high dilution, its activity becomes directly proportional (but not equal) to its concentration, as shown by the straight dashed line in figure 9.6. In other words,

$$a_i = h_i X_i,$$

where h_i is a proportionality constant. This is a statement of *Henry's Law.*

These two laws can be used to explain the mixing behavior of the components of many minerals. Raoult's Law behavior is a common assumption for many major components of solid solution series, as in the case of the Mg_2SiO_4 (forsterite) and Fe_2SiO_4 (fayalite) components of olivine. Henry's Law is commonly used to describe the behavior of many trace components, such as nickel in olivine.

STANDARD STATES AND ACTIVITY COEFFICIENTS

Chemical potentials are functions of temperature, pressure, and composition. It is common practice to define the chemical potential relative to a chemical potential at

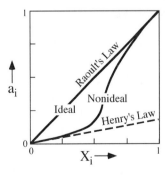

FIG. 9.6. Relations between activity and mole fraction for solutions obeying Raoult's Law and Henry's Law. Raoult's Law may be a good approximation for nonideal solutions in which X_i is large, and Henry's Law is applicable to highly diluted components.

some reference value of temperature and pressure (called the *standard state*), adding a term that corrects this value for deviations in temperature, pressure, and composition:

$$\mu_i = \mu_i^0 + RT \ln a_i, \tag{9.16}$$

where μ_i is the chemical potential of component i at T and P in the phase of interest, μ_i^0 is its value at some standard set of conditions, and a_i is the activity of the component. Note that when the component of interest is in its standard state, its activity a_i must equal one, so that its logarithm is zero and the second term drops out.

The standard state can be chosen arbitrarily, because it has no effect on the final result as long as we are consistent in its use. It might seem most logical to choose the standard state as a pure end-member component at 298 K and 1 bar. However, because of the availability of tabulated thermodynamic data, it is often more convenient to select the standard state as a pure component at the temperature and/or pressure of interest.

The relationships between chemical potential and the activity term in equation 9.16 can be most readily seen in graphical form. Figure 9.7a shows μ_i versus $\ln X_i$. The line illustrated is not straight (because it is for a real substance that does not mix ideally at all concentrations): it has two linear segments (shown extended with dashed lines). In the straight line segment extending from pure component i ($X_i = 1$, so that $\ln X_i = 0$) to slight dilution with another component, component i obeys Raoult's Law ($a_i = X_i$). The slope of this segment is RT and its intercept on the μ_i axis is μ_i^0. In the region where Raoult's Law holds, chemical potential can be expressed as:

$$\mu_i = \mu_i^0 + RT \ln X_i. \tag{9.17}$$

The Henry's Law region extends from infinite dilution of component i to some modest value and, because $a_i = h_i X_i$, it is also represented by a straight line in figure 9.7a. Its slope is also RT, but its intercept on the μ_i axis is not μ_i^0. Instead, this intercept is now μ_i^0 plus the added term $RT \ln h_i$, which is a function of temperature, pressure, and the composition of the component with which i is diluted, but is not a function of X_i. The expression for chemical potential in the Henry's Law region, therefore, is:

$$\mu_i = \mu_i^0 + RT \ln X_i + RT \ln h_i,$$

where h_i is the Henry's Law constant for component i in this phase. Combining the two logarithmic terms in this equation gives:

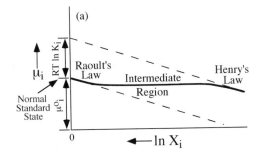

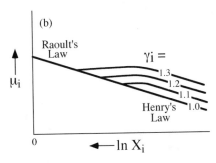

FIG. 9.7. (a) Plot of μ_i versus $\ln X_i$ showing the dependence of chemical potential on composition. In the Raoult's Law region, $\mu_i = \mu_i^0$; in the Henry's Law region, $\mu_i = \mu_i^0 + RT \ln h_i$. The slope of the Raoult's and Henry's Law regions is RT. The normal standard state for component i in a pure substance at P and T is μ_i^0. (b) Plot of μ_i versus $\ln X_i$ showing how the intermediate region changes with activity coefficient. With ideal mixing ($\gamma_i = 1$), the Raoult's and Henry's Law regions lie along the same trend. As γ_i becomes larger, the inflection appears and the intermediate region becomes larger.

$$\mu_i = \mu_i^0 + RT \ln h_i X_i. \tag{9.18}$$

The intermediate region between Henry's and Raoult's Law behavior is characterized by a curved segment that connects the two straight line segments (fig. 9.7a). This region is thus expressed by an equation that is intermediate between equations 9.17 and 9.18. To formulate such an equation, we can describe the degree of nonideality by using γ_i, the activity coefficient, introduced in chapter 4. The activity coefficient is a function of composition, such that in the Raoult's Law region,

$$\gamma_i \to 1, \text{ so that } a_i \to X_i,$$

and in the Henry's Law region,

$$\gamma_i \to h_i, \text{ so that } a_i \to h_i X_i.$$

Thus we can formulate a general equation for chemical potential of a pure component i in any phase at any temperature and pressure:

$$\mu_i = \mu_i^0 + RT \ln \gamma_i X_i. \qquad (9.19)$$

We can see that as component i becomes more similar to the other component with which it is mixing, $h_i \rightarrow 1$. Moreover, greater similarities in these components will increase the size of the Raoult's and Henry's Law regions in figure 9.7a, to the point where the intermediate region disappears and the Raoult's and Henry's Law regions connect. In such a case, we would then have one straight mixing line in this diagram (equivalent to the line that extrapolates to μ_i^0), and the mixing would be ideal. The effects of changing activity coefficients on the various mixing regions are illustrated in figure 9.7b.

Besides the standard state we have just considered, there are others that could be used, and in fact, it is advantageous to do so under certain conditions. Obviously, we must use another standard state if it is not possible to make the phase of interest out of pure component i. For example, we cannot synthesize a garnet of lanthanum silicate, so μ_{La}^0 in garnet based on the pure phase cannot be determined directly. In this case, the standard state could be a hypothetical garnet with properties obtained by extrapolating the Henry's Law region to the pressure and temperature of interest. Such a situation is illustrated in figure 9.8a. This is the standard state we assumed without any explanation in chapter 4. In fluids, it is not possible to measure the activity of pure Na^+ ions, for example, so we must extrapolate from the dilute salt solution.

We might also envision a case where component i cannot be studied directly in the phase of interest, but its chemical potential can be analyzed in another phase. This could occur, for example, if we were interested in determining the chemical potential of Al_2O_3 in a magma but found it easier to study glass quenched from the magma. We could use the activity of Al_2O_3 in a glass of pure Al_2O_3 composition to define the standard state. Such a standard state is illustrated in figure 9.8b. The expressions for chemical potentials using other standard states can be derived from equation 9.19.

SOLUTION MODELS: ACTIVITIES OF COMPLEX MIXTURES

Activity coefficients are functions of temperature, pressure, and composition. They can be greater or less than one. The values for activity coefficients cannot be determined from thermodynamic properties but must be obtained experimentally from measurements of a_i and X_i or estimated empirically from mineral data. Many different expressions for activity coefficients as a function of composition—often called *solution models*—have been formulated. The only restrictions on them are that they must obey the constraints given above for the Raoult's and Henry's Law regions; that is, they must explain mixing at the extreme ends of the compositional spectrum, where $X_i \rightarrow 1$ and $X_i \rightarrow 0$.

One solution model that is often used in solving geochemical problems assumes ideal mixing. This is a trivial model, because the expression for $\gamma_i = 1$. From figure 9.7, we can see that the assumption will be correct in the Raoult's Law region, but will become increasingly less accurate in moving toward the intermediate and Henry's Law regions. Thus, this solution model works best in cases for which X_i is large.

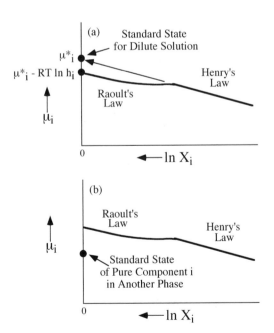

FIG. 9.8. Diagrams of μ_i versus $\ln X_i$ illustrating (a) the standard state for component i in a dilute solution extrapolated from the Henry's Law region, and (b) the standard state for component i taken as the activity of i in another phase at P and T.

Another solution model that is commonly used is that for a *symmetric regular solution*. For this model,

$$RT \ln \gamma_i = WX_j^2 = W(1 - X_i)^2, \qquad (9.20)$$

where W is an interaction parameter that is a function of temperature and pressure but not composition, and X_j is the mole fraction of the other component with which component i is mixing. Substitution of this expression in equation 9.20 gives the chemical potential for component i using a symmetric regular solution model:

$$\mu_i = \mu_i^0 + RT \ln X_i + WX_j^2. \qquad (9.21)$$

Note that this equation obeys the necessary restrictions in the Raoult's and Henry's Law regions. In the case of Raoult's Law, as $X_i \to 1$, $WX_j^2 \to 0$ and $\gamma_i \to 1$. For Henry's Law, as $X_i \to 0$, $WX_j^2 \to W$ and $\gamma_i \to h_i$.

In the regular solution model, the inflection point in the intermediate region on the μ_i-ln X_i plot is flanked by symmetric limbs, as illustrated in figure 9.9. However, in real systems, mixing behavior need not be this regular. The *asymmetric regular solution* model is a somewhat more complex formulation that is also very useful in some situations. This model involves two interaction parameters, W_i and W_j, rather than one as in the regular solution model. For the asymmetric regular solution model,

$$RT \ln \gamma_i = X_j^2(W_i + 2X_i(W_j - W_i)). \qquad (9.22)$$

This model can be used in situations in which mixing is not symmetric, as illustrated in figure 9.9. Note that when $W_i = W_j$, the asymmetric regular solution model reduces

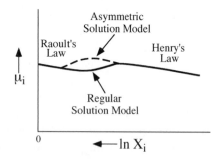

FIG. 9.9. The difference in mixing behavior for the symmetric and asymmetric regular solution models is illustrated by the shapes of the intermediate regions. The symmetric solution model produces symmetrical limbs flanking the inflection point, and the asymmetric model does not.

to the symmetric regular solution model. This mixing model also obeys the necessary restriction in the Raoult's and Henry's Law regions. However, in the Henry's Law region, $RT \ln \gamma_i \to W_i$ and $RT \ln \gamma_j \to W_j$.

Worked Problem 9.9

Calculate the relationship between activity and mole fraction of $Fe_3Al_2Si_3O_{12}$ (almandine) in a garnet solid solution. Although some geochemists have argued that almandine mixes nearly ideally with $Mg_3Al_2Si_3O_{12}$ (pyrope), others have advocated a symmetric regular solution model with a temperature-dependent interaction parameter of the form:

$$W_{Fe-Mg} = 3480 - 1.2T,$$

where T is the temperature in °C and W has units of cal g^{-1} atom^{-1}. Strictly speaking, regular solution models do not allow for variation of W with temperature, but let's ignore this minor problem and use this solution model to illustrate how interaction parameters work.

The garnet we consider has the following composition: $X_{Fe} = 0.710$ and $X_{Mg} = 0.088$, along with minor amounts of Ca and Mn. Let's calculate the activity of Fe in garnet at 827 K or 554°C (the reason this temperature was chosen will become clear shortly). The value of the interaction parameter is therefore:

$$W_{Fe-Mg} = 3480 - 1.2(554) = 2815 \text{ cal g}^{-1} \text{ atom}^{-1}.$$

Applying this value to equation 9.20, we have:

$$RT \ln \gamma_{Fe} = 2815(1 - X_{Fe})^2,$$

or

$$\ln \gamma_{Fe} = (2815)(1 - 0.710)^2/(1.987)(827) = 0.144$$

and

$$\gamma_{Fe} = 1.155.$$

Fe mixes on three crystallographic sites in garnet, so $a_{Fe} = \gamma_{Fe}X_{Fe}^3$. Therefore, the activity of Fe in this garnet is:

$$a_{Fe} = 1.155(0.710)^3 = 0.413.$$

For comparison, for an ideal solution ($\gamma_{Fe} = 1$), the activity of Fe in this garnet would be:

$$a_{Fe} = 1(0.710)^3 = 0.446.$$

THERMOBAROMETRY: APPLYING WHAT WE HAVE LEARNED

The formulation and application of geothermometers and geobarometers involves many of the concepts just introduced. We use one rather simple example of each of these techniques to illustrate how useful the concepts can be in solving geologic problems. For a particular reaction to be used as a geothermometer, it must be a strong function of temperature but be nearly independent of pressure. Conversely, a geobarometer should be sensitive to pressure but not to temperature. We have already seen that at equilibrium:

$$\Delta G^0 = -RT \ln K_{eq} = \Delta \bar{H} - T\Delta \bar{S} + P\Delta \bar{V}.$$

This equation assumes that $\Delta \bar{V}$ is independent of temperature and pressure; otherwise we would have to include expressions for thermal expansion and isothermal compressibility. The form of this equation indicates that if K_{eq} is constant, there is only one equilibrium pressure at any given temperature, and fixing K_{eq} determines the position of the equilibrium reaction curve in P-T space. The factors that control the slopes of equilibrium curves in P-T space can be evaluated by differentiating the above equation with respect to temperature at constant pressure and with respect to pressure at constant temperature:

$$(\partial \ln K_{eq}/\partial T)_P = \Delta \bar{H}/RT^2 \qquad (9.23)$$

and

$$(\partial \ln K_{eq}/\partial P)_T = -\Delta \bar{V}/RT. \qquad (9.24)$$

If the right side of equation 9.23 is larger than that of 9.24, then the equilibrium depends more on temperature than pressure and would be a suitable geothermometer. Thus, we can see that geothermometer reactions should have large values of $\Delta \bar{H}$ and, conversely, that geobarometer reactions should be characterized by large $\Delta \bar{V}$ values. It will also be necessary to have accurate standard-state thermodynamic data for the reaction available, as well as formulations for activity coefficients.

Worked Problem 9.10

The partitioning of Fe and Mg between coexisting biotite and garnet serves as a very useful geothermometer. The exchange reaction is:

$$\underset{\text{almandine}}{Fe_3Al_2Si_3O_{12}} + \underset{\text{phlogopite}}{KMg_3AlSi_3O_{10}(OH)_2} \rightleftarrows$$
$$\underset{\text{pyrope}}{Mg_3Al_2Si_3O_{12}} + \underset{\text{annite}}{KFe_3AlSi_3O_{10}(OH)_2}.$$

The enthalpy change for this reaction is large but $\Delta \bar{V}$ is quite small, as appropriate for a geothermometer. Available data suggest that Fe and Mg mix almost ideally in biotite and garnet, so we assume that they obey Raoult's Law and $a_i = X_i$. The Fe-Mg

partitioning between these phases is for the most part only a function of T (and, to a lesser extent, P) so long as we assume ideal mixing in both garnet and biotite. No formulation for their activity coefficients is necessary in this case. (Recall that some workers have suggested that a symmetric regular solution model provides a better expression for garnet mixing, as we explored in worked problem 9.9.)

There are quite a few expressions for the garnet-biotite geothermometer; we will consider the experimental calibration of John Ferry and Frank Spear (1978). They measured the progress of this exchange reaction at a series of temperatures, and obtained the following expression for T (in kelvins):

$$\ln K_D = -2109/T + 0.782, \qquad (9.25)$$

where K_D equals $(X_{Mg}/X_{Fe})_{garnet}/(X_{Mg}/X_{Fe})_{biotite}$. (For equilibria in which the same phase or phases appear on both sides of the equation, K_{eq} is commonly called K_D, the *distribution coefficient*.) The coefficients were determined by a linear least-squares fit of the experimental values of $\ln K_D$ versus $1/T$, determined at a pressure of ~2 kbar. (To see how this is done, refer to appendix A.) Using the Clapeyron equation, Ferry and Spear were also able to calculate dP/dT and, therefore, to estimate the dependence of the reaction on pressure. Their expression for this Fe-Mg exchange reaction at any pressure and temperature is:

$$\ln K_D = -(2089 + 0.0096P)/T + 0.782 \qquad (9.26)$$

where P is in bars and T is in kelvins. From this equation, we can see that the effect of pressure on this exchange equilibrium is small, and it can be ignored if no pressure estimate is available. To verify that the error generated by ignoring pressure is small, let's solve for temperature in two ways.

First, we use this geothermometer to calculate the temperature of equilibration for a metamorphic rock containing garnet and biotite with these compositions:

garnet: $X_{Fe} = 0.710$, $X_{Mg} = 0.088$

biotite: $X_{Fe} = 0.457$, $X_{Mg} = 0.323$

The value of K_D is then:

$$K_D = (0.088/0.710)/(0.323/0.457) = 0.1743$$

Since we do not know the pressure, we can use equation 9.25 to obtain T:

$$\ln 0.1743 = -2089/T + 0.782;$$

$$T = 827 \text{ K} = 554°C.$$

Let's now say that we have an independent measure of pressure, determined to be 5 kbar. What effect does including the pressure term have on the calculated temperature? Applying equation 9.26:

$$-1.743 = -(2089 + 0.0096[5000])/T + 0.782.$$

Solving this expression for T (our second temperature estimate), we obtain:

$$T = 846 \text{ K} = 573°\text{C},$$

a difference of only 11°C, or about 2°C per kbar of pressure.

Worked Problem 9.11

A useful geobarometer is based on the reaction:

$$\underset{\text{almandine}}{Fe_3Al_2Si_3O_{12}} + \underset{\text{rutile}}{3TiO_2} \rightleftarrows \underset{\text{ilmenite}}{3FeTiO_3} + \underset{\substack{\text{Al} \\ \text{silicate}}}{Al_2SiO_5} + \underset{\text{quartz}}{2\,SiO_2}.$$

This reaction is almost independent of temperature, but has a large $\Delta \bar{V}$ that varies slightly depending on which aluminosilicate polymorph (kyanite, andalusite, or sillimanite) is produced by the reaction. The equilibrium constant for this reaction is:

$$K_{eq} = [(a_{ilm})^3(a_{Al_2SiO_5})(a_{qz})^2]/[(a_{Fe\;gar})(a_{ru})^3]. \quad (9.27)$$

The great advantage of this particular geobarometer is that so many of the activity terms in equation 9.27 are equal to unity (rutile, quartz, and the aluminosilicate polymorphs form essentially pure phases). Ilmenite forms a solid solution with hematite, but it rarely contains more than 15 mol % Fe_2O_3 and can be treated as an ideal solution or, in some cases, a pure phase. Fe in garnet can be treated as an ideal solution or as a symmetrical regular solution, as we saw in worked problem 9.8. Consequently, the activity terms for garnet and ilmenite depend on their compositions.

The end-member reaction (involving pure almandine garnet) was calibrated experimentally by Steven Bohlen and coworkers (1983). They then calculated the relationship between P, T, and K_{eq}, which of course takes into account other compositions of garnet and ilmenite. To do this, Bohlen and his collaborators used available data on molar volumes, thermal expansion, and compressibility, and the relationships we have already derived in this chapter. They presented their results in graphical form, contouring log K_{eq} for the reaction on a P-T diagram, as shown in figure 9.10. You can see that the slopes of the log K_{eq} contours in this diagram are very shallow, demonstrating that the reaction is very sensitive to pressure but almost independent of temperature. Also note that the slopes change slightly when contours move from the kyanite field into the sillimanite field or when they cross the α-quartz–β-quartz transition, reflecting changes in $\Delta \bar{V}$ of the reaction.

Let's use this diagram to determine the equilibration pressure of a metamorphic rock containing rutile, kyanite, ilmenite, and quartz, as well as garnet and biotite with the compositions as in worked problem 9.10. Note that we have already used this same garnet-biotite pair to calculate the temperature (554°C). To keep the problem simple, when calculating K_{eq}, we stipulate that ilmenite is a pure phase. Bohlen and coworkers recom-

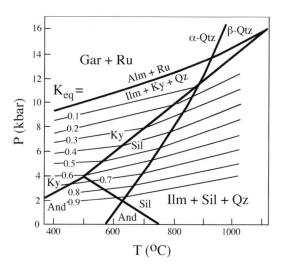

FIG. 9.10. *P-T* diagram showing the experimentally determined reaction almandine + rutile $\rightleftarrows$ ilmenite + Al_2SiO_5 + quartz, and contours of log K_{eq} for the reaction with impure garnet and ilmenite compositions, after Bohlen et al. (1983). The slopes of the contours show that the reaction is strongly dependent on pressure but not on temperature (as appropriate for a geobarometer). Boundary curves for the polymorphic transformation of andalusite, kyanite, and sillimanite, and of α- and β-quartz are also shown.

mended a symmetric regular mixing model for Fe in garnet, and we have conveniently already used this model (and a temperature of 554°C) to determine a_{Fe} in worked problem 9.9. Substituting this value into equation 9.27 gives:

$$K_{eq} = 1/0.413 = 2.421.$$

Therefore,

$$\log K_{eq} = 0.384.$$

Applying this value and a temperature of 554°C to figure 9.10, we obtain a pressure of ~6 kbar.

Before leaving this problem, let's see how much difference it would make if instead we had used an ideal solution model for Fe in garnet. From equation 9.27, the value of K_{eq} becomes:

$$K_{eq} = 1/(0.710)^3 = 2.793,$$

and

$$\log K_{eq} = 0.446.$$

The log K_{eq} value corresponds to a pressure of about 5 kbar, a difference of 1 kbar.

SUMMARY

We have reaffirmed that equilibrium is a condition of minimum free energy for a system. The phase rule provides a useful test for equilibrium. Most geologic systems at equilibrium have variances of two or higher (reflecting the fact that most rocks are stable over a range of temperatures and pressures). The effects of changing temperature and pressure on the free energy of a system can be determined, and the direction of a particular reaction under different conditions can be predicted from appropriate thermodynamic data. The Clapeyron equation illustrates the usefulness of thermodynamic data in assessing the slopes of reactions on *P-T* diagrams.

Systems containing more than one component require analysis of the way such components interact in the phases that constitute the system. Components that mix ideally obey Raoult's Law; that is, their activities equal their concentrations. Components at high dilution obey Henry's Law; that is, their activities are directly proportional to (but not equal to) their concentrations. The chemical potentials of components are generally expressed in the form $\mu_i = \mu_i^0 + RT \ln \gamma_i X_i$ (the chemical potential at some standard state plus an activity term that corrects this value for deviations in conditions and composition). Activity coefficients are equal to one for Raoult's Law behavior and h_i for Henry's Law behavior. Various solution models have been formulated for activity coefficients for components exhibiting intermediate (nonideal) behavior. Solution models find practical application in the formulation of geothermometers and geobarometers.

With this thermodynamic background, we are now ready to consider phase diagrams for multicomponent systems. These are introduced in chapter 10.

SUGGESTED READINGS

There are a number of excellent textbooks that provide an introduction to thermodynamics, as well as various geological applications. These generally contain significantly more detail than has been presented here, and are highly recommended for the serious student.

Denbigh, K. 1971. *The Principles of Chemical Equilibrium*. London: Cambridge University Press. (A classic thermodynamic text containing rigorous derivations but no examples of geologic interest; a superb reference, but not for the fainthearted.)

Ernst, W. G. 1976. *Petrologic Phase Equilibria*. San Francisco: Freeman. (A very readable introductory text; chapter 3 illustrates the computational approach to phase diagrams, and chapter 4 contains descriptions of $\bar{G}$-*T* diagrams.)

Essene, E. J. 1982. Geologic thermometry and barometry. In J. M. Ferry, ed. *Characterization of Metamorphism through Mineral Equilibria*. Reviews of Mineralogy 10. Washington, D.C.: Mineralogical Society of America, pp. 153–206. (An excellent review of the calibration, assumptions, and precautions in determining metamorphic temperatures and pressures.)

Kretz, R. 1994. *Metamorphic Crystallization*. New York: Wiley. (Chapter 2 provides a good description of relations between thermodynamic properties and metamorphic equilibria.)

Philpotts, A. R. 1990. *Principles of Igneous and Metamorphic Rocks*. Englewood Cliffs: Prentice-Hall. (Chapter 9 of this comprehensive textbook gives an especially good section on solution models.)

Powell, R. 1978. *Equilibrium Thermodynamics in Petrology*. London: Harper and Row. (An excellent introductory text that dispenses with proofs; chapter 2 describes $\bar{G}$-*T* diagrams, and chapter 3 gives useful formulations for various standard states not presented in the present book.)

Saxena, S. K. 1973. *Thermodynamics of Rock-Forming Crystalline Solutions*. New York: Springer-Verlag. (A more advanced book that provides a wealth of information on mixing models and solution behavior for real mineral systems.)

Wood, B. J., and D. G. Fraser. 1977. *Elementary Thermodynamics for Geologists*. Oxford: Oxford University Press. (A clear and concise introduction to thermodynamics, with numerous worked examples of geologic interest; chapter 3 treats regular solution models, and chapter 4 provides a much more detailed look at geothermometers and geobarometers than is given in the present book.)

These papers were also referenced in this chapter.

Bohlen, S. R., V. J. Wall, and A. L. Boettcher. 1983. Experimental investigations and geological applications of equilibria in the system FeO-TiO_2-Al_2O_3-SiO_2-H_2O. *American Mineralogist* 68:1049–1058.

Eugster, H. P., and D. R. Wones. 1962. Stability relations of the ferruginous biotite, annite. *Journal of Petrology* 3:82–125.

Ferry, J. M., and F. S. Spear. 1978. Experimental calibration of the partitioning of Fe and Mg between biotite and garnet. *Contributions to Mineralogy and Petrology* 66:113–117.

Helgeson, H. C., J. M. Delany, H. W. Nesbitt, and D. K. Bird. 1978. Summary and critique of the thermodynamic properties of the rock forming minerals. *American Journal of Science* 278A:1–229.

Holdaway, M. J., and B. Mukhopadhyay. 1993. A reevaluation of the stability relations of andalusite: Thermochemical data and phase diagram for the aluminum silicates. *American Mineralogist* 78:298–315.

Kerrick, D. M., ed. 1990. The Al$_2$SiO$_5$ polymorphs. Reviews of Mineralogy 22. Washington, D.C.: Mineralogical Society of America.

Zen, E. 1969. The stability relations of the polymorphs of aluminum silicates: A survey and some comments. *American Journal of Science* 267:297–309.

PROBLEMS

(9.1) Using figure 9.2, construct qualitative $\bar{G}$-P diagrams for the Al$_2$SiO$_5$ system at temperatures T_1 and T_2.

(9.2) From the understanding that $(\partial\Delta\bar{G}/\partial T)_P = -\Delta\bar{S}$ and $(\partial\Delta\bar{G}/\partial P)_T = \Delta\bar{V}$, construct a schematic $\bar{S}$-T diagram at pressure P_1 and a $\bar{V}$-P diagram at temperature T_2 for the Al$_2$SiO$_5$ system (fig. 9.2).

(9.3) Consider an isolated system containing the following phases at equilibrium: (K,Na)Cl, (Na,K)AlSi$_3$O$_8$, and aqueous fluid. (a) Choose a set of components for this system. (b) State the conditions necessary for the system to be in complete equilibrium.

(9.4) Clinoenstatite (MgSiO$_3$) melts incongruently to forsterite (Mg$_2$SiO$_4$) plus liquid at 1557°C at atmospheric pressure. (a) Using thermodynamic data from some reliable source, calculate the chemical potential of SiO$_2$ in the liquid, when these three phases coexist at equilibrium. (b) What will happen to the system if μ_{SiO_2} is increased while T and P are held constant?

(9.5) Using the data provided,
(a) Calculate $\Delta\bar{H}^0_{800}$, $\Delta\bar{S}^0_{800}$, and $\Delta\bar{G}^0_{800}$ for the following reaction:

NaAlSi$_3$O$_8$ → NaAlSi$_2$O$_6$ + SiO$_2$.
  albite   jadeite   quartz

	$\bar{H}^0_{298}$ (cal mol^{-1})	$\bar{S}^0_{298}$ (cal deg^{-1} mol^{-1})	$\bar{C}_p$ (cal deg^{-1} mol^{-1})
albite	−937.146	50.20	$61.7 + 13.9T/10^3 - 15.01 \times 10^5/T^2$
jadeite	−719.871	31.90	$48.2 + 11.4T/10^3 - 11.87 \times 10^5/T^2$
quartz	−217.650	9.88	$11.2 + 8.2T/10^3 - 2.7 \times 10^5/T^2$

(b) Calculate $\Delta\bar{G}^0_{800}$ for the reaction assuming $\Delta\bar{C}_p = 0$. Compare the two values and comment.

(9.6) In worked problem 9.8, we calculated the slope of the andalusite-kyanite boundary in P-T space. Its location was fixed using the experimentally determined triple point of Holdaway and Mukhopadhyay discussed in a highlighted box. Using the Clapeyron equation and data in worked problem 9.8, construct the remainder of the aluminosilicate phase diagram. Assume that $\Delta\bar{C}_p = 0$ and that $\bar{V}$ is independent of P and T.

(9.7) For the reaction:

Mg(OH)$_2$ → MgO + H$_2$O,
  brucite   periclase

(a) Calculate the equilibrium temperature for the reaction at 1000 bars pressure using thermodynamic data from some appropriate source. Assume that $\Delta\bar{C}_p = 0$, $\Delta\bar{V}_{solids}$ = constant, and that H$_2$O behaves as an ideal gas.

(b) Calculate the equilibrium temperature under the same conditions as before, but this time with $P_{H_2O} = 0.5$.

(9.8) Using the values for molar volume, thermal expansion, and isothermal compressibility given below, calculate $\bar{V}$ for each of these phases at 800°C and 3 kbar.

(a) Almandine:

$$\bar{V}_{298,1} = 118.^2 \text{ cm}^3 \text{ mol}^{-1};$$

$$\alpha_p = 25 \times 10^{-6} \text{ deg}^{-1};$$

$$\beta_T = 1.27 \times 10^{-6} \text{ bar}^{-1}.$$

(b) Quartz:

$$\bar{V}_{298,1} = 22.69 \text{ cm}^3 \text{ mol}^{-1};$$

$$\alpha_p = 69 \times 10^{-6} \text{ deg}^{-1};$$

$$\beta_T = 26.71 \times 10^{-6} \text{ bar}^{-1}.$$

(9.9) Calculate a_i as X_i increases, using Raoult's Law and a symmetric regular solution model with $W = 10$ kJ, and determine how small X_i can become before Raoult's Law is no longer a good approximation.

(9.10) Given the reactions below, discuss what standard states you would choose for the chemical potentials:

(a) leucite + SiO_2 (melt) → orthoclase, at 1 bar and 1170°C;

(b) $EuAl_2Si_2O_8$ (melt) → $EuAl_2Si_2O_8$ (plagioclase), at 1300°C and 1 bar;

(c) gypsum → anhydrite + water, at 100°C and 1 kbar.

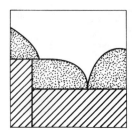

CHAPTER TEN

PICTURING EQUILIBRIA

Phase Diagrams

OVERVIEW

Even after years of experience, many geologists find thermodynamics difficult to understand unless they can draw pictures that relate their equations to tangible systems. Phase diagrams provide a convenient and powerful way to picture equilibria. In this chapter, we introduce the principles of $\bar{G}$-X_2 diagrams for binary systems and illustrate graphically the minimization of free energy. We then construct T-X_2 diagrams by stacking $\bar{G}$-X_2 sections atop one another, using examples of real geochemical systems. These diagrams illustrate simple crystallization, the formation of chemical compounds, solid solution, and exsolution. We review equilibrium crystallization paths and show how binary phase diagrams can be constructed from thermodynamic data. In addition, we examine phase diagrams that employ fluid composition or partial pressure as variables. With this background, we then describe P-T diagrams for binary systems. Finally, we briefly introduce graphical relationships for systems with three components.

$\bar{G}$-X_2 DIAGRAMS

In chapter 9, we learned that a phase diagram is simply a graphical representation of the stability fields of one or more phases, drawn in terms of convenient thermo-

dynamic variables. In a one-component (unary) system, the equilibrium state is determined when we specify two thermodynamic variables, such as some combination of P, T, $\bar{S}$, or $\bar{V}$. In our earlier discussion of the unary system H_2O, we considered a phase diagram with pressure and temperature as Cartesian axes, which showed single-phase fields for water, ice, and vapor, as well as lines along which two phases coexisted stably. To determine the state of a two-component (binary) system completely, an additional compositional variable must be specified. Thus, a complete phase diagram for a binary system is drawn in three dimensions. Once again, we are free to choose any three thermodynamic functions as Cartesian axes, but for most geochemical applications, it is most convenient to select P, T, and composition, normally expressed as X_2 (the mole fraction of component 2). Of course, we could just as easily choose X_1 as our compositional variable, because $X_1 = 1 - X_2$.

Molar Gibbs free energy is a function of all three of these variables (that is, $\bar{G} = \bar{G}(P, T, X_2)$), so let us consider a diagram (fig. 10.1) in which a single phase with a continuous compositional range between $X_2 = 0$ and $X_2 = 1$ is represented. The curved line on this diagram is a graph of the function $\bar{G} = \bar{G}(P, T, X_2)$, considered at a fixed P and T. Now let's examine some significant features of this diagram.

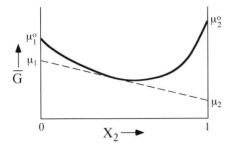

FIG. 10.1. $\bar{G}$-X_2 diagram at fixed T and P for a phase with a continuous compositional range from $X_2 = 0$ to 1. At any particular composition, the ends of a tangent to the $\bar{G}$ curve define the values for μ_1 and μ_2 in that phase, as illustrated by the dashed line.

We can, of course, define the value of $\bar{G}$ for this phase at any specific composition X_2 by reading it directly from the ordinate on the graph. It is more instructive, however, to consider the individual contributions of the chemical potentials μ_1 and μ_2 to $\bar{G}$. We do this by constructing a tangent to the $\bar{G}$-X_2 curve at X_2, as illustrated in figure 10.1 by the dashed line. Because the first derivative of $\bar{G}$ with respect to X_2 equals μ_2, the point at which this tangent intersects the ordinate at $X_2 = 1$ is the value for μ_2 in the phase; similarly, the point at which it intersects the ordinate at $X_2 = 0$ ($X_1 = 1$) is the value of μ_1 in the phase. Note that, although as drawn in figure 10.1, $\bar{G}$ reaches a minimum value at some point between $X_2 = 0$ and $X_2 = 1$, the value of μ_2 increases monotonically (and the value of μ_1 decreases monotonically) as X_2 increases. Also note that the value for either chemical potential approaches a maximum as the composition approaches that of the pure chemical component. Therefore, μ_2^0 represents the chemical potential of component 2 in the phase at its pure $X_2 = 1$ end member.

In this example, we have chosen a phase exhibiting complete solid solution behavior. There are, of course, many pure substances with no compositional variation whatsoever. For such substances, the $\bar{G}$-X_2 curve reduces to a point. (This point could be represented equally well by a sharp vertical spike in the $\bar{G}$-X_2 diagram.)

If several phases are possible in the system, each will have its own $\bar{G}$-X_2 curve or point. The equilibrium state for the system is that phase or set of phases that yields a minimum value of $\bar{G}$ for any specified value of X_2 (still, remember, at constant P and T). We can best visualize the equilibrium states for various values of X_2 by imag-

ining a tangent line or set of tangent lines that touch the set of $\bar{G}$-X_2 curves or points, forming a sort of floor to the diagram as illustrated by the dashed lines in figure 10.2a. In this example, we have one pure phase B and two phases (A and C) that exhibit solid solution behavior. Note that if we start at the left side of the diagram and move gradually toward the right, the tangent to the left limb of the $\bar{G}$-X_2 curve for phase A changes slope monotonically, which is to say that both μ_{1A} and μ_{2A} vary monotonically. This situation continues until we reach the value of X_2 at which the tangent to the curve for phase A bumps against the $\bar{G}$-X_2 point for phase B. Now $\mu_{1A} = \mu_{2A}$ and $\mu_{2A} = \mu_{1B}$. This is, of course, a condition defined originally in chapter 3 as chemical equilibrium between phases A and B. Beyond this value of X_2, the free energy of the system can be found along the common tangent and is lower than the free energy for either

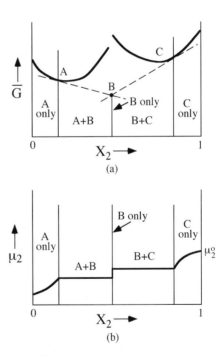

FIG. 10.2. (a) $\bar{G}$-X_2 diagram at fixed T and P for a system containing three phases. Phase B is pure, but phases A and C show limited ranges of composition. The equilibrium states for various values of X_2 are those phases which yield the minimum $\bar{G}$ for the system. These can be readily determined from tangents to the $\bar{G}$ curves, as shown by the dashed lines. (b) The chemical potential for component 2 in the stable phase or phase assemblage in (a) increases with X_2, as illustrated in this diagram. μ_2 is fixed by the coexistence of two phases, but changes monotonically if only one phase is present.

phase A or B alone. The two-phase assemblage A + B continues to be stable until we reach the value of X_2 corresponding to the composition of B. At this point, only phase B is stable, and the tangent pivots instantaneously at B to become the tangent to both B and C. Now, B + C becomes the stable phase assemblage. A similar pattern occurs as we continue to the right and encounter a region in which phase C alone is the stable phase.

We see, therefore, that the chemical potential of either component 1 or 2 changes monotonically across single-phase fields as we vary X_2, but both chemical potentials remain fixed in two-phase fields. This can be seen graphically in figure 10.2b, a schematic construction of the variation of μ_2 in the stable phase or phase assemblage as X_2 ranges from 0 to 1 in the system in figure 10.2a. When two phases are stable, a binary system has only two degrees of freedom (T and P), as dictated by the phase rule.

DERIVATION OF T-X_2 AND P-X_2 DIAGRAMS

A $\bar{G}$-X_2 diagram such as those we have just considered is valid only for the fixed values of T and P that were specified when drawing the diagram. To construct a

phase diagram for the system that takes into account changing temperature and pressure conditions, we must recall from chapter 9 how $\bar{G}$ varies with T and with P at constant X_2:

$$(\partial \bar{G}/\partial T)_{P, X_2} = -\bar{S}, \qquad (10.1)$$

and

$$(\partial \bar{G}/\partial P)_{T, X_2} = \bar{V}. \qquad (10.2)$$

Because $\bar{S}$ and $\bar{V}$ are inherently positive quantities, all $\bar{G}$-X_2 curves will shift upward or downward simultaneously with increasing P or T, respectively. However, because the specific $\bar{S}$ and $\bar{V}$ values for each phase are different and because $\bar{S}$ and $\bar{V}$ are themselves functions of composition, relative movements of the $\bar{G}$-X_2 curves and changes in their shapes will generally occur. Consider, for example, a sequence of $\bar{G}$-X_2 diagrams drawn at T_1, T_2, and T_3, a set of arbitrarily decreasing temperatures, illustrated in figure 10.3. With increasing temperature, the $\bar{G}$-X_2 curves for all three phases in the system move upward, but in this example, the curve for phase B shifts more than do the other two curves. As a result, the curves for all three phases have a common tangent at the unique temperature T_2, and we have the condition $\mu_{1A} = \mu_{1B} = \mu_{1C}$; that is, a three-phase field A + B + C exists at

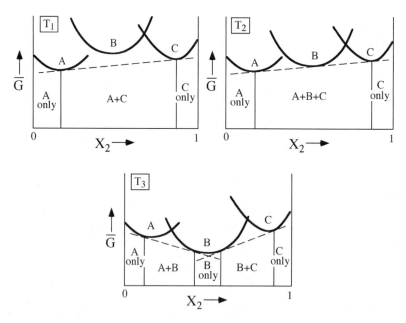

FIG. 10.3. A series of three $\bar{G}$-X_2 diagrams at fixed T and P for a succession of temperatures decreasing from T_1 to T_3. The $\bar{G}$ curves for phases A, B, and C shift downward at different rates with decreasing temperature, and thus different combinations of phases are stable.

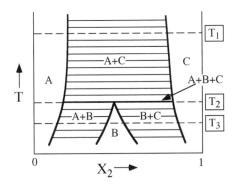

FIG. 10.4. A "stack" of G-X_2 diagrams at various temperatures produces a T-X_2 diagram. This particular diagram results from stacking the three $\bar{G}$-X_2 diagrams shown in figure 10.3. The compositions of coexisting phases are connected by horizontal lines, because at equilibrium, the temperatures of all coexisting phases must be the same.

T_2, replacing the fields A + B, B only, and B + C that were stable at temperatures below T_2. A further increase in temperature lifts the curve for phase B above the common tangent to A and C, and thus phase B is no longer stable either alone or in combination with A or C. The diagram for T_1 illustrates this condition. Phase B is said to be *metastable* with respect to A + C.

Now let us visualize $\bar{G}$-X_2 diagrams drawn over a continuous range from T_1 to T_3, and extending to higher and lower temperatures beyond these limits. The one-, two-, and three-phase fields in these diagrams will be seen to change shapes continuously. If we now envision a "stack" of all such diagrams and rotate it so that it can be viewed from the side with temperature as the vertical axis and X_2 the horizontal one, we have created a T-X_2 diagram for the system. Such a diagram is illustrated in figure 10.4.

Notice that as in the $\bar{G}$-X_2 diagrams, phase fields for single phases and pairs of phases alternate from left to right across the composition axis. The three-phase field A + B + C exists only at temperature T_2. Phase B does not exist for any value of X_2 at temperatures above the line A + B + C. The appearance and disappearance of phases are probably the most complicated and confusing parts of phase diagrams. Other kinds of phase changes are illustrated and explained in the next section.

Note also the set of tie-lines parallel to the X_2 axis in figure 10.4. These connect the compositions of phases A + B, B + C, or A + C that are in equilibrium in the two-phase fields at each temperature. Because it is a necessary

condition of equilibrium that the temperatures of all phases be the same, tie-lines must always be isothermal.

It may not be immediately apparent that a point on any tie-line has a significance in terms of the relative proportions of phases in the two-phase field. We can, in fact, define a *lever rule* using such a point. The proportion of each phase in the field at that point varies inversely as the length of the tie-line from the point to the composition of that phase. For example, consider the two-phase field A + C in figure 10.4. A point on the tie-line in this field at a composition one quarter of the way from the left edge (the composition of phase A) represents a mixture of 75% phase A and 25% phase C, and the tie-line segments to the left and right of the point are 25% and 75% of the total length, respectively. This configuration is illustrated in figure 10.5. Proportions of the phases in a two-phase field can thus be determined by inspection.

The process by which we just developed a T-X_2 diagram from $\bar{G}$-X_2 sections could also be duplicated to generate a P-X_2 diagram. In general, however, the appearance of a P-X_2 diagram will be inverted from that of a T-X_2 diagram, because increases in temperature and pressure have inverse effects on most materials; that is, high temperature favors higher $\bar{V}$ and higher $\bar{S}$, but high pressure favors lower $\bar{V}$ and lower $\bar{S}$. P-X_2 diagrams are not commonly used in solving geochemical problems. Instead, the effects of pressure are generally gauged from P-T diagrams, considered in a later section. For the time being, we focus on the characteristics of T-X_2 diagrams.

T-X_2 DIAGRAMS FOR REAL GEOCHEMICAL SYSTEMS

Now we look at some $\bar{G}$-X_2 and corresponding T-X_2 diagrams for real binary systems, to illustrate the geochemically important features commonly encountered in

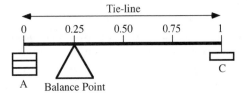

FIG. 10.5. This balance schematically illustrates the lever rule. For the two-phase field A + C in figure 10.4, a point on the tie-line one quarter of the way from the composition of A represents a mixture of 75% phase A and 25% phase C, and the tie-line segments on either side of this point are 25% and 75%, respectively.

phase diagrams for geologic materials. In most of the following examples, a liquid phase is stable at high temperatures and solidifies to form other phases as the temperature is lowered. Each of these systems will be considered only at atmospheric pressure, at least for the moment. We present a series of $\bar{G}$-X_2 diagrams for various temperatures at constant pressure, and then "stack" these to construct the appropriate T-X_2 diagram for that system. We can then explore the characteristics of each T-X_2 diagram in detail.

Simple Crystallization in a Binary System: $CaMgSi_2O_6$-$CaAl_2Si_2O_8$

The simplest case we consider is an example of a two-component system that is liquid at high temperature and crystallizes to form two pure solid phases, diopside $CaMgSi_2O_6$ and anorthite $CaAl_2Si_2O_8$, as shown in figure 10.6. First let us examine some $\bar{G}$-X_{An} diagrams for this system at T_1 through T_4, representing a progression of decreasing temperatures. At T_1, a tangent to the $\bar{G}$-X_{An} curve for liquid (L) defines the lowest free energy for the system at any X_{An}, so that liquid is the only stable phase. By T_2, the $\bar{G}$ point for anorthite has dropped farther than the curves and points for the other phases, so that a tangent to anorthite and liquid offers the lowest free energy configuration for the system at high X_{An} values. By T_3, the $\bar{G}$ for diopside has lowered to the point at which Di + L or An + L are the stable phases over nearly the entire compositional range, but at one special composition, all three phases are stable. At T_4, liquid is metastable, because a tangent to the two solid phases lies below the liquid $\bar{G}$ curve.

By combining the information on phase stability in these $\bar{G}$-X_{An} diagrams, we can construct a T-X_{An} diagram for the system diopside-anorthite, as illustrated at the bottom of figure 10.6. Examination of this diagram reveals some interesting features, probably already familiar to petrology students. The curve defining the upper boundary of the area in which liquid coexists with a solid phase, called the *liquidus,* represents the onset of crystallization. The lower boundary of this area denotes a region in which liquid is not a stable phase. This boundary, called the *solidus,* defines the temperature below which the system is fully crystallized. The point at which the liquidus touches the solidus is called a *eutectic.* It represents the composition at which the liquid can persist to the lowest temperature, as well as the point at which

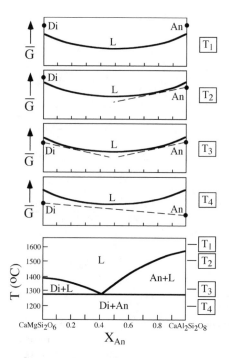

FIG. 10.6. $\bar{G}$-X_{An} diagrams for the system diopside-anorthite at $P = 1$ atm and at various temperatures can be stacked to produce the T-X_{An} diagram at the bottom. Phases and their abbreviations are: diopside (Di), anorthite (An), and liquid (L). This diagram contains a eutectic, representing the three-phase equilibrium between Di, An, and L.

both diopside and anorthite crystallize together with falling temperature.

Worked Problem 10.1

What is the equilibrium crystallization sequence for a liquid in the diopside-anorthite system with composition $X_{An} = 0.2$? The system, initially all liquid, begins to crystallize diopside as it cools and intercepts the diopside liquidus. From figure 10.6, we can see that at $X_{An} = 0.2$, this occurs at a temperature of approximately 1350°C. During subsequent cooling, the composition of the remaining liquid follows the liquidus down slope, changing as diopside is removed. Finally, the liquid reaches the eutectic, at which anorthite and diopside crystallize together. The system can cool no further until the liquid is fully crystallized. Following solidification of the last drop of liquid, the mixture of diopside and anorthite crystals cools to a lower temperature without further change. The final proportions of phases can be determined from the position of this composition along the horizontal axis of the diagram, using the lever rule: 80% diopside and 20% anorthite.

Formation of a Chemical Compound in a Binary System: KAlSi$_2$O$_6$-SiO$_2$

Now we examine a system in which the $\bar{G}$-X_2 curves for two phases having the same composition intersect. Figure 10.7 illustrates $\bar{G}$-X_{silica} diagrams for the system leucite KAlSi$_2$O$_6$-tridymite SiO$_2$ at various temperatures. At T_1, liquid is the stable phase at any composition. First the $\bar{G}$ point for cristobalite, and then that for leucite, drop with lowering temperature to produce the situation at T_2. So far, this system does not differ in its behavior from the diopside-anorthite system we have just examined; however, this changes at T_3. The composition of orthoclase KAlSi$_3$O$_8$ can be represented as a mixture of KAlSi$_2$O$_6$ and SiO$_2$, as shown by its position on the horizontal axis of figure 10.7. At T_3, the orthoclase $\bar{G}$ point coincides with the $\bar{G}$ curve for liquid (that is, orthoclase and liquid have the same free energy), and by T_4, the orthoclase point has pierced the liquid curve to replace liquid as the stable phase over part of the compositional range. Liquid has become metastable by T_5.

The T-X_{silica} diagram produced from this information is shown at the bottom of figure 10.7. The liquidus is indicated by the curves touching the liquid (L) field, and the solidus is defined by the lines that separate liquid-bearing fields from fields with only solid phases. The right side of the diagram contains a eutectic and is similar to the diopside-anorthite system discussed earlier. However, cristobalite transforms to tridymite at 1470°C, because the tridymite $\bar{G}$ point descends more rapidly than that for cristobalite and eventually overtakes it with lowering temperature. The left side of the T-X_{silica} diagram contains a new feature, called a *peritectic* (a point of reaction between solid and liquid), produced when the orthoclase $\bar{G}$ point intersects the $\bar{G}$ curve for liquid. Note that in the $\bar{G}$-X_{silica} section at T_3, the tangent between leucite and liquid has been replaced by tangents between each of these phases and orthoclase. This means that leucite and liquid are no longer stable together at this temperature; thus, these phases react to produce a new phase with intermediate composition, orthoclase.

Worked Problem 10.2

What is the equilibrium crystallization sequence for a composition $X_{silica} = 0.3$ in the system leucite-silica? The first phase to crystallize from a liquid of this composition is leucite, as the temperature drops to ~1425°C. With continued cooling, the liquid follows the liquidus down slope until intersecting the peritectic. At this point, already crystallized leucite reacts with liquid to produce orthoclase (at T_3, a tangent joining the compositions of orthoclase and liquid replaces that between leucite and liquid), and the temperature of the system cannot decrease until all of the leucite is consumed. When the reaction is complete, the liquid resumes its descent along the liquidus, all the while crystallizing more orthoclase. When it reaches the orthoclase-tridymite eutectic, these two phases crystallize together, and after final solidification, the mixture of orthoclase and tridymite crystals cools to a lower temperature. The final proportion of phases can be ascertained from the position of the initial liquid composition along the horizontal axis of figure 10.7. Application of the lever rule, this time using tridymite

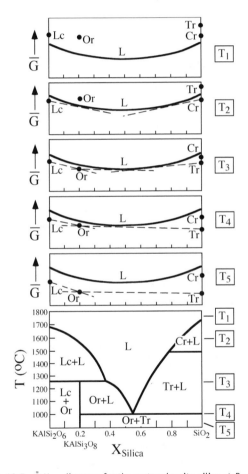

FIG. 10.7. $\bar{G}$-X_{SiO_2} diagrams for the system leucite-silica at $P = 1$ atm and at a progression of temperatures are stacked to make the T-X_{SiO_2} diagram at the bottom. Phases and their abbreviations are: leucite (Lc), orthoclase (Or), cristobalite (Cr), tridymite (Tr), and liquid (L). This diagram contains a eutectic representing equilibrium between Or, Tr, and L, and a peritectic representing the reaction between Lc and L to form Or.

and orthoclase (rather than leucite) as the ends of the tie-line, gives ~87% orthoclase and the remainder tridymite.

Let's now try the crystallization sequence for a composition $X_{silica} = 0.1$ in this system. Minerals crystallize in the same sequence as before, but in this case, the liquid will never be able to persist past the peritectic. The final product for a liquid of this composition must be a mixture of leucite + orthoclase, because that is the mixture of solids in the two-phase field at $X_{silica} = 0.1$ below temperature T_5. Therefore, the liquid phase must be consumed before all of the leucite reacts to form orthoclase.

Solid Solution in a Binary System: NaAlSi$_3$O$_8$-CaAl$_2$Si$_2$O$_8$

Albite (NaAlSi$_3$O$_8$) and anorthite (CaAl$_2$Si$_2$O$_8$) exhibit complete solid solution in plagioclase. Some $\bar{G}$-X_{An} sections for this system are shown in figure 10.8. In this case, the plagioclase composition that is stable at any temperature is determined by the interaction of the $\bar{G}$-X_{An} curves for plagioclase and liquid. At temperatures above

T_1, only liquid is stable. At T_1, the plagioclase $\bar{G}$ curve intersects the liquid $\bar{G}$ curve at $X_{An} = 1$. As the temperature decreases, the plagioclase $\bar{G}$ curve sweeps downward across the liquid $\bar{G}$ curve, intersecting it obliquely at lower and lower X_{An} values. Thus, at T_2 and T_3, compositions with high X_{An} have only plagioclase stable, intermediate compositions have both plagioclase and liquid, and those with low X_{An} contain only liquid. At some value between T_3 and T_4, the plagioclase and liquid $\bar{G}$ curves intersect at $X_{An} = 0$. At T_4, liquid is metastable at any composition.

The corresponding T-X_{An} diagram for the plagioclase system is shown at the bottom of figure 10.8. This diagram contains a loop that defines the boundaries for the liquid-plagioclase field. The upper curve is thus the liquidus, and the lower curve is the solidus. The liquidus gives the composition of the liquid coexisting with a solid whose composition is specified by the solidus at the same temperature.

Worked Problem 10.3

Describe the equilibrium crystallization path for a liquid composition $X_{An} = 0.65$. The final product will obviously be crystals with composition An$_{65}$, as this is a solid solution series. The path by which this product is formed is illustrated in figure 10.9. Crystallization begins when the liquid intersects the liquidus at point a, and the first crystals to form have composition b. As the liquid cools further to point c, the intersections of a horizontal line through this point with the liquidus (point d) and solidus (point e) give the compositions of the liquid and solid, respectively, that are in equilibrium at this temperature. Recall that two phases in equilibrium must have the same temperature, so their compositions are connected by a horizontal tie-line in this diagram. As cooling continues, the composition of the liquid follows the liquidus down slope, and that of crystallizing plagioclase follows the solidus. The intersection of liquid and solid $\bar{G}$ curves seen in the previous diagram indicates that a reaction takes place between liquid and already crystallized solid, so that anorthite-rich plagioclase continually re-equilibrates to form more albite-rich plagioclase that is stable at lower temperatures. Finally, the liquid reaches point f, its farthest point of descent along the liquidus. A horizontal line through this point passes through point g, the composition of the solid at this temperature. Because this composition is $X_{An} = 0.65$, the required final composition for plagioclase, reaction with liquid must end here and the liquid disappears. Crystals of An$_{65}$ composition then cool to low temperature without further change. (Actually, there are some subsolidus reactions, similar to those that take place in alkali feldspars discussed in the next example, which may take place in slowly cooling plagioclases, but we will ignore those here.)

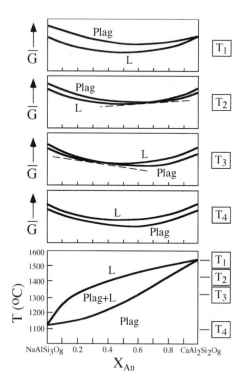

FIG. 10.8. $\bar{G}$-X_{An} diagrams for the system albite-anorthite at $P = 1$ atm and at various temperatures are combined to make the T-X_{An} diagram at the bottom. Plagioclase shows a complete solid solution between these end members, forming a loop.

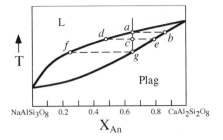

FIG. 10.9. An example of an equilbrium crystallization path in the system albite-anorthite. The liquid cools to point *a*, where crystallization begins with crystals of composition *b*. With falling temperature, liquid changes composition progressively from *a* to *f*, and solid from *b* to *g*.

Unmixing in a Binary System: $NaAlSi_3O_8$-$KAlSi_3O_8$

Finally, we consider a system with an inflection in a $\bar{G}$-X_2 curve. Such inflections cause a single homogeneous phase to separate into two new phases as it cools. A natural example is the system albite $NaAlSi_3O_8$-

orthoclase $KAlSi_3O_8$, a solid solution at high temperature that unmixes into discrete alkali feldspars at lower temperature. For our purposes here, we are only interested in what happens below the solidus in this system, so we ignore temperatures above the solidus.

Several $\bar{G}$-X_{Or} sections are illustrated in figure 10.10. At T_1, the alkali feldspars exhibit complete solution at any composition. At T_2, a small inflection in the alkali feldspar $\bar{G}$ curve appears. Notice that a tangent providing the lowest system free energy touches the curve at two places, indicating that two stable feldspar phases should exist, at least for a small compositional range near the middle of the diagram. With decreasing temperature, the inflection widens and thereby increases the compositional range affected, as shown at T_3.

The T-X_{Or} diagram generated from this phenomenon is illustrated at the bottom of figure 10.10. The solidus is shown at the top of this diagram to emphasize the point that we are looking at what happens in the subsolidus region. The gradually widening tangent points in the $\bar{G}$-X_{Or} diagrams define a concave-downward curve, called a *solvus*, which identifies a region of unmixing or *exsolution*. Above the solvus, one homogeneous phase is stable, and below it, this phase separates into two phases with distinct compositions.

Worked Problem 10.4

What is the subsolidus crystallization path for the composition $X_{Or} = 0.6$ in the alkali feldspar system? With cooling, a homogeneous feldspar of composition $X_{Or} = 0.6$ reaches the solvus at approximately 630°C (fig. 10.10). At this point, the homogeneous feldspar begins to unmix in the solid state, producing two coexisting feldspars. At any temperature below the solvus, intersections of a horizontal (isothermal) line with the two limbs of the solvus give the compositions of the two stable phases. For example, at 550°C, the coexisting phases will have compositions of $X_{Or} = 0.14$ and 0.72. Notice that the compositions of these phases become further separated as temperature decreases and the solvus widens. At some low temperature, the diffusion of ions becomes so sluggish that unmixing ceases, and the compositions of the exsolution products are frozen in at that point. A photograph of an exsolved alkali feldspar formed in this way is illustrated in figure 10.11.

We have now examined the common features of binary T-X_2 diagrams. Simple crystallization of pure phases produces a eutectic. Reactions between liquid and solid either form a compound at a peritectic, or result in solid

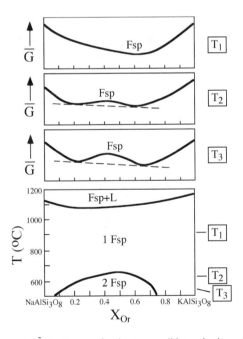

FIG. 10.10. $\bar{G}$-X_{Or} diagrams for the system albite-orthoclase at $P = 1$ atm and at various temperatures are stacked to produce the T-X_{Or} diagram at the bottom. An inflection in the $\bar{G}$ curve with lowering temperature creates a region of unmixing in the solid state, called a *solvus*. Abbreviations for phases are: feldspar (Fsp) and liquid (L).

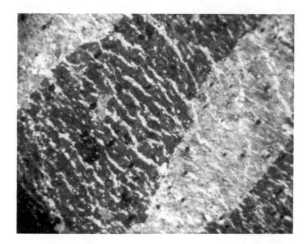

FIG. 10.11. Photomicrograph of perthite, an intergrowth of sodic and potassic alkali feldspars, formed by exsolution. The white lamellae are albite, and the twinned host phase is K-feldspar. This sample is from the Winnsboro granite, South Carolina. Width of the photomicrograph is ~2 mm.

solution described by a loop. Unmixing produces a solvus. Although binary systems are relatively simple, they nevertheless provide important insights into the crystallization of multicomponent systems of geologic interest.

THERMODYNAMIC CALCULATION OF PHASE DIAGRAMS

Up to this point, we have presented without proof a series of qualitative $\bar{G}$-X_2 diagrams and their corresponding T-X_2 diagrams. We confess that to do this, we first worked the problems backwards, by using experimentally determined T-X_2 diagrams to construct qualitatively correct $\bar{G}$-X_2 sections. However, if we had enough good thermodynamic data, we could do each of the examples in the preceding pages just as we presented them; that is, we could calculate rigorous $\bar{G}$-X_2 diagrams, and from them, construct T-X_2 diagrams. In practice, it is often easier to omit the $\bar{G}$-X_2 diagrams and to calculate the T-X_2 sections directly. This problem is similar to the calculation of phase boundaries in the P-T diagram for the unary system Al_2SiO_5 that was introduced in chapter 9, with the added complication that the mixing properties of the two binary components must be modeled.

Worked Problem 10.5

Let's see how accurately we can construct the albite-anorthite phase diagram from thermodynamic data. We can write four sep-

arate equations to express the chemical potentials of albite and anorthite in the plagioclase solid solution and in the liquid:

$$\mu_{Ab,S} = \mu^0_{Ab,S} + RT \ln X_{Ab,S}, \tag{10.3a}$$

$$\mu_{Ab,L} = \mu^0_{Ab,L} + RT \ln X_{Ab,L}, \tag{10.3b}$$

$$\mu_{An,S} = \mu^0_{An,S} + RT \ln X_{An,S}, \tag{10.3c}$$

$$\mu_{An,L} = \mu^0_{An,L} + RT \ln X_{An,L}. \tag{10.3d}$$

The two phases, liquid (L) and solid (S), are indicated by the appropriate subscripts above. In each case, the chemical potentials with the 0 superscripts are standard state values (using the conventional standard state defined by pure plagioclase or pure liquid of end-member compositions). For this exercise, we assume that both the liquid and the solid are ideal solutions, so that $a_i = X_i$.

Within the loop (the two-phase field) we know that $\mu_{Ab,S}$ must equal $\mu_{Ab,L}$ (in other words, equations 10.3a and 10.3b must be equal), and likewise that equations 10.3c and 10.3d must also be equal. The consequence of two-phase equilibrium, therefore, is that:

$$\mu^0_{Ab,S} - \mu^0_{Ab,L} = RT \ln(X_{Ab,L}/X_{Ab,S}) \tag{10.4a}$$

and

$$\mu^0_{An,S} - \mu^0_{An,L} = RT \ln(X_{An,L}/X_{An,S}). \tag{10.4b}$$

The chemical potentials with the 0 superscripts are simply the free energy values for pure albite, pure anorthite, and the pure liquids of end-member composition. The left sides of equations 10.4a and 10.4b therefore correspond to the free energies of melting pure albite and pure anorthite. Because $\Delta \bar{G} = \Delta \bar{H} - T\Delta \bar{S}$ ($\Delta \bar{H}$ and $\Delta \bar{S}$ in this context refer to the enthalpy or entropy changes associated with the melting process), we now have:

$$\mu^0_{Ab,S} - \mu^0_{Ab,L} = \Delta \bar{H}_{Ab} - T\Delta \bar{S}_{Ab} \tag{10.5a}$$

and

$$\mu^0_{An,S} - \mu^0_{An,L} = \Delta \bar{H}_{An} - T\Delta \bar{S}_{An}. \tag{10.5b}$$

These can be substituted into equations 10.4a and 10.4b, respectively, to give

$$RT \ln(X_{Ab,L}/X_{Ab,S}) = \Delta \bar{H}_{Ab} - T\Delta \bar{S}_{Ab} \tag{10.6a}$$

and

$$RT \ln(X_{An,L}/X_{An,S}) = \Delta \bar{H}_{An} - T\Delta \bar{S}_{An}. \tag{10.6b}$$

Let us now take the exponential of both sides of equations 10.6a and 10.6b to solve for the mole fractions. From 10.6a we get:

$$(X_{Ab,L}/X_{Ab,S}) = \exp((\Delta \bar{H}_{Ab} - T\Delta \bar{S}_{Ab})/RT),$$

or

$$X_{Ab,L} = X_{Ab,S} \exp((\Delta \bar{H}_{Ab} - T\Delta \bar{S}_{Ab})/RT). \tag{10.7a}$$

Equation 10.6b similarly yields:

$$X_{An,L} = X_{An,S} \exp((\Delta \bar{H}_{An} - T\Delta \bar{S}_{An})/RT). \tag{10.7b}$$

Now recall that the mole fractions X_{Ab} and X_{An} are not independent, but are related by:

$$X_{An,L} = 1 - X_{Ab,L} \tag{10.8a}$$

and

$$X_{An,S} = 1 - X_{Ab,S}. \tag{10.8b}$$

From equations 10.7a, 10.7b, and 10.8a, we can derive:

$$X_{An,S} \exp((\Delta \bar{H}_{An} - T\Delta \bar{S}_{An})/RT) = \\ 1 - X_{Ab,S} \exp((\Delta \bar{H}_{Ab} - T\Delta \bar{S}_{Ab})/RT).$$

We can eliminate $X_{An,S}$ from this equation by substituting for it from equation 10.8b:

$$(1 - X_{Ab,S}) \exp((\Delta \bar{H}_{An} - T\Delta \bar{S}_{An})/RT) = \\ 1 - X_{Ab,S} \exp((\Delta \bar{H}_{Ab} - T\Delta \bar{S}_{Ab})/RT).$$

If we rearrange this result to solve for $X_{Ab,\bar{S}}$,

$$X_{Ab,S} = [1 - \exp((\Delta \bar{H}_{An} - T\Delta \bar{S}_{An})/RT)]/[\exp((\Delta \bar{H}_{Ab} \\ - T\Delta \bar{S}_{Ab})/RT) - \exp((\Delta \bar{H}_{An} - T\Delta \bar{S}_{An})/RT)]. \tag{10.9}$$

When the temperature is equal to the melting temperature of albite (that is, when $T = T_{Ab}$), we know that equation 10.5a is equal to zero. In other words, the liquidus and solidus curves coincide for pure albite. Thus:

$$\Delta \bar{H}_{Ab} = T_{Ab}\Delta \bar{S}_{Ab}.$$

With this relationship, we can recast equation 10.9 into:

$$X_{Ab,S} = \frac{1 - \exp\left(\dfrac{\Delta \bar{H}_{An}}{R(1/T - 1/T_{An})}\right)}{\exp\left(\dfrac{\Delta \bar{H}_{Ab}}{R(1/T - 1/T_{Ab})}\right) - \exp\left(\dfrac{\Delta \bar{H}_{Ab}}{R(1/T - 1/T_{An})}\right)}. \tag{10.10}$$

This equation contains the enthalpies of melting for albite and anorthite, which have been determined calorimetrically. Numerically, $\Delta \bar{H}_{Ab}$ and $\Delta \bar{H}_{An}$ are equal to 13.1 kcal mol^{-1} and 29.4 kcal mol^{-1}, respectively. With these values in hand, we can calculate the mole fraction of albite in the solid solution at any temperature above the melting temperature of pure albite from equation 10.10. The calculated set of compositions and temperatures defines the solidus. The liquidus curve can be determined from these values with the help of equation 10.8a. These calculated values are tabulated below and plotted graphically in figure 10.12. This phase diagram constructed from thermodynamic data is an almost perfect fit to the plagioclase phase diagram determined by experiment.

T (°C)	Solidus (X_{Ab})	Liquidus (X_{Ab})
1100	1.000	1.000
1150	0.983	0.830
1200	0.955	0.689
1250	0.913	0.579
1300	0.852	0.463
1350	0.767	0.366
1400	0.650	0.275
1450	0.491	0.185
1500	0.280	0.095
1550	0.000	0.000

In the example above, we calculated a T-X_2 phase diagram that was already known from experimental petrology. This is an interesting academic exercise, but it could be considered a waste of time insofar as we have learned nothing new about plagioclase crystallization. However, it illustrates an important point. There are other geochemically important systems for which experimentally derived phase equilibria are difficult to obtain on any reasonable time scale. Calculation of equilibrium relationships in these systems from thermodynamic data may be the most efficient way of attacking such problems.

There is another cogent reason for calculating phase diagrams. If the phase relationships in a system can be modeled accurately under one set of physical conditions, then these relationships can be calculated under other physical conditions. In the example above, the phase diagram was determined at atmospheric pressure. Now that we have demonstrated that we can model this system, we could calculate the phase diagram at elevated pressures by using appropriate thermodynamic data and the relationships (solution models, corrections for compressibility, etc.) derived in the previous chapter.

BINARY PHASE DIAGRAMS INVOLVING FLUIDS

We might also wish to show on phase diagrams the effects of parameters other than temperature, pressure, and X_2. The nature of the fluid phase in a system can be

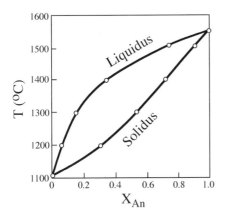

FIG. 10.12. The plagioclase solid solution loop calculated in worked problem 10.5.

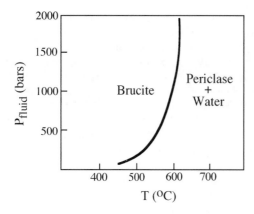

FIG. 10.13. P_{fluid}-T diagram for the system MgO-H_2O, showing how the reaction brucite $\rightleftarrows$ periclase + H_2O depends on P_{fluid}, even though confining pressure remains constant at 1 kbar.

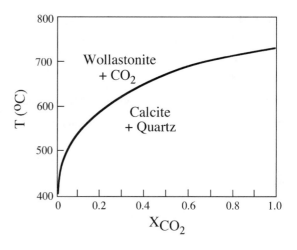

FIG. 10.14. T-X_{CO_2} diagram at constant confining pressure (2 kbar), showing the effect of fluid composition on the reaction calcite + quartz $\rightleftarrows$ wollastonite + CO_2. X_{CO_2} in the fluid can be altered by mixing with H_2O.

a particularly important variable. For example, fluid pressure may be decidedly less than confining (total) pressure in some metamorphic systems because fluids have been progressively driven off. In such a case, P_{fluid} can be considered an independent variable. Employing the same technique we used earlier to derive T-X_2 diagrams, we could envision $\bar{G}$-P_{fluid} diagrams at various values of T and at constant P_{total}, and "stack" these to construct a T-P_{fluid} phase diagram. However, we will spare you the derivation and go directly to a real geochemical example. Phase equilibria in the system MgO-H_2O are illustrated in figure 10.13. This particular diagram is for a fixed confining pressure of 1000 bars and a fixed system composition such that X_{H_2O} equals that in brucite. We can readily see that the temperature at which the equilibrium

$$Mg(OH)_2 \rightleftarrows MgO + H_2O$$
$$\text{brucite} \quad \text{periclase}$$

occurs is controlled by P_{fluid}, even though P_{total} remains constant. Notice that the slope of this reaction curve, like most involving a fluid phase, becomes markedly steeper at high fluid pressures, as we learned from consideration of the Clapeyron equation in chapter 9.

Even in systems in which P_{fluid} equals P_{total}, it is sometimes necessary to treat the fluid composition as a separate parameter. For example, the fluid in contact with a marble undergoing metamorphism may be a mixture of H_2O and CO_2. But only one of these fluid components (CO_2) exerts an influence on reactions in the marble, such as:

$$CaCO_3 + SiO_2 \rightarrow CaSiO_3 + CO_2. \quad (10.11)$$
$$\text{calcite} \quad \text{quartz} \quad \text{wollastonite}$$

In this case, we might want to devise a phase diagram that shows the relationship of this equilibrium to the composition of the fluid. The reaction as written above is univariant. Adding H_2O to the system increases the number of components by one, but does not change the number of phases, because CO_2 and water are miscible. Consequently, the variance is two when calcite, quartz, and wollastonite coexist with a mixed CO_2-H_2O fluid. But by specifying one of the variables (in this example, we fix P_{total} at 2000 bars), we can represent this equilibrium by a univariant curve on a plot of T versus X_{CO_2}. Such a diagram is illustrated in figure 10.14. In this example, we have specified fluid composition in terms of X_{CO_2}, but we could have used P_{CO_2} as well, since $P_{CO_2} = P_{total} X_{CO_2}$.

Worked Problem 10.6

To illustrate the usefulness of phase diagrams involving fluid components, compare the conditions under which wollastonite forms in a closed and an open system. From figure 10.14, we can see that siliceous limestone heated to approximately 740°C will form wollastonite, but the same reaction occurs at much lower temperatures if the resulting CO_2 fluid is diluted with H_2O (thus lowering X_{CO_2}).

First, we assume that the siliceous limestone contains a pore fluid that is a mixture of water and CO_2 with $X_{CO_2} = 0.3$. What happens during metamorphism under closed-system conditions? Inspection of figure 10.14 indicates that wollastonite will begin to form at 625°C. However, this reaction releases CO_2, thereby changing the composition of the fluid in the pore space. The temperature must then increase for the reaction to continue. Provided that the rocks are hot enough, the equilibrium follows the reaction curve, and the fluid becomes progressively richer in CO_2. The fluid composition is controlled by the mineral assemblage in the rock itself; in other words, the system is *internally buffered*. At some point (say $X_{CO_2} = 0.6$, corresponding to $T = 695°C$), one of the reactants is finally consumed, and the reaction ceases.

Now let us consider the same reaction in an open system, in which an infiltrating fluid with $X_{CO_2} = 0.3$ is derived from adjacent rocks (outside of the system as we have defined it). We will stipulate that this fluid permeates through the limestone in such large quantities that it can flush away any CO_2 generated by local reactions. In this case, the fluid composition is *externally buffered* and does not change during metamorphism. So, the wollastonite-producing reaction occurs at 625°C, and the temperature remains constant during the course of the reaction.

P-T DIAGRAMS

So far, we have been dealing mostly with isobaric slices through P-T-X_2 space, largely because we are constrained by the two-dimensional nature of a sheet of paper. It is possible to extend our results into three dimensions, however. As an example, let us consider the system Fe-O, a sketch of which is illustrated in figure 10.15a.

What happens to the various univariant equilibria that we have already defined in such a three-dimensional diagram? A eutectic (the equilibrium between three phases) traces out a ruled surface in P-T-X_2 space. The surface is everywhere parallel to the X_2 axis because we have the equilibrium condition that P and T must be identical in all phases. A peritectic traces out a three-dimensional curve in P-T-X_2 space. An equilibrium involving a pure substance defines a two-dimensional curve lying completely in the plane of constant composition (X_2). These univariant curves and surfaces divide P-T-X_2 space into one- and two-phase volumes (with variances of two or higher) as they intersect at invariant points or lines.

We have already examined some T-X_2 slices (and briefly mentioned the form of P-X_2 slices) through such a P-T-X_2 diagram, so let us now consider the remaining two-dimensional section, a P-T diagram with fixed com-

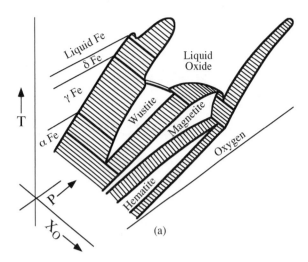

(a)

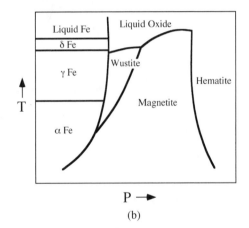

(b)

FIG. 10.15. (a) P-T-X_2 diagram for the system Fe-O. (b) P-T diagram for the system Fe-O obtained by projecting the three-dimensional diagram in part (a) onto the P-T plane.

position. Of all the diagrams we have discussed, P-T diagrams are perhaps the most commonly used and certainly the most difficult to understand.

Inspection of the Fe-O diagram in figure 10.15a suggests the key features of such diagrams. The P-T diagram is obtained by collapsing the three-dimensional diagram onto the P-T plane (that is, projecting parallel to the X_2 axis), as illustrated in figure 10.15b. On this P-T projection, univariant equilibria appear as curves. Invariant equilibria appear as intersections of these curves, although not all apparent intersections are invariant points, because some curves "intersect" only in projection and not in three-dimensional space.

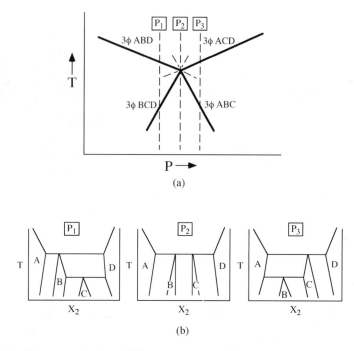

FIG. 10.16. (a) An example of one of the 14 ways in which univariant (three-phase) curves can intersect on a *P-T* diagram. P_1, P_2, and P_3 are three pressures for which representative *T-X$_2$* diagrams are constructed. (b) Isobaric *T-X$_2$* diagrams at the three pressures identified in part (a). Figure out for yourself where the two- and three-phase fields are in these diagrams.

There are many unique topological ways (14, to be precise) in which univariant curves may intersect on a *P-T* diagram. Derivation of these, or even presentation of them, is beyond the scope of this book. Here we consider only one such intersection, just to illustrate how this relates to its corresponding *T-X$_2$* and *Ḡ-X$_2$* sections.

The intersection that we examine is illustrated in figure 10.16a. Each of the univariant lines on this diagram describes an equilibrium between three of the possible phases A, B, C, and D. Each might also be thought of as a line describing a reaction. To see this, refer to the *T-X$_2$* diagrams in figure 10.16b. (It probably isn't obvious that these *T-X$_2$* sections are not topologically unique, but that is the case. The topology of the set of *T-X$_2$* sections that can be drawn from this single invariant point depends on its orientation relative to the *P* and *T* axes. In fact, three distinct topologies can be constructed for different arrangements of isotherms and isobars. We consider only one.)

In the *T-X$_2$* diagram at pressure P_1, there are two univariant three-phase equilibria shown. The first, encoun-

tered at low temperature, is the one in which phases B, C, and D coexist stably. The second, at a somewhat higher temperature, is the one in which phases A, B, and D are stable. Inspection of the *P-T* diagram (fig. 10.16a) indicates the same sequence of equilibria with increasing temperature along the P_1 isobar. Similarly, the *T-X$_2$* diagram drawn at pressure P_3 (fig. 10.16b) can easily be correlated with the sequence of phase equilibria appearing with increasing temperature on the P_3 isobar. The *T-X$_2$* diagram at P_2, however, contains only one line of interest, the one at which phases A, B, C, and D are all stable together. This corresponds to the unique point where all four three-phase lines on the *P-T* diagram intersect—the invariant point.

Now consider the significance of any of these univariant lines as a reaction. The three-phase line marked ABD, for example, is a line along which A, B, and D coexist. It is also a line separating fields for phases A, B, and D on the low-temperature side from a two-phase field at A + D on the high-temperature side (fig. 10.16b). Therefore, the three-phase line ABD may also be viewed as a description of the reaction:

$$B \rightleftarrows A + D,$$

in which B is consumed in the formation of A and D with increasing temperature. Similarly, the three-phase lines ACD, ABC, and BCD may be seen as reactions in which phases C, B, and D are consumed. The invariant (four-phase) point ABCD represents a reaction:

$$B + C \rightleftarrows A + D,$$

in which both B and C are consumed. These relations should not come as any surprise, of course, because they are what we have already been led to expect from our earlier treatment of $\bar{G}$-X_2 diagrams.

SYSTEMS WITH THREE COMPONENTS

It becomes increasingly difficult to handle graphically systems with three or more components. Although we will not consider this subject in great detail, you should at least be aware of how such systems are presented.

For systems containing a liquid phase, we commonly combine three binary T-X_2 sections into a single triangular diagram, as illustrated for the system diopside-albite-anorthite in figure 10.17a. The three binaries fold up to form a triangular prism, whose vertical axis represents temperature, as illustrated in figure 10.17b. The upper lid of the prism is the liquidus, now a curved, isobaric T-X surface. Three-dimensional perspective sketches such as figure 10.17b are difficult to draw, and it is almost impossible to derive quantitative information from them, so details of the liquidus surface are commonly projected onto the compositional base of the prism, as shown in figure 10.17c. The slopes of the liquidus surfaces with temperature are illustrated by isothermal contour lines, and the heavy line delineates the intersection of the liquidus surfaces for different phases—a low-temperature valley called a *boundary curve* or *cotectic*. Slope on the boundary curve is indicated by the arrowhead. Along such boundary curves, the liquid coexists with two solid phases, either crystallizing both simultaneously (*subtraction curve*) or reacting with one solid phase to produce the other (*reaction curve*). Thus, subtraction and reaction boundary curves are equivalent to eutectic and peritectic points, respectively, in binary systems. We have included a more detailed analysis of the various kinds of boundary curves and points in ternary phase diagrams in an accompanying box for interested readers.

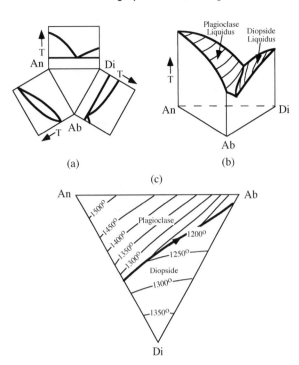

FIG. 10.17. Representations for ternary systems. (a) The three binary phase diagrams for the system albite-anorthite-diopside (Ab-An-Di) oriented around a triangle. (b) These binary systems can be stood upright, and the liquidus surface within the triangle can be extrapolated from the binary systems and contoured, as shown here. (c) A projection of the liquidus surface in (b) onto the bottom triangle is shown in this figure. The arrowhead indicates the direction of slope for the boundary between the primary phase fields of diopside and plagioclase.

SUMMARY

Phase diagrams provide a useful way of visualizing the effects that intensive and extensive variables have on a system. For binary systems, one of the most widely used is the T-X_2 diagram. We have seen how this diagram can be derived from stacking $\bar{G}$-X_2 sections at various temperatures, holding pressure constant. We have explored the characteristics of univariant equilibria in these diagrams, including eutectic and peritectic points, solid solution loops, and solvi. We also considered phase diagrams in which the composition or partial pressure of fluid phases were variables. In addition, we discussed how univariant equilibria appear in P-T diagrams. The various kinds of boundary curves and invariant points in ternary liquidus phase diagrams have also been explored.

INTERPRETATION OF TERNARY LIQUIDUS PHASE DIAGRAMS

Ternary phase diagrams are divided into *primary phase fields* by *boundary curves*. Each primary phase field represents the first phase to crystallize upon intersecting the liquidus. As an example, we take the phase diagram for the system anorthite-leucite-silica, shown in figure 10.18a. Such phase diagrams can be analyzed using Alkemade's Theorem, which encompasses the following statements:

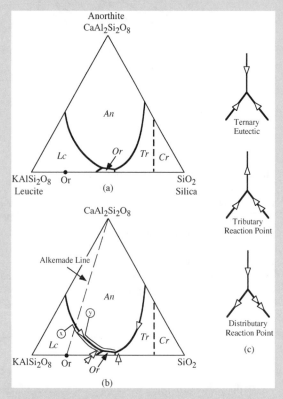

FIG. 10.18. (a) The phase diagram for the system anorthite-leucite-silica ($CaAl_2Si_2O_8$-$KAL_2Si_2O_8$-SiO_2) at $P = 1$ atm. Other phases and their abbreviations are: orthoclase (Or), tridymite (Tr), and cristobalite (Cr). Primary phase fields are identified by abbreviations in italics. The boundary between the primary phase fields for Tr and Cr is dashed to indicate that a phase transition between these two polymorphs occurs at a temperature above the eutectic point. (b) The same diagram with the addition of the Alkemade line between An and Or. Boundaries between primary phase fields are identified as subtraction or reaction curves by single- or double-headed arrows, respectively. (c) These curves may intersect to form three different kinds of invariant points, as illustrated.

1. An *Alkemade line* is a straight line connecting the compositions of two phases whose primary phase fields share a common boundary curve. The first step in the analysis of any ternary phase diagram is to draw in all of the Alkemade lines. In figure 10.18a, most of the Alkemade lines already exist as external boundaries to the triangle. For example, the primary phase fields for leucite and anorthite share a common boundary curve, and the appropriate Alkemade line joining their compositions is the left side of the triangle. In this example, the only additional Alkemade line we must draw is a line joining the compositions of anorthite and orthoclase, as illustrated in figure 10.18b.

2. Alkemade lines divide a triangular diagram into smaller triangles. The ultimate equilibrium crystallization product is defined by the corners of the small Alkemade triangle, in which the original liquid composition lies. For example, a liquid of composition X in figure 10.18b will ultimately crystallize to form leucite + orthoclase + anorthite. If the original liquid lies *on* an Alkemade line, the ultimate assemblage consists of a mixture of the two phases at the ends of the line.

3. The intersection of an Alkemade line with its pertinent boundary curve represents the maximum temperature along that boundary curve. For example, the intersection of the leucite-anorthite boundary curve and Alkemade line occurs along the left side of the triangle. This point represents the highest temperature along that particular boundary curve, and temperature must fall off as one follows the curve to the right.

4. If a boundary curve or its tangent intersects the corresponding Alkemade line, that portion of the boundary curve is a *subtraction curve*. In figure 10.18b, a tangent to the leucite-anorthite boundary curve at any place along its length will intersect the Alkemade line, so the entire curve is a subtraction curve, along which the liquid crystallizes to form two phases, leucite + anorthite. It is conventional to identify subtraction curves with a single arrowhead pointing in the direction of decreasing temperature. If the boundary curve must be extended

to intersect the pertinent Alkemade line, the curve is still a subtraction curve and the point of intersection represents the thermal maximum. The orthoclase-anorthite boundary curve in figure 10.18b must be extended to intersect its Alkemade line, so it is a subtraction curve and temperature decreases away from the Alkemade line, as shown by the arrowhead.

5. If a boundary curve or its tangent does not intersect the appropriate Alkemade line, but merely an extension of that line, then that portion of the boundary curve is a *reaction curve*. For example, a tangent drawn to any point along the leucite-orthoclase boundary curve in figure 10.18b does not intersect the leucite-orthoclase Alkemade line, but only an extension of that line. Thus, this boundary is a reaction curve along its entire length, as indicated by double arrowheads. You might envision a hypothetical boundary curve that changes from subtraction to reaction along its length as its tangent swings outside the limits of its Alkemade line. The intersection of the curve with the extended Alkemade line is the maximum temperature along the boundary curve.

Once the various kinds of boundary curves have been identified using the rules above, we can recognize several types of invariant points formed by intersections of these curves, illustrated in figure 10.18c. A *eutectic point* occurs where three boundary curves form a "dead end," and the liquid has no direction of escape with further cooling. An example of a eutectic in figure 10.18b is the intersection of the orthoclase-anorthite, orthoclase-tridymite, and anorthite-tridymite subtraction curves. A *tributary reaction point* provides one outlet for liquid to follow once the reaction that occurs at that point is completed. In figure 10.18b, the intersection of the leucite-anorthite and orthoclase-anorthite subtraction curves with the leucite-orthoclase reaction curve produces this kind of point. A *distributary reaction point*, shown in figure 10.18c, has two outlets and, after the reaction that occurs at the point is complete, permits the liquid to follow either of two branches, depending on what its final crystallization product must be (deter-

mined by the Alkemade triangle in which it resides). There is no example of a distributary reaction point in the anorthite-leucite-silica system.

We can now describe equilibrium crystallization paths for any liquids in this system. We take two examples, identified by X and Y in figure 10.18b. Liquid X lies in the Alkemade triangle leucite-anorthite-orthoclase, so those phases must be the final product. These phases only coexist at one point in the diagram, the tributary reaction point, which must be where crystallization will end. Upon intersecting the liquidus, X first begins to precipitate leucite, as it lies in the leucite primary phase field. The liquid composition moves directly away from the leucite corner as leucite forms, finally intersecting the leucite-anorthite subtraction curve. At this point, leucite + anorthite crystallize simultaneously, and the liquid follows the curve toward the tributary reaction point. The ratio of leucite to anorthite in the crystallizing assemblage can be found at any point along the subtraction curve by noting the intersection of the tangent to the boundary curve with the Alkemade line and employing the lever rule. When the liquid reaches the tributary reaction point, already crystallized leucite reacts with the liquid to produce orthoclase. We know that the reaction cannot run to completion, because we must have some leucite left in the final product. Therefore, we must run out of liquid before all the leucite can react.

Liquid Y in figure 10.18b lies in the Alkemade triangle anorthite-orthoclase-silica, so this must be the final product. These phases only coexist at the eutectic point. Anorthite is the first phase to crystallize, and the liquid composition moves from Y directly away from the anorthite corner. Upon intersecting the leucite-anorthite subtraction curve, the crystallization sequence is joined by leucite. Both phases crystallize as the liquid moves down the curve until it reaches the tributary reaction point. Then leucite reacts with the liquid to form orthoclase. In this case, the reaction runs to completion, and the liquid is then free to move down the orthoclase-anorthite subtraction curve, all the while crystallizing these two phases. When it finally reaches the eutectic point, orthoclase + anorthite + tridymite continue to precipitate until the liquid disappears.

At every opportunity, we have emphasized the relationships between phase diagrams and thermodynamics, and we hope that this treatment will aid you in comprehending these relationships. So far, we have considered only systems at equilibrium. The next chapter explores nonequilibrium conditions.

SUGGESTED READINGS

Phase diagrams are usually introduced more rigorously in igneous/metamorphic petrology texts than they are in geochemistry texts, although few relate these diagrams to thermodynamic functions. Some of the best explanations of phase diagrams are noted below.

Edgar, A. D. 1973. *Experimental Petrology, Basic Principles and Techniques*. London: Oxford University Press. (A techniques book that describes the equipment, procedures, and pitfalls in experimentally determining phase diagrams.)

Ehlers, E. G. 1972. *The Interpretation of Geological Phase Diagrams*. San Francisco: Freeman. (A complete and very readable introduction to the subject that provides much more detail than we do in this book; chapter 2 describes binary phase diagrams and chapter 3 discusses ternary systems.)

Ernst, W. G. 1976. *Petrologic Phase Equilibria*. San Francisco: Freeman. (An exceptionally good reference on phase diagrams; chapter 2 describes the experimental determination of phase equilibria, chapter 3 provides a look at the computational approach, and chapter 4 introduces unary, binary, and ternary systems.)

Ferry, J. M., ed. 1982. *Characterization of Metamorphism through Mineral Equilibria*. Reviews in Mineralogy 10. Washington, D.C.: Mineralogical Society of America. (This book contains numerous applications of phase diagrams to metamorphic rocks; chapter 1 by J. B. Thompson describes the relationship between thermodynamics and metamorphic phase diagrams.)

Hess, P. C. 1989. *Origins of Igneous Rocks*. Cambridge: Harvard University Press. (Chapter 2 presents a thorough explanation of phase diagrams, although without reference to thermodynamics.)

Morse, S. A. 1994. *Basalts and Phase Diagrams*. Malabar: Krieger Publishing. (A thorough and very readable book about phase diagrams. As in this book, Morse's discussions of binary diagrams end with $\bar{G}$-X diagrams.)

Philpotts, A. R. 1990. *Principles of Igneous and Metamorphic Rocks*. Englewood Cliffs: Prentice-Hall. (The relationship between free energy and phase diagrams is introduced in chapter 8 of this excellent text, and chapter 10 treats binary and ternary phase diagrams.)

Ottonello, G. 1997. *Principles of Geochemistry*. New York: Columbia University Press. (Chapter 2 of this thick book provides one of the few systematic treatments of $\bar{G}$-X diagrams that we are aware of in a geochemistry text.)

Yoder, H. S., Jr., ed. 1979. *The Evolution of the Igneous Rocks, Fiftieth Anniversary Perspectives*. Princeton: Princeton University Press. (An updated version of Bowen's treatise that uses phase diagrams throughout. Chapter 4 by A. Muan gives a thorough description of phase diagrams.)

PROBLEMS

(10.1) Using figure 10.6, reconstruct the equilibrium crystallization sequence for a liquid with a composition that corresponds exactly to the eutectic in this system.

(10.2) Equilibrium melting is the reverse process of equilibrium crystallization. From figure 10.7, determine the melting sequence for a mixture of 90 mol% orthoclase and 10% tridymite.

(10.3) The liquid in problem 10.2 is now cooled again from 1200°C to room temperature under equilibrium conditions. In this experiment, you continuously measure the temperature of the system over the four hours it takes to reach room temperature. Sketch a qualitative plot of how temperature in this system changes with elapsed time, and explain why temperature does not fall continuously with time.

(10.4) Using figure 10.9, describe the melting under equilibrium conditions of plagioclase with composition $X_{An} = 0.2$.

(10.5) Using the experimental results tabulated below, construct a T-X_{Fa} diagram for the system forsterite-fayalite. The glass composition represents the liquid.

Bulk Composition (X_{Fa})	T(°C)	Olivine Composition (X_{Fa})	Glass Composition (X_{Fa})
0.60	1800	—	0.60
0.60	1700	—	0.60
0.60	1600	0.34	0.60
0.60	1500	0.42	0.75
0.60	1400	0.58	0.91
0.60	1300	0.60	—
0.20	1800	0.08	0.25
0.20	1700	0.17	0.47
0.20	1600	0.20	—

(10.6) (a) Construct $\bar{G}$-X_2 diagrams for a system that is a regular solution with $W = 15$ kJ at 800, 900, 1000, 1100, and 1200 K. Then construct the corresponding T-X_2 diagram for this system. (b) At 1000 K, what is the effect of changing W from 2 to 5 kcal? Show your work on a $\bar{G}$-X_2 diagram.

(10.7) Draw a series of topologically reasonable isothermal P-X_2 sections that describe the phase relations shown in figure 10.16a.

(10.8) Calculate a T-X_{CO_2} diagram for the reaction in worked problem 10.2 at 1 atm for a CO_2-H_2O fluid that behaves as an ideal gas.

(10.9) Using the T-X_{SiO_2} diagram on the bottom of figure 10.7, determine the relative proportions of stable phases at 900, 1100, and 1300°C for a bulk composition of $X_{SiO_2} = 0.3$.

CHAPTER ELEVEN

KINETICS AND CRYSTALLIZATION

OVERVIEW

Not all geochemical processes make the final adjustment to equilibrium conditions. Even those processes that eventually go to completion, such as crystallization of magmas and recrystallization of metamorphic rocks, are sometimes controlled by kinetic factors. In this chapter, we discover what these factors are and how they can be understood. We first examine the effect of temperature on rate processes, which leads to the concept of activation energy. We then consider three specific rate processes: diffusion, nucleation, and growth. Differences between volume diffusion and grain boundary diffusion are explained, and the importance of crystal defects is considered. We introduce nucleation by discussing the extra free energy term related to surface energy, and then consider rates for homogeneous and heterogeneous nucleation. Crystal growth may be dominated by interface- or diffusion-controlled mechanisms, so we discuss each briefly. Finally, we illustrate the importance of determining the rate-limiting step with several examples, as well as discuss several empirical methods for estimating nucleation and growth rates without reference to kinetic theory.

EFFECT OF TEMPERATURE ON KINETIC PROCESSES

You may not have been surprised to learn in chapter 5 that diagenetic processes are kinetically controlled. Most of us are intuitively aware of how sluggish reactions are at low temperature. It may not be so obvious that kinetics also plays a role in many geochemical processes at high temperatures, such as crystallization of magma or recrystallization of metamorphic rocks.

The rate r of a chemical reaction $X \rightarrow Y + Z$ can be expressed as:

$$-dX/dt = k(X)^n,$$

where X is the reactant's concentration at any time t, k is the rate constant, and the exponent n indicates the order of the reaction. The minus sign denotes that the rate decreases with time, as the reactant is used.

For rate-controlled processes in magmatic or metamorphic systems, it is critical that we take into account the temperature dependence of rate constants. The relationship between any rate constant k and temperature follows the classic equation proposed by Arrhenius more than a century ago:

$$k = A \exp(-\Delta\bar{G}^*/RT), \tag{11.1}$$

where A is the frequency factor, $\Delta\bar{G}^*$ is the activation energy, R is the molar gas constant, and T is the absolute temperature. Because a reaction requires the collision of two molecules, the reaction rate is proportional to the frequency of such collisions. The *frequency factor*, then, indicates the number of times per second that atoms are close enough to react. Not all collisions are comparable, though. A very gentle collision between molecules may be insufficient to cause a reaction. The reaction rate, therefore, is assumed to depend on the probability that collisions will have energies greater than some threshold value. The *activation energy* is an expression of this free energy hurdle that must be overcome for the reaction to proceed at an appreciable rate (fig. 11.1). The term $\exp(-\Delta\bar{G}^*/RT)$ expresses the fraction of reacting atoms that have energy higher than the average energy of atoms in the system and are thus most likely to participate in chemical reactions. This probability term resembles the Boltzmann distribution law of statistics, and is called the *Boltzmann factor*. The negative exponential form of the Boltzmann factor explains why this term must be considered in high-temperature systems: depending on the size of $\Delta\bar{G}^*$, rate constants may change by several orders of magnitude over a temperature range of a few hundred degrees.

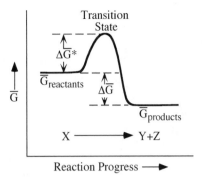

FIG. 11.1. Representation of activation energy needed for the reaction $X \rightarrow Y + Z$. $\Delta\bar{G}$ is the free energy for the reaction under equilibrium conditions, and $\Delta\bar{G}^*$ is the activation energy necessary to allow the reaction to proceed.

plot), we can graphically obtain values for the activation energy and frequency factor.

A plot of Mathews's data is shown in figure 11.2. The intercept of the least squares regression line with the ln k axis corresponds to a value for A of 1.2×10^{11}. The slope of the line corresponds to a $\Delta\bar{G}^*$ value of 44.5 kcal mol^{-1}.

Unfortunately, there are too few experimental data on reaction rates of geochemical interest, such as those given in worked problem 11.1. One notable exception is a 1963 study of the system $MgO\text{-}SiO_2\text{-}H_2O$ by Hugh

Worked Problem 11.1

How can we determine activation energy and frequency factor for a real geochemical system? As an example, consider the reaction analcite + quartz $\rightarrow$ albite + water. Experimental data for the rate of this reaction in NaCl at various temperatures were reported by A. Mathews (1980):

k (kcal/RT)	T (°C)
9.02×10^{-4}	419
2.55×10^{-3}	435
5.95×10^{-3}	454
9.80×10^{-3}	474

First, we recast equation 11.1 in a linear form by taking the logarithm of both sides:

$$\ln k = \ln A - [\Delta\bar{G}^* R \,(1/T)]. \tag{11.2}$$

This expression has the form of a straight line ($y = b + mx$), in which the slope is $-\Delta\bar{G}^*$ and the y intercept is ln A. Therefore, by plotting ln k versus reciprocal temperature (an Arrhenius

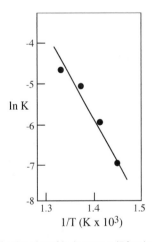

FIG. 11.2. Arrhenius plot of ln k versus $1/T$ for the reaction analcite + quartz $\rightarrow$ albite + water. The activation energy for the reaction can be obtained from the slope of the regression line, and the frequency factor is given by the y intercept.

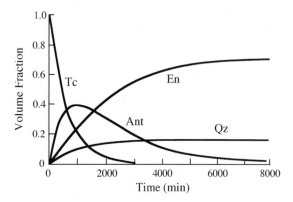

FIG. 11.3. Volume fractions of talc (Tc), anthophyllite (Ant), enstatite (En), and quartz (Qz) as a function of time produced by the talc dehydration reaction at 830°C and 1 kbar. (After Mueller and Saxena 1977.)

Greenwood. Experiments show that at 830°C and 1 kbar, talc dehydrates to form enstatite plus quartz. However, a multistep process involving the breakdown of talc to form anthophyllite as a reactive intermediate is kinetically more favorable than the direct transformation of talc into enstatite plus quartz. Using Greenwood's measured rate constants, Robert Mueller and Surendra Saxena (1977) calculated the proportional volumes of talc, anthophyllite, enstatite, and quartz in a reacting assemblage as functions of time. We summarize their results in figure 11.3. Even though the stable assemblage under these conditions is enstatite plus quartz, anthophyllite is the most abundant phase during the time interval from 1000 to 1500 minutes. Thus, experiments of short duration might give the false impression that anthophyllite is the stable phase under these conditions.

DIFFUSION

Although diffusion may be easily thought of as a physical phenomenon, it resembles a chemical reaction in being governed by an activation energy. Components can migrate over free surfaces of grains (surface diffusion), along the boundaries between grains (grain-boundary diffusion), and through the body of grains or liquids (volume diffusion). The latter two styles of diffusion are of most interest in high-temperature systems. When we consider volume diffusion, it is also useful to distinguish between *self diffusion*, by which we mean the one-directional movement of similar atoms or ions (for example, when

oxygen diffuses through the oxygen framework of feldspar), and *interdiffusion*, in which the diffusion of one atom or ion is dependent on the opposing diffusion of another (for example, Mg-Fe exchange in olivine). Generally, the requirement of local charge balance implied by interdiffusion results in a decrease in the rate of diffusion.

In chapter 5, when we first discussed diffusion, we presented a macroscopic or empirical view, in which Fick's Laws were used to describe the bulk transport of material through a medium that was considered to be continuous in all its properties. That treatment involved the definition of diffusion coefficients (D), which come in a bewildering array of forms. We can now add two more. Self diffusion is commonly described by *tracer diffusion coefficients* (D^*), which are measured experimentally by following the progress of isotopically tagged ions (for example, ^{18}O moving through the oxygen framework of a silicate). *Interdiffusion coefficients* (D_{1-2}), describing the interaction of ions of species 1 and species 2, are measured empirically or can be calculated from the tracer coefficients of each species, weighted by their relative abundance in the system:

$$D_{1-2} = (n_1 D_1^* + n_2 D_2^*)/(n_1 + n_2).$$

In geologic systems of even moderate complexity, it can be difficult to calculate D_{1-2}, because we first need to collect a large amount of experimental data to estimate tracer coefficients in systems that react sluggishly and are hard to analyze.

Geochemists are adopting an alternate approach, which considers diffusion from a mechanistic perspective at the atomic level and has been used with some success by metallurgists and ceramists. It offers the advantage of describing diffusive transport in terms of forces between ions in a mineral lattice or other medium. In theory, these can be predicted from conventional bonding models, given detailed information about crystal structures. This approach holds great promise but has been applied to very few materials of geologic significance so far. It is beyond the scope of this chapter, although we can use the atomistic perspective to make some broad generalizations to help visualize the behavior of diffusing ions.

Volume diffusion in solids usually involves migration of atoms or ions through imperfections in the periodic structures of crystals. Point defects can arise from the absence of atoms at lattice sites (vacancies). Two kinds

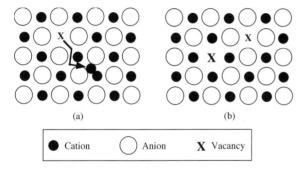

FIG. 11.4. Volume diffusion is facilitated by defects in crystals. (a) Frenkel defects result when ions leave normal crystal sites for interstitial positions, leaving vacancies. (b) Schottky defects occur when cation and anion vacancies preserve charge balance.

of *intrinsic* vacancies in ionic crystals maintain electrical neutrality: *Frenkel defects* are caused by an ion abandoning a lattice site and occupying an interstitial position, and *Schottky defects* involve equal numbers of vacancies in cation and anion positions. These imperfections are illustrated in figure 11.4. Impure crystals may also contain *extrinsic* defects, which arise because foreign ions may have a different charge from the native ions they replace. The requirement for overall charge balance leads to vacancies in the crystalline structure. For example, the incorporation of Fe^{3+} into olivine normally requires a cation vacancy somewhere in the crystal.

Nonstoichiometric crystals, by definition, also have vacancies. Real crystals commonly contain both intrinsic and extrinsic defects, each of which dominate the overall diffusion at different temperatures. Intrinsic mechanisms tend to be most effective at high temperatures, when thermal vibrations of atoms are rapid. As temperatures are lowered, the number of vacancies induced by impurities becomes greater than those generated intrinsically. This can be seen in figure 11.5, an Arrhenius plot of experimentally determined diffusion coefficients for Fe-Mg along the *c* axis in olivine as functions of temperature. In the high-temperature region (left side of fig. 11.5), diffusion occurs mostly by an intrinsic mechanism, which has a steeper slope (larger $\Delta \bar{G}^*$) than for the extrinsic region. The transition between diffusion mechanisms occurs at ~1125°C. The difference in slope tells us, qualitatively, that it takes more energy to move ions between intrinsic defects than between extrinsic ones (30 kcal mol^{-1} in the intrinsic region, 62 kcal mol^{-1} in the extrinsic region, when extrapolated to pure forsterite).

Extrinsic diffusion mechanisms probably predominate in natural systems, because most minerals contain many impurities, and temperatures for most geologic processes are less than required for intrinsic mechanisms.

Extended defects also play important roles in volume diffusion. These include dislocations, which are linear displacements of lattice planes, and various kinds of planar defects, such as twin boundaries and stacking faults. Irregular microfractures also fall into this category. Extended defects provide very effective conduits (sometimes called *short circuits*) for diffusion, and where they are abundant, they may control this process. This accounts in part for the high reactivity of strained versus unstrained crystals. However, extended defects are difficult to model quantitatively.

Grain boundaries are sometimes considered to be extended defects, but this blurs the distinction between volume diffusion and grain-boundary diffusion. Most geochemists believe that the most efficient path for diffusive mass transfer in metamorphic rocks is along grain boundaries. This idea is grounded in the notion that metamorphic fluids along such boundaries provide faster diffusion paths than volume diffusion through solid grains. Although this idea is probably correct in many cases, the situation is not as straightforward as it may seem. The diffusion cross section offered by grain

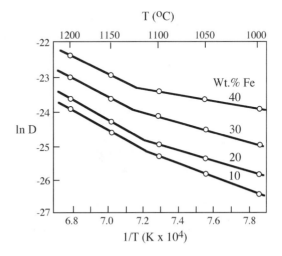

FIG. 11.5. Arrhenius plot of ln *D* versus 1/*T*, summarizing data on interdiffusion of Fe and Mg in olivines of different composition (Buening and Buseck 1973). The kink in these lines at ~1125°C corresponds to a change from an intrinsic to an extrinsic diffusion mechanism with different activation energy (and hence, slope).

boundaries is quite small, the tortuosity of these paths may be appreciable, and the amount of fluid may decrease considerably at high metamorphic grades. Attempts have been made to place limits on grain-boundary diffusion rates in metamorphic rocks, but there are few data with which to test numerical models of this process.

Worked Problem 11.2

Can diffusion be used to constrain the rate at which a magma cools? Larry Taylor and coworkers (1977) developed a "speedometer" applicable to basalts, using frozen diffusion profiles of Fe/Mg in olivines. Let's examine how this system works.

From the experimental data of Buening and Buseck (1973), illustrated in figure 11.5, we can derive an expression for the diffusion coefficient along the olivine c axis, D_c, for temperatures below 1125°C:

$$D_c = 10^2 f_{O_2}^{1/6} \exp(-0.0501\, X_{Fa} - 14.03)$$
$$\exp([-31.66 + 0.2191 X_{Fa}]/RT), \qquad (11.3)$$

where X_{Fa} is the mole fraction of Fe_2SiO_4 in olivine. Buening and Buseck also showed that olivine diffusion is highly anisotropic, with transport being fastest in the c direction. The measured profiles of olivine grains, however, reflect reactions limited by the slowest diffusive rate, which is along the a axis. Taylor and coworkers estimated that D_c was greater than D_a by a factor of about four at 1050°C, so that:

$$D_a = D_c/4.$$

One uncertainty in boundary conditions for this problem is that the concentration profile in olivine as it solidified is unknown; that is, some compositional zoning may have been present originally. Taylor and coworkers treated the forsterite content of olivine as a step function, so that each grain originally had no compositional gradient other than a sharp step between the magnesian core and iron-rich rim. Such a simplification means that the calculated cooling rate will be a minimum rate. In a later paper, Taylor and coworkers (1978) corrected this problem by estimating an "as-solidified" concentration profile. For simplicity, we ignore this later refinement.

In the case of olivine, the limiting diffusion coefficient D_a is a function of both temperature (which is, in turn, a function of time t) and composition. We can use a one-dimensional diffusion equation to calculate the compositional profile:

$$\frac{\partial C}{\partial t} = \frac{\partial}{\partial x}\left\{ D_a[C(x), T(t)] \frac{\partial C}{\partial x} \right\}, \qquad (11.4)$$

where C is the Fe concentration. Unlike the worked problem at the beginning of chapter 5, which yielded easy, analytical solutions, this equation is nonlinear and must be solved by numerical iteration to approach the correct solution.

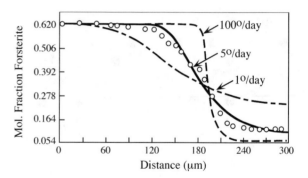

FIG. 11.6. Calculated compositional profiles in olivine as a function of distance from the core-rim interface and cooling rate. These profiles are produced by interdiffusion of Fe and Mg during cooling. Circles are electron microprobe data for zoned olivines in lunar basalt 15555. (After Taylor et al. 1977.)

Taylor and his collaborators applied their solution to determination of the cooling rates for lunar basalts. Oxygen fugacities (which are temperature dependent) appropriate for lunar rocks obey the relation:

$$\log f_{O_2} = 0.015 T\ (K) - 34.6.$$

This relation provides the information we need to determine D_a as a function of f_{O_2} for substitution into equation 11.4. The measured compositional profile in an olivine grain in lunar basalt 15555, obtained by electron microprobe traverse, is illustrated by circles in figure 11.6. Calculated diffusion profiles for various cooling rates are shown for comparison. The minimum cooling rate for this basalt is ~5°C/day.

NUCLEATION

It is a wonder that crystals ever form. The most difficult step, from a thermodynamic perspective, is *nucleation*—the initiation of a small volume of the product phase. Crystal nuclei form because local thermal fluctuations in the host phase result in temporary, generally unstable clusters of atoms. This is an extremely difficult process to study, because no experimental technique has yet been devised that allows direct observation of the formation of crystal nuclei.

The driving force for nucleation, like other rate processes, is the deviation of the system from equilibrium conditions. A common expression for this driving force is *supersaturation*, the difference between the concentration of the component of interest and its equilibrium concentration. In studying nucleation in geochemical systems, we often define deviation from equilibrium in another way; that is, in terms of *undercooling* (ΔT). This parameter is the difference between the equilibrium temperature for the first appearance of a phase and the temperature at which it actually appears. For example, in magmas, crystallization commonly fails to occur at the liquidus temperature of a melt, where it should be expected. The magma is therefore said to be *undercooled* if it is at a temperature below its liquidus and crystallization is not yet initiated. The value of ΔT is equal to the liquidus temperature minus the actual temperature.

Nucleation in Melts

As we learned in earlier chapters, the appearance of a new phase is associated with a decrease in the free energy of the system. This change can be expressed as a total differential:

$$dG = \left(\frac{\partial G}{\partial T}\right)_{P,n_i} dT + \left(\frac{\partial G}{\partial P}\right)_{T,n_i} dP + \left(\frac{\partial G}{\partial n_i}\right)_{P,T,n_j \neq n_i} = dn_i,$$

in which the partial derivatives are written more familiarly as $-\bar{S}$, $\bar{V}$, and μ_i. In practice, we apply this equation by declaring that each phase in the system consists of a material whose thermodynamic properties are continuous. The boundary between one phase and another is marked by a discontinuity in thermodynamic properties, but we assume that it does not contribute any energy to the system itself. Calculations performed on this basis are generally consistent with observations of real systems.

We know, however, that the boundaries between phases are not simple discontinuities. The atoms on the surface of a droplet or a crystal are not surrounded by the same uniform network of bonds found in its interior. This irregular bonding environment leads to structural distortions that must increase the free energy of the system. The *surface free energy* contribution is generally quite small, because the number of atomic sites near the boundary of a phase is much less than the number in its interior. We can predict, however, that this contribution increases in importance if the mean size of crystals or droplets in the system is vanishingly small, as it is during nucleation. Surface free energy, therefore, accounts for the kinetic inhibition we have described as undercooling.

The extra energy associated with the formation of an interface is minimized if the surface area of the new par-

ticle is also minimized. Consequently, a growing particle must perform work to stretch the interface. This work, for a liquid droplet, is commonly defined in terms of a tensional force, σ, applied perpendicular to any line on the droplet surface. This concept of *surface tension* is only strictly applicable to liquid surfaces, but no adequate alternative has been formulated for solids. It is common practice to refer loosely to σ even when the nucleating particle is a crystal.

Surface tension can be viewed as a measure of the change in free energy due to an increase in particle surface area A; that is,

$$\sigma = \left(\frac{\partial G}{\partial A}\right)_{T,P,n}.$$

The total differential of G thus can be modified to the form:

$$dG = (dG)_{vol} + \left(\frac{\partial G}{\partial A}\right)_{T,P,n} dA$$
$$= (dG)_{vol} + \sigma dA,$$

in which $(dG)_{vol}$ is a shorthand notation for the familiar temperature, pressure, and composition terms that apply throughout the volume of the phase. For spherical nuclei with radius r, this becomes:

$$\Delta G_{total} = \frac{4\pi r^3}{3} \Delta G_v + 4\pi r^2 \sigma, \tag{11.5}$$

where ΔG_v is the free energy change per unit volume. The surface energy term dominates when the radius is small, and the volume term dominates when it is large. Because σ is always positive, equation 11.5 implies that the total free energy of a nucleating particle increases with r until some critical radius (r_{crit}) is reached, after which the free energy decreases with further increase in r. Small nuclei with $r < r_{crit}$ are unstable and tend to redissolve, whereas nuclei with $r > r_{crit}$ persist and grow. We can illustrate the importance of this observation with the following problem.

Worked Problem 11.3

How does the free energy of a nucleus vary with radius? To answer this, let's consider the formation of spherical nuclei of forsterite in Mg_2SiO_4 liquid. Assume an undercooling of 10°C and a surface energy of 10^3 ergs cm^{-2} (= 2.39×10^{-5} cal cm^{-2}). Calculate the critical radius for olivine nuclei under these conditions.

ΔG_v for olivine at a temperature of 2063 K (10 K below the forsterite liquidus) is -1.738 cal cm^{-3}. Substitution of these values into equation 11.5 gives:

$$\Delta G_{total} = \frac{4\pi}{3}(-1.738)r^3 + 4\pi(2.39 \times 10^{-5})r^2$$

$$= -7.276r^3 + 3.0 \times 10^{-4}r^2.$$

For $r = 2 \times 10^{-5}$ cm, ΔG_{total} is thus 6.18×10^{-14} cal. Substitution of other values for r gives a data set from which the solid line in figure 11.7 is plotted. From this figure, we can readily see that increasing r initially causes an increase in total free energy of the system because the surface energy term dominates. At the point where the volume energy term begins to dominate, there is a downturn in total free energy, and further crystal growth can take place spontaneously with a net decrease in free energy.

The peak of the free energy hump in this diagram corresponds to r_{crit}. We can calculate this value directly by recognizing that the reversal in slope corresponds to the point where the first derivative of equation 11.5 with respect to r equals zero. Solving for r under this condition yields:

$$r_{crit} = -2\sigma/\Delta G_v. \tag{11.6}$$

Substitution of the values given above for σ and ΔG_v gives r_{crit} for olivine = 2.75×10^{-5} cm under these conditions.

The total energy barrier to nucleation ΔG_{homo} (for homogeneous nucleation, defined below) can be found by combining

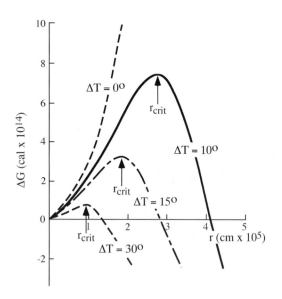

FIG. 11.7. Calculated total free energy for nucleation of forsterite in a liquid of its own composition, as a function of nucleus radius r and undercooling ΔT. The critical nucleus at each undercooling is r_{crit}.

the conditions necessary to form clusters of the critical radius (equation 11.6) with equation 11.5 to give:

$$\Delta G_{homo} = (16\pi\sigma^3)/(3\Delta G_v^2). \tag{11.7}$$

Also illustrated in figure 11.7 by dashed lines are energy barriers to nucleation at $\Delta T = 0$, 15, and 30°C. We can see that this barrier, as well as the critical radius for nuclei, decreases as undercooling increases.

What we have presented so far is an apparent paradox. A particle is not stable until its radius is greater than r_{crit} but it can never reach r_{crit} because it must always begin as a nucleus too small to be stable. Worked problem 11.3, however, suggests one way to avoid this paradox—by undercooling.

If nuclei persist once they reach critical radius, then the nucleation rate can be expressed by an Arrhenius-type equation as the product of the concentration of these nuclei and the rate at which atoms attach themselves to critical nuclei. The concentration of nuclei having the critical radius (N_r) at equilibrium is given by an expression of the Boltzmann distribution:

$$N_r = N_v \exp(-\Delta G_{total}/RT), \tag{11.8}$$

where N_v is the number of atoms per unit volume of the reactant phase, and the exponential term is the probability that they have sufficient energy to react. The attachment frequency of atoms is the product of the number of atoms (n) located next to the critical cluster, the frequency (ν) with which atoms try to overcome the barrier, and the probability (f) of atoms having sufficient energy to succeed. This is expressed by:

$$f_{attach} = n\nu \exp(-\Delta G^*/RT), \tag{11.9}$$

where ΔG^* is the activation energy needed for attachment. The forms of equations 11.8 and 11.9 are similar to that of equation 11.1, the expression for the rate constant in terms of frequency factor and activation energy. We will see equations of this form over and over again in this chapter, because kinetic processes are governed by statistical probability. Multiplying equations 11.8 and 11.9 gives the nucleation rate J:

$$J = n\nu N_v \exp(-\Delta G_{total}/RT)$$
$$\exp(-\Delta G^*/RT). \tag{11.10}$$

For nuclei that form free in a liquid, ΔG_{total} in this equation is identical to ΔG_{homo} in equation 11.7. We can make the further approximation that $\Delta G_v = -\Delta S \Delta T$, where

ΔS is the entropy change associated with nucleation and ΔT is the undercooling. By combining equations 11.7 and 11.10, therefore, we find that:

$$J = n\nu N_v \exp\left(\frac{16}{3}\pi\sigma^3\Delta S^2\Delta T^2 RT\right)\exp(-\Delta G^*/RT). \tag{11.11}$$

We can use equation 11.11 to make some generalizations about nucleation rate. Not surprisingly, undercooling exerts a major influence, and there is a rapid rise in nucleation rate as undercooling increases; that is, as the system moves farther away from equilibrium. As ΔT increases and T decreases, the two exponential terms compete for dominance, with the result that the nucleation rate reaches a maximum at some critical undercooling and then decreases thereafter. In physical terms, the free energy advantage gained by undercooling is finally overwhelmed by the increasing viscosity of the liquid. If we were to plot J against ΔT, the result would be an asymmetrical bell-shaped curve.

Undercooling, then, reduces the kinetic inhibition to nucleation and offers a solution to the paradox. Nature provides another, more common solution as well. Equations 11.5–11.7 and equation 11.11 are written in terms of a free-standing spherical particle. This geometrical condition is known as *homogeneous nucleation*. Observation tells us, however, that nucleation more commonly takes place on a substrate. Frost forms on a windowpane, dew on spiderwebs, and crystals on other crystals. There must be an energetic advantage to this style of *heterogeneous nucleation* that makes it so widespread.

To see how nucleation on a substrate helps avoid the problem of a critical radius, consider figure 11.8. Here, a particle of a phase C with an exceedingly small volume has been allowed to form from liquid phase L against a foreign surface S. Note that the radius of this particle, if it had condensed homogeneously, would have been $<r_{crit}$. Here, however, the same volume of phase C forms a spherical cap with an effective radius that is considerably larger than r_{crit}. Because the surface tension on the particle is a function of its degree of curvature, this nucleus now has a low surface free energy and is stable.

If the interfaces between phases C, L, and S are all mechanically stable, then the tensional forces perpendicular to any line along the phase interfaces must be perfectly balanced. This must also be true at the three-phase contact (point P), where there are three force vectors to resolve. Each is proportional to a surface tension in the

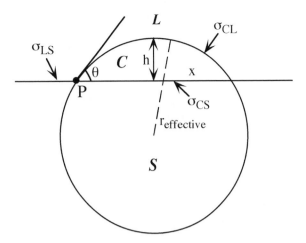

FIG. 11.8. Schematic cross section through a cluster of atoms of phase C nucleating heterogeneously from phase L on a substrate S. The contact angle between the surface of the nucleus and S is θ. The effective radius of this nucleus is much greater than the critical radius of a spherical nucleus of the same volume formed by homogeneous nucleation. Distances x and h are used in problem 11.5 at the end of this chapter.

system. σ_{LS} and σ_{CS} oppose each other along the flat boundaries L-S and C-S, respectively. The remaining vector, represented by σ_{CL}, is tangential to the curved C-L interface at point P. Because the sum of these three must be zero, we see that:

$$\sigma_{LS} = \sigma_{CS} + \sigma_{CL}\cos\theta, \tag{11.12}$$

where θ is the angle between the tangent and interface C-S.

From equation 11.12, it is apparent that nucleation of a new particle is particularly favorable if σ_{CL} is nearly equal to σ_{LS}, and σ_{CS} is very small. This condition might easily be met if the substrate were a seed crystal of the nucleating phase itself. Under these conditions, θ is minimized, and the effective radius of the nucleus is at its largest.

By combining equation 11.12 with appropriate expressions for the surface area and volume of the spherical cap in figure 11.8, we can express the energy barrier to nucleation ΔG_{hetero} as:

$$\Delta G_{hetero} = \left(\frac{4\pi}{3}\sigma_{CL}^3\Delta G_v^2\right)(2 - 3\cos\theta + \cos^3\theta). \tag{11.13}$$

Comparing this to equation 11.7, which describes the total energy barrier to homogeneous nucleation, we see that by nucleating against a substrate, the barrier has been reduced by a factor of $(2 - 3\cos\theta + \cos^3\theta)/4$. Clearly, then, nucleation against foreign objects is favored over nucleation of unsupported particles.

The rate of heterogeneous nucleation can be expressed by an equation similar to equation 11.11, with appropriate energy values substituted. We have not spent much time on these rate equations, for the simple reason that they do not appear to work very well for geologic systems. Theoretical models for nucleation qualitatively reproduce the general shapes of nucleation curves, but the temperatures for the maximum nucleation rate and peak widths are commonly different from experimentally determined curves. This difference between theoretical and experimental curves might be due to heterogeneous nucleation, but some researchers think that classic nucleation theory is flawed. Therefore, these equations may require significant revision before they can be applied successfully to understand nucleation in silicate melts.

Nucleation in Solids

The principles we have just discussed are also applicable to solid phases, but the process of nucleation in solids is more complicated. A phase transformation in the solid state may involve a significant volume change that cannot be accommodated by flow of the reactant phase. This produces a "room problem" that induces strain energy into the system. This extra energy must be added to the energy barrier to nucleation. The free energy for nucleation thus becomes:

$$\Delta G_{\text{total}} = 4\pi/3 r^3 (\Delta G_v + \Delta G_s) + 4\pi r^2 \sigma,$$

where ΔG_s is the strain energy.

Nucleation in metamorphic rocks is probably always heterogeneous. Grain boundaries, dislocations, and strained areas provide favorable nucleation sites. The formation of a nucleus at a boundary or dislocation involves the destruction of part of an existing surface, thereby lowering the free energy of the system. Nucleation at grain corners and triple points is also energetically favored.

An interesting example of solid-state nucleation is provided by exsolution, which commonly occurs in slowly cooled pyroxenes and feldspars. The upper part of figure 11.9 shows the experimentally determined solvus in the system $NaAlSi_3O_8$-$KAlSi_3O_8$. In chapter 10, we showed that subsolidus cooling of a homogeneous alkali

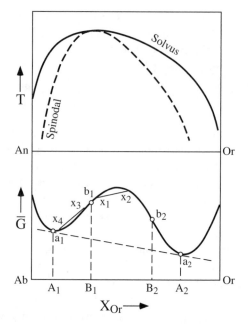

FIG. 11.9. The top shows a T-X_{Or} diagram of the subsolidus portion of the alkali feldspar system showing locations of the solvus and spinodal, as determined by Waldbaum and Thompson (1969). Below that is a schematic $\bar{G}$-X_{Or} diagram for the alkali feldspar system representing some temperature below the crest of the solvus. Tangent points a_1 and a_2 define the locations of solvus limbs, and inflection points b_1 and b_2 define spinodal limbs at this temperature.

feldspar below the solvus will result in its separation into two phases with compositions defined by the limbs of the solvus. The exsolved phases nucleate, homogeneously or heterogeneously, in the solid state. The lower part of figure 11.9 shows a schematic $\bar{G}$-X_{Or} diagram for some temperature below the crest of the alkali feldspar solvus. Points a_1 and a_2 define the locations of the limbs of the solvus at this particular temperature, corresponding to phase compositions A_1 and A_2.

Another style of nucleation, which may be kinetically more favorable, results in the formation of umixed feldspars with compositions that are metastable relative to A_1 and A_2. To show how this occurs, we have illustrated another curve, called the *spinodal*, in the upper part of figure 11.9. Inside the limits of the solvus, the free energy curve at the fixed temperature in the lower diagram has two inflections ($\partial^2\bar{G}/\partial X^2 = 0$), labeled b_1 and b_2, which define the positions of the spinodal limbs. If any feldspar having an overall composition between B_1 and B_2 contains small local fluctuations in composition (as,

INTERNAL PRESSURE, BEER BUBBLES, AND PELE'S TEARS

If you have ever opened a shaken can of beer or soda pop, you have experienced the fountain effect that follows rapid pressure release. Why does this happen? If not shaken, the same can emits only a gentle hiss when it is opened. To answer this question, we need to explore the relationship between surface tension and pressure.

The work required to transfer a small volume dV of material from liquid to the inside of a bubble, $(P_{int} - P_{ext})dV$, must equal the work required to extend its surface, σdA:

$$(P_{int} - P_{ext})dV = \sigma dA.$$

Differentiating the expressions for the volume and area of a sphere gives:

$$dV = 4\pi r^2 dr$$

and

$$dA = 8\pi r dr.$$

Substituting these values into the above equation results in:

$$(P_{int} - P_{ext})\, 4\pi r^2 dr = \sigma 8\pi r dr,$$

or

$$P_{int} - P_{ext} = 2\sigma/r. \tag{11.14}$$

Thus, the internal pressure of a vapor bubble must be greater than external pressure by an amount $2\sigma/r$.

In an unshaken can of beer, all of the vapor phase comprises a large "bubble" at the top. Its interface with the liquid is flat, so r in equation 11.14 is infinitely large. Under these circumstances, $P_{int} = P_{ext}$. If the can is shaken, however, the same volume of vapor is distributed among a very large number of small bubbles. The internal pressure on each is significantly higher than the external pressure because r is a fraction of a millimeter. It is this vapor pressure that causes the explosive fountain.

The actual pressure, of course, depends on the magnitude of σ as well as the radii of the vapor bubbles. It can be particularly high in some systems of geochemical interest. For example, the surface tension of basaltic magma is ~350 dynes cm^{-1} at 1200°C and 1 bar. Highly fluidized basaltic ejecta sometimes form tiny glassy droplets and filaments called *Pele's tears*. The internal pressure of such droplets, which have radii on the order of 10^{-3} cm, can be calculated using equation 11.13:

$$P_{int} = 1 + (2 \times 350)/(10^{-3} \times 10^6) = 1.7 \text{ bars.}$$

The internal pressure is nearly double the external (atmospheric) pressure!

---

for example, x_1 and x_2), its free energy will be less than that for a homogeneous feldspar. This situation also results in unmixing, but in this case, the process is called *spinodal decomposition*. Within the limits of the spinodal (b_1 to b_2), coexisting compositions such as x_1 and x_2 are stable relative to a homogeneous phase. Outside the spinodal, a homogeneous phase is stable with respect to unmixed phases such as x_3 and x_4. The energy difference between the homogeneous and exsolved phases is very small in either case, but unmixing is only favored inside the spinodal. Because the compositional difference between x_1 and x_2 is small, the structural mismatch between them is small and the strain energy is minimal. Therefore, spinodal decomposition involves no major nucleation hurdle and the interface between the unmixed phases is diffuse. In this way, exsolution lamellae with

compositions between A_1 and A_2 can form metastably. Feldspars with bulk compositions that lie between the solvus and the spinodal (that is, between A_1 and B_1 or A_2 and B_2) can unmix only by solvus-controlled exsolution. In this case, the unmixed compositions are relatively far apart, requiring the accommodation of significant structural mismatch, so there is a high kinetic barrier to nucleation. For this reason, nucleation may be sluggish, and exsolution phase boundaries are likely to be sharp.

The kinetics of exsolution processes can be rather complex, and it is convenient to summarize such behavior with the aid of a time-temperature-transformation (sometimes called TTT) plot, as illustrated in figure 11.10. These plots provide a synthesis of kinetic data obtained from experiments carried out at different temperatures and cooling rates. The two sets of curves in figure 11.10

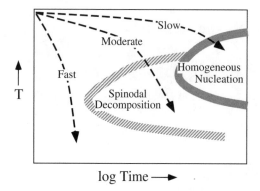

FIG. 11.10. Time-temperature-transformation (TTT) plot illustrating how various cooling rates might control whether homogeneous nucleation or spinodal decomposition occurs. Spinodal decomposition takes place in more rapidly cooled systems because nucleation is not required.

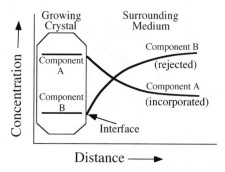

FIG. 11.11. During diffusion-controlled growth, compositional gradients are established in the medium surrounding a growing crystal. Components rejected by the crystal cannot diffuse away fast enough, and components used in making the crystal cannot diffuse to the interface fast enough, so that gradients result.

correspond approximately to the beginning (1% product) and completion (99% product) of exsolution. Completion of exsolution occurs when equilibrium is reached.

The behavior of a mineral under various thermal conditions can be determined using cooling curves such as those illustrated. For example, at slow cooling rates, homogeneous nucleation is favored over spinodal decomposition, whereas the reverse is true for moderate cooling rates. During fast cooling, even spinodal decomposition will be incomplete, and quenched samples show no unmixing at all.

GROWTH

Crystal growth is a complex activity involving a number of processes: (1) chemical reaction at the interface between a growing crystal and its surroundings; (2) diffusion of components to and from the interface; (3) removal of any latent heat of crystallization generated at the interface; and (4) flow of the surroundings to make room for the growing crystal. For most silicate minerals, only processes (1) and (2) are thought to be significant factors. These lead, then, to two end-member situations. When diffusion of components in the surroundings is fast relative to their attachment to the crystal nucleus, the rate of growth is controlled by the interface reaction. Conversely, when attachment of components to the crystal nucleus is faster than transport of components to it, the rate of growth is controlled by diffusion. For the special case in which crystals and surroundings have the

same composition, no diffusion is necessary and growth is interface-controlled, although removal of latent heat of crystallization may be important in this situation. Buildup of latent heat may also be important where two growing crystals converge. Interface-controlled growth may dominate in igneous systems and diffusion-controlled growth is likely to be more important in metamorphic systems, but intermediate situations in which both processes play a role are also common in both environments.

In diffusion-controlled growth, migration of components to and from the area immediately surrounding the growing crystal cannot keep pace with their uptake or rejection at the interface, so that concentration gradients are formed, as illustrated in figure 11.11. The formation of these gradients obviously slows down the rate of growth. Fluid advection, if it occurs, tends to destroy the gradients and causes growth rate to increase. In interface-controlled growth, concentrations of components in the zone around the growing crystal are similar to those of the bulk surroundings, because the limitation on growth is not movement through the surroundings. As a consequence, growth rate under these conditions is unaffected by fluid movement.

Interface-Controlled Growth

The rate at which interface-controlled growth takes place is the difference between the rates at which atoms are attached to and detached from the crystal. The rate of attachment, r_a, can be expressed by a probability relationship similar in form to equations we have already seen:

$$r_a = \nu \exp(-\Delta G^*/RT),$$

where ν is the frequency at which the atoms vibrate and ΔG^* is the activation energy for attachment. Similarly, the rate of detachment, r_d, is given by:

$$r_d = \nu \exp(-[\Delta G' + \Delta G^*]/RT),$$

where $\Delta G'$ is the thermodynamic driving force. The growth rate, Y, is $(r_a - r_d)$ times the thickness per layer, a_0, and the fraction of surface sites to which atoms may successfully attach, f:

$$Y = a_0 f \nu \exp(-\Delta G^*/RT)[1 - \exp(-\Delta G'/RT)]. \tag{11.15}$$

Equation 11.15 exhibits the same qualitative variation with undercooling as that for nucleation rate, equation 11.11. At zero undercooling, the rate is zero because $\Delta G'$ is zero. As the system cools below the equilibrium temperature, growth rate increases because $\Delta G'$ increases. However, at large degrees of undercooling, the growth rate begins to decrease because ΔG^* becomes significant, reflecting the fact that atoms are less mobile as fluid viscosity increases. Therefore, with increasing undercooling, the growth rate first increases, then passes through a maximum, and finally decreases, just as nucleation rate does.

Chemists recognize several mechanisms for interface-controlled growth, each dominant in a different growth environment with different proportions of surface sites available for growth (f in equation 11.15). *Continuous* models dominate where the growth interface is rough and atoms can attach virtually anywhere. Consequently,

$f = 1$ for all undercoolings and growth is relatively fast. In *layer-spreading* models, the interface is flat except at steps, which provide the only locations where atoms can attach. Thus, $f < 1$. Two commonly observed layer-spreading mechanisms are illustrated in figure 11.12. Surface nucleation forms one-atom-thick layers that spread laterally over the interface by adding atoms at the edges. If screw dislocations occur on the interface, atoms are attached to form spiral-shaped steps. One way to recognize these growth mechanisms is illustrated in worked problem 11.4.

Worked Problem 11.4

Jim Kirkpatrick and coworkers (1976) measured the growth rate of diopside from diopside melt by making movies of experiments in which a seed crystal was introduced into melt at various undercoolings. Using their data, presented in figure 11.13a, can we deduce the growth mechanism for diopside?

Because diopside is growing from a melt of its own composition, we know that the growth rate will not be diffusion-controlled. The interface-controlled growth rate expression is difficult to solve in the form of equation 11.15 because we do not have appropriate values for ΔG and $\Delta G'$. A theoretical alternative, known as the Stokes-Einstein relation, however, allows us to replace the exponential terms in equation 11.15 with a viscosity (η) term. We do not produce the mathematical derivation for this step, because it is rather complex, but the step should be intuitively reasonable. For any fixed degree of undercooling, the growth of crystals from a low-viscosity melt (low ΔG and $\Delta G'$) should be more rapid than from a high-viscosity melt. Thus, we should be able to define a term $Y\eta/T$ which is, like f, a measure of available attachment sites on the crystal surface. A plot of $Y\eta/T$ versus undercooling can be used to determine the growth mechanism. This yields a horizontal line

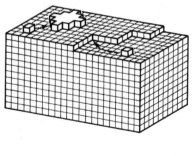

(a)

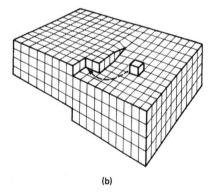

(b)

FIG. 11.12. Diagrams of layer-spreading mechanisms for crystal growth. (a) In surface nucleation, atoms are added only at the edges of a spreading layer of atoms. (b) In screw dislocations, atoms attached to the crystal surface form spiral stairs.

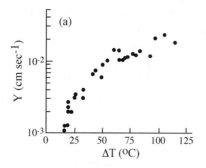

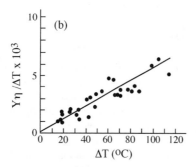

FIG. 11.13. Experimental data on the growth of diopside from melt of the same composition. (a) Relationship between growth rate Y and undercooling ΔT. (b) Calculated $Y\eta/\Delta T$ versus undercooling for data in (a), using viscosity measurements of Kirkpatrick (1974). The linear trend with positive slope indicates a screw dislocation growth mechanism. (After Kirkpatrick et al. 1976.)

for continuous growth, a straight line with positive slope for growth by a screw dislocation mechanism, and a curve with positive curvature for growth by a surface nucleation mechanism.

The first step is to determine the parameter $Y\eta/\Delta T$ for each data point. Kirkpatrick (1974) tabulated viscosity data for liquid diopside at various degrees of undercooling. For example, at an undercooling of 50°C, η equals 19 poise (g cm^{-1} sec^{-1}). Substituting this value and the measured growth rate of 9×10^{-3} cm sec^{-1} at this undercooling,

$$Y\eta/\Delta T = 3.4 \times 10^{-3} \text{ g sec}^{-2} \text{ deg}^{-1}.$$

A plot of similarly calculated values for all the data points in figure 11.13a against ΔT is presented in figure 11.13b. These data define a nearly straight line with positive slope, so diopside must grow by a screw dislocation mechanism.

We can apply this formulation to other types of problems as well. For example, let's consider the following problem.

Worked Problem 11.5

Crystallization under plutonic conditions may occur at a very small degree of undercooling relative to volcanic rocks. How long will it take to grow a 1-cm diopside crystal at an undercooling of only 1°C?

The only growth rate data on diopside are those we presented in worked problem 11.3. It is not possible to measure the growth rate at small degrees of undercooling accurately, but we can extrapolate a rate from experimental data at larger undercoolings. The straight line passing through the point (0, 0) in figure 11.13b allows us to estimate Y at $\Delta T = 1°$. The equation for this line is:

$$Y\eta/\Delta T = 6.6 \times 10^{-5} \ T.$$

Therefore, at $\Delta T = 1°$, $Y\eta = 6.6 \times 10^{-5}$ g sec^{-2}. Because $\eta = 15$ poise at 1° undercooling (Kirkpatrick 1974), Y is on the order of 10^{-6} cm sec^{-1}. At this rate, it would take 10^6 sec, ~10 days, to grow a 1-cm crystal.

For a number of reasons, this calculated rate must be higher than the growth of diopside in a real plutonic system. The experimental temperature (1390°C) is considerably higher than the liquidus temperatures of most magmas, and the viscosity of the experimental system is lower. Also, diffusion might play a role in a real, multicomponent magma.

Diffusion-Controlled Growth

Formulation of an equation for the rate of diffusion-controlled growth must start with either Fick's First or Second Law, depending on whether a steady state is reached. The growth rate is controlled by how rapidly atoms can diffuse to the interface. Consequently, growth rate decreases as time passes, because compositional gradients such as those shown in figure 11.11 extend farther into the surroundings with time. The expressions for growth rate under these conditions involve combining a diffusion equation with a mass balance equation, because the flux of atoms reaching the interface by diffusion must equal the atoms attached to the interface. Derivation of such expressions is rather complex and is not attempted here. Equations for growth rate Y under diffusion-controlled conditions take the general form:

$$Y = k(D/t)^{1/2}, \tag{11.16}$$

where k is a constant involving concentration terms, D is the diffusion coefficient of the slowest diffusing species, and t is time.

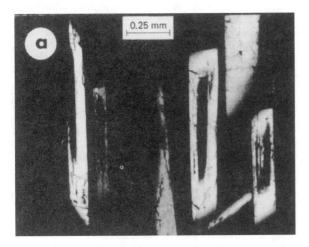
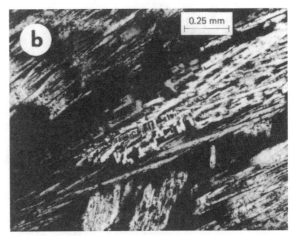

FIG. 11.14. Photomicrographs of (a) skeletal and (b) dendritic plagioclase crystals formed in rapidly cooled experimental melts. Sluggish diffusion at large undercoolings causes breakdown of planar crystal surfaces. (Courtesy of G. Lofgren.)

The compositional or thermal gradients produced by diffusion can affect growth by causing crystal faces to break down. When a faceted crystal becomes unstable, protuberances begin to develop. Heat and components rejected during crystal growth flow away from these protuberances, hindering the development of additional structures in their vicinity. This can produce skeletal and dendritic crystals, such as those in figure 11.14.

With increasing undercooling, D decreases and Y increases, so the ratio D/Y decreases. Gary Lofgren (1974) illustrated by experiment how increasing undercooling (that is, decreasing D/Y) controls crystal morphologies. He studied the shapes of plagioclase crystals growing at 5 kbar in plagioclase melts at various degrees of undercooling, as shown in figure 11.15. At nearly constant temperature intervals below the liquidus, various plagioclase compositions crystallized to form tabular, skeletal, dendritic, and spherulitic morphologies. The onset of diffusion-controlled growth in this system (that is, the condition under which faceted crystals become unstable) is defined approximately by the boundary separating tabular and skeletal crystals in figure 11.15. The periodicity of protuberances increases as D/Y decreases. The skeletal morphologies are instabilities with protuberances spaced far apart. With increasing undercooling, these give way to finer dendritic structures spaced closer together. Similar morphological sequences are observed in olivine crystals in the interior and exterior parts of pillow lavas. Even more familiar, of course, are the fine, feathery structures of snowflakes, which grow rapidly at high degrees of undercooling.

SOME APPLICATIONS OF KINETICS

Many geochemical systems reflect the interplay of several kinetic processes acting concurrently. In such cases, understanding reaction rates requires identifying the slowest step in the process, known as the *rate-limiting step*. If we can do this, it may be possible to model the

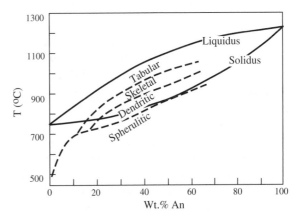

FIG. 11.15. T-X_{An} diagram for the system albite-anorthite, summarizing how plagioclase crystal morphologies change with undercooling. Skeletal and dendritic morphologies are illustrated in figure11.14. (After Lofgren 1974.)

kinetics of crystallization, as illustrated in two examples below. However, the application of kinetic theory to experimental systems with controlled rates of nucleation and growth has often produced disappointing results. Impure crystals and other complexities can frustrate even the most determined theorist. Empirical methods that bypass theory but still provide information on nucleation and growth rates have been developed. We also consider several of these.

Aragonite ⇌ Calcite: Growth as the Rate-Limiting Step

The polymorphic transformation aragonite ⇌ calcite has been studied extensively. Calcite is the thermodynamically stable phase under all conditions at the Earth's surface, as we demonstrated in worked problem 9.5. Although aragonite is only stable at high pressures, it is sometimes exposed on the surface, where high-pressure metamorphic rocks have been uplifted. If we understood the kinetic pathway by which aragonite inverts to calcite, we could constrain the rates and conditions under which such terrains were uplifted.

Metastable aragonite in shells is readily replaced by calcite during diagenesis, so the preservation of metamorphic aragonite must imply unusual conditions. Hydrothermal experiments indicate that at temperatures as low as 200° C, there is significant replacement of aragonite in only a few days, although the reaction is slower in the absence of water. In these experiments, however, aragonite was dissolved and reprecipitated as new grains of calcite rather than being replaced. In contrast, petrologists observe that metamorphic aragonite has often been partly converted to calcite, indicating that the calcite replaced aragonite in situ. We can then conclude that metamorphic rocks containing aragonite cannot have had a pore fluid present when they were uplifted.

Experiments to study the replacement reaction under dry conditions were carried out by William Carlson and John Rosenfeld (1981). They concluded that nucleation of calcite was not the rate-limiting step, because partially replaced aragonite is commonly surrounded by numerous small calcite grains. Consequently, they set about to measure calcite growth rates by observing the grains on a heating stage attached to a microscope. An Arrhenius plot of the calcite growth rate (in different crystallographic directions) versus temperature is shown in figure 11.16. Extrapolating these growth rates to lower

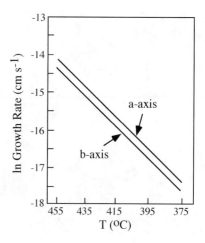

FIG. 11.16. Arrhenius plot showing the rate of growth of calcite crystals replacing aragonite, as a function of crystallographic orientation and temperature. (After Carlson and Rosenfeld 1981.)

temperatures allows inferences to be made about the transformation of aragonite to calcite under natural conditions. At 250° C, calcite replaces aragonite at a rate of 1 m/10^6 yr, so even large aragonite crystals would be completely transformed. However, at 100° C, calcite grows at only 10^{-3} mm/10^6 yr, allowing aragonite to persist over geologic time periods. This work clearly demonstrates that the uplifted rocks must have been cool and dry, as well as illustrating the importance of temperature in controlling reaction rates.

Iron Meteorites: Diffusion as the Rate-Limiting Step

Iron meteorites are thought to be samples of the differentiated cores of small bodies (asteroids). This interpretation derives from the rate of exsolution in metal alloys, in which interdiffusion of Fe and Ni is the rate-limiting step.

The subsolidus phase diagram for Fe-Ni is illustrated in figure 11.17, in which a two-phase region separates the fields of kamacite and taenite (both iron-nickel phases). The lines that bound this region are actually limbs of a solvus, like that encountered in the alkali feldspar system in chapter 10, although this one has an unusual shape. On cooling, an initially homogeneous alloy X containing 10% Ni encounters the solvus at 680° C, unmixing to form plates of kamacite within taenite. Exsolution continues as temperature decreases, and the

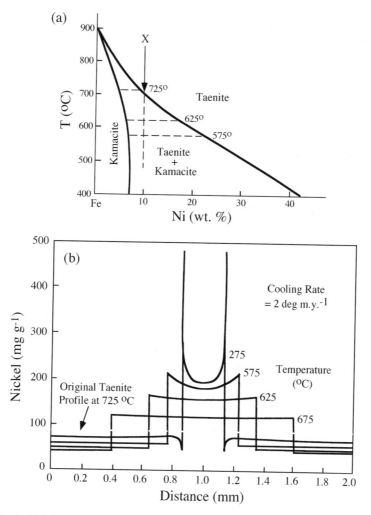

FIG. 11.17. (a) Subsolidus phase diagram for the Fe-Ni system, showing the solvus for unmixing of kamacite and taenite. (b) Calculated Ni diffusion profiles in taenite, as a function of cooling rate.

compositions of the coexisting kamacite and taenite at any temperature are given by intersections of an isothermal line with the solvus limbs. Because both solvus limbs slope in the same direction (fig. 11.17a), both phases become more nickel-rich with decreasing temperature. The only way this is possible is if the proportion of kamacite (the low-Ni phase) increases at the expense of high-Ni taenite. Thus, the kamacite plates become thicker with decreasing temperature.

Exsolution is made possible by diffusion of Fe and Ni in the metal alloys. Measurements show that diffusion is slower in taenite than in kamacite. Because nickel is expelled from kamacite more rapidly than it can diffuse into the interior of adjacent taenite grains, the nickel

distribution in taenite develops an "M"-shaped profile (fig. 11.17b). The more rapidly the metal is cooled, the more pronounced is the dip in the nickel diffusion profile. Calculations based on Fick's Second Law, using diffusion rates determined from experiments, give nickel profiles at various cooling rates (fig. 11.17b).

Iron meteorites commonly exhibit intergrowths of kamacite and taenite (called a Widmanstätten pattern and pictured in fig. 11.18). A nickel profile in meteoritic taenite analyzed by electron microprobe can be compared with calculated nickel diffusion profiles for taenite plates having the same width to estimate the cooling rate of the iron meteorite. These estimates, generally a few tens of degrees per million years, correspond to cooling rates

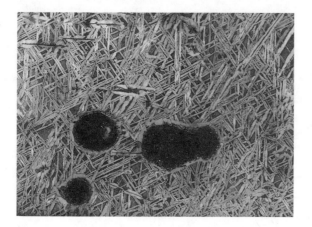

FIG. 11.18. Etched surface of the Mount Edith (Australia) iron meteorite, showing Widmanstätten texture formed by exsolution of kamacite and taenite. Dark blobs are sulfide (troilite). (Courtesy of the Smithsonian Institution.)

expected at the centers of asteroids with diameters <200 km. The interiors of larger bodies would cool more slowly, because rocks provide good insulation.

Bypassing Theory: Controlled Cooling Rate Experiments

The challenge of unraveling kinetic controls on complex geologic systems may be more amenable to experiment than to calculation. Early in this chapter, we summarized some experimental results on the rate of dehydration of talc. There are very few kinetic data for such metamorphic systems. Experiments in which magma compositions are cooled and solidified at predetermined rates are more common.

An example is a programmed cooling rate study of a lunar basalt by David Walker and others (1976), the results of which are summarized in figure 11.19. Experimental charges were cooled at different rates and quenched from different temperatures within the crystallization range to construct the liquidus lines shown in the diagram. Liquidus temperatures from equilibrium experiments are illustrated by arrows on the left side of the figure.

Several interesting features emerge from these data. First, the crystallization sequence is different under equilibrium and dynamic cooling conditions; specifically, the

order of appearance of plagioclase and ilmenite reverses. Walker and coworkers inferred from the texture of the rock itself that ilmenite began to form before plagioclase, suggesting that the controlled cooling rate experiments may be more relevant than equilibrium crystallization experiments to understanding the cooling history of this basalt. A second feature seen in figure 11.19 is that all phases begin crystallizing below their liquidus temperatures in the programmed cooling experiments, and the amount of temperature suppression increases with cooling rate. Another conclusion, not apparent from this figure, is that the compositions of pyroxenes, in terms of both major and minor elements, are distinct in the equilibrium and controlled cooling rate experiments. Based on order of phase appearance, texture, and mineral compositions, Walker and coworkers estimated the actual cooling rate for this lunar rock decreased from ~1°C/hr when olivine first crystallized to ~0.2°C/hr during crystallization of pyroxene. Decreasing cooling rate is consistent with its location near the margin of a lava flow.

Programmed cooling experiments provide an empirical method by which to assess the mineralogical and textural features that characterize rapidly cooling igneous rocks and to quantify the physical conditions that produced them.

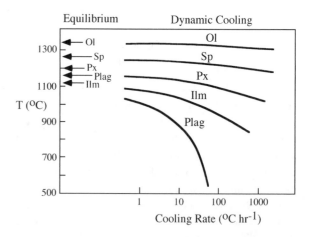

FIG. 11.19. Results of controlled cooling rate experiments on lunar basalt 12002 by Walker et al. (1976). Equilibrium liquidi for various phases are shown by arrows on the left-hand side. The curves are contours of the first appearance of various phases at different cooling rates.

Bypassing Theory Again: Crystal Size Distributions

Crystal size distribution (CSD) analysis provides a way to assess the kinetics of crystallization in rocks. The method depends on a population balance equation that describes the change in numbers and sizes of crystals as they nucleate and grow in either a liquid or solid environment. This relationship is expressed as:

$$n = n^0 \exp(-L/Yt). \qquad (11.17)$$

In this equation, n is the crystal population density (the number of crystals of a given size class per unit volume), and n^0 is a constant that represents the population density of crystal nuclei. The specific size range for crystals is L, Y is the linear growth rate (the growth rate along one crystallographic direction), and t is the average residence time for crystals in the system. A plot of $\ln n$ versus L is a straight line with slope $-1/Yt$ and intercept n^0. If either the growth time or the growth rate is known, the other may be determined. The nucleation rate J can be calculated by recognizing that:

$$J = n^0 Y. \qquad (11.18)$$

In practice, one must count the number density of crystals as a function of size. This is expressed as a cumulative distribution. The population density n is the number of crystals in any size range divided by the size interval.

Katherine Cashman and Bruce Marsh (1988) applied CSD analysis to drill core samples from Makaopuhi lava lake, Kileauea volcano, Hawaii. The lengths of plagioclase crystals were measured in rock thin sections and binned by size, and the population bins were converted to volume proportions. A typical CSD plot (n versus L) for plagioclase is illustrated in figure 11.20. The data define a straight line, implying that crystal nucleation and growth were continuous throughout the crystallization interval. Crystal fractionation or other magmatic processes would have disturbed this pattern and produced more complex CSD plots. The nuclei population density n^0 for this basalt, determined from the y intercept, is 2.68×10^7 cm^{-4}. The value for the slope, $-1/Yt = -544$ cm^{-1}, can be used to calculate the plagioclase growth rate Y, since t has been estimated at various depths for this cooling lava lake. Using a time t of 3.075×10^7 sec, $Y = 6.0 \times 10^{-11}$ cm/sec (~0.02 mm/yr). Applying equation 11.18, the nucleation rate J of plagio-

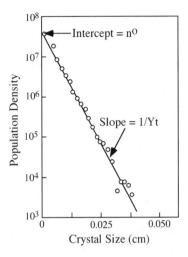

FIG. 11.20. Crystal size distribution (CSD) plot for plagioclase grains in basalt sample from Makaopuhi lava lake. The slope of the linear regression line is related to the growth rate Y and crystal residence time t, and the y intercept gives the nuclei population density n^0. (After Cashman and Marsh 1988.)

clase is 1.6×10^{-3} nuclei/cm^3 sec (~50,000 nuclei/yr). This example shows that the role of kinetics in the crystallization of rocks can be quantified from analysis of completely solidified samples, even without recourse to kinetic theory.

SUMMARY

In this chapter, we have seen that kinetic factors can be very important to understanding some geologic systems, even at high temperatures. The temperature dependence of rate processes is commonly expressed in terms of an activation energy barrier that must be overcome to initiate the process. This is reflected in supersaturation and undercooling, both ways of overstepping equilibrium crystallization conditions.

Three rate processes are particularly important in high-temperature systems. The first of these, diffusion, occurs in response to a gradient in chemical potential. Migration of components may occur along grain boundaries or through phases. Volume diffusion uses naturally occurring defects in crystal structures as pathways for movement. Diffusion rates can be calculated from Fick's First and Second Laws.

Nucleation, the second major rate-controlling process, is inhibited by surface free energy; this extra energy requirement is eliminated after the nucleus reaches a certain critical size. Nucleation may occur by homogenous or heterogeneous mechanisms. This process is very difficult to study experimentally, and theoretical models provide only qualitative agreement with experimental data. Exsolution requires nucleation and thus is rate sensitive, but unmixing by spinodal decomposition occurs without nucleation of a new phase.

Growth of crystals, the third process, may be controlled by the rate of diffusion of components to the interface or their rate of attachment to the new phase. Interface-controlled growth may occur by continuous, surface nucleation, or screw dislocation mechanisms, each of which is characterized by a different relationship between growth rate and undercooling. The rate of diffusion-controlled growth slows with time because of chemical gradients that form in the medium surrounding the growing crystal. Compositional or thermal gradients near the growing interface also cause planar crystal surfaces to break down into skeletal or dendritic morphologies.

Several examples (polymorphic transformation of aragonite into calcite; unmixing of kamacite and taenite in iron meteorites) illustrate the importance of determining the rate-limiting step in kinetic processes. Because kinetic theory is often difficult to apply in understanding natural systems, geochemists sometimes use other methods that bypass theory and directly determine nucleation and growth rates. These include controlled cooling rate experiments and analysis of crystal size distributions.

SUGGESTED READINGS

The literature on geochemical kinetics is not as voluminous as that for equilbrium thermodynamics, and most references are portions of books on other subjects.

Carmichael, I. S. E., F. J. Turner, and J. Verhoogen. 1974. *Igneous Petrology.* New York: McGraw-Hill. (Chapter 4 provides a discussion of kinetic factors applicable to magmatic systems.)

Gill, R. 1996. *Chemical Fundamentals of Geology,* 2nd ed. London: Chapman & Hall. (Chapter 3 provides an easily understandable summary of kinetically controlled geochemical processes.)

Hoffmann, A. W., B. J. Gilletti, H. S. Yoder Jr., and R. A. Yund, eds. 1974. *Geochemical Transport and Kinetics.* Washington, D.C.: Carnegie Institution of Washington. (A collection of rather technical papers relating to diffusion.)

Kirkpatrick, J. R. 1975. Crystal growth from the melt: A review. *American Mineralogist* 60:798–814. (An excellent review paper on the kinetics of crystal growth in silicate melts.)

Lasaga, A. C., and R. J. Kirkpatrick, eds. 1981. *Kinetics of Geochemical Processes.* Reviews in Mineralogy 8. Washington, D.C.: Mineralogical Society of America (Probably the best available reference on geochemical kinetics; contains excellent chapters on rate laws and dynamic treatment of geochemical cycles, as well as applications to metamorphic and igneous rocks.)

Lofgren, G. E. 1980. Experimental studies of the dynamic crystallization of silicate melts. In Hargraves, R. B., ed. *Physics of Magmatic Processes.* Princeton: Princeton University Press. (A comprehensive review of controlled cooling rate experimentation.)

Marsh, B. D. 1988. Crystal size distribution (CSD) in rocks and the kinetics and dynamics of crystallization. *Contributions to Mineralology and Petrology* 99:277–291. (An excellent review of the theory behind CSD analysis of rocks and the means of extracting kinetic information from them.)

Thompson, A. B., and D. C. Rubie, eds. 1985. *Metamorphic Reactions, Kinetics, Textures, and Deformation.* Advances in Physical Chemistry 4. New York: Springer-Verlag. (This reference includes chapters on the kinetics of metamorphic reactions, as well as the importance of crystal defects and deformation on rate processes.)

Other papers referenced in this chapter are:

Buening, D. K., and P. R. Buseck. 1973. Fe-Mg lattice diffusion in olivine. *Journal of Geophysical Research* 78:6852–6862.

Carlson, W. D., and J. L. Rosenfeld. 1981. Optical determination of topotactic aragonite-calcite growth kinetics: Metamorphic implications. *Journal of Geology* 89:615–638.

Cashman, K. V., and B. D. Marsh. 1988. Crystal size distribution (CSD) in rocks and the kinetics and dynamics of crystallization, II: Makaopuhi lava lake. *Contributions to Mineralogy and Petrology* 99:292–305.

Greenwood, H. J. 1963. The synthesis and stability of anthophyllite. *Journal of Petrology* 4:317–351.

Kirkpatrick, R. J. 1974. Kinetics of crystal growth in the system $CaMgSi_2O_6$-$CaAl_2SiO_6$. *American Journal of Science* 274: 215–242.

Kirkpatrick, R. J., G. R. Robinson, and J. F. Hays. 1976. Kinetics of crystal growth from silicate melts: Anorthite and diopside. *Journal of Geophysical Research* 81:5715–5720.

Lofgren, G. 1974. An experimental study of plagioclase morphology. *American Journal of Science* 273:243–273.

Mathews, A. 1980. Influences of kinetics and mechanism in metamorphism: A study of albite crystallization. *Geochimica et Cosmochimica Acta* 44:387–402.

Mueller, R. F., and S. K. Saxena. 1977. *Chemical Petrology.* New York: Springer-Verlag.

Taylor, L. A., P. I. K. Onorato, and D. R. Uhlmann. 1977. Cooling rate estimations based on kinetic modelling of Fe-Mg diffusion in olivine. *Proceedings of the Lunar Scientific Conference* 8:1581–1592.

Taylor, L .A., P. I. K. Onorato, D. R. Uhlmann, and R. A. Coish. 1978. Subophitic basalts from Mare Crisium: Cooling rates. In R. B. Merrill and J. J. Papike, eds. *Mare Crisium: The View from Luna 24.* New York: Pergamon, pp. 473–481.

Waldbaum, D. R., and J. B. Thompson Jr. 1969. Mixing properties of sanidine crystalline solutions, part IV: Phase diagrams from equations of state. *American Mineralogist* 54: 1274–1298.

Walker, D., R. J. Kirkpatrick, J. Longhi, and J. F. Hays. 1976. Crystallization history of lunar picritic basalt sample 12002: Phase equilibria and cooling rate studies. *Bulletin of the Geological Society of America* 87:646–656.

PROBLEMS

(11.1) Given the following data, determine the activation energy for the reaction analcite + quartz → albite + water in sodium disilicate solution.

k	T (°C)
2.41×10^{-4}	345
2.68×10^{-3}	379
1.84×10^{-2}	404

(11.2) Sketch schematic plots of concentration of element x in the medium surrounding a growing crystal as functions of distance from the interface for the following conditions: (a) diffusion-controlled growth, (b) interface-controlled growth, (c) growth controlled by both mechanisms.

(11.3) Using equations 11.11 and 11.15, construct a plot that illustrates how the rates of nucleation and growth vary with undercooling.

(11.4) (a) Given that $\Delta H = \Delta G + T\Delta S$, show that the surface enthalpy of a liquid droplet is correctly described by $\Delta H = \sigma - T(\partial\sigma/\partial T)_p$. (b) The surface tension of water against air at 1 atm has the following values at various temperatures:

			T (°C)		
	20	22	25	28	30
σ (dynes cm^{-1})	72.75	72.44	71.97	71.50	71.18

What is the surface enthalpy (in cal cm^{-2}) of water at 25°C?

(11.5) (a) The volume of the spherical cap shown in figure 11.8 can be calculated from:

$$V = \tfrac{1}{6}\pi h \, (3x^2 + h^2).$$

Using this formula and equations 11.6 and 11.1, derive expressions to calculate the volume (V) and height (h) of the smallest stable nucleus that can be formed on a substrate S, assuming that the quantities σ_{CL}, ΔG_v, and ($\sigma_{LS} - \sigma_{CL}$) can be measured experimentally. (b) What is the volume of the smallest stable nucleus of forsterite for which ($\sigma_{LS} - \sigma_{CS}$) = 2.378×10^{-5} cal cm^{-2} at 2063°C? Use the values for σ (=σ_{CL}) and ΔG_v for forsterite at 2063°C from worked problem 11.3. How does this minimal volume compare to the volume of the smallest homogeneously nucleated forsterite particle?

CHAPTER 12

THE SOLID EARTH AS A GEOCHEMICAL SYSTEM

OVERVIEW

In this chapter, we explore how the various parts of the Earth's interior can be integrated into a grand geochemical system. First, we estimate the compositions of the reservoirs—that is, crust, mantle, and core—in terms of both chemistry and phase assemblages. We then investigate how these reservoirs interact through the exchange of heat and matter. The continental crust was extracted from the upper mantle, which appears to be geochemically isolated from the lower mantle. Generally, convection within the upper mantle and crust is manifested in plate tectonics. During certain episodes, however, whole-mantle convection may occur, resulting in sinking of lithospheric plates into the lower mantle and return flow of matter to the surface via large mantle plumes. These episodes are driven by thermal convection in the liquid outer core, resulting from exothermic crystallization of the solid inner core. Many geochemical interactions between mantle and crust involve basaltic magmatism, so we discuss the thermodynamics of melting and the geochemical characteristics of partial melts of mantle rocks generated under various conditions. Most magmas experience significant chemical changes en route to the surface. Differentiation of magmas by fractional crystallization and liquid immiscibility is considered, and we attempt to quantify their geochemical consequences. The

behavior of compatible and incompatible trace elements during melting and differentiation is also examined. As we shall see, trace elements can be used to quantify geochemical models of magma evolution. We then consider the compositions, reservoirs, and cycling of volatile elements within the mantle and crust.

RESERVOIRS IN THE SOLID EARTH

Composition of the Crust

The crust is the outer shell of the Earth, in which P-wave seismic velocities are <7.7 km sec^{-1}. The lower boundary of the crust is the Mohorovicic discontinuity (the Moho), a layer within which seismic velocities increase rapidly or discontinuously from crustal values to mantle values of >7.7 km sec^{-1}. The average thickness of the crust defined in this way is 6 km in ocean basins, ~35 km in stable continental regions, and ranges up to 70 km under mountain chains. Even so, the crust is only a minuscule portion of the mass of the planet, constituting only about 0.5% of it by weight.

Oceanic crust shows no major lateral and vertical changes in P-wave velocity. As a first approximation, it can be considered to have basaltic composition throughout, although studies of ophiolites reveal that it has a veneer of siliceous sediments and contains ultramafic rocks at depth. In contrast, the lateral heterogeneity of continental crust, so evident in geologic maps, apparently persists to great depth as revealed by variations in subsurface P-wave velocities. This heterogeneity complicates any attempt to estimate the bulk composition of the crust as a whole.

Continental crust is conventionally thought to be subdivided into a "granitic" upper layer and a "gabbroic" lower layer, separated by a seismic discontinuity (the Conrad discontinuity) that is not everywhere laterally continuous. This view is certainly too simplistic and may be wrong altogether. The common mineral assemblage of gabbro, for example, is apparently not stable at the temperatures and pressures that occur in the lower crust. Rocks of this composition should be transformed to eclogite, a mixture of garnet and clinopyroxene. The density of eclogite, however, is too high to fit the measured seismic velocities of the lower crust. If there is enough water in the lower crust, "gabbroic" rocks could instead consist of amphibolites (amphibole + plagioclase rocks), which do have the appropriate seismic properties. Another idea consistent with seismic constraints is that much of the lower crust under continents is composed of granulites of intermediate composition.

There are various ways to approach the thorny problem of determining an average composition for the crust, and all depend on assumptions that we make in describing crustal heterogeneity. We might choose to take the weighted average of the compositions of various crustal rocks in proportion to their occurrence, using, for example, geologic maps as a basis. Less direct approaches utilize the analyses of clays derived from glaciers that have sampled large continental areas, or estimate the relative proportions of granitic and basaltic end members by mixing their compositions in the proper ratio to reproduce the compositions of sediments derived by crustal weathering. All of these methods have been tried and generally give similar results.

The most widely accepted crustal compositions usually involve averaging the compositions of rocks assigned to hypothetical crustal models. This approach was adopted by A. B. Ronov and A. A. Yaroshevsky (1969), using the crustal model illustrated in figure 12.1. Their calculated crustal composition is given in table 12.1. Ross Taylor and Scott McLennan (1985) estimated the composition of the crust by assuming that it was composed of 75% average Archean crustal rocks and 25%

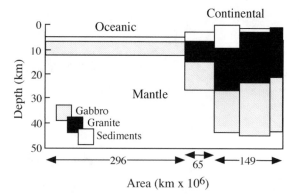

FIG. 12.1. A model for the Earth's crust used in the calculation of its average composition. (Adapted from Ronov and Yaroshevsky 1969.)

andesite. The Taylor and McLennan model (table 12.1) is somewhat more mafic than that of Ronov and Yaroshevsky, with lower concentrations of SiO_2 and alkalis and higher concentrations of FeO, MgO, and CaO. However, both of these calculated crustal compositions are remarkably similar to the composition of average andesite, also shown in table 12.1. The mineralogy of the average crust is thought to be similar to that of andesite, which is dominated by feldspars, quartz, pyroxenes, and amphibole.

The crust is strongly enriched in *incompatible* elements (those that partition into magmas because they do not fit easily into the crystal structures of the minerals that remain unmelted). The incompatible elements in the crust are *lithophile*, so-called because they prefer

TABLE 12.1. Estimates for the Average Composition of the Earth's Crust

Wt. % Oxide	Ronov and Yaroshevsky (1969)	Taylor and McLennan (1985)	Average Andesites[1]
SiO_2	59.3	57.3	58.7
TiO_2	0.9	0.9	0.8
Al_2O_3	15.9	15.9	17.3
Fe_2O_3	2.5	—	3.0
FeO	4.5	9.1	4.0
MnO	0.1	—	0.1
MgO	4.0	5.3	3.1
CaO	7.2	7.4	7.1
Na_2O	3.0	3.1	3.2
K_2O	2.4	1.1	1.3
P_2O_5	0.2	—	0.2

[1]Average of 89 andesites from island arcs compiled by McBirney (1969).

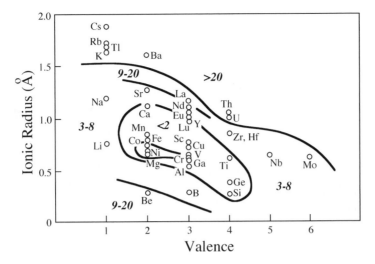

FIG. 12.2. Enrichment of lithophile elements in the continental crust relative to abundances in the Earth's primitive mantle. Ions of large size and/or charge are incompatible in common mineral structures and so are concentrated in melts. Over time, these melts have enriched the crust in incompatible elements. Contours group elements with similar degrees of enrichment, from <2, 3–8, 9–20, and >20. (After Taylor and McLennan 1985.)

silicate phases to metal or sulfide, as noted in chapter 2. Included in this category are elements with large ionic size (K, Rb, Cs, U, Th, Sr, Ba, and some rare earth elements, such as La and Nd) and high field strength elements that have high ionic charge (Nb, Ta, Zr, Hf, Ti). Figure 12.2 summarizes the enrichment of lithophile elements in the crust relative to abundances in the primitive mantle, plotted against ionic radius and valence. Contours in this figure group elements with similar enrichment factors. Enrichments of lithophile elements in the continental crust are so extreme that multiple periods of melting are required, resulting in a refining process that further sequesters these elements at each melting step. Lithophile elements are thought to have been extracted from the mantle during crust formation.

Composition of the Mantle

The mantle, that region of the Earth's interior between the Moho and the core, is of course not directly accessible for study. What we know about it comes from indirect evidence, mostly geophysical. Seismic wave velocities as a function of depth can be used to calculate elastic parameters, which depend on the chemical and mineralogic composition of the mantle, as well as temperature and pressure. Comparison of measured seismic data with the results of laboratory measurements of the ultrasonic properties of various phases provides the information necessary for inferring the composition of the upper mantle. For the lower mantle, we must use theoretical relationships between velocity and density and deduce compositions from shock wave experiments. Despite the inherent uncertainties in interpreting mantle seismic data, the mineralogic composition of the mantle seems reasonably well determined.

P-wave seismic velocities of 7.8–8.2 km sec^{-1} just below the Moho are consistent with an upper mantle having the bulk density of peridotite (olivine + orthopyroxene + clinopyroxene) or eclogite (defined above as clinopyroxene + garnet). Although both of these rock types occur in the mantle, peridotite is thought to predominate. As we shall see, one of the constraints on petrologic modeling is that mantle rocks must be capable of generating basaltic liquids on partial melting. Peridotite can do this, but eclogite cannot. The simple reason is that eclogite already has a basaltic composition, and thus must melt completely to produce basaltic magma. Clearly then, the mantle, consisting mostly of ultramafic

TABLE 12.2. Estimates for the Average Composition of the Earth's Mantle

Wt. %	White (1967)[1]	Hutchison (1974)[2]	Ringwood (1975)[3]	Anderson (1980)[4]
SiO_2	44.5	45.0	45.7	47.3
TiO_2	0.15	0.09	0.09	0.2
Al_2O_3	2.6	3.5	3.4	4.1
Fe_2O_3	1.5	—	—	—
Cr_2O_3	—	0.41	0.4	0.2
FeO	7.3	8.0	8.0	6.8
MnO	0.14	0.11	0.14	0.2
Ni	0.2	0.25	0.27	0.2
MgO	41.7	39.0	38.4	37.9
CaO	2.3	3.25	3.1	2.8
Na_2O	0.25	0.28	0.4	0.5
K_2O	0.02	0.04	—	0.2

[1]From frequency histograms of 168 ultramafic rocks.
[2]From mantle peroditite nodules.
[3]Pyrolite.
[4]80% garnet peridotite + 20% eclogite.

rocks, has a profoundly different bulk composition from the crust.

Some crustal occurrences of tectonically emplaced ultramafic rocks may have been derived directly from the mantle. The compositions of abundantly exposed ultramafic rocks may be representative of the mantle. A preferred mantle composition, calculated from frequency histograms of ultramafic rocks, is presented in column 1 of table 12.2.

Another constraint on the composition of the mantle comes from xenoliths of mantle material carried upward by erupting magmas. Those that have not been altered by extraction of basaltic magma consist of either peridotite or eclogite. We focus on the peridotites, which we have already concluded are the best candidates for the bulk of the mantle. There are essentially two types: spinel peridotites, containing olivine, enstatite, diopside, and Cr-Al spinel; and garnet peridotites, containing olivine, enstatite, diopside, and pyrope garnet. In both cases, olivine and pyroxenes predominate, but the presence of spinel or garnet is very important. Even though the pyroxenes in peridotite xenoliths are commonly aluminous, they do not contain enough aluminum to account for the plagioclase component of basalts produced by partial melting. In rocks of ultramafic composition, plagioclase itself is the Al-bearing phase at low pressures, but this gives way to spinel and garnet at higher pressures. The spinel- and garnet-forming reactions can be represented by:

$$2CaAl_2Si_2O_8 + 2Mg_2SiO_4 \rightarrow$$
anorthite   forsterite

$$CaMgSi_2O_6 \cdot xCaAl_2SiO_6$$
aluminous clinopyroxene

$$+ 2Mg_2SiO_3 \cdot xMgAl_2SiO_6 + (1-x)CaAl_2Si_2O_8$$
aluminous orthopyroxene   unreacted anorthite

$$+ (1-x)MgAl_2O_4 \rightarrow 2CaMg_2Al_2Si_3O_{12}.$$
spinel   garnet

One example of the use of peridotite xenoliths as a measure of the composition of the mantle is illustrated in column 2 of table 12.2.

One of the problems we encounter in estimating the mantle composition has to do with partial melting. A peridotite assemblage without spinel or garnet (and often without clinopyroxene as well) is said to be *depleted*, because basaltic magma has already been extracted from it. Conversely, peridotite that can still produce basaltic melt is called *undepleted*. But how do we determine what proportion of the mantle is depleted and what proportion is undepleted? This dilemma forms the basis for a class of hypothetical mantle models known as *pyrolite*, originally devised in 1962 by Ted Ringwood. He postulated a primitive mantle material that was defined by the property that on fractional melting it would yield basaltic magma and leave behind a residual refractory dunite-peridotite. The composition of pyrolite ("pyroxene–olivine" rock) can be derived, therefore, by combining depleted peridotite and the complementary basalt in the proper proportions. This is a rather flexible model for the mantle composition. Ringwood and his collaborators at various times have presented pyrolite compositions with peridotite:basalt ratios varying from 1:1 to 4:1, although the 3:1 ratio is most often quoted as a reasonable value. A mantle composition based on pyrolite is compared with others already discussed in column 3 of table 12.2.

Not only is basaltic magma removed from the mantle, but it is also recycled back into the mantle (in the form of eclogite) by subduction. Consequently, we might also envision a model composition for the mantle that mixes average eclogite and undepleted peridotite in the correct proportions. The model presented in column 4 of table 12.2 does just that.

All of these methods give comparable results, and the composition of the mantle seems reasonably well constrained, at least for the major elements. We should keep in mind, however, that these estimates are for the upper mantle, although they are commonly used to describe the composition of the entire mantle. Is there any basis for this extrapolation?

Ringwood (1975) provided a wealth of information about the stability of phases at various depths in a mantle of pyrolite composition. Down to a depth of 600 km, the assumed mineral assemblages are based on the results of high-pressure experiments, but below this depth, Ringwood had to infer what phases would be stable from indirect evidence, such as experimental data on germanate (GeO_4^{-4}) analog systems. If germanium is substituted for silicon, phase transformations that were experimentally inaccessible for silicates occur within pressure ranges that could be studied in the laboratory. Since that time, multianvil high-pressure apparatus and shock wave experiments have confirmed the proposed phase relationships for the equivalent silicates. These phase changes (see the accompanying box) correspond very well with observed seismic discontinuities in the mantle, so it seems plausible that, to a first approximation, the mantle has a pyrolite composition throughout. This would mean that most discontinuities within the mantle result from phase changes, in contrast to the Moho, which is due to a change in chemical composition.

It is convenient to distinguish the upper and lower mantle, which are separated by a seismic discontinuity at 650–700 km depth. These two reservoirs apparently have different abundances of incompatible trace elements. The upper mantle is strongly depleted in incompatible elements, which were presumably extracted during crust formation, whereas the lower mantle has higher abundances of these elements. There is also some evidence that the upper and lower mantle may have different compositions in terms of major elements, expressed in their inferred Mg/Si, Fe/Si, and Ca/Al ratios. Such variations might have resulted from crystallization in a largely molten mantle during its formation. If this is correct, then the 650-km discontinuity (sometimes called the 670-km discontinuity) is also a compositional boundary, and not just a phase change.

The lowermost 200 km of the mantle, sometimes called region D, has peculiar seismic properties and may have a distinct chemical composition from the rest of the lower mantle. This region may be a mixture of mantle material with subducted slabs of ancient crustal rocks. The term *piclogite* is sometimes used to describe this mixture. Piclogite differs from pyrolite in the upper mantle in that its ultramafic component is undepleted and thus rich in incompatible elements.

Composition of the Core

The mean density of the Earth is consistent with a large, metallic iron core, and its moment of inertia and free oscillations indicate that this mass is concentrated at the planet's center. A major seismic discontinuity at 2900 km depth marks the mantle-core transition, so the core is approximately half the radius of the Earth. Variations in compressional wave velocities through the core have led to the conclusion that the outer core is liquid and the inner core is solid. In fact, the outer core is the largest magma chamber inside the Earth, forming a layer 2260 km thick between the crystalline lower mantle and the solid inner core. Most of what we know about the core is derived from indirect seismic observations. Despite the rigorous aspects of this discipline, one geophysicist has described the inversion of geophysical data to yield geological information as "like trying to reconstruct the inside of a piano from the sound it makes crashing down stairs." Can geochemistry shed any light on this complex problem?

Geophysical methods applied to study of the core are capable of detecting elements with very different atomic weights, but they cannot distinguish among elements similar in weight to iron. One such element, nickel, is thought to be an important constituent of the core, based on its *siderophile* character (its geochemical affinity for metal) and its high *cosmic abundance* (its abundance on average in the solar system). To first order, the core can thus be visualized as a mass of Fe-Ni alloy, part liquid and part solid. We can also infer that other siderophile elements, such as cobalt and phosphorus, might be important constituents.

The density of the outer core is about 8% less than that of the inner core. This is interpreted as reflecting the presence of an additional component, an element of low atomic weight, in the outer core. Considerable support has developed for the hypothesis that this element is sulfur. The seismic velocities of sulfide and metal at high pressures indicate that ~10 wt % S would be necessary to account for the lower density of the outer core. Iron meteorites, which formed as cores in asteroids, also contain significant amounts of sulfur as FeS (the mineral

PHASE CHANGES IN THE MANTLE

Ted Ringwood's suggested mantle mineralogy as a function of depth is summarized in figure 12.3. The wavy diagonal line from upper left to lower right represents the density distribution with depth, inferred from P-wave velocities and corrected for the effect of compression due to the overlying rock. Starting near the top, just below the Moho, we first encounter the low-velocity zone, a region of inferred partial melting in which seismic velocity decreases. The stable assemblage at this point is olivine + orthopyroxene + clinopyroxene + garnet. Just below 300 km, pyroxenes transform to a garnet crystal structure. One of the two major discontinuities in the mantle occurs at 400 km; this corresponds to the conversion of olivine, volumetrically the most important phase to this point, into a more tightly packed structure called the β phase (also called wadsleyite). At this point, the mantle consists of β-olivine and a complex garnet solid solution. Another less pronounced seismic wiggle at ~500 km may be due to the transformation of the β phase into a spinel (γ) structure (also called ringwoodite), as well as a second reaction in which the calcium component of garnet, possibly with some iron, separates as $(Ca,Fe)SiO_3$ in the dense perovskite structure.

The second major seismic discontinuity in the mantle is seen at 650–700 km. At this point, γ-olivine is thought to disproportionate into its constituent oxides, MgO (periclase) + SiO_2 (stishovite). The $(Mg, Fe)SiO_3$-Al_2O_3 components of the garnet solid solution presumably also transform into the ilmenite structure. Below this transition, the mantle consists of periclase + stishovite + $MgSiO_3 \cdot Al_2O_3$ (ilmenite) + $(Ca,Fe)SiO_3$ (perovskite).

The density changes in mantle materials below ~700 km could be caused by further phase transformations to more tightly packed structures with Mg and Fe in higher coordination, but it is very difficult to specify exactly what transformations may take place. Some additional phases that have been suggested to be stable in the lower mantle include $(Ca, Fe, Mg)SiO_3$ in the perovskite structure, $(Mg, Fe)O$ in the halite structure, and $(Mg, Fe)(Al, Cr, Fe)_2O_4$ in the calcium ferrite structure. From the generality of these formulas, we can see how speculative these phases are.

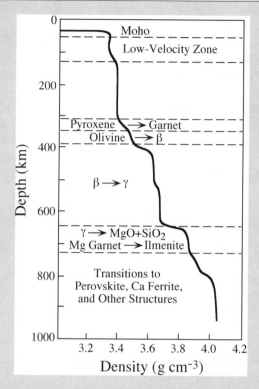

FIG. 12.3. Summary of possible phase transformations in the mantle, as determined by Ringwood (1975). The densities shown by the solid line are zero-pressure densities (corrected for depth) calculated for pyrolite; density discontinuities correspond to phase transformations inferred from seismic studies.

Another possible explanation for the higher densities in the deep mantle is an increased Fe/Mg ratio. Such a change in mantle composition would have important geochemical implications, but it is difficult to confirm because the behavior of iron in the mantle is so uncertain. It has been suggested that Fe^{2+} ions in silicates may undergo a contraction in radius due to spin-pairing of electrons below 1200 km depth. This is potentially important because low-spin Fe^{2+} probably would not substitute for Mg in solid solutions. Another hypothesis, with some experimental justification, is that Fe-bearing silicates might disproportionate to form some metallic iron, even at modest depths of ~350 km. This metal might then sink, stripping Fe from some parts of the mantle and enriching others.

troilite). Other geochemists have argued that oxygen is the light element in the outer core. Bonding in FeO is expected to become metallic, rather than ionic, at very high pressures, and an outer core with a composition of about Fe_2O (50 mol % oxygen) would satisfy the density constraint. Although oxygen is not alloyed with metal in iron meteorites, the cores in asteroids would not have formed at the high pressures necessary for this reaction. Another possibility is silicon; the density of Fe-Si alloys at core conditions requires ~17 wt % Si.

One way to estimate the composition of the core directly is to assume that siderophile elements are present in cosmic relative abundances to iron. (We discuss cosmic abundances in chapter 15.) An example of this approach is shown in the following worked problem.

Worked Problem 12.1

What is the composition of the Earth's core? One way to estimate core composition is to assume that siderophile elements accompany iron in cosmic proportions. Of course, we must also include major amounts of a light element to account for the density of the outer core, and we have just seen that this element may not be in cosmic proportion relative to iron.

Let's assume a value of 9.0 wt % sulfur in the core, based on sulfur abundance from density constraints. The first step, then, is to determine how much iron must be combined with sulfur to make FeS:

$$9.0/32.05 \text{ g mol}^{-1} = x/55.85 \text{ g mol}^{-1},$$

so that:

$$x = 15.68 \text{ wt } \% \text{ Fe.}$$

The wt % FeS in the core is then $15.68 + 9.0 = 24.68$. By difference, the core must consist of 75.32 wt % metal.

The metal phase also contains other siderophile elements in addition to iron. Siderophile elements with the highest cosmic abundances are nickel, cobalt, and phosphorus, and these are prominent constituents of iron meteorites. A mass balance equation for metal is thus:

$$Fe_{metal} + Ni + Co + P = 75.32\%,$$

or

$$Fe_{metal} = 75.32 - Ni - Co - P, \qquad (12.1)$$

where the abundance of each element is expressed in wt %. Nickel can be determined from the cosmic wt ratio of Ni and Fe:

$$Ni = (Ni_{cosmic}/Fe_{cosmic}) \, Fe_{core}$$
$$= (4.93 \times 10^4/58.77 \text{ g mol}^{-1})/$$
$$(55.85 \text{ g mol}^{-1}/9 \times 10^5)Fe_{cor} = 0.0521 \, Fe_{core},$$

where Fe_{core} is the total abundance of iron in the core, equal to $Fe_{metal} + 15.68$. Similarly, $Co = 0.0024 \, Fe_{core}$ and $P = 0.0208 \, Fe_{core}$. Substitution of these values into equation 12.1 gives:

$$Fe_{metal} = 75.32 - 0.0521 \, Fe_{core} - 0.0024 \, Fe_{core}$$
$$- 0.0208 \, Fe_{core} = 68.94 \text{ wt } \%.$$

Thus, the total amount of iron in the core is $68.94 + 15.68 = 84.62$ wt %. Nickel, cobalt, and phosphorus wt percentages are obtained by multiplying their cosmic wt ratios by 84.62; the respective values are 4.41, 0.20, and 1.76 wt %.

FLUXES IN THE SOLID EARTH

Cycling between Crust and Mantle

The mantle and crust are coupled through plate tectonics, which provides a mechanism for adding mantle material and heat to the crust (through volcanic activity) and recycling crust back into the mantle (through subduction). Plate tectonics reflects mantle convection, although the relationship is not as clear as was once thought. At present, ocean floor is being created and subducted at a rate of ~3 km^3 yr^{-1}. If the average thickness of a plate is 125 km, then 375 km^3 of rock is subducted each year. At that rate, the entire mantle (having a volume of 9×10^{11} km^3) could be processed through the plate tectonic cycle in 2.4 billion years (approximately half the age of the Earth). Mantle plumes, also called hot spots, are sites of volcanism that also must relate to convection. Mantle convection plays a dominant role in cooling the Earth, but there is no consensus on how convective flow extends to depth.

The main controversy centers on whether the whole mantle is involved in convective overturn, or the upper and lower mantles convect separately. The most prominent convection is associated with the sinking of plates at subduction zones. These plates are metamorphosed to dense eclogite on subduction, which adds to the gravitational imbalance caused by their cooler temperatures. On approaching 650 km depth, however, the high calcium and aluminum contents of eclogite stabilize garnet and hinder transformation to the denser perovskite that occurs in the surrounding mantle peridotite, so the density contrast decreases. Studies of earthquakes show that the subducted plates descend to depths of 650 km, below which seismicity is absent (fig. 12.4). Some high-resolution seismic images indicate that subducted slabs bounce off the 650-km discontinuity and accumulate in

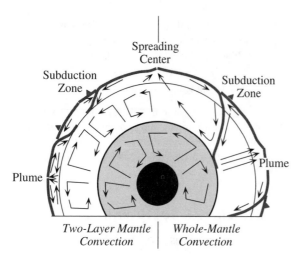

FIG. 12.4. Sketch of a cross-section of the Earth, showing convection in the core, mantle, and crust. Mantle plumes and subduction zones allow interaction of matter between mantle and crust. The left side of the diagram illustrates two-layer mantle convection and the right side shows whole-mantle convection.

the upper mantle. However, other seismic studies have documented that subducted slabs sometimes sink below this boundary, in some cases all the way to the interface between the mantle and core (fig. 12.4). The subducted crust is colder than the lower mantle into which it descends, so differences in seismic velocity allow the slabs to be imaged using seismic tomography. If subducted slabs are used as a tracer for convection, then the upper

and lower mantles are sometimes isolated from each other and sometimes not.

If slabs sink to the core-mantle boundary, an equivalent mass of material must be transferred back into the upper mantle or crust. Such material would be relatively hot and thus have lower seismic velocities. Seismic tomography reveals that two regions of low velocity occur within the lower mantle: one underlies the Pacific basin and the other is situated beneath southern Africa (fig. 12.5). Within both these areas lie the majority of the Earth's plumes, and both areas are highs on the geoid (the geometrical form that describes the planet's shape). The return flow from the lower mantle thus appears to consist of scattered upwellings that form clusters of hot spots.

The existence of separate mantle reservoirs for various kinds of magmas supports the idea that the upper and lower mantles do not convect as a single unit. The midocean ridge basalts (MORB) that erupt at spreading centers are strongly depleted in incompatible elements. Melting experiments suggest that MORB is generated at pressures corresponding to depths within the upper mantle. The depletion of incompatible elements in this source region constitutes evidence that the crust was extracted from the upper mantle, as illustrated in worked problem 12.2. Conversely, magmas thought to be generated within the lower mantle and erupted at hot spots have higher abundances of incompatible elements. The lower mantle may thus be geochemically isolated from

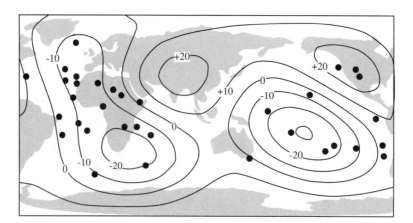

FIG. 12.5. Contours of average lower mantle seismic velocity perturbations define two large regions of high temperature (low velocity, indicated by negative contours). These regions define the locations of mantle plumes, and they correspond to clusters of hot spots (solid circles). (Redrawn from Castillo 1988.)

the upper mantle, although plumes generated at the core-mantle boundary obviously must pass through both the lower and upper mantle on their way to the surface.

Worked Problem 12.2

The extraction of the incompatible-element-enriched crust from the mantle has produced a complementary incompatible-element-depleted reservoir in the mantle. Can we use incompatible element concentrations to estimate what fraction of the mantle was involved in producing the crust?

An approximate upper limit for mantle involvement can be calculated from mass balance considerations. Rubidium is a highly incompatible element, strongly enriched in the crust and depleted in MORB. Using the data in table 12.3, we can calculate the effect of extracting the crustal complement of incompatible Rb and relatively compatible Yb from varying amounts of primitive mantle. The relative masses of crust and mantle are 0.56% and 99.44%, respectively. The mass balance equation is:

Rb in primitive mantle =

Rb in depleted mantle + Rb in crust,

or

$$m_{mantle} (cpm) = nm_{mantle} (cdm) + m_{crust} (cc), \qquad (12.2)$$

where m_{mantle} and m_{crust} are the mass fractions of the mantle and crust, n is the fractional degree of mantle depletion, and cpm, cdm, and cc are the concentrations of Rb in the primitive mantle, depleted mantle, and crust, respectively. Substituting the values in table 12.3 for 100% mantle depletion ($n = 1$) and solving for depleted mantle abundance of Rb (cdm) gives:

$$0.9944(0.55) = 1(0.9944)(cdm) + 0.0056(32),$$

or

cdm = 0.37 ppm Rb at 100% mantle depletion.

Similarly, we can use equation 12.2 to calculate the concentration of Rb and Yb at other degrees of mantle depletion (with n expressed as a fraction of total mantle), as shown in table 12.3. Also shown in table 12.3 is the measured weight ratio of Rb/Yb in MORB, which should reflect the ratio of these

elements in the depleted mantle source region. By comparing the calculated Rb/Yb ratios at different degrees of mantle depletion with the measured MORB Rb/Yb ratio, we can estimate the degree of mantle involvement in crust formation. These simple calculations bracket the amount of mantle that has been affected by extraction of the crust between 33% and 50%. A similar result can be obtained by doing the same calculation with other incompatible elements, such as barium.

If we accept the base of the upper mantle as the 650 km discontinuity, then the upper mantle volume corresponds almost exactly to one-third of the entire mantle. Thus, it seems plausible that the upper mantle constitutes the part that has been partially melted to make the crust.

The contrasting ideas about mantle convection might be reconciled by a model that allows the Earth to lose heat by different mechanisms at different times. During normal periods (sometimes called Wilson cycles), such as we are experiencing now, upper and lower mantle convection cells are distinct, with most heat loss occurring at spreading centers through plate tectonics. Periodically, however, the Earth experiences whole-mantle convection, as large plumes of hot material reach the surface. During these MOMO (mantle overturn and major orogeny) episodes, accumulated cold slabs descend from the 650-km boundary into the lower mantle, and huge plumes rise from the core-mantle boundary. Eruption from the heads of such plumes produce large igneous provinces characterized by vast outpourings of basaltic magma (flood basalts). These contrasting styles are illustrated in figure 12.6. Flood basalt provinces form over relatively short time periods and may have caused significant environmental catastrophies, including some mass extinctions. After the plume head becomes depleted, the plume tail continues to be a source of magmas on a smaller scale. This usually produces a chain of volcanoes, as in Hawaii, as the overlying plate rides over the stationary hot spot.

TABLE 12.3. Rubidium and Ytterbium Mass Balance in the Mantle and Crust

	Primitive Mantle	Crust	Predicted Depleted Mantle[1]				MORB
			100%	50%	33%	25%	
Rb (ppm)	0.55	32	0.37	0.19	0.003	—	2.2
Yb (ppm)	0.37	2.2	0.36	0.35	0.34	0.33	5.1
Rb/Yb	1.49	14.5	1.03	0.55	0.01	—	0.43

From Taylor and McLennan (1985).
[1]Predicted concentrations at various degrees of mantle involvement.

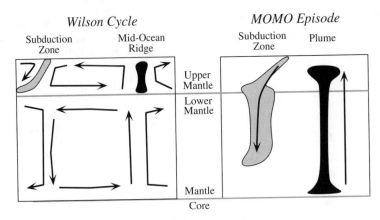

FIG. 12.6. Comparison of convection during Wilson cycles and MOMO episodes. During most periods of the Earth's history, including the present day, plate tectonics prevails and convection cells within the upper and lower mantles are largely isolated. During MOMO (mantle overturn and major orogeny) episodes, subducted slabs that have accumulated at the base of the upper mantle sink into the lower mantle, and multiple plumes rise from the core-mantle boundary.

Heat Exchange between Mantle and Core

The metallic liquid in the Earth's outer core is constantly crystallizing, so that the solid inner core is growing larger with time. Variations in the velocity of seismic waves traveling at different orientations within the inner core have been interpreted as evidence that the inner core may be one huge crystal. Exothermic crystallization yields large amounts of heat that stirs the outer core. Convection of this metallic liquid is responsible in part for the Earth's magnetic field. Heat carried upward through the core is eventually transferred to the bottom of the mantle.

Although most of the flux across the core-mantle boundary is heat, there is some suggestion of mass transfer as well. Reactions between liquid metal and perovskite in the lower mantle have been postulated. If such reactions occur, this might provide a mechanism for adding a light element to the outer core, so in this case, mass transfer would be from the mantle downward into the core. Small amounts of core material might also intrude into the lowermost mantle and eventually be entrained in mantle convection cells. The evidence for this, too, is speculative, based on small anomalies in the isotopic composition of the siderophile element iridium. Heat transfer through the core-mantle boundary, though, is unambiguous. It is a major driving force in mantle convection and must be responsible for mantle plumes.

Fluids and the Irreversible Formation of Continental Crust

Production of continental crust occurs primarily above subduction zones, where melting of the mantle wedge above the subducted slab occurs. At the time it is subducted, oceanic lithosphere has been hydrated by reactions with seawater, so that water is also carried downward into the upper mantle. When the slab reaches a depth appropriate for transformation to eclogite, the water is driven off and migrates into the overlying wedge of mantle peridotite. Melting under hydrous conditions can produce andesitic magmas with higher silica contents than basalt. Andesitic magmas normally contain a few percent of dissolved water, which is cycled back to the surface when the magmas outgas on eruption. Magmas that are erupted at subduction zones beneath continents immediately become continental crust. Where subduction occurs beneath oceanic lithosphere, magmatic arcs form and may be eventually become accreted to continents.

The production of continental crust appears to be unidirectional. The aggregate density of "granitic" continental crust is too low for this material to be subducted. Consequently, once continental crust forms, it is likely to remain at the Earth's surface. This is in marked contrast to oceanic crust, which is consumed in subduction zones at virtually the same rate that it is generated at midocean ridges. In fact, the only way that oceanic crust can escape being reincorporated into the mantle is by its accidental accretion onto continents by tectonic processes (producing ophiolites).

As the continental crust accumulated over time, its composition apparently changed. Scott McLennan (1982) found that Archean clastic sedimentary rocks were depleted in silicon and potassium and enriched in sodium, calcium, and magnesium relative to post-Archean rocks. These data indicate that the sources of Archean sediments were more mafic than for Proterozoic sediments. In an exhaustive survey of 45,000 analyses of common crustal rocks, A. B. Ronov and his collaborators (1988) determined that progressive changes occurred in the compositions of shales, sandstones, basalts, and granitic rocks. Each of these lithologies showed decreases in magnesium, nickel, cobalt, and chromium and increases in potassium, rubidium, and other lithophile elements with increasing time. There is disagreement about whether these secular changes are continuous or a sharp break occurs at the Archean-Proterozoic boundary 2.5 billion years ago.

Models for the growth of continental crust fall into two basic categories: early formation of virtually the entire crust with subsequent recycling, or continuous or episodic increase in the amount of crust throughout geologic time. Although this remains a contentious subject, continuous or quasicontinuous growth models are most popular. Ross Taylor and Scott McLennan (1985) argued that 90% of the continental crust was in place by the end of the Archean. They also proposed that most (60–75%) of the crust was produced episodically during the period 3.2 to 2.5 billion years ago. This transition might explain, in part, the compositional variations between Archean and post-Archean crust.

From our discussion of these interactions between crust, mantle, and core, we can see that the interior parts of the Earth constitute a grand geochemical system. The fluxes of materials and energy between these reservoirs are summarized in figure 12.7. Heat and possibly some matter is exchanged between core and mantle. Heat and matter in the form of ascending magmas or subducted lithosphere are exchanged between mantle and crust. The Earth is continuing to differentiate irreversibly in such a way that the mantle is depleted in fusible components, which are ultimately added to continental crust. Although continental crust, once formed, cannot be recycled, the fluids that play an important role in its formation are cycled between mantle and crust at subduction zones.

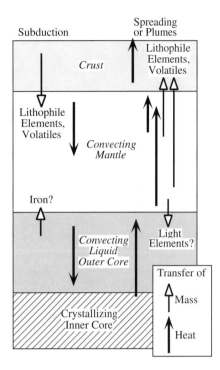

FIG. 12.7. Cartoon summarizing the fluxes of heat and matter between the core, mantle, and crust reservoirs.

All of these fluxes are driven by thermal convection. In the following sections, we examine in more detail how these geochemical interactions take place.

MELTING IN THE MANTLE

Because mass transfer from the mantle to the crust commonly involves magmas (although hot, solid materials also migrate in rising plumes), we consider the melting process in some detail. Melting is probably never complete, and partial melts and residues usually exhibit significant chemical differences.

The transformation from the crystalline to the liquid state is not very well understood at the atomic or molecular level. One (largely chemical) description of the melting process is that it results from increasing the vibrational stretching of bonds between atoms to the point at which the weakest bonds are severed. Another (primarily structural) view of melting is that it parallels a change from long-range crystallographic order to the short-range order characteristic of melts. X-ray diffraction studies of minerals at low and high temperatures have shown no significant distortions of crystal structures as melting temperatures are approached, but unit cell volumes increase. However, more work needs to be done before the physics of the melting process can be detailed.

Thermodynamic Effects of Melting

Melting, like other phase transformations in the mantle, results in discontinuous changes in the thermodynamic properties of the system. A sudden increase in entropy, reflecting the higher state of atomic disorder in liquids relative to crystals, is probably the most obvious thermodynamic change. More important from the geochemist's point of view, however, is the increase in enthalpy that accompanies melting. The thermal energy necessary to convert rock at a particular temperature to liquid at that same temperature is the *enthalpy of melting,* also called the *latent heat of fusion.* This is the thermal energy required beyond that necessary to raise the temperature of the rock to its melting point.

The enthalpy of melting can be measured by dropping crystals or glass into a calorimeter at temperatures just above and just below the melting point. (Calorimetry was mentioned in chapter 3.) From the rise in temperature of the calorimeter bath, or change in the proportion of crystals and liquid, the energy released by each state of the material can be evaluated. It is possible to calculate the enthalpy of melting for more complex systems that consist of several crystalline phases, provided that the phase diagram relating these phases has been determined and the values of ΔH_m for the individual phases are known, as illustrated in worked problem 12.3.

Worked Problem 12.3

How much thermal energy is required to melt a mixture of diopside and anorthite, initially at 0°C? Let's assume that the proportions of these two minerals correspond to the eutectic composition in the system diopside-anorthite (you may wish to review this phase diagram, illustrated at the bottom of fig. 10.6).

This calculation is performed in two steps. First, we must determine the energy required to raise the temperature of this material from 0°C to the melting point, which is 1274°C for a eutectic composition in this system (see fig. 10.6). Recall that the eutectic represents the composition of the last liquid to crystallize or the first liquid to form during partial melting. From the phase diagram, we find that the eutectic composition (in wt %) is $Di_{58}An_{42}$. The heat capacity ($\Delta \bar{C}_p$) for diopside is 100.40 J mol^{-1} K^{-1} = 23.99 cal mol^{-1} K^{-1}, and that for anorthite is 146.30 J mol^{-1} K^{-1} = 34.97 cal mol^{-1} K^{-1} (Robie et al. 1978). We convert these heat capacities into units of cal g^{-1} K^{-1} be-

cause the eutectic composition is given in units of weight. Division by the gram formula weights for diopside (216.55 g mol^{-1}) and anorthite (278.21 g mol^{-1}) gives $\Delta \bar{C}_p = 0.1109$ cal g^{-1} K^{-1} for diopside and 0.1257 cal g^{-1} K^{-1} for anorthite. The energy required to raise the temperature of this mixture by 1274° is therefore:

(0.1109 cal g^{-1} K^{-1})(1274 K)(0.58)
+ (0.1257 cal g^{-1} K^{-1})(1274 K)(0.42) = 149.21 cal g^{-1}.

The second step is to calculate the enthalpy of melting of this eutectic mixture. $\Delta \bar{H}_m$ for diopside is 77.40 kJ mol^{-1} = 18,345 cal mol^{-1} and for anorthite is 81.00 kJ mol^{-1} = 19,359 cal mol^{-1}. Division by the appropriate gram formula weights gives $\Delta \bar{H}_m$ for diopside of 84.71 cal g^{-1} and for anorthite of 69.58 cal g^{-1}. Multiplying these values by the appropriate fractions for the eutectic composition yields:

(84.71 cal g^{-1})(0.58) + (69.58 cal g^{-1})(0.42) =
78.36 cal g^{-1}.

The heats of mixing of these liquids are so small that they can be neglected. Thus, the total thermal energy necessary for this melting process is the sum of the energy required to raise the temperature of a eutectic mixture of diopside and anorthite to the melting point plus the energy required to transform this mixture into a liquid at that same temperature:

149.21 cal g^{-1} + 78.36 cal g^{-1} = 227.51 cal g^{-1}.

A realistic value for the enthalpy of melting of mantle peridotite is difficult to obtain, because of the unknown effects of high pressure and of other solid solution components in pyroxenes, olivine, and garnet or spinel. One commonly quoted value for garnet peridotite at 40 kbar pressure is 135 cal g^{-1}. For melting to occur in the mantle, this amount of thermal energy must be supplied in excess of the heat necessary to bring mantle rock to the melting temperature.

Types of Melting Behavior

If the temperature of mantle peridotite is increased to the melting point and some extra thermal energy is added for the enthalpy of melting, magma is produced. As in the case of eutectic melting in the diopside-anorthite system discussed in worked problem 12.3, fusion begins at some invariant point, so that several phases melt simultaneously. The melting process can be described in several ways. *Equilibrium melting* (also called *batch melting*) is a relatively simple process in which the liquid remains at the site of melting in chemical equilibrium with the solid residue until mechanical conditions allow it to escape as a single "batch" of magma. *Fractional melting* involves continuous extraction of melt from the system as it forms, thereby preventing reaction with the solid residue. Fractional melting can be visualized as a large number of infinitely small equilibrium melting events. *Incremental batch melting* lies between these two extremes, with melts extracted from the system at discrete intervals.

Worked Problem 12.4

To illustrate the distinction between equilibrium and fractional melting, let's examine fusion processes in the two-component system forsterite-silica, shown in figure 12.8. Describe melting in this system under equilibrium and fractional conditions.

This system contains a peritectic point between the compositions of forsterite (Fo) and enstatite (En) and a eutectic between the compositions of enstatite and cristobalite (Cr), seen more clearly in the schematic insets. If enough heat is available to raise the temperature of the system to 1542°C and offset the enthalpy of melting, any composition between En and Cr

begins melting at the eutectic, with both solid phases entering the melt. For small degrees of partial melting under equilibrium conditions, the melt will have the eutectic composition. With further melting, one phase usually becomes exhausted from the solid residue before the other does. Begin, for example, with a composition x that plots between the peritectic and the eutectic, shown in the upper inset in figure 12.8. As this mixture of En + Cr melts, Cr will be exhausted from the residue first. As melting continues under equilibrium conditions, the composition of the liquid leaves the eutectic and follows the liquidus upward, all the while melting enstatite. Total melting occurs when the liquid reaches the starting composition. This path is the reverse of equilibrium crystallization.

Fractional melting produces a sequence of distinct melts rather than one evolving melt composition. Again, consider the same mixture x of enstatite and cristobalite as before, shown in the upper inset. The initially formed liquid, which has the eutectic composition, is removed from the system as it forms, so the composition of the remaining solid shifts toward En. After Cr is exhausted from the residue, the temperature must increase to the peritectic temperature for any further melting to occur. When the temperature reaches 1557°C, En melts incongruently to Fo plus liquid, and the liquid is immediately removed. After all the En is converted to Fo, no further melting

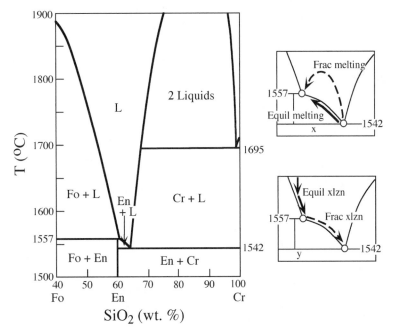

FIG. 12.8. Phase diagram for the system forsterite (Mg_2SiO_4)-silica (SiO_2). The insets illustrate the eutectic and peritectic in this system; the upper inset (melting paths) is used in worked problem 12.4, and the lower inset (crystallization paths) in worked problem 12.5. The silica-rich side of the diagram contains a region of silicate liquid immiscibility ("2 Liquids").

can occur until the liquid reaches the melting point for pure Fo. Fractional melting, if carried to completion, thus produces three batches of melt having distinct compositions, whereas equilibrium melting results in one evolving but (at any time) homogeneous melt.

Some details of the melting relationships of mantle peridotite are illustrated schematically in figure 12.9. Phase changes for the stable aluminum-bearing mineral in fertile peridotites are illustrated at temperatures below the solidus. Melting of peridotite begins as its temperature is raised just above the solidus. As temperature increases further, clinopyroxene, spinel, and garnet are completely melted and thereby exhausted from the solid residue (as indicated by lines labeled Cpx-out, Sp-out, and Gar-out, respectively). This all occurs within ~50°C of the solidus, producing a wide *P-T* field for the coexis-

tence of olivine + orthopyroxene with liquid. The olivine + orthopyroxene residue is the depleted peridotite that was discussed earlier.

The composition of the basaltic liquid produced by partial melting of peridotite changes with pressure, becoming less silica rich at higher pressure. Lower silica contents of liquids increase the proportion of olivine that can ultimately crystallize from the melt on cooling; the percentage of olivine in the melt is contoured as dashed lines in figure 12.9. Variation in pressure (or depth) of magma generation thus results in basaltic melts ranging from tholeiite (the common lava composition at mid-ocean ridges) at modest pressures to more olivine-rich alkali basalt and basanite (which occur at hot spots) at higher pressures.

Causes of Melting

The *geothermal gradient* (also called the *geotherm*) is the rate of increase of temperature with pressure, or depth. In an earlier section of this chapter, we noted that the elastic properties of the low-velocity zones in the mantle are consistent with partial melting. However, from an examination of the relative positions of geotherms and the peridotite solidus in figure 12.9, there is no apparent reason for incipient melting in this zone or, for that matter, anywhere else in the mantle. How can we then account for magma generation? There is no simple answer to this question, but there are at least three intriguing possibilities: localized temperature increase, decompression, and addition of volatiles. Let's see how each of these could cause melting.

Simply raising the temperature of a rock (the path labeled +*T* in fig. 12.10) is the most obvious mechanism by which melting can occur; this has sometimes been called the *hot plate model*. However, in considering this model, we should keep in mind the difference in magnitude between the enthalpy of melting and the heat capacity for silicate minerals. $\Delta \bar{C}_P$ values are generally several hundred times lower than corresponding $\Delta \bar{H}_m$ values (compare the values for diopside and for anorthite in worked problem 12.3), so it is much easier to raise the temperature of a rock than to produce a significant amount of partial melt from it. Melting consumes large quantities of thermal energy, moderating temperature variations within the system in its melting range. Even so, it may be possible to have localized mantle heating that produces magma. Frictional heating of subducted

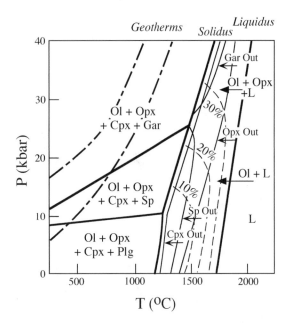

FIG. 12.9. Schematic phase diagram for mantle peridotite, illustrating the effects of partial melting. The solidus and liquidus are indicated by heavy lines. The aluminous phase changes from plagioclase (Plg) to spinel (Sp) and then garnet (Gar) with increasing pressure. With increasing temperature, first clinopyroxene (Cpx) and then spinel or garnet are exhausted from the residue, leaving olivine (Ol) + orthopyroxene (Opx) or olivine alone. The dashed lines are contours of the percentage of dissolved olivine in the melt; they show that with increasing pressure, magmas become more olivine rich. Also illustrated are typical geothermal gradients for oceanic and continental regions.

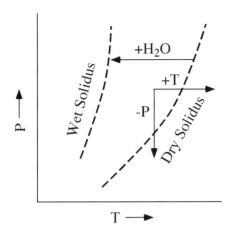

FIG. 12.10. *P-T* diagram illustrating three possible ways that mantle rocks may melt. Increase in temperature (+*T* path) could result from localized concentration of heat producing radionuclides or underplating by other magmas. Decrease in confining pressure (–*P* path) may occur because of diapiric upwelling of ductile mantle rock. Addition of a fluid phase (+H_2O path) depresses the melting curve to lower temperature for the same pressure.

lithospheric slabs or dissipation of tidal energy due to the gravitation attraction of the Sun and Moon have been suggested as possible causes of localized heating in the mantle. More plausible is local heating of the lower crust in response to underplating by mantle-derived magmas.

The most important mechanism for melting mantle rocks is almost certainly the release of pressure, caused by the convective rise of hot, plastic mantle rocks. The effect of decreasing confining pressure is generally to lower the solidus temperature. One example, in the system diopside-anorthite, is illustrated in figure 12.11. The eutectic temperature, which is the point at which partial melting begins, is decreased by >100°C in going from 10 kbar to 1 atm pressure. The position of the eutectic point also shifts, because the melting point of diopside increases much more for a given pressure change than does that of anorthite. Therefore, the Di field is enlarged with increasing pressure at the expense of the An field. The effect of decreasing pressure on the solidus of mantle peridotite can be seen in figure 12.9. Because of the positive slope of this melting curve in *P-T* space, decompression of solid rock can induce melting, as illustrated by the path labeled –*P* in figure 12.10. Because decompression melting occurs at constant temperature (we used the term *adiabatic* to describe such a process in

chapter 3) over some interval of pressure, conventional *T-X* phase diagrams (which are drawn for a fixed pressure) cannot really be used to understand this process. A recent innovation in visualizing polybaric melting is discussed in an accompanying box.

Changes in the composition of a rock system due to gain or loss of volatiles, principally H_2O and CO_2, are also potentially important in facilitating melting. Water lowers the solidus of peridotite, as illustrated in figure 12.13 by the wet solidus curve. The addition of water to a dry peridotite that is already above the wet solidus will cause spontaneous melting, as illustrated by the +H_2O path in figure 12.10. We have already discussed how an influx of water into the mantle can be caused by dehydration reactions in a subducted slab of hydrated lithosphere. The effect of the addition of CO_2 to mantle peridotite is less drastic than for H_2O, but still significant. The solidus of peridotite in the presence of CO_2 is illustrated in figure 12.13 by the line labeled CO_2. When the fluid phase contains both H_2O and CO_2, the effect on the peridotite melting curve is complicated. A solidus for a mixed fluid composition of $X_{CO_2} = 0.6$ is illustrated in figure 12.13. At modest pressures of 10 to 20 kbar, water predominates, because it is most soluble in the melt; but at higher pressures, the effect of CO_2 becomes marked as its solubility increases.

Both of these volatile components also strongly influence the compositions of the magmas that are generated. Water selectively dissolves silica out of mantle phases

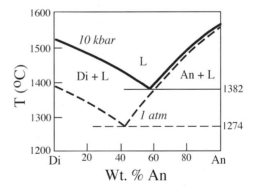

FIG. 12.11. Phase diagram for the system diopside (CaMgSi$_2$O$_6$)-anorthite (CaAl$_2$Si$_2$O$_8$), demonstrating the effect of increasing confining pressure on melting relationships. Lowering pressure from 10 kbar to 1 atm results in a decrease in the eutectic melting temperature of >100°. (Based on experiments by Presnall et al. 1978.)

DECOMPRESSION MELTING OF THE MANTLE

T-X phase diagrams of the type considered in chapter 10 are appropriate for understanding how magmas and crystals behave when temperature and composition are variables and pressure is constant. We have already seen that the Gibbs free energy is minimized at equilibrium under such conditions. However, decompression melting, which is the most important process by which magmas are generated, cannot be visualized using the conventional *T-X* phase diagrams. What are the appropriate variables when melting occurs over a range of pressures?

If decompression melting is adiabatic ($q = 0$) and reversible ($dq = TdS$, where S is the entropy), the process is isentropic ($dS = 0$). The appropriate variables for visualizing polybaric melting are thus P and S, with composition (X) held constant. At equilibrium under these conditions, enthalpy rather than free energy is minimized.

Figure 12.12 illustrates a schematic *P-S* diagram for a one-component system with two phases—solid and liquid. The diagram resembles the familiar *T-X* phase loop bounded by a solidus and a liquidus for a binary system, and it can be read in much the same way. Isentropic decompression is indicated on this diagram by a vertical (decreasing *P*, constant *S*) line. At point 1, the system is entirely solid, but it begins to melt at point 2, where it passes into the two-phase field. At any given pressure within the two-phase field, the relative proportions of solid and liquid can be determined by applying the lever rule to a horizontal line passing through the point. Under equilibrium melting conditions, melting is complete at point 3.

The same diagram can also be used to model fractional melting. Strictly speaking, this process is neither adiabatic nor isentropic, because melt leaving the system carries entropy with it. However, each small increment of melting can be considered isentropic. A solid having a value of entropy given by point 1 undergoes pressure release and begins to melt when it reaches point 2. As fractional melting proceeds, the system constantly changes composition and the solid residue follows the solidus, that is, the path from 2 to 3′. At the beginning of melting at point 2, the first increment of melt is identical, whether melting is equilibrium or fractional. However, for fractional melting,

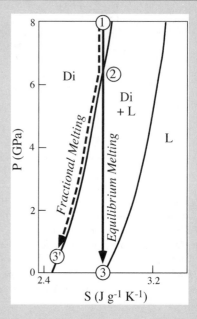

FIG. 12.12. Pressure-entropy diagram illustrating melting of diopside under equilibrium (path 1-2-3) and fractional (path 1-2-3′) conditions. This diagram allows us to visualize how melting takes place over a range of decreasing pressures typical of those experienced in an ascending mantle plume.

although the mass of the solid residue continuously decreases, it is not possible to melt the solid completely by this process. As a consequence, fractional melting produces less melt than equilibrium melting for a given decompression interval.

One subtlety of the *P-S* diagram is that the amount of melt generated per decrement of pressure *increases* with decreasing pressure. Using figure 12.12, you can demonstrate this for yourself by comparing the pressure intervals required to produce 50% melting and 100% melting under equilibrium melting conditions. This effect results from the curvature of the liquidus and the solidus. The conclusion that melting accelerates with decompression cannot be worked out by studying this process on a *P-T* diagram. The effect is appreciable in natural rock systems and probably accounts for most melts generated within the mantle. Complications result from phase changes in the mantle, which produce nonuniform zones of melting during upwelling, as discussed by Paul Asimow and his collaborators (1995).

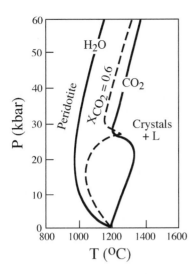

FIG. 12.13. Phase diagram showing the solidus in the system peridotite-CO_2-H_2O. Melting in the presence of pure water occurs at lower temperatures than for pure CO_2. The dashed line labeled $X_{CO_2} = 0.6$ corresponds to peridotite melting with a mixed fluid of that composition. Water is the most important volatile species at pressures of less than ~20 kbar, but CO_2 becomes increasingly important at higher pressures because of its increased solubility in the melt.

before melting occurs; as soon as the melt forms, it effectively swallows this fluid and thereby becomes richer in silica (sometimes producing andesites rather than basalts). However, CO_2 favors the production of alkalic magmas poor in silica (nephelinites and kimberlites).

Concluding this section on melting, we summarize some of these concepts by relating current ideas about how mantle-derived magmas are formed in various tectonic settings: (1) Partial melting of relatively dry peridotite accounts for the vast outpourings of tholeiitic basalt at midocean ridges. This style of melting is initiated by decompression resulting from the rise of mantle under spreading centers. (2) Hydrated slabs of oceanic lithosphere subducted at convergent plate boundaries transform into eclogite at depths of ~100 km, thereby releasing water into the overlying wedge of mantle peridotite. This addition of H_2O may cause melting to produce siliceous magmas, such as andesites. (3) At substantially deeper levels in the mantle, localized heating (probably by heat generated through crystallization of the inner core) at hot spots generates alkalic magmas. Magmas underplating the crust can also induce melting in the overlying crustal rocks by local increase in temperature.

DIFFERENTIATION IN MELT-CRYSTAL SYSTEMS

In the preceding section, we painted a broad picture of melting processes in the mantle, gleaned mostly from studies of volcanic rocks. In reality, the picture is much fuzzier, because very few mantle-derived magmas arrive at the Earth's surface in pristine condition. Most have been altered in composition (*differentiated*) to various degrees. We will now see how that happens.

Fractional Crystallization

The most common mechanism by which magmas are differentiated is *fractional crystallization*: the physical separation of crystals and liquid so that these phases can no longer maintain equilibrium. Crystals may be isolated from their parent magma through gravity settling (or rarely, floating), or they may be carried to the bottom of a magma chamber and deposited by convection currents. A pile of accumulated crystals may be purified further by squeezing out of some of the interstitial liquid as the overburden of additional crystals accumulates. Rapid crystallization, in which the reaction of melt with solid phases cannot keep up with falling temperature and advancing solidification, can produce compositionally zoned crystals. This kinetic isolation of the interiors of crystals prevents their equilibration with the liquid and thus can also be considered as fractional crystallization.

Unlike equilibrium crystallization and melting, fractional crystallization is not the reverse of fractional melting. To see how fractional crystallization differs from equilibrium crystallization, let's consider two binary phase diagrams that involve olivine.

Worked Problem 12.5

Fractional melting in the system forsterite-silica has already been discussed in worked problem 12.4. Using the phase diagram for this system in figure 12.8, trace the fractional crystallization paths of melts having compositions *x* and *y*.

Cooling of any melt composition to the liquidus causes the appropriate solid phase to appear. If we begin with a liquid of composition *x*, removal of enstatite (En) crystals as they form drives the residual liquid composition toward the eutectic. When this point is reached, both En and crystobalite (Cr) separate from the melt. The liquid trajectory is, in this case, no different from that for equilibrium crystallization, but in fractional crystallization, the bulk composition of the magma changes because

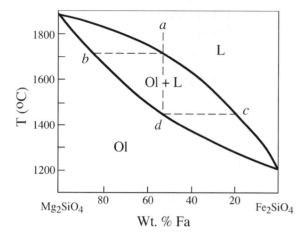

FIG. 12.14. Phase diagram for the system forsterite (Mg_2SiO_4)-fayalite (Fe_2SiO_4), illustrating how fractional crystallization affects a solid solution series. Olivine crystals forming from a melt with initial composition *a* will range in composition from *b* to some point past *d*.

crystals are removed. Under equilibrium conditions, these crystals remain suspended in the magma and become part of the resulting erupted lava and igneous rock.

Now consider a liquid having composition *y*, shown in the lower inset of figure 12.8. The first solid phase to form is now forsterite (Fo). Under equilibrium conditions, the liquid should follow the Fo liquidus to intersect the peritectic, at which point it reacts with already crystallized Fo to produce En. Under fractional crystallization conditions, Fo is physically removed from the system as it forms, so that there is none of this phase to react with the melt at the peritectic temperature. Thus, there is no Fo + L → En reaction to cause the liquid to delay at the peritectic. The cooling liquid slides through the peritectic and immediately begins to crystallize En rather than Fo. The melt composition continues toward the eutectic. Under conditions of perfect fractional crystallization, the liquid reaches the eutectic, but such ideal conditions do not exist in nature, so that the liquid may be exhausted at some point before the eutectic is reached.

Olivine that crystallizes from real magmas is, of course, not pure forsterite, but a solid solution of Mg- and Fe-rich end members. The phase diagram for the system forsterite (Fo)-fayalite (Fa), shown in figure 12.14, is very similar to one for the plagioclase system that was used in chapter 10 to illustrate solid solution behavior. Crystallization of a liquid of composition *a* in figure 12.14 initially produces olivine of composition *b*. If equilibrium were achieved, this olivine would then react continuously with the melt to form more Fe-rich compositions, ultimately having composition *d*, as the last drop of liquid is exhausted. Under fractional crystallization conditions, however, the initially formed Mg-rich olivine is removed from the melt,

so that this reaction cannot take place. Hence, the liquid crystallizes a range of olivines that are progressively more Fe-rich, from composition *b* to *d* and beyond. Again, under conditions of perfect fractional crystallization, the liquid should ultimately reach the composition of pure fayalite, but in real systems we cannot specify the exact point at which the last liquid will solidify, only that the liquid composition continues past point *c*.

Fractional crystallization affects ternary systems in an analogous manner. Reaction curves, the ternary equivalents of binary peritectic points, no longer define the compositions of fractionating liquids, because there are no solid phases to react. Fractional crystallization in a more complex system is described in worked problem 12.6.

Worked Problem 12.6

Phase relationships in the system forsterite-anorthite-silica are shown in figure 12.15. This problem will not make sense to you, unless you have studied the box on ternary phase diagrams in chapter 10. (Fig 12.15 actually represents a pseudoternary rather than a ternary system, because the primary phase field of spinel appears in the diagram, but the composition of spinel lies outside of this plane. The only complication that this situation presents is that we cannot specify the crystallization path of liquids within the spinel field, because we do not know where

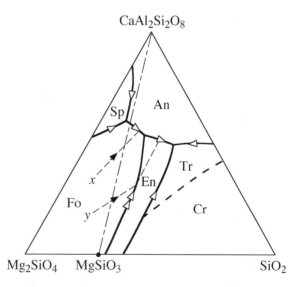

FIG. 12.15. Phase diagram for the pseudoternary system forsterite (Mg_2SiO_4)-anorthite ($CaAL_2Si_2O_8$)-silica (SiO_2). The compositions of liquids derived by fractional crystallization of melt compositions *x* and *y* are illustrated by dashed lines.

the projection of spinel plots on this diagram.) We have already applied Alkemade's Theorem to identify the various subtraction and reaction curves (single- and double-headed arrows, respectively) in this diagram. Describe fractional crystallization paths for liquids of composition x and y in figure 12.15.

Liquid x cools to the Fo liquidus and then changes in composition away from the forsterite apex as this phase is removed. When the liquid composition intersects the subtraction curve, Fo + An crystallize and are separated out together. The liquid moves to and through the tributary reaction point, because there is no Fo left in contact with melt to react to form En. The liquid continues toward the ternary eutectic point between En, An, and Tr (tridymite). We cannot specify the exact point where the liquid phase will be exhausted.

A liquid with composition y will fractionate Fo until it reaches the curve for the reaction Fo + L → En. Because there is no Fo left to react, the melt steps over this curve into the En primary phase field. At this point, En begins to crystallize, and the liquid composition moves directly away from the enstatite composition. Continued En fractionation drives the melt to the subtraction curve, and An joins the fractionating assemblage. The liquid then follows the subtraction curve toward the ternary eutectic.

Consideration of these experimentally derived phase diagrams provides some indication of the ways in which magmas *might* differentiate, but what is the evidence that fractional crystallization actually occurs as a widespread phenomenon in magmatic systems? To answer this question, we must examine real igneous rocks for the effects of this process. Experimental studies of most basaltic rocks indicate that they are *multiply saturated* at low pressures; that is, several minerals begin to crystallize at virtually the same temperature. We would expect that kind of behavior for partial melts formed at low pressure, because melting begins at invariant points (such as eutectics, the lowest temperature points at which liquids are stable). However, basaltic magmas are mantle derived and could not have formed at low pressures, so multiple saturation must have another explanation. That most basalts are already multiply saturated at low pressure suggests that they have experienced fractional crystallization.

It has also been recognized that basalts cluster at certain preferred compositions. This observation is difficult to interpret if these are primitive magmas, unless the preferred compositions represent those of liquids in equilibrium with mantle peridotite at invariant points. An alternative (and more plausible) view is that these

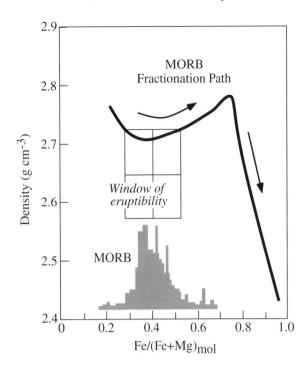

FIG. 12.16. The density of fractionating basaltic magma as a function of composition. MORB compositions cluster at the minimum density, suggesting that only those compositions can ascend through the crust and erupt. (After Stolper and Walker 1980.)

compositions represent points along a fractionation path that have special properties that allow the magmas to erupt. Edward Stolper and David Walker (1980) observed that the density of basaltic liquids first decreases and then increases during fractional crystallization, as illustrated in figure 12.16. In this diagram, fractionation is indicated by the progressive increase in the Fe/(Fe + Mg) ratio of the lavas. The commonly observed composition for MORB corresponds to that which has the minimum density. This fractionated composition would be most likely to ascend through the less dense crust and reach the planet's surface.

Widespread fractional crystallization has obvious ramifications for what we can hope to learn about the mantle from igneous rocks. Crystallization experiments on fractionated rocks serve no purpose, except to understand the fractionation path itself. Luckily, a few unfractionated lavas somehow reach the surface, so judicious selection of samples for experiments can still provide useful information on mantle composition and melting.

Chemical Variation Diagrams

The chemical changes wrought by fractional crystallization can be modeled quantitatively by using mixing calculations, which are most easily visualized from element *variation diagrams*. Figure 12.17 illustrates a magma's chemical evolution, called the *liquid line-of-descent*, as it undergoes fractional crystallization. In figure 12.17a, crystals of composition A separate from liquid L_1, with the result that the liquid composition is driven toward L_2. In figure 12.17b,c, mixtures of two or three minerals crystallize simultaneously. The bulk composition of the extracted mineral assemblage is given in each case by A, and the liquid line-of-descent follows the path $L_1 \rightarrow L_2$. In figure 12.17d, magma L_1 first crys-

tallizes mineral D and then D plus E, with the bulk composition of the fractionating assemblage represented by A. In this case, the liquid line-of-descent contains a kink, where the additional phase joins the crystallization sequence. A more complex case in which the fractionating phase is a solid solution is illustrated in figure 12.17e. Mineral AB is a solid solution of end members A and B, and the proportions of these end members change as temperature drops during crystallization. The trajectory of the liquid at any point is defined by a tangent to the liquid composition drawn from the composition of crystallizing AB. For example, the liquid path when $A_{80}B_{20}$ and $A_{50}B_{50}$ are extracted is shown by the tangents intersecting these compositions. In this case, the liquid line-of-descent traces out a curve. In all of these diagrams, the relative proportions of fractionating phases can be determined by application of the lever rule.

In actual practice, the problem must be worked in reverse. A suite of volcanic rocks representing various points along a liquid line-of-descent is analyzed, and from these analyses, we must determine the fractionated phases. To test possible solutions to this problem, we require rock bulk compositions and the identities and compositions of possible extracted minerals. The latter may be obtained from petrographic observations and electron microprobe analyses of phenocrysts in the rocks. An example of such an exercise is given in worked problem 12.7.

Worked Problem 12.7

How can we test for fractional crystallization in real rocks? Assume that we have collected a suite of volcanic rock samples that we believe to be a fractionation sequence. The first step is to obtain bulk chemical analyses of the rocks. Petrographic observations indicate that three phases—olivine (Ol), plagioclase (Plg), and clinopyroxene (Cpx)—exist as phenocrysts in this suite of these rocks, and we have also chemically analyzed these minerals. Because these three phases were obviously on the liquidus at some point, we determine whether fractional crystallization of any one or some combination of them could have produced the observed liquid line of descent.

Figure 12.18 shows three different variation diagrams with the liquid line-of-descent ($L_1 \rightarrow L_2$) defined by the analyzed bulk-rock compositions. Also shown are the compositions of the three phenocryst minerals. The information gained from any individual diagram is ambiguous, but we can compare diagrams to obtain a unique answer. In figure 12.18a, we see that the liquid line-of-descent is consistent with fractionation of Ol + Plg, Ol + Cpx, or Ol + Plg + Cpx. This diagram thus rules out Cpx

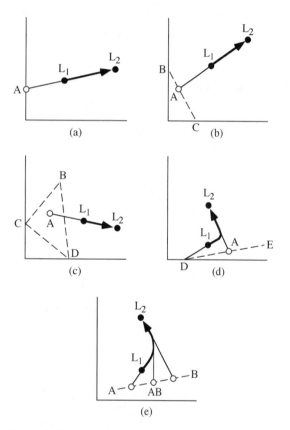

FIG. 12.17. Schematic chemical variation diagrams (axes can be any two elements or oxides) showing the liquid line-of-descent ($L_1 \rightarrow L_2$) during fractional crystallization of one (a), two (b), or three (c) phases. In (d), the line-of-descent has an inflection due to late entry of a second phase. In (e), the line-of-descent is curved because of continuous change in the composition of an extracted solid solution phase.

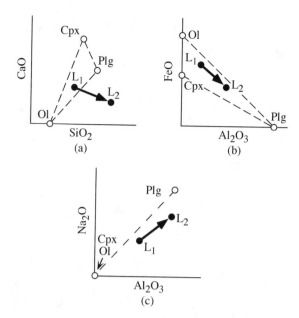

FIG. 12.18. Schematic chemical variation diagrams for a suite of basalts containing phenocrysts of olivine (Ol), clinopyroxene (Cpx), and plagioclase (Plg). The liquid line-of-descent defined by bulk chemical compositions of these basalts can be derived by fractional crystallization of Ol + Cpx.

+ Plg as a possibility. In figure 12.18b, Ol + Cpx or Ol + Plg + Cpx are permissible assemblages that could be extracted to produce this sequence, but Ol + Plg (allowed in fig. 12.18a), is not allowed. The final distinction between Ol + Cpx and Ol + Plg + Cpx is made using figure 12.18c. This diagram indicates that fractionation of Ol + Cpx is the only internally consistent answer. The graphical testing of all interrelationships between elements is an important part of this kind of study; otherwise, we could have incorrectly assumed that all of the observed phenocrysts can fractionate to produce this liquid line-of-descent. Notice also that the relative proportions of olivine and clinopyroxene (approximately 1:1) in all the diagrams are the same, as deduced from application of the lever rule to the point at which the extrapolated liquid line-of-descent crosses the Ol-Cpx tie-line. Not only the identity of the fractionating phases but also their relative proportions should be consistent in this set of variation diagrams.

As the number of possible fractionating phases or analyzed elements increases, it becomes more difficult to apply graphical methods. Numerical methods offer a real advantage in such situations. Several computer programs, commonly called *petrologic mixing programs,* are available for performing calculations of this type.

These programs compute the composition of the parent magma by mixing the residual liquid with fractionating phases in various proportions. The answer is a best fit to the least-squares regression for the analyzed liquid line-of-descent.

Liquid Immiscibility

Fractional crystallization is not the only mechanism by which magmas can differentiate. Experimental petrologists have recognized for half a century that the binary systems CaO-SiO_2, FeO-SiO_2, and MgO-SiO_2 contain fields in which silicate melts separate spontaneously into two liquids. The phase diagram for the forsterite-silica system is shown in figure 12.8; a stability field for two liquids is shown near the SiO_2 compositional end member. This region of unmixing behaves exactly like the solvus we introduced earlier in chapter 10, except that it occurs above the solidus. Liquid phase separation, or *immiscibility,* can cause magmatic differentiation, provided that the two melts have different densities. In such a case, globules of the more dense liquid tend to coalesce and sink to the bottom of the magma chamber, in a manner similar to the segregation of crystals by gravity fractionation.

Liquid immiscibility obviously indicates highly nonideal mixing of melt components. However, it may be helpful to recognize that the separation of an immiscible liquid fraction is really no different than the formation of a crystal, at least in the thermodynamic sense. In both cases, the original melt becomes saturated with respect to a new phase as the temperature decreases. However, there is a difference in the ease with which equilibrium can be achieved in these two situations. Crystal growth and separation of an additional liquid phase both proceed by diffusion of ions through the melt to the sites of growth and across the interface. Diffusion in liquids is generally many orders of magnitude faster than in crystals, so immiscible melts are more likely to maintain equilibrium with falling temperature. Consequently, a cooling magma may form zoned plagioclase crystals under nonequilibrium conditions while at the same time segregating an immiscible liquid fraction under equilibrium conditions.

Although examples of silicate liquid immiscibility are known from melting experiments on simple systems, most of these experiments take place at very high temperatures and have geologically unreasonable compositions (see, for example, the very high temperature and high SiO_2 content of immiscible liquids in the system forsterite-silica in fig. 12.8). An experimentally known example of silicate liquid immiscibility that may have a natural analog is in the system fayalite (Fe_2SiO_4)-leucite ($KAlSi_2O_6$)-silica. This two-liquid field is a reasonable compositional analog for immiscibility observed in some uncommon dike rocks (lamprophyres) and some residual liquids formed during the last stages of basalt crystallization (*mesostasis*). It is unlikely, however, that silicate liquid immiscibility produces large volumes of differentiated magmas, because these compositions are only produced near the end of fractional crystallization sequences, when little liquid is left.

Liquid immiscibility seems to be most effective when the structures of the two melts are radically different. This can be achieved most readily if melt components with nonsilicate compositions are considered. One important example involves the separation of carbonate and silicate melts in CO_2-bearing systems. A photomicrograph of carbonatite globules suspended in alkali basalt is illustrated in figure 12.19. Immiscibility in such systems has been documented experimentally and almost certainly accounts for the production of the carbonatite magmas that are sometimes associated with alkalic silicate magmas. Another example is the separation of immiscible sulfide melts from basaltic magmas on cooling. The sulfide melts are more dense than silicate magma and readily sink. This may be an important mechanism for producing some sulfide ore deposits.

THE BEHAVIOR OF TRACE ELEMENTS

Much of the gemstone industry depends on the chance incorporation of minute quantities of certain elements into otherwise "ho-hum" minerals. Small proportions of foreign impurities can turn ordinary corundum into valuable rubies or sapphires. For the geochemist, elements present in trace concentrations can serve another purpose—to test and quantify geochemical models.

Trace elements are commonly defined as those occurring in rocks in concentrations of a few tenths of a percent or less by weight. The mixing behavior of trace components in crystals or melts is often highly nonideal; consequently, different minerals may concentrate or exclude trace elements much more selectively than they do major elements. Trace element distributions can there-

FIG. 12.19. Photomicrograph of immiscible carbonatite globules in alkali basalt. Two globules are coalescing. This sample is from the Kaiserstuhl volcanic complex, Germany. Width of the photograph is ~4 mm.

fore provide quantitative constraints on the processes of partial melting and differentiation that cannot be deduced from consideration of major elements.

Trace Element Fractionation during Melting and Crystallization

In melt-crystal systems under equilibrium conditions, elements are partitioned among phases according to their activities in those phases. For trace elements, we normally assume Henry's Law behavior, $a_i = h_i X_i$. The *distribution coefficient*, K_D, derived in chapter 9, is given by K_D = concentration in mineral/concentration in liquid. Trace element distribution coefficients are based on the assumption that the trace component obeys Henry's Law in the phase of interest. Experimental studies suggest that the linear Henry's Law region for many trace elements extends at least to several wt %, so it is applicable for the concentration ranges measured in natural rocks. In principle, K_D can be measured from partially crystalline experimental charges or from natural lavas containing phenocrysts and glass. Many of the available distribution coefficients were determined from natural rocks, but there is unfortunately no assurance that these rocks reached equilibrium. There are also analytical problems in obtaining concentrates of pure phases (that is, crystals with no adhering or included glass, and vice versa), because trace elements are sometimes below the detection limits for in situ measurements. Another problem with K_D values is that they are dependent on the compositions of the phases. In other words, their distributions are nonideal, and enough information to determine the appropriate activity coefficients is usually not in hand. Despite these potential pitfalls, distribution coefficients are widely used to solve geochemical problems, and their compositional dependence can be minimized by selecting K_D values for phases similar in composition to those in the problem of interest.

The *bulk distribution coefficient* D is the K_D value adjusted for systems containing more than one solid phase. It is calculated from the weight proportions ω of each mineral ϕ by:

$$D = \sum \omega_\phi K_{D\phi}. \qquad (12.3)$$

D must be calculated if we intend to handle models for partial melting of mantle peridotite or for fractionation of several minerals. Using the bulk distribution coefficient, we can predict trace element distributions during various magmatic processes.

We now derive expressions that describe changes in the concentrations of trace elements during solidification of a magma. We consider two ideal cases: equilibrium crystallization and perfect fractional crystallization (the latter is sometimes called *Rayleigh fractionation*). We assume that more than one phase is crystallizing.

In the case in which equilibrium between crystals and liquid is maintained,

$$X_{i\,\text{solid}}/X_{i\,\text{melt}} = D. \qquad (12.4)$$

If a fraction of magma crystallizes, the fraction of melt remaining F is given by:

$$F = n_{\text{melt}}/n^0_{\text{melt}}, \qquad (12.5)$$

where n^0_{melt} is the number of moles of all components in the original magma and n_{melt} is the total number of

moles remaining after crystallization commences. The ratio of solid to melt remaining is therefore:

$$(n^0_{\text{melt}} - n_{\text{melt}})/n_{\text{melt}} = 1/F - 1. \qquad (12.6)$$

If we define y_i as the number of moles of trace component i in the system, equations 12.4 and 12.6 can be combined to give:

$$X_{i\,\text{solid}}/X_{i\,\text{melt}} = [y_{i\,\text{solid}}/(n^0_{\text{melt}} - n_{\text{melt}})]/(y_{i\,\text{melt}}/n_{\text{melt}})$$
$$= D.$$

Because $y_{i\,\text{solid}} = y^0_{i\,\text{melt}} - y_{i\,\text{melt}}$,

$$D = [(y^0_{i\,\text{melt}}/y_{i\,\text{melt}}) - 1][(n_{\text{melt}}/(n^0_{\text{melt}} - n_{\text{melt}})].$$

Rearranging and dividing by $n^0_{\text{melt}}/n_{\text{melt}}$ gives:

$$(y^0_{i\,\text{melt}}/n^0_{\text{melt}})/(y_{i\,\text{melt}}/n_{\text{melt}}) =$$
$$\{D[(n^0_{\text{melt}} - n_{\text{melt}})/n_{\text{melt}}] + 1\}n/n^0.$$

Substituting equation 12.5 into the expression above yields:

$$X_{i\,\text{melt}}/X^0_{i\,\text{melt}} = 1/(D - FD + F). \qquad (12.7)$$

Expression 12.7 gives the concentration of trace element i in the liquid as a function of the original concentration, the bulk distribution coefficient, and the fraction of liquid remaining for a system undergoing equilibrium crystallization. The same expression also holds for equilibrium melting.

Now let's see how this differs from fractional crystallization. Using the same symbols as above, the mole fraction of trace component i in the system is $X_i = y_i/n$. If fractionation of a mineral containing i occurs, then n becomes $n - dn$, and y becomes $y - dy$. The compositions of solid and melt are therefore:

$$X_{i\,\text{solid}} = dy/dn \qquad (12.8a)$$

and

$$X_{i\,\text{melt}} = (y - dy)/(n - dn). \qquad (12.8b)$$

If trace component i obeys Henry's Law, then:

$$X_{i\,\text{solid}} = dy/dn = D\,X_{i\,\text{melt}}. \qquad (12.9)$$

However, dy/dn can also be expressed in terms of $X_{i\,\text{melt}}$. If we neglect dy in comparison with much larger y and dn in comparison with n in equation 12.8b, we see that $X_{i\,\text{melt}}$ can be approximated as y/n and y as $n(X_{i\,\text{melt}})$.

Differentiating the latter expression with respect to n gives:

$$dy/dn = n(dX_{i\,melt}/dn) + X_{i\,melt}.$$

Substitution of this expression for dy/dn into equation 12.9 gives:

$$DX_{i\,melt} = n(dX_{i\,melt}/dn) + X_{i\,melt},$$

which can be recast into the form:

$$(1/X_{i\,melt}[D-1])\, dX_{i\,melt} = (1/n)\, dn. \qquad (12.10)$$

Because X_i changes during crystallization, but not necessarily linearly, the way we obtain the total change in concentration is to sum all the individual steps. In other words, we integrate equation 12.10 between X_i^0 and X_i and between the initial and final moles of melt n^0 and n:

$$\ln(n/n^0) = [1/(D-1)]\ln(X_{i\,melt}/X_{i\,melt}^0),$$

or

$$X_{i\,melt}/X_{i\,melt}^0 = (n/n^0)^{D-1}.$$

Because (n/n^0) equals F, the fraction of liquid remaining:

$$X_{i\,melt}/X_{i\,melt}^0 = F^{D-1}. \qquad (12.11)$$

Thus equations 12.7 and 12.11 describe the behavior of trace elements under the extreme conditions of equilibrium and fractional crystallization, respectively.

Using the same principles, analogous expressions can be derived for partial melting models. These are given here without derivation. For equilibrium melting:

$$X_{i\,melt}/X_{i\,solid}^0 = 1/(D - FD + F), \qquad (12.12)$$

and for fractional melting:

$$X_{i\,melt}/X_{i\,solid}^0 = 1/D(1-F)^{(1/D)-1}. \qquad (12.13)$$

Because fractional melting involves continuous removal of melt as it is generated, this process probably does not occur in nature, but it can be considered as an end member for intermediate melting processes. Geochemical models for magma generation have become very complex and are beyond the scope of this book; Shaw (1977) gives an excellent account of this topic.

The equations we have just derived can be expressed equally well in units of concentration other than mole fractions, as long as the distribution coefficients are defined in the same terms. In fact, most trace element analyses are expressed in units of weight, for example, parts per million (ppm), and we can substitute these concentration units for mole fractions in equations 12.7 and 12.11 if D is expressed as a weight ratio.

The changes in trace element concentration for residual liquids under equilibrium and fractional crystallization are illustrated graphically in figure 12.20. Curves

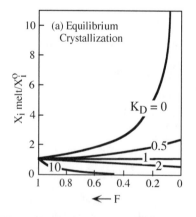 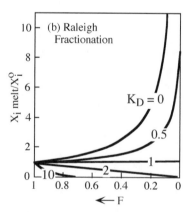

FIG. 12.20. Changes in trace element concentrations in residual liquids for different values of K_D during (a) equilibrium crystallization and (b) Rayleigh fractionation. Element concentrations are given in terms of mole fractions of each element i relative to its concentration in the parent magma. F is the fraction of liquid remaining. (After Wood and Fraser 1977.)

labeled $K_D = 0$ are limiting cases, and trace element enrichments greater than these cannot be produced by crystallization processes alone.

Compatible and Incompatible Elements

Earlier, we noted that it is useful to distinguish between trace elements that are compatible in the crystallographic sites of minerals and those that are incompatible with such minerals. Compatible elements have $K_D \gg 1$, whereas incompatible elements have $K_D \ll 1$. Consequently, compatible elements are preferentially retained in the solid residue on partial melting or extracted from the liquid during fractional crystallization; incompatible elements exhibit the opposite behavior. This can readily be seen in figure 12.20. Of course, as new minerals begin to melt or crystallize during the evolution of a magma, a particular trace element may alter its behavior. For example, phosphorus may act as an incompatible element during magmatic crystallization until the point at which the phosphate mineral apatite starts to form, after which it behaves compatibly. For applications involving mantle-derived magmas, it is often convenient to define compatible and incompatible behavior in terms of the minerals that make up mantle rocks—olivine, pyroxenes, garnet, and spinel. Incompatible elements are generally those with ionic radii too large to substitute for the more abundant elements, or those with charges of +3 or higher.

PREDICTING TRACE ELEMENT BEHAVIOR: TRANSITION METALS AND RARE EARTH ELEMENTS

Chemical physics provides powerful ways to predict trace element behavior. We illustrate this approach by considering several types of *transition elements*, which are characterized by inner d or f atomic orbitals that are incompletely filled by electrons. Unpaired electrons in transition elements are responsible for the distinctive colors of the minerals that contain them, as well as the magnetism of minerals. First we see how chemical bonding models can predict relative distribution coefficients.

Elements of the first transition series (Sc, Ti, V, Cr, Mn, Fe, Co, Ni, and Cu) have incompletely filled 3d orbitals, and cations are formed when the 4s and some 3d electrons are removed. Divalent ions in this series are compatible in common ferromagnesian minerals because of similarities to Mg^{2+} in ionic size and charge. However, the degree of compatibility differs markedly and can be explained by the effect of crystal structure on d orbitals. Crystal-field theory assumes that electrostatic forces originate from the anions surrounding the crystallographic site in which a transition metal ion may be located, and that these control its substitution in the structure.

A transition metal ion has five 3d orbitals, which have different spatial geometries (illustrated in fig. 2.7). In an isolated ion, the d electrons have equal probability of being located in any of these orbitals, but they attempt to minimize electron repulsion by oc-cupying orbitals one electron at a time. For ions with more than five d electrons, this is obviously not possible, so a second electron can be added to each orbital by reversing its spin.

When a transition metal ion is placed in a crystal (for example, in octahedral coordination with surrounding anions), the five 3d orbitals are no longer degenerate; that is, they no longer have the same energy. Electrons in all five orbitals are repelled by the negatively charged anions, but electrons in the d_{z^2} and $d_{x^2-y^2}$ orbitals, which are oriented parallel to bonds between the cation and the anions around it, are affected to greater extent than those in the other three. In octahedral coordination, therefore, these two orbitals will have a higher energy level. The reverse is true for ions in tetrahedral coordination.

The energy separation between the two groups of orbitals is called *crystal-field splitting*. This is illustrated in figure 12.21. Ions with electrons in only the lower-energy orbitals for a particular site geometry are stable in that site, whereas electrons in the higher-energy orbitals destabilize the ion. Of course, the number of electrons depends on which transition metal the cation is made from and on its oxidation state, so some transition elements prefer octahedral sites more than others do. For example, the predicted octahedral site preference energies for divalent cations decrease in the order Ni > Cu > Co > Fe > Mn, and for

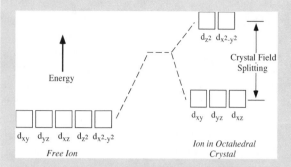

FIG. 12.21. Sketch of the energies of the five d orbitals in transition metal ions (refer to fig. 2.7 for geometric representations of these orbitals). When the free ion is placed in an octahedral mineral site, its total energy increases. Some orbitals have higher energy because they experience greater electron repulsion; the magnitude of the energy difference between the two groups of orbitals is the crystal-field splitting.

trivalent cations, Cr > V > Fe. As olivine or spinel (which offer octahedral sites) crystallizes from a magma, we expect to see rapid uptake of Ni and Cr in these minerals relative to other compatible transition metals. Tetrahedral sites in minerals produce a different but predictable order of cation site preferences. Thus, crystal-field effects provide an explanation for the differing K_D values for transition metals. The geochemical and mineralogical applications of crystal-field theory were treated comprehensively by Roger Burns (1970).

The lanthanide or *rare earth elements,* commonly abbreviated as REE, include La, Ce, Pr, Nd, Pm, Sm, Eu, Gd, Tb, Dy, Ho, Er, Tm, Yb, and Lu. The REEs find wide application in metallurgy, the coloring of glass and ceramics, and the production of magnets. In this part of the periodic table, increasing atomic number adds electrons to the inner 4f orbitals, so this is a somewhat similar case to the transition metals just discussed. The f electrons in REE, however, do not participate in bonding. The valence electrons occur in higher-level orbitals, and loss of three of these results in a +3 charge for most REE cations. These are large ions; their size, coupled with their high charges, makes them incompatible.

Europium (Eu) is unique in that its third valence electron does not enter the 5d orbital as in other REEs, but a 4f orbital, which then becomes exactly half full.

This is a particularly stable configuration, and more energy is required to remove the third electron. A consequence of this is that under reducing conditions, such as on the Moon or within the Earth's mantle, Eu may exist in both the divalent and trivalent states (the ratio of Eu^{2+} and Eu^{3+} depends on the oxidation state of the system). Eu^{2+} has about the same size and charge as Ca^{2+}, for which it readily substitutes, so europium, at least in part, may behave as a compatible element.

Under oxidizing conditions, such as in the marine environment, cerium (Ce) is oxidized to Ce^{4+}. This results in a significant contraction in its ionic radius; thus, Ce^{4+} has a very different behavior from trivalent ions of the other REEs. Cerium in the oceans precipitates in manganese nodules, consequently leading to a dramatic Ce depletion in seawater. Marine carbonates formed in this enviroment mimic this geochemical characteristic. Metalliferous sediments deposited at midocean ridges also exhibit Ce depletions, pointing to a seawater source for the hydrothermal solutions that produced them.

The natural abundances of REEs vary by a factor of thirty. To simplify comparisons, REE concentrations in rocks are normalized by dividing them by their concentrations in another rock sample. Chondritic meteorites, which are thought to contain unfractionated element abundances, are the most commonly used standard for REE normalization. The geochemical significance of these meteorites is considered in chapter 15. Shales also generally have uniform lanthanide patterns, and several sets of shale REE abundances are used for normalization in some applications. As we have already discussed earlier in this chapter, shales have geochemical significance because their compositions may represent that of the average crust. Normalization of REE data using chondritic meteorites or shales produces relatively smooth patterns that enable comparisons between samples, accentuating relative degrees of enrichment or depletion and contrasting behavior between neighboring elements in the lanthanide series. Normalized REE concentrations are conventionally shown in diagrams such as that shown in figure 12.22, plotted by increasing atomic number. The elements on the left side of the diagram thus have lower atomic weights and are called the light rare earth elements (LREE) to distinguish them from their heavier counterparts (HREE) on the right. What is important from the standpoint of geochemical behavior, however, is that the lanthanides show a regular contraction in ionic radius from La^{3+} to Lu^{3+}.

Except for Eu^{2+} in reduced systems and Ce^{4+} in oxidized systems, the REEs are all incompatible but, like the first series transition metals, their behavior varies in degree. Some minerals can tolerate certain REEs more easily than others, as shown in figure 12.22. For example, apatite can readily accommodate all rare earth elements. Some minerals discriminate between different rare earth elements, resulting in their fractionation. Garnet has a phenomenal preference for HREEs because of their smaller ionic radii, and pyroxenes show a similar but less pronounced affinity. The extra Eu in plagioclase, accounting for the spike or *positive europium anomaly*, is Eu^{2+} substituting for calcium. Thus K_D values for the REEs in various phases are controlled by size rather than by bonding characteristics.

Studies of REEs have found many uses in geochemistry. They constitute important constraints on models for partial melting in the mantle and magmatic fractionation. The lanthanides are insoluble under conditions at the Earth's surface, so their abundances in sediments reflect those of the average crust. The short residence times for REEs in seawater also allow them to be used to track oceanic mixing processes. More exhaustive treatments of REE geochemistry can be found in Henderson (1983) and Taylor and McLennan (1987).

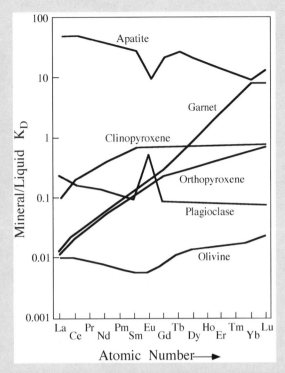

FIG. 12.22. Typical distribution coefficients for REEs between various minerals and basaltic melt. REE concentrations are normalized to values in chondritic meteorites. All of these phases are intolerant of REE (that is, K_D <1) except apatite (and garnet for the HREEs). The positive Eu anomaly in plagioclase occurs because some of this element is in the divalent state and can substitute for Ca^{2+} ions. (After Zielinski 1975.)

A knowledge of compatible and incompatible trace element distribution coefficients can provide a powerful test for geochemical models involving partial melting or crystallization. Some representative K_D values for trace elements in minerals in equilibrium with basaltic magma are presented in table 12.4. An example of such a test is given in worked problem 12.8.

We are now in a position to evaluate how the crust-mantle system has become stratified in terms of its heat-producing radioactive elements. Potassium, uranium, and thorium are all very large ions and behave as incompatible elements. Consequently, any partial melts produced in the mantle scavenges these elements, and over time, they become depleted from mantle residues and concentrated in the crust by ascending magmas. This differentiation of heat-producing elements has had a profound effect on the Earth's thermal history.

Worked Problem 12.8

How can trace element abundances in basalts be used to constrain models for their origin? A geochemical study of volcanic

TABLE 12.4. Representative Trace Element K_D Values for Minerals in Equilibrium with Basaltic Magma

Mineral	Ni	Rb	Sr	Ba	Ce	Sm	Eu	Y
Olivine	10	0.001	0.001	0.001	0.001	0.002	0.002	0.002
Clinopyroxene	2	0.001	0.07	0.001	0.1	0.3	0.2	0.3
Orthopyroxene	4	0.001	0.01	0.001	0.003	0.01	0.01	0.05
Garnet	0.04	0.001	0.001	0.002	0.02	0.2	0.3	4.0
Amphibole	3	0.3	0.5	0.4	0.2	0.5	0.6	0.5
Plagioclase	0.01	0.07	2.2	0.2	0.1	0.07	0.3	0.03

From Cox et al. (1979).

rocks from Reunion Island in the Indian Ocean by Robert Zielinski (1975) provides an excellent example. Measured trace element abundances in eight samples are shown in figure 12.23. All of these samples contain phenocrysts of olivine, clinopyroxene, plagioclase, and magnetite in a groundmass of the same minerals plus apatite. Zielinski used the major element compositions of whole rocks and phenocrysts to construct a fractionation model. Using a petrologic mixing program, he found that lava samples 3–8 could be formed by fractionation of these phenocrysts from assumed parental magma 2. Basalt 1 was not a suitable parental magma for this suite, even though it has the highest Mg/Fe ratio; this rock probably contains cumulus olivine and pyroxene. Zielinski then tested this model using trace elements.

The chondrite-normalized rare earth element patterns (fig. 12.23a) are parallel and increase in the order sample 1 to 8. This is just as expected; incompatible REEs should have higher concentrations in residual liquids, and none of the solid phases fractionates LREEs from HREEs appreciably. Using the K_D values shown in figure 12.22 and the proportions of fractionating phases from the mixing program, it is possible to calculate the bulk distribution coefficient (equation 12.3) for each sample. The mixing program also gives the fraction of liquid F. Assuming equilibrium between the entire separated solid and melt, substitution of these values into equation 12.7 gives the ratio $C_{i\,melt}/C^0_{i\,melt}$. Assume, as Zielinski did, that rock 2 is the parental magma; its trace element concentrations are then $C^0_{i\,melt}$. Thus, we can calculate the expected REE pattern for liquids derived from sample 2. When Zielinski did this exercise, he was able to duplicate the measured REE patterns fairly accurately. The negative Eu anomaly in rock 8 is caused by the removal of large quantities of plagioclase at the end of the crystallization sequence.

The abundances of some other trace elements in these rocks are shown in figure 12.23b. Uranium and thorium are even more incompatible than REEs and increase progressively in the sequence. Barium is also incompatible, but it and strontium can substitute in sodic plagioclase, so both drop towards the end of the sequence as feldspar fractionation becomes dominant. In

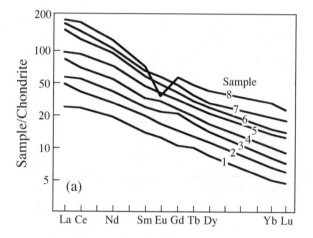

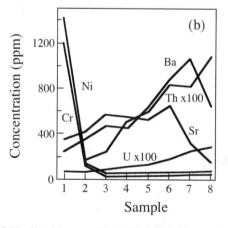

FIG. 12.23. Chondrite-normalized REE patterns (a) and other trace element variations (b) for a suite of volcanic rocks analyzed by Zielinski (1975). The patterns are consistent with a model whereby samples 3–8 were formed by progressive fractional crystallization of a liquid having the composition of sample 2. Sample 1 appears to be a cumulate rock.

contrast, nickel and chromium are compatible and show marked depletions in fractionated rocks. Their contents in rock 1 are much higher than expected for a primary magma, and this observation lends support to the idea that this rock contains cumulus phases. Using appropriate D and F values, we could also predict the abundances of these elements in the hypothetical fractionation sequence. Zielinski was also able to model these fairly well.

Even though Zielinski's modeling may appear convincing, we should note that this is not a unique solution to this problem. The same geochemical patterns might also be produced by various degrees of partial melting of mantle peridotite. Rock 8 would represent a magma formed during a small amount of melting. As melting advanced, the incompatible elements in the melt phase would be progressively diluted, ultimately to produce magma 2. Actually, Zielinski recognized this possibility, but argued against it on the basis of field evidence for shallow differentiation. Like many other tools of geochemistry, trace element modeling may give ambiguous answers, and is at its best when used in combination with other investigative methods.

VOLATILE ELEMENTS

The crust and mantle also interact geochemically by the exchange of volatile elements. Hydrogen, oxygen, carbon, sulfur, and other volatiles in the Earth's interior may be bound into crystal structures, dissolved in magmas, or exist as a discrete fluid phase. We use the term *fluid* because P-T conditions in the mantle and crust are such that this phase is generally above its critical point, where the distinction between liquid and gas is meaningless. Before we can investigate how fluids are cycled within the planet's interior, we should examine their compositions.

Crust and Mantle Fluid Compositions

The compositions of fluids in the crust and mantle can be ascertained from several different lines of evidence. The first is direct chemical analysis of waters from deep wells in geothermal fields such as the Salton Sea, California. In chapter 7, we showed how computer programs such as EQ6 can be used to model the chemical evolution of natural waters. Such sophisticated methods have been used to verify that the measured compositions of these fluids are in equilibrium with the metamorphic minerals in core samples, supporting the idea that these are contemporary metamorphic fluids. It is often difficult, however, to assess the extent to which such fluids may have been contaminated by meteoric water or drilling muds, and sampling of fluids from depths of more than a few kilometers is out of the question.

Fluid inclusions trapped in the minerals of crustal and mantle rocks (both igneous and metamorphic) provide another method of direct observation. Although these inclusions contain only minute quantities of fluids, they can sometimes be extracted for chemical analysis. More commonly, their compositions are inferred from heating and freezing experiments in situ. The experimental manipulation of fluid inclusions in considered further in an accompanying box.

EXTRACTING INFORMATION FROM FLUID INCLUSIONS

Fluids are commonly stranded in interstitial spaces as minerals crystallize. They may be trapped along old crystal boundaries, in fractures, or within pore spaces. With time, these interstitial spaces gradually change shape to minimize their surface energies. The result is a population of more or less equally dimensioned spherical fluid inclusions. Because the minimal surface energy may be best attained by adapting inclusion shape to the structure of the host mineral, some inclusions have *negative crystal* shapes. At the time of trapping, the fluid is usually a single phase liquid or supercritical fluid. The fluid shrinks on cooling, because the coefficient of thermal expansion for the host mineral is much less than for the fluid. As the pressure in inclusions drops below the vapor pressure of the fluid components, vapor bubbles nucleate and grow. Most fluid inclusions, therefore, contain both a liquid and a vapor phase. Solid phases are also commonly encountered in inclusions that have become saturated with respect to one or more salts on cooling. These *daughter crystals* may be valuable clues to the compositions of the parent fluid. It is also common, particularly in sedimentary environments, to find inclusions containing a second, immiscible fluid (usually composed mostly of organic matter). Many of these features are illustrated in the inclusion

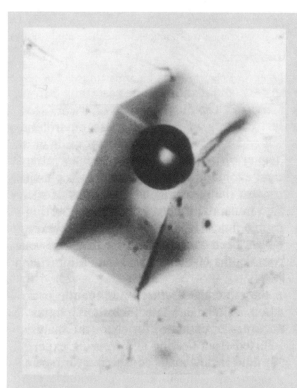

FIG. 12.24. Photomicrograph of a fluid inclusion in fluorite from southern Illinois. The inclusion measures ~200 mm in its longest dimension, and consists of a NaCl-rich brine and a gas bubble. Note that the inclusion has a shape that mimics the host fluorite crystal.

geothermometry is based on the assumption that the original sealed fluid was a single, homogeneous phase; an observational test of this assumption is that all inclusions trapped simultaneously should have the same bulk composition and therefore the same volumetric ratio of phases at room temperature.

The fluids trapped in large inclusions can be sampled by drilling into them and extracting them with a micropipette. Their compositions can be determined directly by various analytical techniques. Fluids in small inclusions present a much greater challenge. If only one generation of inclusions is present, the fluid they contain may be released by crushing, but some information can also be gleaned from nondestructive observations. Like the salt used to melt ice on roads in winter, the salinity of the fluid depresses its freezing point relative to pure water. Salinity can therefore be determined by measuring the fluid's freezing point. This can be done on the same microscope stage used for the heating experiments by passing liquid nitrogen through a cooling jacket on the stage. Freezing point depression is a function of total salinity and the identity of the salt in solution. Generally, it is assumed that inclusions are primarily NaCl solutions, and salinity

in figure 12.24. The evolution of such an inclusion is summarized in figure 12.25, which resembles the phase diagram for water.

Experimental manipulation of fluid inclusions can provide constraints on fluid composition and the conditions of trapping. Phase transformations that occur with changing temperature can be observed by fitting a heating stage to a microscope, in effect observing the cooling process (fig. 12.25) in reverse. As an inclusion is heated, daughter crystals redissolve and the vapor bubble shrinks and finally disappears. The temperature of homogenization is an approximate measurement of the temperature at which the fluid was trapped, provided that the fluid was very near its boiling point. If the trapped fluid was supercritical, its trapping temperature may be estimated from the homogenization temperature and an independent estimate of confining pressure. Fluid inclusion

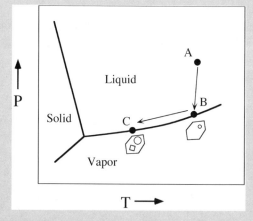

FIG. 12.25. *P-T* diagram showing the evolution of a fluid inclusion. The fluid is trapped at *A*. Because it occupies a space of constant volume, cooling causes the fluid pressure to fall from *A* to *B*. At *B*, the fluid becomes saturated with vapor, and a bubble nucleates and grows as the inclusion cools along the liquid-vapor boundary. The inclusion becomes saturated with a solute at *C*, and a daughter crystal (of, say, halite) nucleates and grows with further cooling.

is calculated in terms of wt % NaCl equivalent. More sophisticated studies involve spectroscopic measurements of inclusions to provide further information on fluid compositions. Daughter crystals may be identified by using optical properties, such as color, pleochroism, and crystal habit, as well as melting points.

As with most geochemical techniques, there are many potential pitfalls in fluid inclusion work. Leakage of fluid into or out of inclusions after trapping is always of great concern. Care must be taken to identify multiple generations of inclusions by optical methods, because each generation will presumably

have its own specific composition and trapping temperature. Recrystallization of the host mineral may cause several generations of inclusions to coalesce into large inclusions, or separation of large inclusions into smaller ones (*necking*). This latter process is critical if inclusions consist of more than one phase, because the phases may not be redistributed uniformly. These potential problems can be overcome in most cases, however, so that fluid inclusions studies offer a unique source of information on fluids from the Earth's interior. Shepherd et al. (1985) provide practical information on how to study fluid inclusions.

A third method of determining fluid compositions is by studying metamorphic assemblages that depend on fluid composition. Characterization of fluids in this case is based on the thermodynamic principles that we considered in chapter 9. For example, for the equilibrium reaction:

$$Al_2Si_4O_{10}(OH)_2 \rightleftarrows Al_2SiO_5 + 3SiO_2 + H_2O,$$
$$\text{pyrophyllite} \quad \text{andalusite} \quad \text{quartz} \quad \text{fluid}$$

we know that:

$$\mu_{Al_2Si_4O_{10}(OH)_2, \text{ pyroph}} = \mu_{Al_2SiO_5, \text{ and}} + 3\mu_{SiO_2, \text{ qtz}} + \mu_{H_2O, \text{ fluid}}.$$

Evaluation of the chemical potentials in the solid phases allows calculation of $\mu_{H_2O, \text{ fluid}}$. This can be converted into fluid composition ($X_{H_2O, \text{ fluid}}$) if the relationship between chemical potential and composition is known from an appropriate equation of state for the fluid. Worked problem 12.9 illustrates how this is done.

Worked Problem 12.9

John Ferry (1976) determined the composition of fluids during regional metamorphism in south-central Maine. Let's examine his calculations for zoisite-bearing rocks, in which the following reaction occurs:

$$2 \, Ca_2Al_3Si_3O_{12}(OH) + CO_2 = 3 \, CaAl_2Si_2O_8 + CaCO_3$$
$$\text{zoisite} \qquad\qquad \text{anorthite} \quad \text{calcite}$$
$$+ H_2O.$$

At equilibrium,

$$3(\Delta\bar{G}_{An}^0 + RT \ln a_{An, \text{ plag}}) + \Delta\bar{G}_{Cc}^0 + RT \ln a_{Cc, \text{ calcite}}$$
$$+ \Delta\bar{G}_{H_2O}^0 + RT \ln a_{H_2O, \text{ fluid}} = \qquad (12.12)$$
$$2(\Delta\bar{G}_{Zo}^0 + RT \ln a_{Zo, \text{ zoisite}}) + \Delta\bar{G}_{CO_2}^0 + RT \ln a_{CO_2, \text{ fluid}}.$$

In this equation, the $\Delta\bar{G}_i^0$ terms refer to standard free energy values at P and T. By recalling that:

$$(\Delta\bar{G}_i^0)_{P,T} = (\Delta\bar{G}_i^0)_{1,T} + \int_1^P \Delta\bar{V}_i^0 \, dP,$$

we can rewrite equation 12.12 as:

$$(3\Delta\bar{G}_{An}^0 + \Delta\bar{G}_{Cc}^0 + \Delta\bar{G}_{H_2O}^0 - 2\Delta\bar{G}_{Zo}^0 - \Delta\bar{G}_{CO_2}^0)$$
$$+ \int_1^P (3\bar{V}_{An}^0 + \bar{V}_{Cc}^0 - 2\bar{V}_{Zo}^0) \, dP$$
$$+ RT \ln [(a_{An, \text{ plag}}^3 \, a_{Cc, \text{ calcite}})/a_{Zo, \text{ zoisite}}^2] \qquad (12.13)$$
$$- RT \ln f_{CO_2} + RT \ln f_{H_2O} = 0.$$

This is an equation of the general form:

$$\Delta\bar{G}^0 + \Delta\bar{V}_{\text{solids}}^0 (P - 1) + RT \ln K - RT \ln f_{CO_2} \qquad (12.14)$$
$$+ RT \ln f_{H_2O} = 0.$$

The first two terms in this equation can be evaluated from:

$$\Delta\bar{G}^0 + \Delta\bar{V}_{\text{solids}}^0 (P - 1) = \Delta H^0 - T\Delta\bar{S}^0 + P\Delta\bar{V}_{\text{solids}}^0,$$

because the difference between P and $P - 1$ is negligible in geologic problems. P and T can be estimated from mineral assemblage stability fields or from geothermometers and barometers. The compositions of coexisting minerals can be used to determine K; however, in this case, we require two reactions to be solved simultaneously for the two unknowns f_{CO_2} and f_{H_2O}.

For zoisite + calcite + plagioclase rocks in Maine, Ferry estimated metamorphic P and T values of 3500 bar and 798–710 K, respectively. The ratio f_{CO_2}/f_{H_2O} can be determined from the reaction discussed above. Using estimated thermodynamic data and K values obtained from coexisting mineral compositions at various temperatures, Ferry determined f_{CO_2}/f_{H_2O} ratios ranging from 0.53 to 5.10 throughout his field area. These rocks coexist with other assemblages that can be used to estimate either f_{CO_2} or f_{H_2O}, and thus both variables can be determined by solving simultaneous equations.

During prograde metamorphism, devolatilization reactions commonly release enough fluid that we can safely assume the condition:

$$P_{fluid} = P_{H_2O} + P_{CO_2} = P_{total}.$$

It follows that:

$$P_{H_2O} + P_{CO_2} = f_{H_2O}/\gamma_{H_2O} + f_{CO_2}/\gamma_{CO_2} = 3500 \text{ bar.} \quad (12.15)$$

If we assume that the gas mixture was ideal,

$$f_i = f_i^0 \, X_i = \gamma_i^0 \, PX_i = \gamma_i^0 \, P_i. \quad (12.16)$$

From these relationships, Ferry determined values for $X_{CO_2 \text{ fluid}}$ ranging from 0.06 to 0.32 for his assemblages. This range of values indicates that gradients in fluid composition existed on an outcrop scale, possibly due to mixing of several fluids.

The compositions of fluids determined from fluid inclusions and calculated from metamorphic equilibria can generally be represented by the system H-O-C-S, although chloride complexes may also be present. The major species in naturally occurring fluids are H_2O, CO_2, CH_4, and H_2S. Although the species O_2 and S_2 actually occur in vanishingly small quantities, it is common practice to characterize fluid compositions by defining the fugacities of these species. These variables can serve as useful monitors for oxidation and sulfidation processes in such fluids. The relative proportions of these species shift constantly as intensive variables for the system change. Figure 12.26 illustrates how the species in a fluid of fixed composition alter in response to changing temperature.

Fluids in the upper crust are typically H_2O-rich except in carbonate rocks, where CO_2 may predominate. Granulites from the middle and lower crust have fluid inclusions dominated by CO_2 and, to a lesser extent, by CO and CH_4. The most abundant volatile species in fluid inclusions from mantle xenoliths is CO_2, although some inclusions from the same specimens may be dominated by H_2O. Magmatic gases in tholeiitic basalts have higher H_2O/CO_2 ratios than those from alkaline lavas that may have formed by melting at deeper levels.

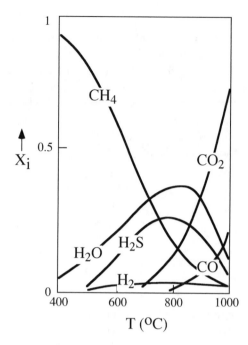

FIG. 12.26. The effect of temperature on speciation of fluids in the system H-C-O-S coexisting with graphite at 2 kbar pressure. The fugacities of O_2 and S_2 are fixed by appropriate buffers. (Based on calculations by Holloway 1981.)

Mantle and Crust Reservoirs for Fluids

The abundances of volatiles in the mantle are difficult to specify, but analyses of dissolved fluids and fluid inclusions in basalts provide some constraints. For example, the water content of the MORB source region has been estimated at 140–350 ppm, whereas the deeper mantle sources for ocean island basalts are slightly wetter, with 450–525 ppm water. The solubility of water and other volatile species in most upper mantle minerals is low or modest. However, it has been shown experimentally that β-olivine (wadsleyite) and γ-olivine (ringwoodite) can accommodate as much as 2 wt % water. Some hydrous magnesian silicates, such as serpentine and chlorite, are also stable at pressures appropriate for the lower reaches of the upper mantle. The Mg-perovskite that dominates the lower mantle cannot contain much water. Thus, it is possible that the mantle is stratified in terms of its water storage capacity, with a relatively wet zone at 350–650 km depth sandwiched between drier upper and lower mantle rocks.

In metamorphic rock systems within the crust, it is common to express the proportion of fluid in terms of a fluid pressure (P_{fluid}) relative to lithostatic pressure ($P_{lithostatic}$). The upper and lower crusts are characterized by different hydrologic regimes that are thought to be determined by rock strength. Strong upper crustal rocks can maintain networks of open pores, so that P_{fluid} is independent of $P_{lithostatic}$ and tends to follow the local hydrostatic gradient. As a consequence, crustal fluids can flow downward or upward between these pores and thus will circulate vertically and horizontally. Lower crustal rocks, however, tend to deform plastically, so that porosity can be maintained only if the pores are filled with fluids at or close to $P_{lithostatic}$. Fluids in the lower crust, then, tend to follow the lithostatic pressure gradient and therefore always migrate upward. Following this logic, fluids in the lower crust are ephemeral.

Cycling of Fluids between Crust and Mantle

Volatile elements are readily transported as dissolved fluid species in magmas. Significant quantities of mantle-derived fluids are expelled from magmas erupting at midocean ridges and subduction zones. The movement of a fluid in this way depends critically on its solubility in magma. H_2O is very soluble for all magma compositions, but CO_2 solubility depends more critically on magma composition and pressure. Most magmas become saturated with respect to volatiles during ascent, causing the formation of a discrete fluid phase at some point. Once a fluid phase has formed, all of the dissolved volatiles partition between it and the magma, and the fluid phase is likely to escape into the surrounding crust or atmosphere. Magmas provide transport for volatiles in only one direction—upward.

Metamorphism occurs on the seafloor near midocean ridges, as warm basaltic lavas react with seawater. Fluids trapped in these rocks are ultimately recycled into the mantle within subducted slabs. At depths appropriate for the eclogite transformation, dehydration reactions occur and fluids are released. These may rise into the mantle wedge above the subducted plate and promote melting. The resultant magmas dissolve these volatiles and carry them again into the crust. Alternatively, some fluids in subducted slabs may endure in stable hydrous minerals, which can be sequestered for a time near the boundary between the upper and lower mantles.

Regional metamorphism is associated with collisions between continents. The sedimentary rocks caught in this predicament contain considerable quantities of crustal fluids. High-grade metamorphism can drive off most of the fluid component, and the effect of recycling these volatiles into the atmosphere may be dramatic. For example, the collision of India and Asia ~50 million years ago was coincident with the warmest epoch in the Cenozoic. Kerrick and Caldeira (1993) suggested that metamorphism liberates huge quantities of CO_2 (>10^{18} moles per million years) by such reactions as $CaCO_3 + SiO_2 \rightarrow CaSiO_3 + CO_2$. Expulsion of CO_2 into the atmosphere then caused global warming by the greenhouse effect. In contrast, Selverstone and Gutzler (1993) observed that high-pressure metamorphic rocks in the Alps contain carbonates and graphite that was buried to depths of >50 km during the Alpine orogeny. In this case, carbon remained sequestered in the lower crust, and they argued that global cooling was the result. Although the net effect of metamorphism on fluid cycling is debated, it is clear that metamorphic reactions at compressional plate boundaries can cycle fluids into and out of the crust.

Near the beginning of this chapter, we noted that granulites may constitute a major part of the lower crust. The CO_2-rich fluid inclusions that characterize granulites, as well as the evidence for low a_{H_2O}, suggest that these rocks may form when the lower crust is invaded by CO_2. It is possible that such CO_2-dominated fluids are due to mantle degassing, possibly providing another mechanism (streaming from mantle to crust, independently of plate tectonics) for the cycling of fluids. During metamorphism, continuous fluxing with CO_2 must take place, because dehydration reactions produce water that would otherwise dilute the fluid. Significant quantities of CO_2 are required over the metamorphic episode, but the actual amount of CO_2 in the rocks at any time may be small, because the fluids occur as films at grain boundaries.

SUMMARY

The compositions of the crust, mantle, and core are so different that it may be tempting to surmise that they do not interact, but we have seen that geochemical exchange does occur on a global scale. Ascending magmas formed by partial melting of mantle peridotite carry mantle components into the crust. Mantle melting is promoted by local temperature increases, depressurization, or fluxing by fluids. Incompatible and volatile elements are partitioned into these melts, so that they are ultimately deposited in the crust. The geochemical record in igneous rocks is commonly obscured by fractional crystallization or other processes that alter magma compositions during ascent to the surface.

Crustal materials are carried downward in subducted plates, where they may be stranded indefinitely. However, much of the volatile element content in these slabs may be recycled because of dehydration or decarbonation reactions during metamorphism in the mantle. Only oceanic crust is subducted; the amount of buoyant continental crust appears to have increased episodically with time.

With this background in the processes and pathways by which materials in the Earth's interior interact, we are now in a position to explore another tool for formulating and testing geochemical interactions in these regions. In the following chapters on isotopes, we build on the information just presented.

SUGGESTED READINGS

The wealth of excellent books and papers on the subjects in this chapter makes selection of a few difficult. We recommend the following literature enthusiastically.

Anderson, D. L. 1989. *Theory of the Earth*. Oxford: Blackwell. (This thoughtful book weaves together numerous constraints from geophysics and geochemistry to provide much detail on the composition of the Earth's mantle.)

Carlson, R. W. 1994. Mechanisms of Earth differentiation: Consequences for the chemical structure of the mantle. *Review of Geophysics* 32:337–361. (A superb review of constraints on mantle geochemistry and the processes that account for its variations.)

Cox, K. G., J. D. Bell, and R. J. Pankhurst. 1979. *The Interpretation of Igneous Rocks*. London: Allen and Unwin. (Chapter 6 gives a superb account of element variation diagrams, and chapter 14 discusses trace element fractionation.)

Ferry, J. M., ed. 1982. *Characterization of Metamorphism through Mineral Equilibria*. American Reviews in Mineralogy 10. Washington, D.C.: American Mineralogical Society. (An up-to-date manual of the quantitative geochemical aspects of metamorphism; chapters 6 [by J. M. Ferry and D. M. Burt] and 8 [by D. Rumble III] provide very readable information on fluids in metamorphic systems.)

Fyfe, W. S., N. J. Price, and A. B. Thompson. 1978. *Fluids in the Earth's Crust*. Amsterdam: Elsevier. (Chapter 2 summarizes the compositions of naturally occurring fluids, and other chapters describe their generation and transport.)

Hargraves, R. B., ed. 1980. *Physics of Magmatic Processes*. Princeton: Princeton University Press. (A high-level but very informative book about the quantitative aspects of magmatism; chapter 4 [by S. R. Hart and C. J. Allegre] presents

trace element constraints on magma genesis, and chapter 5 [by E. R. Oxburgh] describes the relationship between heat flow and melting.)

Henderson, P. 1983. *Rare Earth Element Geochemistry*. Developments in Geochemistry 2. Amsterdam: Elsevier. (A book devoted to research developments in the field of REEs; chapter 4 [by L. A. Haskin] provides an in-depth look at petrogenetic modeling using these elements.)

Jeanloz, R. 1990. The nature of the Earth's core. *Annual Reviews of Earth and Planetary Sciences* 18:357–386. (An interesting review of geophysical research bearing on the properties of the core; this paper contains some controversial ideas about reactions at the core-mantle boundary).

Philpotts, A. R. 1990. *Principles of Igneous and Metamorphic Petrology*. Englewood Cliffs: Prentice-Hall. (Chapter 22 of this superb petrology text gives an up-to-date explanation of how melting occurs in the mantle, and magmatic processes are treated in chapter 13.)

Ringwood, A. E. 1975. *Composition and Petrology of the Earth's Mantle*. New York: McGraw-Hill. (An exhaustive, although now somewhat dated, monograph on the nature of the mantle; chapter 5 describes the pyrolite model, and later chapters describe experimental constraints on mantle mineralogy.)

Silver, P. G., R. W. Carlson, and P. Olson. 1988. Deep slabs, geochemical heterogeneity, and the large-scale structure of mantle convection: Investigation of an enduring paradox. *Annual Reviews of Earth and Planetary Sciences* 16:477–541. (A very thoughtful review of geophysical and geochemical constraints on the mantle.)

Taylor, S. R., and S. M. McClennan. 1985. *The Continental Crust: Its Composition and Evolution*. Oxford: Blackwell Scientific. (Everything you might want to know about the composition and origin of the Earth's crust, presented in a very readable manner.)

Walther, J. V., and B. J. Wood, eds. 1986. *Fluid-Rock Interactions during Metamorphism*. Advances in Geochemistry 5. New York: Springer-Verlag. (Chapter 1 [by M. L. Crawford and L. S. Hollister] provides a summary of fluid inclusion research.)

Wood, B. J., and D. G. Fraser. 1977. *Elementary Thermodynamics for Geologists*. Oxford: Oxford University Press. (Chapter 6 presents derivations of equations for trace element behavior.)

Yoder, H. S. Jr., ed. 1979. *The Evolution of the Igneous Rocks. 50th Anniversary Perspectives*. Princeton: Princeton University Press. (An updated version of the classic treatise by N. L. Bowen; chapters 2 [by E. Roedder] and 3 [by D. C. Presnall] describe liquid immiscibility and fractional crystallization and melting, and chapter 17 [by P. J. Wyllie] considers magma generation.)

The following publications are also cited in this chapter and provide much more detail for the interested student:

Anderson, D. L. 1980. The temperature profile of the upper mantle. *Journal of Geophysical Research* 85:7003–7010.

Asimow, P. D., M. M. Hirschmann, M. S. Ghiorso, M. J. O'Hara, and E. M. Stolper. 1995. The effect of pressure-induced solid-solid phase transitions on decompression melting of the mantle. *Geochimica et Cosmochimica Acta* 59:4489–4506.

Burns, R. G. 1970. *Mineralogical Applications of Crystal Field Theory*. Cambridge: Cambridge University Press.

Castillo, P. 1988. The Dupal anomaly as a trace of the up-welling lower mantle. *Nature* 336:667–670.

Ferry, J. M. 1976. P, T, f_{CO_2}, and f_{H_2O} during metamorphism of calcareous sediments in the Waterville-Vassalboro area, south-central Maine. *Contributions to Mineralogy and Petrology* 57:119–143.

Holloway, J. R. 1981. Compositions and volumes of supercritical fluids in the earth's crust. In L. S. Hollister and M. L. Crawford, eds. *Fluid Inclusions: Applications to Petrology*. Calgary: Mineralogical Association of Canada, pp. 13–38.

Hutchison, R. 1974. The formation of the Earth. *Nature* 250:556–568.

Kerrick, D., and K. Caldeira. 1993. Paleoatmospheric consequences of CO_2 released during early Cenozoic regional metamorphism in the Tethyan orogen. *Chemical Geology* 108:201–230.

McBirney, A. R. 1969. Compositional variations in Cenozoic calc-alkaline suites of Central America. *Oregon Department of Geology Mineral Industries Bulletin* 65:185–189.

McLennan, S. M. 1982. On the geochemical evolution of sedimentary rocks. *Chemical Geology* 37:335–350.

Presnall, D. C., S. A. Dixon, J. R. Dixon, T. H. O'Donnell, N. L. Brenner, R. L. Schrock, and D. W. Dycus. 1978. Liquidus phase relations on the join diopside-forsterite-anorthite from 1 atm to 20 kbar: Their bearing on the generation and crystallization of basaltic magma. *Contributions to Mineralogy and Petrology* 66:203–220.

Ringwood, A. E. 1962. A model for the upper mantle. *Journal of Geophysical Research* 67:857–866.

Robie, R. A., B. S. Hemingway, and J. R. Fisher. 1978. Thermodynamic properties of minerals and related substances at 298.15 K and 1 bar (105 pascals) pressure and at higher temperatures. Geological Survey Bulletin 1452. Washington, D.C.: U.S. Geological Survey.

Ronov, A. B., and A. A. Yaroshevsky. 1969. Chemical composition of the earth's crust. In P. J. Hart, ed. *The Earth's Crust and Upper Mantle*. Monograph 13. Washington, D.C.: American Geophysical Union.

Ronov, A. B., N. V. Bredanova, and A. A. Migdisov. 1988. General compositional-evolutionary trends in continental crust sedimentary and magmatic rocks. *Geochemistry International* 25:27–42.

Selverstone, J., and D. S. Gutzler. 1993. Post-125 Ma carbon storage associated with continent-continent collision. *Geology* 21:885–888.

Shaw, D. M. 1977. Trace element behavior during anatexis. *Oregon Department of Geology Mineral Industries Bulletin* 96:189–213.

Shepherd, T. J., A. H. Rankin, and D.H.M. Alderton. 1985. *A Practical Guide to Fluid Inclusion Studies*. London: Blackie.

Stolper, E., and D. Walker. 1980. Melt density and the average composition of basalt. *Contributions to Mineralogy and Petrology* 74:7–12.

Taylor, S. R., and S. M. McLennan. 1987. The significance of the rare earths in geochemistry and cosmochemistry. *Handbook on the Physics and Chemistry of Rare Earths* 13. Amsterdam: North-Holland.

White, I. G. 1967. Ultrabasic rocks and the composition of the upper mantle. *Earth and Planetary Science Letters* 3:11–18.

Zielinski, R. A. 1975. Trace element evaluation of a suite of rocks from Reunion Island, Indian Ocean. *Geochimica et Cosmochimica Acta* 39:713–734.

PROBLEMS

(12.1) Describe the fractional crystallization path for a liquid with 45 wt. % SiO_2 in the system forsterite-silica (see fig. 12.8).

(12.2) Describe the equilibrium crystallization path for a liquid with 72 wt. % SiO_2 in the system forsterite-silica (see fig. 12.8).

(12.3) Describe the phase assemblages in the melting sequence under equilibrium and fractional melting conditions for a composition of Fo_{50} in the system forsterite-fayalite (fig. 12.14).

(12.4) (a) Sketch qualitative chemical variation diagrams for CaO versus MgO and Al_2O_3 versus SiO_2 for liquid lines-of-descent during fractionation of compositions x and y in figure 12.15. (b) How would these variation diagrams be affected if the system also contained FeO? (Hint: what effect would this have on olivine and orthopyroxene?) (c) Describe the sequence of rocks produced by perfect fractional crystallization of these two compositions.

(12.5) A basaltic magma has the following trace element abundances: Ni, 1100 ppm; Rb, 100 ppm; Sr, 300 ppm; Ba, 220 ppm; Ce, 82 ppm; Sm, 9.4 ppm; Eu, 3.3 ppm; and Yb, 3.3 ppm. Fractional crystallization of 20 wt. % olivine, 12 wt. % orthopyroxene, and 10 wt. % plagioclase from this magma results in a residual liquid that erupts on the surface. Using the distribution coefficients in table 12.4, calculate the trace element contents of this erupted melt.

(12.6) The following are REE abundances in chondritic meteorites: Ce, 0.616 ppm; Sm, 0.149 ppm; Eu, 0.056 ppm; and Yb, 0.159 ppm. Construct a chondrite-normalized REE pattern for the parent magma and residual liquid in problem 12.5, following the example in figure 12.23. How would you explain the general shapes of these patterns?

(12.7) A spinel peridotite has a solidus temperature of 1500°C at 20 kbar. (a) Using figure 12.9, describe the phases involved in its melting as a mantle diapir of this material rises adiabatically to a depth of 40 km. (b) How does the liquid composition, in terms of normative olivine, change over this interval?

CHAPTER 13

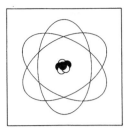

USING STABLE ISOTOPES

OVERVIEW

This is the first of two chapters on isotope geochemistry. In it, we discuss how geochemists use stable isotopes of hydrogen, carbon, nitrogen, oxygen, and sulfur to interpret geologic processes and environments. You may want to review portions of chapter 2 to refresh your familiarity with the language and some of the basic concepts in nuclear chemistry. With the aid of several case studies, we show how those and other concepts can be applied to problems of practical interest. In each case study, the common principle is mass fractionation. We explain how the different stable isotopes of a single element can separate from each other in a variety of thermal or biochemical processes, and how we can trace the chemical history of a system by measuring the abundance ratios of these isotopes in coexisting phases. Geochemists interpret some fractionation processes by using the familiar tools of equilibrium thermodynamics. Among these processes are simple isotopic exchange reactions that have been used successfully as geothermometers. Other isotope fractionations are kinetically controlled. In many cases, abundance patterns produced by geologic and biologic processes can yield insights into the pathways along which isotopic changes have occurred.

HISTORICAL PERSPECTIVE

Some of the most dramatic advances in the modern geologic sciences have been made in the field of isotope geochemistry. It is a discipline with a surprisingly short history, having attracted talented geochemists in large numbers since only about 1950. The roots of the field lie barely half a century before that, in the work of Henri Becquerel and the Curies, Marie and Pierre. Their studies of radioactivity led British geologist Arthur Holmes and others to create the new field of geochronology, which occupies much of our attention in chapter 14. That work was well underway, however, before chemists knew much about the structure of the atom.

Frederick Soddy hypothesized the existence of isotopes in 1913, attempting to explain why some atoms of a single element weighed more than others. Even with that theoretical step forward, it was 1932 before the neutron was discovered and Soddy's hypothesis was confirmed. Not by coincidence, a dramatically new emphasis in geochemistry began that same year with Harold C. Urey's discovery of deuterium (^{2}H). For that discovery and for showing that the isotopes of hydrogen vary in relative abundance from one geologic environment to another, Urey won the 1934 Nobel Prize. Thus opened the field of stable isotope chemistry.

As we noted in chapter 1, Alfred O. C. Nier's refinement of the mass spectrometer led to an explosion of knowledge about isotopes in the years before World War II. By the late 1940s, Urey and his students at the University of Chicago were pioneers in the use of oxygen isotopes for geothermometry and in an increasing number of provenance studies. Samuel Epstein and his students at Cal Tech further refined that work and applied it to the study of ore deposits. Canadian geochemist Harry Thode, working at McMaster University in the 1950s, developed techniques for extracting sulfur isotopes from ore minerals, opening yet another fruitful class of studies in stable isotope geochemistry. In the ensuing half-century, geochemists have used isotopes not only of oxygen and sulfur, but also carbon, nitrogen, silicon, and boron to trace chemical pathways in the Earth. It is difficult nowadays to find a study of ore genesis, crustal evolution, or global geochemical cycles that does not include consideration of stable isotopes.

WHAT MAKES STABLE ISOTOPES USEFUL?

Interpretations based on studies of stable isotopes depend on the ability of various kinetic and equilibrium processes to separate (fractionate) light from heavy isotopes. By understanding the mechanisms involved, we can measure the relative abundances of light and heavy isotopes of an element in natural materials and interpret the pathways and environments they have undergone. Isotopes of H, C, N, O, and S share a set of characteristics that make them particularly well suited for this sort of work:

1. They all have low atomic mass. Nuclides with $Z > 16$ generally do not fractionate efficiently in nature.

2. The relative difference in mass between heavy and light isotopes of each of these nuclides is fairly large. The difference is largest for hydrogen, whose heavy isotope, deuterium (^2H, also given the symbol D) is twice as massive as the light isotope, ^1H. At the other extreme on our short list of stable nuclides, ^{34}S is still 6.25% heavier than ^{32}S. Because isotopes fractionate on the basis of relative mass, this is an important criterion.

3. All five elements are abundant in nature and constitute a major portion of common Earth materials. Oxy-

gen, in particular, makes up almost 47% of the crust by weight (nearly 92% by volume).

4. Carbon, nitrogen, and sulfur exist in more than one oxidation state and may therefore participate in processes over a wide range of redox conditions.

5. Each of the five elements forms bonds with neighboring atoms that may range from ionic to highly covalent. As we show shortly, fractionation of heavy and light isotopes is greatest between phases that have markedly different bond types or bond strengths. For this reason, most other light elements like magnesium or aluminum, which form the same type of bonds in almost all common Earth materials, do not show marked isotopic fractionation.

6. For each of the five elements, the abundance of the least common isotope still ranges from a few tenths of a percent to a few percent of the most common isotope. From an analytical point of view, this means that the precision of measurements is quite high.

7. All are biogenic elements, and thus can be fractionated by biologic processes.

Geochemists commonly use three different types of notation to express the degree of isotopic fractionation. First, they describe the distribution of stable isotopes between coexisting phases A and B in terms of a *fractionation factor*, α_{A-B}, defined by:

$$\alpha_{A-B} = R_A/R_B, \tag{13.1}$$

in which R is the ratio of the heavy to the light isotope in the phase indicated by the subscript. The fractionation of ^{18}O and ^{16}O between quartz and magnetite, for example, would be indicated by the magnitude of:

$$\alpha_{qz-mt} = (^{18}O/^{16}O)_{qz}/(^{18}O/^{16}O)_{mt}.$$

Concentrations, rather than activities, are used in calculating fractionation factors, because activity coefficients for isotopes of the same element are nearly identical and would cancel each other. Values for α_{A-B} are commonly between 1.0000 and 1.0040 for inorganic processes and higher for biologic processes.

Because it is inconvenient to use absolute isotopic ratios (R_A/R_B) and because α_{A-B} values commonly differ only in the third or fourth decimal place, geochemists commonly use a second notation, δ (delta), instead. With this convention, the isotopic ratio in a sample is compared with the same ratio in a standard, using the formula:

$$\delta = 1000 \times (R_{sample} - R_{standard})/R_{standard}. \quad (13.2)$$

The numerical value that results from this procedure is a measure of the deviation of R in parts per thousand, or *per mil* (‰, by analogy with percent) between the sample and the standard. Samples with positive values of δ are said to be *isotopically heavy* (that is, they are enriched in the heavy isotope relative to the standard); those with negative values are *isotopically light*.

For each isotopic system, laboratories have agreed on one or two readily available substances to serve as standards. For hydrogen and oxygen, the universal standard is Standard Mean Ocean Water (SMOW). In older studies, $\delta^{18}O$ values were commonly referred to the isotopic ratio in a Cretaceous marine belemnite fossil from the Pee Dee formation in South Carolina. Today, the PDB scale for ^{18}O is only used in paleoclimatology studies. Carbon isotopic measurements, however, are almost always compared with $^{13}C/^{12}C$ in the Pee Dee Belemnite, so most people think of PDB as a standard for carbon rather than oxygen. (In the literature since the 1980s, SMOW and PDB are called V-SMOW and V-PDB, referring to reference standards available from the International Atomic Energy Agency [IAEA] in Vienna. Comparisons between these standards, with earlier samples of SMOW and PDB, and with other commonly used oxygen standards can be found in Coplen et al. [1983].) Nitrogen measurements are all reported relative to the $^{15}N/^{14}N$ ratio in air, and sulfur measurements are reported relative to the $^{34}S/^{32}S$ ratio in troilite (FeS) from the Canyon Diablo meteorite, whose impact produced Meteor Crater in Arizona. In practice, each laboratory develops its own standards, which are then calibrated against these universal standards.

Finally, geochemists also use the symbol Δ to compare δ values for coexisting substances. We can derive this third quantity from either of the other systems of notation. From the definition of $\alpha_{A\text{-}B}$, we see that:

$$\alpha_{A\text{-}B} - 1 = (R_A - R_B)/R_B.$$

From the definition of δ, we also see that:

$$\delta_A = 1000 \times (R_A - R_{standard})/R_{standard}$$

and

$$\delta_B = 1000 \times (R_B - R_{standard})/R_{standard},$$

so:

$$\alpha_{A\text{-}B} - 1 = (\delta_A - \delta_B)/(1000 + \delta_B).$$

Because δ_B is small compared with 1000, we can write this last result as:

$$1000(\alpha_{A\text{-}B} - 1) \approx \delta_A - \delta_B.$$

For light stable isotope pairs with fractionation factors on the order of 1.000–1.004 (all except D-H), it is a very good approximation to write:

$$\ln \alpha_{A\text{-}B} \approx \alpha_{A\text{-}B} - 1.$$

(The natural logarithm of 1.004, for example, is 0.00399.) With this approximation, we finally see that:

$$1000 \ln \alpha_{A\text{-}B} \approx \delta_A - \delta_B = \Delta_{A\text{-}B}. \quad (13.3)$$

In this way, fractionation factors between coexisting minerals can be calculated from their respective δ values, measured relative to a universal laboratory standard.

Worked Problem 13.1

Atmospheric CO_2 has a $\delta^{13}C$ value of –7‰. If HCO_3^- in a sample of river water assumed to be in equilibrium with the atmosphere has a measured value of +1.24‰, what are the equivalent values of Δ and the fractionation factor $\alpha_{HCO_3^-\text{-}CO_2}$?

We can easily substitute these values into equation 13.3 to see that:

$$\delta_{HCO_3^-} - \delta_{CO_2} = +1.24‰ - (-7‰) = +8.24‰,$$

and

$$\alpha_{HCO_3^-\text{-}CO_2} = \exp(\Delta_{HCO_3^-\text{-}CO_2}/1000) = 1.00827.$$

Fractionation factors and delta values can be derived in several different ways, generally yielding compatible results. Experiments involving isotopic exchange can be performed in a laboratory setting, and fractionation can be determined by analyzing the run products. In some cases, geochemists can also use isotopic analyses from pairs of natural materials if they have independent information about the conditions of isotopic exchange between them. For many situations, however, values of α are calculated from principles of statistical mechanics, using spectroscopic data. Our approach to thermo-dynamics in this textbook has been macroscopic rather than statistical, so we have not laid the foundation for an in-depth discussion of this third technique. Because it helps explain the thermodynamic driving force for isotopic fractionation, however, let's take a brief detour into processes at the molecular level.

MASS FRACTIONATION AND BOND STRENGTH

Fractionation among isotopes occurs because some thermodynamic properties of materials depend on the masses of the atoms of which they are composed. In a solid or liquid, the internal energy is related largely to the stretching or vibrational frequency of bonds between the atom and adjacent ligands. The vibrational frequency of a molecule, in turn, is inversely proportional to the square root of its mass (see accompanying box). There-fore, if two stable isotopes of a light element are distrib-uted randomly in molecules of the same substance, the vibrational frequency associated with bonds to the lighter isotope will be greater than that with bonds to the heav-ier one. As a result, molecules formed with the lighter isotope will have a higher internal energy and be more easily disrupted than molecules with the heavier one.

The effect of isotopic composition on the physical properties of a phase can be surprisingly large, as you can see in table 13.1. Molecules containing the lighter isotope are more readily extracted from a material dur-ing such processes as melting or evaporation. In general, then, the lighter isotope of pairs such as D-H or ^{18}O-^{16}O tends to fractionate into a vapor phase rather than a liq-uid or into a liquid rather than a solid. This is the basis for many kinetic mass fractionation processes.

By inverting this logic, we can see that an isotope, placed in an environment with coexisting phases, will fractionate between the phases on the basis of bonding

characteristics. Bonds to small, highly charged ions have higher vibrational frequencies than those to larger ions with lower charge. (Note that we are now comparing bonds that have higher vibrational frequencies because they have different "spring" constants, not because they connect atoms with different masses.) Substances with such bonds tend to accept heavy isotopes preferentially, because doing so reduces the vibrational component of their internal energy. This, in turn, reduces the energy available for chemical reactions—the free energy of the system—and makes the phase more stable.

The effect of substituting heavy isotopes into a sub-stance with fewer small, highly charged ions is generally less pronounced. There is a tendency, consequently, for substances such as quartz or calcite, in which oxygen is bound to small Si^{4+} or C^{4+} ions by covalent bonds, to incorporate a high proportion of ^{18}O. These minerals, therefore, have characteristically large positive values of $\delta^{18}O$. In feldspars, this tendency is less dramatic, largely because some of the Si^{4+} ions in the silicate framework have been replaced by Al^{3+}. The contrast is even greater in magnetite, where oxygen forms ionic bonds to Fe^{2+} and Fe^{3+}, which are much larger and less highly charged than Si^{4+}. These bonds, therefore, have a lower vibra-tional frequency. Magnetite, therefore, tends to prefer the lighter isotope, ^{18}O. At the extreme, uraninite (UO_2) is commonly among the most ^{18}O-depleted minerals found in nature, as we might predict from the large radius and mass of the U^{4+} ion.

GEOLOGIC INTERPRETATIONS BASED ON ISOTOPIC FRACTIONATION

The distribution of stable isotopes among minerals and fluids has proven to be a highly fruitful area of research in geochemistry. In this section, we examine some of the more prominent problems to which isotopic interpre-tations have been applied. This survey is not intended to be comprehensive, but rather to suggest the wide variety of geologic questions that can be addressed from the perspective of stable isotope fractionation.

Thermometry

We can describe fractionation of stable isotopes be-tween minerals by writing simple chemical reactions in which the only differences between the reactants and the

TABLE 13.1. Physical Properties of Water as a Function of Isotopic Composition

	$H_2{}^{16}O$	$D_2{}^{16}O$
Melting point	0.0°C	3.81°C
Boiling point	100.0°C	101.42°C
Maximum density	at 3.98°C	at 11.23°C
Viscosity	8.9 mpoise	10 mpoise

CHEMICAL BONDS AS SPRINGS

Recall from chapter 2 that bond strength is primarily a function of the electronic interaction between atoms. A bond is in many ways analogous to a simple spring. Hooke's Law tells us that a mass (m) attached to a spring and moved a distance (x) experiences a restoring force:

$$F = -kx,$$

in which k is a *spring constant* that describes the elasticity of the spring itself. The elasticity of a chemical bond, similarly, is an inherent property. From basic physics, we also know that if we displace a mass on a spring, it will oscillate with a frequency:

$$f = \left(\frac{1}{2}\pi\right)\sqrt{(k/m)}.$$

Isotopes of the same element have the same electronic configuration, so the bonds they form with other atoms should have the same "spring" constant, k. Electronic contributions to the strength of a Ca-^{16}O bond and a Ca-^{18}O bond, for example, are virtually identical, so the bonds are equally elastic. But because ^{18}O is heavier, the Ca-^{18}O bond has a lower vibrational frequency, f.

Finally, elementary physics allows us to calculate the maximum velocity of an oscillating mass on a spring from:

$$V_{max} = 2\pi x_{max}f = 2\pi x_{max}\sqrt{(k/m)}.$$

Imagine, then, that we place two different masses (^{16}O and ^{18}O) on the ends of two springs (Ca-^{16}O and Ca-^{18}O) with the same spring constant, k, and stretch the two springs to the same length, x_{max}, before releasing them (fig. 13.1). A little algebra shows that the one with the greater mass and lower vibra-

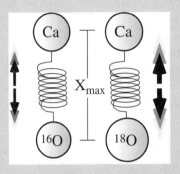

FIG. 13.1. The bond between atoms in a molecule can be modeled as a spring. The elasticity of a Ca-O bond is analogous to the spring constant, k, for a mechanical spring. It depends on the electronic interaction between atoms and is therefore insensitive to the mass of either atom. The vibration frequency (f) of a Ca-^{16}O bond is higher than the vibration frequency of a Ca-^{18}O bond, because f is proportional to $\sqrt{(k/m)}$.

tional frequency, ^{18}O, will also have the lower maximum velocity:

$$V_{max}(^{18}\text{O})/V_{max}(^{16}\text{O}) = f_{18_O}/f_{16_O} = \sqrt{(m_{16_O}/m_{18_O})}.$$

When the velocity of an oscillating mass is maximized, the internal energy of the spring-mass system is all in kinetic energy ($K = \frac{1}{2}mV_{max}^2$) rather than potential energy. It is easy to see that the kinetic energy associated with the heavier isotope will be lower than with the lighter one:

$$\begin{aligned}
K_{\text{Ca-}^{18}\text{O}} &= \tfrac{1}{2}m_{18_O}V_{max}^2(^{18}\text{O}) \\
&= \tfrac{1}{2}m_{18_O}V_{max}^2(^{16}\text{O})(m_{16_O}/m_{18_O}) \\
&= \tfrac{1}{2}V_{max}^2(^{16}\text{O})(m_{16_O}/m^2_{18_O}) \\
&= K_{\text{Ca-}^{16}\text{O}}/m^2_{18_O}.
\end{aligned}$$

products are their isotopic compositions. Water and calcite, for example, may each contain both ^{16}O and ^{18}O, so the isotopic exchange between them is given by:

$$\text{CaC}^{16}\text{O}_3 + \text{H}_2{}^{18}\text{O} \rightleftarrows \text{CaC}^{18}\text{O}_3 + \text{H}_2{}^{16}\text{O}.$$

The equilibrium constant, K_{eq}, for this reaction is equal to:

$$\begin{aligned}
K_{eq} &= a_{\text{CaC}^{18}\text{O}_3}a_{\text{H}_2{}^{16}\text{O}}/a_{\text{CaC}^{16}\text{O}_3}a_{\text{H}_2{}^{18}\text{O}} \\
&= (^{18}\text{O}/^{16}\text{O})_{\text{CaCO}_3}/(^{18}\text{O}/^{16}\text{O})_{\text{H}_2\text{O}} \\
&= \alpha_{\text{CaCO}_3\text{-H}_2\text{O}}.
\end{aligned}$$

The free energy change for this reaction is due only to the very slight difference between the vibrational energy states of a ^{16}O-C bond and a ^{18}O-C bond in calcite and between a ^{18}O-H bond and a ^{16}O-H bond in water. Unlike the mineral reactions we have discussed so far, in which $\Delta \bar{G}_r^o$ is on the order of tens of kilocalories per mole, the free energy associated with this exchange is therefore only a few calories per mole. As written, the exchange reaction has a small negative $\Delta \bar{G}_r^o$, so $\alpha_{CaCO_3-H_2O}$ has a positive value.

In 1947, Harold Urey recognized that the isotopic exchange between ^{16}O and ^{18}O could be used as the basis of a paleothermometer to estimate the temperature of the ancient ocean. At 25°C, the fractionation factor for the calcite-water system has a value of 1.0286, which means that $\delta^{18}O$ is equal to +28.6‰, if the water we have been talking about is SMOW. The value of α is a function of temperature, such that with decreasing temperature, calcite becomes more ^{18}O-enriched relative to seawater. Urey hypothesized that if accurate determinations of $\delta^{18}O$ could be made on marine limestones, it should be possible to calculate the ocean temperature at their time of limestone formation.

Despite early promise, this particular paleothermometer has only proven reliable for Cenozoic sediments, and even then with great difficulty. Its most serious limitation is that the oxygen isotopic composition of seawater has varied considerably through time, so that the isotopic fluctuations in carbonate sediments do not reflect changes in temperature alone. Later in this chapter, for example, we discuss kinetic effects that cause polar ice to be isotopically lighter than mean seawater. As glacial epochs have come and gone and the balance of ^{16}O and ^{18}O has shifted between ice and the oceans, the $\delta^{18}O$ of the oceans may have varied by as much as 1.4‰. The diagram in figure 13.2 attempts to correct for this complication by offering two temperature scales: one for ice-free periods and a second for periods (like the present) when there is glacial ice on the Earth. Sea surface temperatures within the past 50,000 years can be validated independently by methods such as the alkenone U_{37}^K index that we discussed in chapter 6, but relating these temperatures to the isotopic record in bottom sediments remains a challenge. Other practical problems related to diagenetic changes in the isotopic composition of limestone and the biochemical fractionation of oxygen by shell-forming organisms (vital effects) have contributed further complications to carbonate paleothermometry.

Although Urey's original idea for a marine carbonate thermometer has turned out to be of limited value in

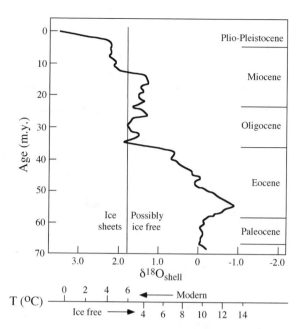

FIG. 13.2. Oxygen isotopic data from benthic foraminifera collected in the North Atlantic Ocean. The process of forming glaciers fractionates ^{16}O from the oceans, leaving them relatively richer in ^{18}O when ice is present on the Earth than in ice-free times (see fig. 13.6 and the accompanying text). Therefore, an isotopic temperature scale based on $\delta^{18}O$, shown at the bottom, has to include a correction for this fractionation for ice-free times. For large portions of the mid-Tertiary, it is not easy to determine how great a correction to apply. (Modified from Miller et al. 1987.)

practice, the basic concept is a good one. Especially at metamorphic or igneous temperatures, kinetic complications are rarer than at 25°C, and a large number of silicate and oxide mineral pairs have been shown to yield valid temperature estimates. Oxygen isotope thermometry is particularly attractive in studies of deep crustal processes, because isotopic fractionation takes place almost independently of pressure. This is a substantial advantage over the geothermometers that we examined in chapter 9.

Figure 13.3 shows how a number of fractionation factors for oxygen isotopes between quartz and other rock-forming minerals vary with temperature. These have been calculated from spectroscopic data, using a quantum mechanical model for lattice dynamics. More commonly, plots of this type are constructed from experimental data. At temperatures >1000 K, it has been shown that the natural log of α is proportional to $1/T^2$, so it is con-

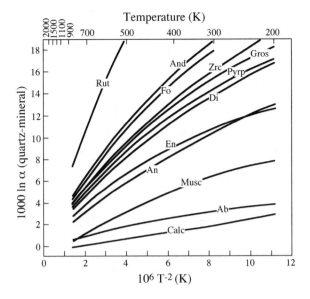

FIG. 13.3. Temperature dependence of oxygen isotope fractionation factors between quartz and common rock-forming minerals, calculated from spectroscopic data. Abbreviations: Ab = albite; An = anorthite; And = andradite; Calc = calcite; Di = diopside; En = enstatite; Fo = forsterite; Gros = grossular; Musc = muscovite; Rut = rutile; Pyrp = pyrope; Zrc = zircon. (Modified from O'Neil 1986.)

venient to construct plots like figure 13.3 on which equations of the form:

$$1000 \ln \alpha = A \, (10^6 T^{-2}) + B \qquad (13.4)$$

plot as straight lines. A and B are empirical constants. For many pairs of geologic materials, this proportionality can be used to extrapolate to temperatures much lower than 1000 K. Deviations are most obvious below 500 K, but are often small enough to make extrapolation reasonable. The following problem is typical of many in isotope geothermometry.

Worked Problem 13.2

The Crandon Zn-Cu massive sulfide deposit in northeastern Wisconsin is a volcanogenic ore body within a series of intermediate to felsic volcanic rocks. Munha and coworkers (1986) have sampled andesitic flows and tuffs adjacent to the ore body and determined the oxygen isotope compositions of quartz and plagioclase in them. How can these be used to estimate the temperature at which the ore body formed?

Matsuhisa and colleagues (1979) performed a series of isotopic exchange experiments between quartz and water, and other

experiments between plagioclase and water. By performing least-squares fits of equation 13.4 to their data, they determined that:

$$1000 \ln \alpha_{qz\text{-}H_2O} = 3.13 \times 10^6 T^{-2} - 2.94,$$

$$1000 \ln \alpha_{ab\text{-}H_2O} = 2.39 \times 10^6 T^{-2} - 2.51,$$

and

$$1000 \ln \alpha_{an\text{-}H_2O} = 1.49 \times 10^6 T^{-2} - 2.81,$$

where T is in kelvins. These equations are strictly valid only over the range 673–773 K, where the experiments were performed. Considering the likely uncertainties encountered when using these curves to determine temperatures from natural samples, however, it is probably safe to apply them somewhat outside this range.

If we assume, as Matsuhisa and his coworkers did, that the fractionation factor for an intermediate plagioclase varies in a linear fashion with its anorthite content, then we can combine the two feldspar-water equations to yield:

$$1000 \ln \alpha_{pl\text{-}H_2O} = (2.39 - 0.9 \, An) \times 10^6 T^{-2}$$
$$- (2.51 + 0.3 An).$$

To describe the fractionation between quartz and plagioclase, we subtract this result from the quartz-water equation above. The final expression is:

$$1000 \ln \alpha_{qz\text{-}pl} = (0.74 - 0.9 An) \times 10^6 T^{-2}$$
$$- (0.43 - 0.3 An),$$

which we can solve for temperature to get:

$$T = [(0.74 - 0.9 An) \times 10^6/(\delta^{18}O_{qz} - \delta^{18}O_{pl})$$
$$+ 0.43 - 0.3 An]^{1/2} \qquad (13.5)$$

Munha and his colleagues collected two samples from the footwall rocks at the Crandon deposit. Both show petrographic evidence of propylitic alteration (characterized by the appearance of chlorite, serpentine, and epidote in hydrothermally altered andesitic rocks), which commonly causes marked albitization of plagioclase. The mole fraction of anorthite in feldspars from these rocks is ~0.15. In one sample, the measured $\delta^{18}O$ for quartz is +9.38‰ and for plagioclase is +6.71‰. By inserting these values in equation 13.5, we find that $T = 533$ K (260°C). In the other sample, $\delta^{18}O_{qz} = +9.53$‰ and $\delta^{18}O_{pl} = +6.44$‰, from which we calculate that $T = 503$ K (230°C). These are compatible with temperatures from 240° to 310°C measured by independent techniques.

Isotope geothermometry is perhaps most heavily used by economic geologists, for problems like the one shown above. Sulfur and oxygen isotopes are commonly measured to determine the temperature of ore formation. A few of the most popular sulfur isotope thermometers are indicated in figure 13.4 and table 13.2. Where sulfide

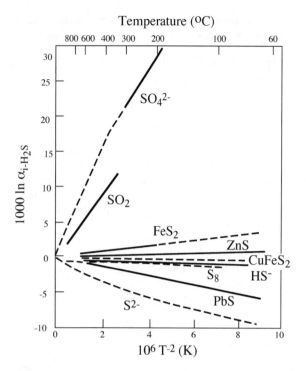

FIG. 13.4. Temperature dependence of sulfur isotope fractionation factors between minerals or fluid species and H_2S. Solid curves are experimentally determined, and dashed curves are estimated or extrapolated. (Modified from Ohmoto and Goldhaber 1997.)

minerals can be shown to have formed in isotopic equilibrium with a hydrothermal fluid, these can provide valuable insights into the conditions of mineralization. Unfortunately, however, sulfur isotopic systems are easily upset by kinetic factors. The oxidation state and pH of ore-forming fluids, for example, can seriously affect the rates of isotopic exchange between aqueous species and ore minerals, particularly at low temperature.

TABLE 13.2. Selected Sulfur Isotope Thermometers

Isotope Thermometer	Temperature (K)
Pyrite-galena	$[(1.01 \pm 0.04) \times 10^3]/\Delta^{1/2}$
Sphalerite-galena or Pyrrhotite-galena	$[(0.85 \pm 0.03) \times 10^3]/\Delta^{1/2}$
Pyrite-chalcopyrite	$[(0.67 \pm 0.04) \times 10^3]/\Delta^{1/2}$
Pyrite-pyrrhotite or Pyrite-sphalerite	$[(0.55 \pm 0.04) \times 10^3]/\Delta^{1/2}$

$\Delta = \delta^{34}S_A - \delta^{34}S_B = 1000 \ln \alpha_{A-B}$. Data from Ohmoto and Goldhaber (1997).

Isotopic Evolution of the Oceans

In chapter 8, we found that SO_4^{2-}, at an average concentration of 2700 ppm, is the second most abundant anion in seawater. The total mass of sulfur in the oceans is $\sim 3.7 \times 10^{21}$ g. These values have probably not changed significantly since the Proterozoic era. The isotopic composition of oceanic sulfate, however, has varied considerably. Today, marine SO_4^{2-} is very nearly homogeneous, with a $\delta^{34}S$ value of almost +20‰. In the past 600 million years, that value has ranged from ~10 to almost +30‰. As illustrated in figure 13.5, these systematic variations, reflected in the isotopic composition of sulfate evaporites, offer promise as a stratigraphic tool.

The specific causes of variation are not well understood, but they are widely interpreted to represent a dynamic balance between redox processes that supply and remove sulfur from the ocean. Sulfur enters the ocean primarily as sulfate derived from a variety of continental sources and dissolved in river water. Among these sources are sulfidic shales and carbonate rocks, volcanics and other igneous rocks, and evaporites. Sulfur leaves the

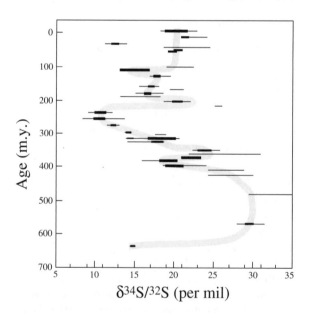

FIG. 13.5. The variation of the sulfur isotopic composition of sulfate evaporites through the Phanerozoic era has been used to estimate the isotopic composition of the oceans through time. Most measured values lie within the heavy lines. Thin lines indicate the range of extreme values. (Modified from Holser and Kaplan 1966.)

ocean as sulfate evaporites or in reduced form as pyrite and other iron sulfides in anoxic muds.

In general, sulfides are isotopically light compared to sulfates. This is due, in part, to equilibrium fractionation (see fig. 13.4), but is also a product of kinetic fractionation by bacteria. Metabolic pathways in sulfate-reducing bacteria involve enzyme-catalyzed reactions that favor ^{32}S over ^{34}S, because ^{32}S-O bonds are more easily broken than ^{34}S-O bonds. The degree of fractionation depends on sulfate concentration as well as temperature. Some sedimentary sulfide is >50‰ lighter than contemporaneous gypsum in marine evaporites.

The isotopic composition of sulfate entering the oceans, therefore, may vary with time, because of fluctuations in global volcanic activity, or because different proportions of evaporates and sulfidic shales are exposed to continental weathering and erosion. In the same way, changes in the rate of deposition of pyritic sulfur in marine sediments may affect the $\delta^{34}S$ value in seawater sulfate. Berner (1987) has shown that such changes may be the indirect result of fluctuations in the mass and diversity of plant life during its evolution on the continents, or may reflect changes in the relative dominance of euxinic and "normal" marine depositional environments. By studies of this type, the grand-scale processes that have moderated the global cycle for sedimentary sulfur (and carbon and oxygen, which mimic or mirror the history of sulfur) are becoming well understood. The events that have given rise to specific fluctuations in the marine $\delta^{34}S$ record, however, remain largely unknown.

Worked Problem 13.3

What is the maximum change in $\delta^{18}O$ that could have occurred in ocean water due to the weathering of igneous and metamorphic rocks to form sediments during the history of the Earth? To answer this question, we need a few basic pieces of information. The average $\delta^{18}O$ value for igneous rocks is ~+8‰, and that for sedimentary and metamorphic rocks is ~+14‰. These numbers are not very well constrained, but they are good enough for a rough calculation. The mass of the oceans (appendix C) is 1.4×10^{24} g, and the mass of sedimentary and metamorphic rocks in the crust is ~2.1×10^{24} g.

Assume that all of the sedimentary and metamorphic rocks were once igneous, and therefore once had a $\delta^{18}O$ value of +8‰ (8/1000 heavier than the present ocean). Their $\delta^{18}O$ value, then, has since increased by +6‰. The complementary change in the composition of the hydrosphere must have been:

$$(+6‰)(2.1 \times 10^{24} \text{ g})/(1.4 \times 10^{24} \text{ g}) = +9‰.$$

The oxygen isotopic mix in the oceans, in other words, is gradually becoming lighter through geologic time as a result of crustal weathering reactions.

Fractionation in the Hydrologic Cycle

Because there are two stable isotopes of hydrogen (^{1}H and ^{2}D) and three of oxygen (^{16}O, ^{17}O, and ^{18}O), there are nine different ways to build isotopically distinct water molecules. Masses of these molecules range from a low of 18 for $^{1}H_2^{16}O$ to a high of 22 for $^{2}D_2^{18}O$. Earlier in this chapter, we found that the vibrational frequency associated with bonds to a light isotope is greater than that for bonds to a heavy isotope. Consequently, "light" water molecules escape more readily from a body of water into the atmosphere than do molecules of "heavy" water. When water evaporates from the ocean surface, it has a δD value of about −8‰ and a $\delta^{18}O$ value around −9‰. The degree of fractionation is, of course, a function of temperature. The result at any place on the Earth, however, is that atmospheric water vapor always has negative δD and $\delta^{18}O$ values relative to SMOW.

The reverse process occurs when water condenses in the atmosphere. Condensation is nearly an equilibrium process, favoring heavy water molecules in rain or snow. The first rain to fall from a "new" cloud over the ocean, therefore, has values for both δD and $\delta^{18}O$ that are ~0‰. Water vapor remaining in the atmosphere, however, is systematically depleted in deuterium and ^{18}O by this process. Subsequent precipitation is derived from a vapor reservoir that has delta values even more negative than freshly evaporated seawater. The isotopic ratio (R) in the remaining vapor is given by:

$$R = R_o f^{(\alpha-1)}, \tag{13.6}$$

where R_o is the initial $^{18}O/^{16}O$ value in the vapor, f is the fraction of vapor remaining, and α is the fractionation factor. This is the *Rayleigh distillation equation*, which also appears in chapter 14 when we consider element fractionation during crystallization from a melt. Let's use it in an example.

Worked Problem 13.4

Suppose that rain begins to fall from an air mass whose initial $\delta^{18}O$ value is −9.0‰. If we assume that the fractionation factor is 1.0092 at the condensation temperature, what should we expect the isotopic composition of the air mass to be after 50% of the vapor has recondensed? What happens when 75% or 90% has been recondensed? If we were to collect all of the rain that fell as the air mass approached total dryness, what would its bulk $\delta^{18}O$ value be?

First, we convert the R values to equivalent delta notation:

$$(\delta + 1000) = 1000(R/R_{SMOW}).$$

The Rayleigh distillation equation (13.6), therefore, can be written in the form

$$R/R_o = (\delta^{18}O + 1000)/([\delta^{18}O]_o + 1000) = f^{(\alpha-1)},$$

or

$$\delta^{18}O = [(\delta^{18}O)_o + 1000]f^{(\alpha-1)} - 1000.$$

After 50% recondensation,

$$\delta^{18}O = [-9.0‰ + 1000](0.5)^{(0.0092)} - 1000 = -15.3‰.$$

After 75%,

$$\delta^{18}O = [-9.0‰ + 1000](0.25)^{(0.0092)} - 1000 = -21.6‰.$$

After 90%,

$$\delta^{18}O = [-9.0‰ + 1000](0.1)^{(0.0092)} - 1000 = -29.8‰.$$

What happens to the isotopic composition of rainwater during this evolution? To find out, we write:

$$\alpha = R_{rain}/R_{vapor} = (\delta^{18}O_{rain} + 1000)/(\delta^{18}O_{vapor} + 1000).$$

By rearranging terms, we find that:

$$\delta^{18}O_{rain} = \alpha(\delta^{18}O_{vapor} + 1000) - 1000.$$

We use the results of the first calculation to find that after 50% condensation, $\delta^{18}O$ of rain is:

$$1.0092(-15.3 + 1000) - 1000 = -6.2‰.$$

After 75%, $\delta^{18}O_{rain} = -12.69‰$. After 90%, it reaches −20.9‰. With continued precipitation, therefore, rainwater becomes lighter. Notice, however, that if we collected all of the rain that falls, conservation of mass would require that its bulk $\delta^{18}O$ value be the same as the initial $\delta^{18}O$ value of the water vapor.

─────────────────────────

Because of this fractionation process, $\delta^{18}O$ values for rainwater are always negative compared with values for seawater. A similar fractionation affects hydrogen isotopes. For both isotopic systems, the separation between rainwater and seawater becomes more pronounced as air masses move farther inland or are lifted to higher elevations. Also, because the fractionation factors for both oxygen and hydrogen isotopes become larger with

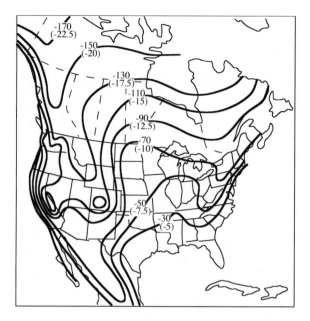

FIG. 13.6. Contours of δD and $\delta^{18}O$ (values in parentheses) in rainwater become increasingly negative as moist air moves inland, to higher elevations, or to higher latitudes. For this reason, ice sheets during glacial periods lock up a disproportionate amount of the Earth's ^{16}O, leaving the oceans relatively ^{18}O-enriched. (After Sheppard et al. 1969.)

decreasing temperature, the difference between seawater and precipitation increases toward the poles. The lightest natural waters on the Earth are in snow and ice at the South Pole, where $\delta^{18}O$ values less than −50‰ and δD less than −45‰ have been measured. These various trends are illustrated for the North American continent in figure 13.6.

These observations make it possible to recognize the isotopic signature of meteoric waters in a particular region and to use that information to study the evolution of surface and subsurface waters. Analyses of $\delta^{18}O$ and δD are commonly displayed on a two-isotope plot like figure 13.7, originally proposed by Cal Tech chemists Sam Epstein and Toshiko Mayeda (1953) and refined through the work of Harmon Craig (1961) and others. Meteoric waters on such a plot lie along a straight line given by:

$$\delta D = \delta^{18}O + 10.$$

Two-isotope plots have been used for many interesting purposes. Shallow groundwaters tend to retain the

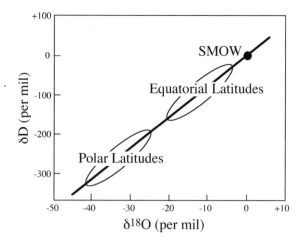

FIG. 13.7. Oxygen and hydrogen isotopic values in meteoric water vary with temperature and hence latitude, but always lie along the diagonal meteoric water line in this two-isotope plot. (Modified from Craig 1961.)

isotopic pattern they inherit from rain. Water in deep aquifers, however, can deviate from local rainwater in many ways. Because of climatic changes, for example, the isotopic composition of modern rainwater may be significantly different from that of water that fell thousands of years ago and is now buried in deep aquifers, even though both lie on the meteoric water line. Siegel and Mandle (1984) have sampled well waters drawn from the Cambrian-Ordovician aquifer system in Minnesota, Iowa, and Missouri. They find that $\delta^{18}O$ values in the north, where the aquifer is exposed, are close to modern rainwater values. To the south, however, $\delta^{18}O$ becomes steadily more negative. In Missouri, well water is as much as 11‰ lighter than present precipitation. Siegel and Mandle conclude, therefore, that a contour map of $\delta^{18}O$ values (fig. 13.8) can serve in this case as a qualitative guide to rates of groundwater movement during the past 12,000 years.

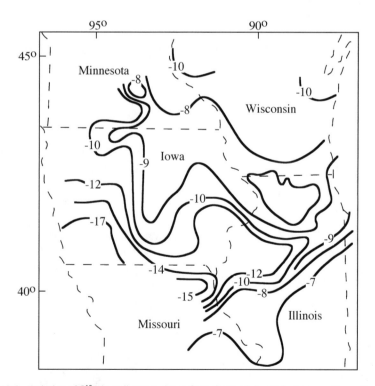

FIG. 13.8. Variation of $\delta^{18}O$ in well waters drawn from the Cambrian-Ordovician aquifer system in the north-central midwestern United States. Values are more negative to the southwest, where the aquifer is deepest. These waters may have been introduced from glacial meltwater sources during the late Pleistocene epoch. (Modified from Siegel and Mandle 1984.)

MASS-INDEPENDENT FRACTIONATION

For studies of the type we are highlighting in this chapter, geochemists use the ratio of the two most abundant isotopes of an element. When three or more stable isotopes exist, however, they could just as easily choose to use a less abundant pair. One reason that they usually do not is that the various isotopes of any element tend to fractionate in direct proportion to their masses, so that a measurement of $\delta^{17}O$, for example, offers no more information than a measurement of $\delta^{18}O$. To see why, take a look at figure 13.9. Because ^{18}O is two mass units heavier than ^{16}O and ^{17}O is only one unit heavier, any process that alters a mixture containing both isotopes should change $\delta^{18}O$ twice as much as it does $\delta^{17}O$. For this reason, all samples of oxygen-bearing materials on the Earth should lie on a mass fractionation line through the composition of SMOW ($\delta^{17}O = \delta^{18}O = 0.0‰$) for which $\delta^{17}O = 0.5\delta^{18}O$.

Imagine the surprise when geochemists at the University of Chicago (Robert Clayton et al. 1973) discovered that some meteorites contain high-temperature

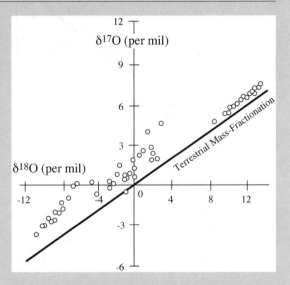

FIG. 13.10. Gypsum from the central Namib Desert and from dry valleys in Antarctica is relatively enriched in ^{17}O, so that its composition plots above the terrestrial mass fractionation line. This gypsum is deposited directly from the atmosphere, where its sulfate is produced by reactions between biogenic dimethyl sulfide and ^{17}O-rich O_3 and H_2O_2. (After Thiemens et al. 2001.)

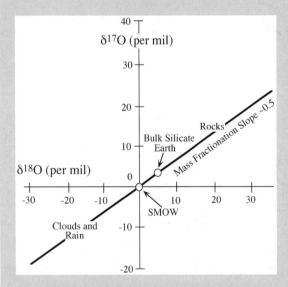

FIG. 13.9. On this "three-isotope plot," values of $\delta^{17}O$ and $\delta^{18}O$ for almost all Earth materials plot on a mass fraction line with a slope of ~0.5, passing through the isotopic composition of SMOW. Most natural processes fractionate isotopes according to their masses, so that the resulting compositions move up or down the line, but not off of it.

inclusions that do not plot on the mass fractionation line. Because no mass fractionation process could account for this anomaly, they concluded that some unusual nucleosynthetic event—a supernova—must have sprayed pure ^{16}O into the early Solar System. It was incorporated in the inclusions, enriching their mixture of oxygen isotopes by as much as 40‰ in ^{16}O without also adding ^{17}O or ^{18}O.

Although still attractive for other reasons we will discuss at greater length in chapter 15, this interpretation lost some of its unique appeal following the early 1980s, when other examples of mass-independent fractionation began to appear in the chemical literature. Mark Thiemens and coworkers at the University of California in San Diego have found that atmospheric ozone, CO_2, CO, H_2O_2, aerosol sulfate, nitrous oxide, and even O_2 all possess mass-independent isotopic compositions. The fractionation mechanism is still not understood, although it evidently involves molecular reactions triggered by ultraviolet radiation in the upper atmosphere.

This discovery opens the possibility that the meteoritic anomalies may have had a chemical origin rather than a nucleosynthetic one. As importantly, perhaps, it has also given geochemists a new tool for tracing atmospheric gases through terrestrial pathways. Bao and colleagues (2000) and Thiemens and coworkers (2001), for example, have reported relative $\delta^{17}O$ enrichments between +0.5 and +3.4‰ in extensive gypsum deposits in the central Namib Desert and in dry valleys in Antarctica. In both cases, the sulfur source is dimethyl sulfide (DMS), released as a gaseous product of microbial activity in adjacent oceans and then oxidized to sulfate by interaction with atmospheric O_3 and H_2O_2. The ^{17}O-enriched isotopic signatures of these two gases is retained as the sulfate is deposited (see fig. 13.10), making it possible to distinguish the Namibian and Antarctic deposits from more typical sulfate deposits formed by chemical weathering.

Fractionation in Geothermal and Hydrothermal Systems

Where water and rocks combine chemically, shifts in isotopic values can be much larger and more obvious. Exchange reactions between groundwater and rocks generally cause isotopic values to leave the meteoric water line. This tendency is perhaps most easily seen in analyses of geothermal brines. The amount of hydrogen in most minerals is small, so most of the hydrogen in a water-saturated rock is in the water. Exchange reactions, therefore, produce only negligible shifts in the deuterium content of geothermal brines. Particularly at elevated temperatures, however, large changes in $\delta^{18}O$ are possible. More than half of the oxygen in a typical water-rock system is in the rock. Reactions involving silicates or carbonates commonly enrich geothermal waters in ^{18}O relative to rainwater. Four horizontal arrows on figure 13.11 indicate the trends measured at Steamboat Springs (Colorado), and Lassen Park, the Salton Sea, and The Geysers (all in California) by Ellis and Mahon (1977). Trends like the four shown here should converge on the composition of juvenile (mantle-derived) water if any appreciable mixing with deep fluids had taken place. Because they do not, geochemists conclude that water in geothermal systems is derived almost entirely from local precipitation.

This conclusion appears reasonable for any water-dominated hydrothermal system, but what about those in which water must circulate through plutonic rocks or through country rock with a very low permeability? In these, the water:rock ratio should be very low, and there may be comparable amounts of exchangeable hydrogen in the rock and in circulating fluid. In rocks with low permeabilities, therefore, we should observe that both $\delta^{18}O$ and δD shift in the fluid phase.

This is an exciting possibility, because it offers a way to solve a problem that has interested economic geologists for a long time. Particularly where circulating meteoric waters have mobilized and concentrated economically important metals, economic geologists want to know how much water has been involved in exchange reactions. Without making some assumptions about the nature of the subsurface plumbing in an area, it is difficult to calculate the ratio of water to rock during a period of hydrothermal alteration. Suppose, though, that we observed shifts in δD and $\delta^{18}O$ for the mineralizing fluid in a given deposit. Maybe we could use those observations to estimate how much water had passed through.

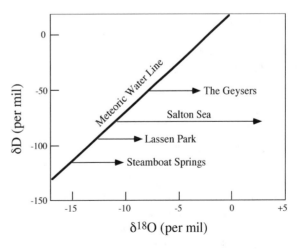

FIG. 13.11. Isotopic compositions of geothermal waters. (Data from Ellis and Mahon 1977.)

The following problem illustrates one attempt of this type.

Worked Problem 13.5

The San Cristobal ore body in the Peruvian Andes is a wolframite-quartz vein deposit associated with the intrusion of Tertiary quartz monzonite into older metamorphic rocks. Oxygen and hydrogen isotopic data were gathered in the area by economic geologist Andrew Campbell and his colleagues in 1984. How can these data be used to estimate the water to rock ratio during hydrothermal mineralization?

Campbell began by noting that isotopic mass balance for either oxygen or hydrogen in a water-rock system can be expressed by:

$$W\delta_{H_2O}(i) + R\delta_{rock}(i) = W\delta_{H_2O}(f) + R\delta_{rock}(f), \quad (13.7)$$

in which W and R are the total atomic percents of exchangeable oxygen or hydrogen in water and rock, respectively, and i and f stand for initial and final states in the system. The δD or $\delta^{18}O$ values for fresh meteoric water ($\delta_{H_2O}(i)$) and unaltered rock ($\delta_{rock}(i)$) are easy to measure in the lab. Determining the final value of $\delta_{H_2O}(f)$, however, takes more work.

Assuming that altered rock and fluid were in equilibrium at the temperature of mineralization, we can calculate 1000 ln δ_{rock-H_2O} ($= \Delta = \delta_{rock}(f) - \delta_{H_2O}(f)$). If $\delta_{rock}(f)$ is known, we can then use equation 13.3 to recast equation 13.7. With a little rearranging, we find that:

$$W/R = [\Delta + \delta_{H_2O}(f) - \delta_{rock}(i)]/[\delta_{H_2O}(i) - \delta_{H_2O}(f)]. \quad (13.8)$$

Now, a brief diversion: In this problem, we are interested in finding out how final values of δD_{H_2O} and $\delta^{18}O_{H_2O}$ in fluid are affected by relative masses of water and rock. In writing equation 13.7, however, we defined W and R as the *atomic* proportions of oxygen or hydrogen in water and rock. To convert from W and R to new variables w and r, which will refer to *mass* quantities of water and rock, we introduce a new quantity z ($= wR/Wr$) to adjust for the different amounts of hydrogen or oxygen in water and rock, so that:

$$w/(w + zr) = W/(W + R).$$

So much for the diversion. To find the isotopic composition of water in equilibrium with altered granite, then, we convert from W and R to w and r and solve equation 13.8 for $\delta_{H_2O}(f)$:

$$\delta_{H_2O}(f) = \delta_{H_2O}(i)[w/(w + zr)] \\ + (\delta_{rock}(i) - \Delta)[zr/(w + zr)]. \quad (13.9)$$

Water is 89 wt % oxygen, and granite contains ~45 wt % oxygen, so when we use equation 13.9 to calculate the $\delta^{18}O$ for water in equilibrium with granite, z will be 45/89, or ~0.5. Bulk rock analyses of granites typically contain ~0.6 wt % water, so when we calculate δD, z will be ~0.006.

Campbell and his coworkers analyzed fresh granite at San Cristobal and found that it has a $\delta^{18}O$ value of +7‰ and

$\delta D = -70‰$. They estimated values of $\delta_{H_2O}(i)$ for hydrogen and oxygen by analyzing fluids trapped at low temperature in late-stage barite. These values are $\delta D = -140‰$ and $\delta^{18}O = -20‰$. They estimated the value of Δ for hydrogen, calculated from a fractionation equation for biotite and water (similar to the method used in worked problem 13.2), to be -49.8‰. The value of Δ for oxygen, calculated from a fractionation equation for plagioclase (An_{30}) and water, was estimated to be +2.4‰.

Campbell and colleagues generated the plot in figure 13.12 by inserting these various parameter values into equation 13.9 and varying the relative magnitudes of w and r. Figure 13.12 shows that, in theory, hydrothermal fluids produced by exchange between meteoric water and rock at high w/r ratios (>0.2) experience only a shift toward heavier $\delta^{18}O$ without any change in δD. As the water:rock ratio decreases below 0.2, however, $\delta^{18}O$ only changes slightly, whereas δD becomes increasingly positive.

To estimate the actual water to rock ratio at San Cristobal, Campbell and coworkers measured $\delta^{18}O$ and δD in trapped fluids and plotted them on this diagram for comparison with

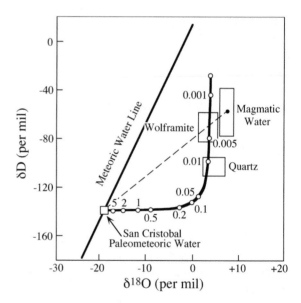

FIG. 13.12. Isotopic compositions of waters in the San Cristobal, Peru, hydrothermal system. δD and $\delta^{18}O$ values along the solid curve are calculated for various ratios of water to rock (w/r), as indicated at various points along the curve. By comparing these calculated isotopic values to analyses of water in isotopic equilibrium with quartz and wolframite in the deposit (boxes), Campbell and colleagues (1984) estimated that the water:rock ratio during ore-formation was between 0.01 and 0.003. Solid circle indicates the calculated value for magmatic water. Wolframite could also have been deposited from a mixture of meteoric and magmatic waters, as indicated by the dashed line. (Modified from Campbell et al. 1984.)

the theoretical values. These measurements, also indicated on figure 13.12, tell us that *w/r* was probably between 0.1 and 0.003.

As this worked example indicates, the isotopic composition of hydrothermal fluids and coexisting rocks depends greatly on the relative amounts of water and rock in a system. Because of this, it is not always clear that an interpretation like the one by Campbell and colleagues (1984) is appropriate. How can we distinguish between fluids that are meteoric waters modified by exchange with rock, on the one hand, and fluids that are simply mixtures of meteoric and juvenile water, on the other? In the San Cristobal study, Campbell and company used equations for Δ_{pl-H_2O} and Δ_{bio-H_2O} to calculate the composition of water in equilibrium with granite at 800°C. This "magmatic" water, plotted as a filled circle on figure 13.12, is colinear with fresh meteoric water and the calculated hydrothermal fluid in the box labeled "wolframite." The fluid in equilibrium with quartz, however, has a much lower δD value. Campbell concludes from this that wolframite at San Cristobal may have been deposited by a mixed meteoric-magmatic fluid, but that quartz probably was not.

Another approach has been used to distinguish between meteoric and magmatic fluids in large-scale porphyry metal deposits. Mineralization in porphyry copper and molybdenum deposits characteristically occurs at the border between a porphyritic granite or granodiorite intrusion and country rock. Economically extractable sulfide minerals are disseminated in a broad, highly altered zone and veins and fractures that cut across both the intrusion and its host. It has long been recognized that this mineralization was promoted by the migration of heated water. Before isotopic data began to accumulate on these systems, however, the source of this water was widely believed to be magmatic.

Several research teams during the 1970s (summarized by Cal Tech geochemist Hugh Taylor [1997]) investigated the oxygen and hydrogen systematics in sericite, pyrophyllite, and hypogene clay minerals from North American porphyry metal deposits. As shown in figure 13.13, δ values for each of these minerals lie in a belt parallel to the meteoric water line. As we compare the deposits to each other, we see also that there is a clearly recognizable latitude trend among them. Both δD and $\delta^{18}O$ decrease consistently as we sample northward from Arizona and New Mexico (Santa Rita, Safford, Copper Creek, and Mineral Park) through Utah, Colorado, and

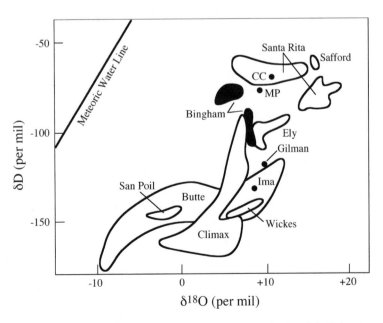

FIG. 13.13. Isotopic compositions of OH-bearing minerals from a selection of North American porphyry metal deposits. Values are more negative for northern deposits, indicating that the hydrothermal fluid was probably circulating meteoric water. Compare with figure 13.7. (Data from Sheppard et al. 1969, and Taylor 1997.)

Nevada (Gilman, Bingham, Ely, and Climax) to Montana, Washington, and Idaho (Butte, San Poil, Ima, and Wickes). This trend looks so much like what we see along the meteoric water line itself (recall fig. 13.7) that it is now commonly believed that porphyry metal deposits are generated largely by recycling local meteoric water.

Fractionation in Sedimentary Basins

Basin brines present yet another class of trends on a two-isotope plot for oxygen and hydrogen. It was once widely believed that pore waters in deep sedimentary basins were *connate fluids* (that is, samples of seawater trapped at the time of sediment accumulation). Researchers working with Robert Clayton (1966) and others more recently have shown that waters within a given basin generally have a distinct isotopic signature. Values for both $\delta^{18}O$ and δD show considerable scatter, but the least saline brines in any basin tend to lie close to the meteoric water line, whereas more saline fluids generally contain heavier oxygen and hydrogen. Furthermore, as with porphyry copper and molybdenum deposits, isotopic values display a latitudinal variation similar to that for surface waters. As indicated on figure 13.14, Gulf Coast basin brines are much heavier than those in the Alberta Basin. The primary source for these brines, therefore, seems to be local rainwater. The degree of scatter and the disparate slopes of isotopic trends from basin to basin, however, suggest that other sources (silicate or carbonate rocks, true connate water, clays, or hydrocarbons) are also important.

Fractionation among Biogenic Compounds

Two stable isotopes of carbon, ^{12}C and ^{13}C, exist in nature. On a global scale, $\delta^{13}C$ values generally range from about −30‰ to +25‰ relative to PDB, although methane in some natural gas fields has been found with values as low as −100‰. In general, carbon in the reduced compounds found in living and fossil organisms is isotopically light, whereas carbonate minerals (particularly in marine environments) are isotopically heavy. This twofold distinction itself is useful in characterizing the sources of secondary carbonate in sediments. The carbonate cap rocks on salt domes, for example, have $\delta^{13}C$ values that are very negative (about −36‰) compared with marine carbonates (near +0‰). This has been taken as evidence that they are produced by oxidation of

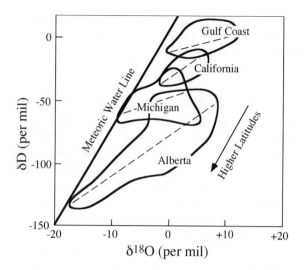

FIG. 13.14. Isotopic compositions of basin brines are increasingly negative from south to north, indicating the influence of meteoric waters. The scatter in values and the variable trends within basins (indicated by dashed lines) are due to the influence of other sources such as true connate waters or magmatic fluids, and to isotopic exchange with host rocks. (Based on data from Taylor 1997.)

methane in adjacent hydrocarbon-rich strata, rather than from recrystallized marine limestones. In a similar way, fresh-water carbonates and dripstone in caves are characteristically light, because they are formed from waters that are charged with soil CO_2. As we saw in chapter 7, soil CO_2 is largely the product of plant respiration and decay, so it should be ^{12}C-rich relative to atmospheric CO_2.

The complex biochemical pathways in living organisms offer many opportunities to fractionate carbon isotopes. Some of these have been investigated, with limited success, for use as paleoenvironmental indicators or as guides in provenance studies. Green plants fractionate carbon isotopes in at least three metabolic steps during photosynthesis, each of which favors the retention of ^{12}C rather than ^{13}C. In broad terms, these involve (1) the physical assimilation of CO_2 across cell walls, (2) the conversion of that CO_2 into intermediate compounds by enzymes, and (3) the synthesis of large organic molecules. The degree of fractionation in each step is kinetically controlled by such factors as the atmospheric or intracellular partial pressure of CO_2.

In addition, terrestrial plants can be divided into two large groups on the basis of the enzyme that dominates in step 2 and the types of large organic molecules that form in step 3. Those in one group, which includes trees, bushes, and grains like wheat and rice, are known as C_3 plants because the initial product of CO_2 fixation is a molecule with three carbon atoms. In these, lignin, cellulose, and waxy lipids constitute a large portion of the total organic matter. By contrast, the C_4 plants, which include corn, sugar cane, many aquatic plants, and some grasses, contain relatively high proportions of carbohydrate and protein. As summarized in table 13.3, C_3 plants have $\delta^{13}C$ values between about $-22‰$ and $-33‰$, whereas C_4 plants range from $-9‰$ to $-16‰$. Only a few plants such as algae, lichens, and cacti occupy an isotopic middle ground between these two major plant groups. Marine plants have $^{13}C/^{12}C$ ratios that are $\sim7.5‰$ less negative than C_3 terrestrial plants, possibly because they assimilate carbon from marine HCO_3^- ($0‰$) rather than atmospheric CO_2 ($-7‰$). Plankton vary from $17‰$ to about $-27‰$.

Sackett and Thompson (1963) measured $\delta^{13}C$ values in Gulf Coast sediment and found that they decrease systematically shoreward from about $-21‰$ to $-26‰$. Using the patterns we introduced in the previous paragraph as a guide, they inferred that the isotopic trend reflects a near-shore mixing of terrestrial and marine plant debris. In more recent studies, isotopic signatures of individual species of phytoplankton have been used to account for the spatial variability of ^{13}C in estuarine and shallow marine sediments. Without painstakingly detailed work, studies of this sort can yield highly ambiguous results because of the large number of hidden source materials and processes that may have contributed to present isotopic compositions.

The transition from primary organic debris to kerogen causes an enrichment in ^{12}C, which becomes more pronounced as kerogen is converted to petroleum or coal. Kerogen ranges from $-17‰$ to $-34‰$. Despite the degrading effects of diagenesis, source materials can still be recognized isotopically in some kerogen and some Cenozoic coals and crude oils. Marine biogenic matter and C_3 terrestrial plant remains retain their relatively light $\delta^{13}C$ values most effectively, presumably because the double bonds associated with polycyclic hydrocarbons resist thermal degradation well. The single bonds in hydrocarbon chains of C_4 plant debris are more vulnerable; yet they, too, resist mild diagenesis and retain their heavier $\delta^{13}C$ signatures well enough that geochemists can infer some source characteristics. Ultimately, of course, diagenesis renders even the most obvious trends useless. Pre-Tertiary coals all have $\delta^{13}C$ values very near $-25‰$ and values in petroleum cluster very close to $-28‰$.

Isotopic Fractionation around Marine Oil and Gas Seeps

In chapter 5, we discussed the bacterially mitigated digenesis of organic matter in the presence of dissolved sulfate, using the schematic reaction:

$$7CH_2O + 4SO_4^{2-} \rightleftarrows 2S^{2-} + 7CO_2 + 4OH^- + 5H_2O.$$

Depending on whether we want to emphasize the end production of sulfide or soluble bicarbonate, an equally plausible overall reaction might be:

$$2CH_2O + SO_4^{2-} \rightleftarrows H_2S + 2HCO_3^-.$$

"CH_2O," in either context, is a generic placeholder for the complex mix of hydrocarbons derived from C_3 and C_4 plants. At places where crude oil and gas seep from the cold ocean floor, however, the organic food for bacteria can instead be methane, the by-product of petroleum maturation deep below the seafloor. The process of bacterial sulfate reduction using CH_4 as the energy source can be described by the overall reaction:

$$CH_4 + SO_4^{2-} \rightleftarrows H_2S + CO_3^{2-} + H_2O.$$

Does it make a difference whether CH_2O or CH_4 is the dominant carbon source at a particular site?

TABLE 13.3. Ranges of $\delta^{13}C$ and $\delta^{15}N$ Values of Inorganic Nutrients, Higher Plants, Photoautotrophs, and Particulate Organic Matter from Terrestrial and Marine Environments

Source	$\delta^{13}C$ (‰)	$\delta^{15}N$ (‰)
Atmospheric CO_2	-7 to -8	
Dissolved CO_2	-7 to -8	
Bicarbonate (HCO_3^-)	0	
Atmospheric N_2		0
Terrestrial C_3 plants	-30 to -23	-7 to 6
Marsh C_3 plants	-29 to -23	3 to 5
Marsh C_4 plants	-15 to -6	1 to 8
CAM plants	-33 to -11	-15 to -1
Seagrasses	-16 to -4	0 to 6
Mangroves	-19 to -18	6 to 7
Macroalgae	-27 to -10	-1 to -10
Temperate marine POM	-24 to -18	-2 to 10
Temperate estuarine POM	-30 to -15	2 to 19

Data from Ostrom and Fry (1993) and Hoefs (1997).

AN ORGANIC ODYSSEY: FROM KANSAS WHEAT FIELDS TO THE MISSISSIPPI DELTA

Isotopic analysis is particularly useful for distinguishing among possible sources of organic matter in aquatic sediments, each of which has a distinct isotopic composition. For example, vascular plants such as trees and shrubs ($\delta^{13}C = -25‰$ to $-28‰$) that utilize the C_3 metabolic pathway can be distinguished from corn, bamboo and many grasses ($\delta^{13}C = -10‰$ to $-14‰$) that employ the C_4 pathway. In addition, freshwater phytoplankton are typically depleted in ^{13}C ($\delta^{13}C = -30‰$ to $-40‰$) relative to C_3 land plants, and marine phytoplankton are usually more enriched ($-15‰$ to $-20‰$). As we discovered in chapter 6, however, organic matter in aquatic and sedimentary environments is a highly altered, complex mixture of organic components from many sources. Isotopic values of preserved organic matter also record the isotopic history of all components from production to bacterial alteration and diagenesis, further complicating things.

Fortunately, as we also found in chapter 6, some biomolecules survive alteration during deposition and burial. The $\delta^{13}C$ values of biomarkers such as lipids (alkanes and sterols) and lignin phenols provide a more detailed picture of individual sources than the isotopic compositions of bulk organic matter do. For example, the $\delta^{13}C$ values of lignin phenol derived from C_3 plants are typically between $-28‰$ and $-34‰$, whereas carbon isotopic values of C_4 lignin phenols are roughly between $-14‰$ and $-20‰$. Compare each of these to the relatively lower $\delta^{13}C$ values for bulk C_3 and C_4 plants that we gave above. In contrast, other biomarkers such as amino acids are enriched in ^{13}C relative to the whole organism. Using the $\delta^{13}C$ values of individual compounds, organic geochemists can determine the sources of multiple components in organic matter in complex systems and trace the fate of these compounds through complex reactions and processes.

Organic geochemists Miguel Goñi, Kathy Ruttenberg, and Tim Eglinton (1998) used the $\delta^{13}C$ values of bulk organic matter and lignin phenols from marine sediments to investigate the input of terrestrial organic matter to the Gulf of Mexico. The $\delta^{13}C$ values of bulk organic matter sampled from depths between 100 and 2250 m ranged from $-19.7‰$ to $-21.7‰$. At first glance, these bulk values seemed to reflect significant contribution from marine phytoplankton and a minor input from C_3 vascular plants. Goñi and his colleagues were skeptical, however, because they knew that a significant portion of Mississippi River sediment is derived from the central United States, which contains extensive grasslands (C_4 plants).

A more detailed picture of the potential sources was needed, so Goñi's research team investigated the distribution of lignin phenols in sediments. In general, the S/V (1.6) and C/V (0.5) ratios (which we introduced in chapter 6) suggested input from nonwoody angiosperms (see table 6.4). Fine-grained, deep sediments (>100 meters), however, had much higher S/V ratios than the coarser grained sediments in shallower samples. This suggested that organic matter in the fine-grained, off-shore sediments was most likely derived from nonwoody angiosperms, such as grasses, which dominate the central United States. The organic matter in near-shore sediments presumably originated from nearby estuarine environments, where nonwoody gymnosperms found in swamps, bays and estuaries leave a lower S/V signature.

To confirm the S/V evidence that midcontinent C_4 plants are the dominant source of organic matter in deep sediments, Goñi and his colleagues turned to stable isotopes. The isotopic composition of the lignin phenols provided unequivocal evidence. For the shallow sites, the $\delta^{13}C$ values were between $-21.1‰$ and $-29.0‰$, but in the deep sediments, signature was much lighter, between $-15.1‰$ and $-21.8‰$. These values are consistent with the $\delta^{13}C$ signatures of lignin phenol from C_3 plants and C_4 plants, respectively. These data showed that terrestrial organic matter is an important component of these deltaic sediments, and also support some of the ideas about organic matter preservation in marine sediments that we discussed in chapter 6.

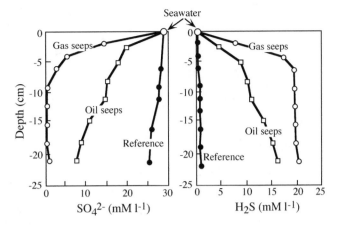

FIG. 13.15. Analyzed SO_4^{2-} and H_2S concentrations in pore waters in sediment associated with oil and gas seeps in the Gulf of Mexico and with sediments at a reference site distant from any seeps. (Modified from Aharon and Fu 2000.)

Geochemists Paul Aharon and Baoshun Fu (2000) have analyzed pore fluids in sediment cores at oil and gas seeps in the Gulf of Mexico. As illustrated in figure 13.15, the decreasing abundance of SO_4^{2-} with depth is accompanied by an increase in H_2S at all sites. Clearly, though, the consumption of SO_4^{2-} is much more complete near seeps, where microbes live on CH_4, than at a distance, where CH_2O is their dominant carbon source. Several previous studies have indicated that the type of source organic matter, rather than the amount of it, controls the rate of sulfate reduction. These data, then, confirm that the microbial diet makes a difference.

Isotopic analyses add a dimension to this conclusion. Generally, both $\delta^{18}O$ and $\delta^{34}S$ increase with depth in each of their sample cores. This is to be expected, because the rate of microbial SO_4^{2-} reduction is faster for $^{32}S^{16}O_4^{2-}$ than for $^{34}S^{18}O_4^{2-}$, consistent with the principles we outlined in a box earlier in this chapter. Therefore, the more SO_4^{2-} is consumed in the closed system environment of the sediment column, the heavier the isotopic composition of any remaining SO_4^{2-} becomes. The surprise for Aharon and Fu was that the degree of isotopic fractionation for both oxygen and sulfur varied dramatically from one sampling site to another, apparently as a function of the mix of CH_2O and CH_4. At a distance from oil or gas seeps (that is, in "normal" marine sediment), both ^{18}O and ^{34}S in residual SO_4^{2-} increase rapidly with depth, even though only a small fraction of the SO_4^{2-} in pore fluid is consumed. Although a greater proportion of SO_4^{2-} is consumed, the relative isotopic enrichments are much

less pronounced at oil seep sites, and significantly less at gas seep sites, as is evident in figure 13.16. That is, microbes degrading CH_2O discriminate between heavy and light isotopes of sulfur and oxygen much more efficiently than do microbes consuming a CH_4-rich diet. Aharon and Fu infer that because temperature and other compositional parameters are the same across all sites, these isotopic differences reflect differences in the metabolic pathways—perhaps involving different enzymatic intermediates—by which sulfate-reducing microbes separate oxygen from sulfur, depending on what they are eating. This may suggest a way to interpret the type of bacterial activity that took place in ancient environments for which we have only an isotopic record in preserved sediments.

SUMMARY

In this chapter, we have discussed some of the basic processes that lead to the fractionation of light stable isotopes from one another in geologic environments. Hydrogen, carbon, nitrogen, oxygen, and sulfur are all abundant in the crust and atmosphere/hydrosphere, and may be found in a wide variety of chemical compounds. Variations in bond strength and site energy among these compounds lead them to favor different proportions of the stable isotopes.

The free energies of isotopic exchange reactions vary with temperature, but not with pressure. This makes them very useful as geothermometers. The most popular

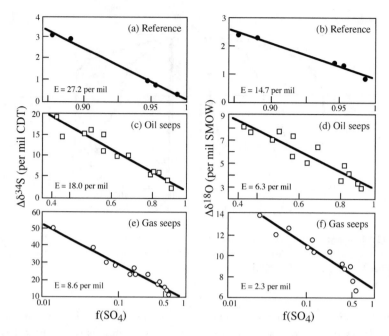

FIG. 13.16. Enrichments of $\delta^{34}S$ (relative to a marine value of 20.3‰) and $\delta^{18}O$ (relative to a marine value of 9.7‰) as a function of the residual fraction (f) of SO_4^{2-} in pore waters after bacterial sulfate reduction. (a, b) At a distance from seeps, (c, d) at oil seeps, and (e, f) at gas seeps. In each case, the enrichment factor (E) is equal to $10^3(\alpha - 1)$, or $(\delta - \delta_0)/\ln f$, where δ is the measured $\delta^{34}S$ or $\delta^{18}O$ value, δ_0 is the corresponding value in seawater, and f is the residual fraction of SO_4^{2-}. (Modified from Aharon and Fu 2000.)

geothermometers involve equilibrium fractionation of oxygen or sulfur isotopes.

Kinetic fractionation effects are also commonly exploited by geochemists. The small difference in vapor pressure between "light" water and "heavy" water leads to a significant fractionation of oxygen and hydrogen isotopes with increasing latitude, altitude, and distance from the ocean. These systematic trends in meteoric water have been useful in recognizing the source of water in aquifers, sedimentary basins, and hydrothermal systems.

The role of living organisms in fractionating light, stable isotopes has made it possible to determine the provenance and, to some degree, the depositional and diagenetic history of organic matter. Isotopic biomarkers in kerogen and young fossil hydrocarbons can be helpful stratigraphic guides, and biomarkers in sedimentary environments can identify primary nutrient sources and biochemical pathways of deposited organic matter.

SUGGESTED READINGS

The following books discuss principles of stable isotope geochemistry at a level appropriate for the beginning student. In most cases, they emphasize the application of methods to specific field problems.

Barnes, H. L., ed. 1997. *Geochemistry of Hydrothermal Ore Deposits,* 3rd ed. New York: Wiley Interscience. (This collection of basic articles by geochemists at the leading edge of research in ore-forming systems has two long chapters devoted to light stable isotopes.)

Clayton, R. N. 1981. Isotopic thermometry. In R. C. Newton, A. Navrotsky, and B. J. Wood, eds. *Thermodynamics of Minerals and Melts.* New York: Springer-Verlag, pp. 85–109. (This concise review paper is a very good way to learn the first principles of stable isotope geothermometry.)

Faure, G. 1986. *Principles of Isotope Geology,* 2nd ed. New York: Wiley. (This is an easy-to-read text dealing largely with radioactive isotopes. Chapters 18–21, however, give a good introduction to light stable isotopes.)

Hoefs, J. 1997. *Stable Isotope Geochemistry*, 4th ed. New York: Springer-Verlag. (This is an excellent text, covering all the basics, with a focus on field examples.)

Jager, E., and J. C. Hunziker. 1979. *Lectures in Isotope Geology*. Berlin: Springer-Verlag. (The lectures in this book were given in 1977 as an introductory short course for geologists. Most deal with radioactive isotopes, but the four final lectures are a useful overview of hydrogen, oxygen, carbon, and sulfur.)

Valley, J. W., H. P. Taylor, and J. R. O'Neil, eds. 1986. *Stable Isotopes in High Temperature Geological Processes*. Mineralogical Society of America Reviews in Mineralogy 16. Washington, D.C.: Mineralogical Society of America. (Papers in this volume are of particular interest to igneous and metamorphic petrologists who use stable isotopes to study processes in the deep crust.)

We refer to the following papers in chapter 13. The interested student may wish to consult them further.

Aharon, P., and B. Fu. 2000. Microbial sulfate reduction rates and sulfur and oxygen isotope fractionations at oil and gas seeps in deepwater Gulf of Mexico. *Geochimica et Cosmochimica Acta* 64:233–246.

Bao, H., M. H. Thiemens, J. Farquhar, D. A. Campbell, C. C.-W. Lee, K. Heine, and D. B. Loope. 2000. Anomalous ^{17}O compositions in massive sulphate deposits on the Earth. *Nature* 406:176–178.

Berner, R. A. 1987. Models for carbon and sulfur cycles and atmospheric oxygen: Application to Paleozoic geologic history. *American Journal of Science* 287:177–196.

Campbell, A., D. Rye, and U. Petersen. 1984. A hydrogen and oxygen isotope study of the San Cristobal mine, Peru: Implications of the role of water to rock ratio for the genesis of wolframite deposits. *Economic Geology* 79:1818–1832.

Clayton, R. N., I. Friedman, D. L. Graf, T. K. Mayeda, W. F. Meents, and N. F. Schimp. 1966. The origin of saline formation waters, I: Isotopic composition. *Journal of Geophysical Research* 71:3869–3882.

Clayton, R. N., L. Grossman, and T. K. Mayeda. 1973. A component of primitive nuclear composition in carbonaceous chondrites. *Science* 182:485–488.

Coplen, T. B., C Kendall, and J. Hopple. 1983. Comparison of stable isotope reference samples. *Nature* 302:236–238.

Craig, H. 1961. Isotopic variations in meteoric waters. *Science* 133:1702–1703.

Deines, P., D. Langmuir, and R. S. Harmon. 1974. Stable carbon isotope ratios and the existence of a gas phase in the evolution of carbonate ground water. *Geochimica et Cosmochimica Acta* 38:1147–1164.

Ellis, A. J., and W.A.J. Mahon. 1977. *Geochemistry and Geothermal Systems*. New York: Academic.

Epstein, S., and T. Mayeda. 1953. Variations of ^{18}O content of waters from natural sources. *Geochimica et Cosmochimica Acta* 4:213–224.

Goñi, M. A., K. C. Ruttenberg, and T. I. Eglinton. 1998. A reassessment of the sources and importance of land-derived organic matter in surface sediments from the Gulf of Mexico. *Geochimica et Cosmochimica Acta* 62:3055–3075.

Hattori, K., and S. Halas. 1982. Calculation of oxygen isotope fractionation between uranium dioxide, uranium trioxide, and water. *Geochimica et Cosmochimica Acta* 46:1863–1868.

Holser, W. T., and I. R. Kaplan. 1966. Isotope geochemistry of sedimentary sulfates. *Chemical Geology* 1:93–135.

Matsuhisa, Y., J. R. Goldsmith, and R. N. Clayton. 1979. Oxygen isotopic fractionation in the system quartz-albite-anorthite-water. *Geochimica et Cosmochimica Acta* 43:1131–1140.

Miller, K. G., R. G. Fairbanks, and G. S. Mountain. 1987. Tertiary oxygen isotope synthesis, sea level history, and continental margin erosion. *Paleoceanography* 2:1–19.

Munha, J., F.J.A.S. Barriga, and R. Kerrich. 1986. High δ^{18}O ore-forming fluids in volcanic-hosted base metal sulfide deposits: Geologic, ^{18}O/^{16}O, and D/H evidence from the Iberian Pyrite belt; Crandon, Wisconsin; and Blue Hill, Maine. *Economic Geology* 81:530–552.

Ohmoto, H., and M. B. Goldhaber. 1997. Isotopes of sulfur and carbon. In H. L. Barnes, ed. *Geochemistry of Hydrothermal Ore Deposits*, 3rd ed. New York: Wiley Interscience, pp. 517–612.

O'Neil, J. R. 1986. Theoretical and experimental aspects of isotopic fractionation. In J. W. Valley, H. P. Taylor, and J. R. O'Neil, eds. *Stable Isotopes in High Temperature Geological Processes*. Mineralogical Society of America Reviews in Mineralogy 16. Washington, D.C.: Mineralogical Society of America, pp. 1–40.

Ostrom, P. H., and B. Fry. 1993. Sources and cycling of organic matter within modern and prehistoric food webs. In M. H. Engel and S. A. Macko, eds. *Organic Geochemistry: Principles and Applications*. New York: Plenum, pp. 785–798.

Rye, R. O. 1974. A comparison of sphalerite-galena sulfur isotope temperatures with filling temperatures of fluid inclusions. *Economic Geology* 69:26–32.

Sackett, W. M., and R. R. Thompson. 1963. Isotopic organic carbon composition of recent continental derived clastic sediments of the eastern Gulf Coast, Gulf of Mexico. *Bulletin of the American Association of Petroleum Geologists* 47:525–531.

Sheppard, S.F.M., R. L. Nielsen, and H. P. Taylor. 1969. Oxygen and hydrogen isotope ratios of clay minerals from porphyry copper deposits. *Economic Geology* 64:755–777.

Siegel, D. I., and R. J. Mandle. 1984. Isotopic evidence for glacial melt-water recharge to the Cambrian-Ordovician aquifer, North-Central United States. *Quaternary Research* 22:328–335.

Taylor, H. P. 1997. Oxygen and hydrogen isotope relationships in hydrothermal mineral deposits. In H. L. Barnes, ed. *Geochemistry of Hydrothermal Ore Deposits,* 3rd ed. New York: Wiley Interscience, pp. 229–302.

Thiemens, M. H., J. Savarino, J. Farquhar, and H. Bao. 2001. Mass-independent isotopic compositions in terrestrial and extraterrestrial solids and their applications. *Accounts of Chemical Research* 34(8):645–652.

Urey, H. C. 1947. *The thermodynamics of isotopic substances. Journal of the Chemical Society (London)* 562–581.

PROBLEMS

(13.1) The relative atomic abundances of oxygen and carbon isotopes are:

$^{16}O{:}^{17}O{:}^{18}O = 99.759{:}0.0374{:}0.2039,$

$^{12}C{:}^{13}C = 98.9{:}1.1.$

Calculate the relative abundances of (a) CO_2 molecules with masses 44, 45, and 46; (b) CO molecules with masses 28, 29, and 30.

(13.2) The fractionation factor for hydrogen isotopes between liquid water and vapor at 20°C is 1.0800. Make a graph to indicate how the value of δD changes in rainwater during continued precipitation from an air mass that initially has a value of −80‰. Show how the δD value of the remaining water vapor changes during this process.

(13.3) Analyses of foraminifera that formed at known temperatures have yielded the following isotopic data:

T (°C)	$\delta^{18}O_{shell}$
30	−3.14
25	−1.97
20	−0.82
15	+0.35
10	+1.51
5	+2.68
0	+3.84

(a) Graph these data and determine the appropriate constants a and b for a linear equation of the form $T = a + b\delta^{18}O_{shell}$.

(b) Seawater did not have its modern $\delta^{18}O$ value of +0.0‰ until ~11,000 years ago. If we know what the isotopic composition of seawater ($\delta^{18}O_{sw}$) was at any earlier time, we can adjust for it by writing $T = a + b\,(\delta^{18}O_{shell} − \delta^{18}O_{sw})$. Between 22,000 and 11,000 years ago, $\delta^{18}O_{sw}$ was +2.0‰. If you find that 13,000-year-old foraminifera in a deep-ocean sample core have the same $\delta^{18}O_{shell}$ values as modern foraminifera in the same core, by how much would you infer that the temperature of ocean water has changed?

(13.4) The following $\Delta^{34}S$ data were collected on samples from Providencia, Mexico, by Rye (1974). On the basis of these data, what is your estimate for the temperature at which these ores formed?

Sample #	Galena	Sphalerite
60-H-67	−1.15‰	+0.95‰
60-H-36-57	−1.11	+0.87
62-S-250	−1.42	+1.03
63-R-22	−1.71	+0.01

(13.5) Hattori and Halas (1982) have determined that the fractionation factor for oxygen exchange between UO_2 and water varies with temperature according to the relationship:

$$1000 \ln \alpha_{UO_2\text{-}H_2O} = 3.63 \times 10^6 T^{-2} - 13.29 \times 10^3 T + 4.42.$$

Using the similar expression for quartz-water fractionation from worked problem 13.2, devise an oxygen isotope geothermometer to estimate temperatures from coexisting quartz and uraninite.

(13.6) The fractionation of carbon isotopes between calcite and CO_2 gas can be calculated from $1000 \ln \delta_{cal\text{-}CO_2} = 1.194 \times 10^6 T^{-2} - 3.63$ (Deines et al. 1974). If the $\delta^{13}C$ value of atmospheric CO_2 is $-7.0‰$, what should be the $\delta^{13}C$ value of calcite precipitated from water in equilibrium with the atmosphere at 20°C?

(13.7) Rain and snow from a large number of sampling sites have been analyzed and found to have the following $\delta^{18}O$ values:

$\delta^{18}O$ (‰)	Mean Annual Air Temperature (°C)
−3.18	15
−8.04	8
−11.52	3
−20.55	−10
−30.98	−25

Using these data, write an equation to predict the oxygen isotopic composition of precipitation at any locality, given its mean annual temperature. Write a second equation based on these data and the meteoric water line, relating δD to mean annual temperature.

(13.8) Use the procedure by Campbell and colleagues (1984), described in worked problem 13.5, to determine what the final values of $\delta^{18}O$ and δD would be in water that initially had values of $-30‰$ and $-230‰$, but that had equilibrated in a closed system with granite at 400°C. Assume the same isotopic values for unaltered granite that were used for the study at San Cristobal, Peru, and a water to rock ratio (w/r) of 0.05.

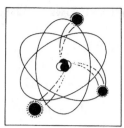

CHAPTER 14

USING RADIOACTIVE ISOTOPES

OVERVIEW

In the geochemist's arsenal of techniques for unraveling geological problems, the study of radioactive isotopes and their decay products has become very prominent. This chapter begins with a discussion of nuclide stability and decay mechanisms. After equations that describe radioactive decay are derived, we examine the utility of certain naturally occurring radionuclide systems (K-Ar, Rb-Sr, Sm-Nd, and U-Th-Pb) in geochronology. The concept of extinct radionuclides that were present in the early solar system is also discussed. The principles behind both mass spectrometry and fission track techniques are introduced. We discuss how induced radioactivity can be used to solve geochemical problems. Examples include neutron activation analysis, ^{40}Ar-^{39}Ar dating, and the use of cosmogenic nuclides (^{14}C and ^{10}Be) for dating geological and archeological materials and for detecting the subduction of sediments. We explore how radionuclides can be used as geochemical tracers in such global problems as determining when the Earth accreted and differentiated, quantifying mantle heterogeneity, assessing the cycling of material between crust and mantle, revealing global oceanic mixing patterns, and understanding degassing of the Earth's interior to produce the atmosphere.

PRINCIPLES OF RADIOACTIVITY

Nuclide Stability

In chapter 13, we learned how stable isotopes can be used to solve geochemical problems. The study of unstable (that is, *radioactive*) nuclides, considered in this chapter, is also an extremely powerful technique. In chapter 2, we saw that only ~260 of the 1700 known nuclides are stable, so we can infer that nuclide stability is the exception rather than the rule. Most of the known radioactive isotopes do not occur in nature. Although some of these may have occurred naturally in the distant past, their decay rates were so rapid that they have long since been transformed into other nuclides. In most cases, radioactive isotopes that are of interest to geochemists require very long times for decay or are produced continually by naturally occurring nuclear reactions.

In chapter 13, we explored how variations in stable isotopes are caused by mass fractionation during the course of chemical reactions or physical processes. With the exception of radioactive ^{14}C and a few other nuclides, the atomic masses of most of the unstable isotopes of geochemical interest are very large, so that mass differences with other nuclides of the same element are minuscule. Consequently, these isotopic systems can be considered to be immune to mass fractionation processes. Thus, all of the measured variations in these nuclides are normally ascribed to radioactive decay.

Decay Mechanisms

Radioactivity is the spontaneous transformation of an unstable nuclide (the *parent*) into another nuclide (the *daughter*). The transformation process, called *radioactive decay*, results in changes in N (the number of neutrons) and Z (the number of protons) of the parent atom, so that another element is produced. Such processes occur by emission or capture of a variety of nuclear particles. Isotopes produced by the decay of other isotopes are said to be *radiogenic*. The radiogenic daughter may be stable or unstable; if it is unstable, the decay process continues until a stable nuclide is produced.

Beta decay involves the emission of negatively charged beta particles (electrons emitted by the nucleus), commonly accompanied by radiation in the form of gamma rays. This is equivalent to the transformation of a neutron into a proton and an electron. As illustrated in figure 14.1, Z increases by one and N decreases by one for each beta particle emitted.

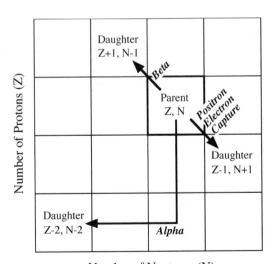

FIG. 14.1. A portion of the nuclide chart illustrating the *N-Z* relationships for daughter nuclides formed from a hypothetical parent by emission of beta particles, positrons, alpha particles, or by electron capture.

Another type of radioactivity occurs as a result of *positron decay*. A *positron* is a positively charged electron expelled from the nucleus. This process can be regarded as the conversion of a proton into a neutron, a positron, and a *neutrino* (a particle with appreciable kinetic energy, but without mass). Positron decay produces a nuclide with N increased by one and Z decreased by one, as illustrated in figure 14.1.

Inspection of figure 14.1 allows us to predict which nuclides tend to transform by beta decay or positron decay. Unstable atoms that lie below the band of stability (as seen in fig. 2.3), and therefore have excess neutrons, are likely to decay by emission of beta particles, so that their N values are reduced. Similarly, nuclides that lie above the band of stability, having excess protons, may experience positron decay. In each case, these processes occur in such a way that the resulting daughter nuclides fall within the band of nuclear stability.

An alternate type of decay mechanism is *electron capture*. In this process, the nuclide increases its N and decreases its Z by addition of an electron from outside the nucleus. A neutrino is also liberated during this process. The daughter nuclide produced during electron capture occupies exactly the same position relative to its parent in figure 14.1 as that produced during positron decay.

Alpha decay proceeds by the emission of heavy alpha particles from the nucleus. An *alpha particle* is composed of two neutrons and two protons. Therefore, the N and Z values for the daughter nuclide both decrease by two, as shown in figure 14.1.

To be complete, we should also add to this list of radioactive decay processes *spontaneous fission*, which is an alternate mode of decay for some heavy atomic nuclei. These atoms may break apart, because electrostatic repulsion between the Z positively charged protons overcomes the strong nuclear binding force. Fission products generally have excess neutrons and tend to decay further by beta emission. For most geochemical applications, fission can be considered a relatively minor side effect.

The decay of natural radionuclides may be much more complex than suggested by the simple decay mechanisms just introduced. A particular transformation may employ several of these decay mechanisms simultaneously, so that the parent atoms form more than one kind of daughter. Such a process is called *branched decay*. As an example, consider the decay of ^{40}K. Most atoms of the parent nuclide decay by positron emission and electron capture into ^{40}Ar, but ^{40}Ca is produced from the remainder by beta decay. Branched decay decreases the yield of daughter atoms of a particular nuclide, requiring more sensitivity in measurement.

Worked Problem 14.1

How many atoms of ^{40}Ar would be produced by the complete decay of ^{40}K in a 1 cm^3 crystal of orthoclase? To answer this question, we must first calculate the number of atoms of parent ^{40}K in the sample. We can convert the volume of orthoclase to weight by multiplying by its density, and weight to moles by dividing by the gram formula weight of orthoclase:

$$1 \text{ cm}^3 (2.59 \text{ g cm}^{-3})/278.34 \text{ g mol}^{-1}$$
$$= 9.30 \times 10^{-3} \text{ mol}.$$

Each mole of orthoclase contains one mole of potassium, so there are 9.30×10^{-3} moles of K in this sample. Multiplication of this result by Avogadro's number gives the number of atoms of potassium:

$$9.30 \times 10^{-3} \text{ mol } (6.023 \times 10^{23} \text{ atoms mol}^{-1})$$
$$= 5.60 \times 10^{19} \text{ atoms}.$$

The natural abundance of ^{40}K in potassium (before decay) is 0.01167%. Therefore, this orthoclase sample contains:

$$5.60 \times 10^{19} \text{ atoms } (1.167 \times 10^{-4})$$
$$= 6.535 \times 10^{15} \text{ atoms } {}^{40}\text{K}.$$

We have already mentioned that ^{40}K undergoes branched decay, and we are concerned here with only one of the daughter nuclides. A total of 11.16% of the ^{40}K atoms decay into ^{40}Ar by electron capture and positron decay, the remainder transforming to ^{40}Ca. Consequently, the number of atoms of ^{40}Ar produced by complete decay is:

$$6.535 \times 10^{15} (0.1116) = 7.293 \times 10^{14} \text{ atoms } {}^{40}\text{Ar}.$$

Rate of Radioactive Decay

In each of the decay mechanisms just discussed, the rate of disintegration of a parent nuclide is proportional to the number of atoms present. In more quantitative terms, the number of atoms (because of convention, we use the symbol N, not to be confused with the symbol for the number of neutrons used in earlier paragraphs) remaining at any time t is:

$$-dN/dt = \lambda N, \tag{14.1}$$

where λ is the constant of proportionality, usually called the *decay constant*. The value of the decay constant is characteristic for a particular radionuclide and is expressed in units of reciprocal time. Rearranging the terms of equation 14.1 and integrating gives:

$$-\int dN/N = \lambda \int dt,$$

or

$$-\ln N = \lambda t + C, \tag{14.2}$$

where C is the constant of integration. If $N = N_0$ at time $t = 0$, then:

$$C = -\ln N_0.$$

By substituting this value into equation 14.2, we obtain:

$$-\ln N = \lambda t - \ln N_0,$$

or

$$\ln N/N_0 = -\lambda t.$$

This is more commonly expressed as:

$$N/N_0 = e^{-\lambda t}. \tag{14.3}$$

Equation 14.3 is the basic relationship that describes all radioactive decay processes. With it, we can calculate the number of parent atoms (N) that remain at any time t from an original number of atoms (N_0) present at time $t = 0$.

We can also express this relationship in terms of atoms of the daughter nuclide rather than the parent. If no daughter atoms are present at time $t = 0$ and none are added to or lost from the system during decay, then the number of radiogenic daughter atoms produced by decay D^* (not to be confused with the same symbol used for tracer diffusion coefficients in chapter 11) at any time t is:

$$D^* = N_0 - N. \tag{14.4}$$

By rearranging equation 14.3 and substituting it into 14.4, we see that:

$$D^* = Ne^{\lambda t} - N,$$

or

$$D^* = N(e^{\lambda t} - 1). \tag{14.5}$$

Equation 14.5 gives the number of daughter atoms produced by decay at any time t as a function of parent atoms remaining. If some atoms of the daughter nuclide were present initially (D_0), then the total number of daughter atoms (D) is:

$$D = D_0 + D^*.$$

Combining this equation with 14.5 produces the useful result:

$$D = D_0 + N(e^{\lambda t} - 1). \tag{14.6}$$

This important equation is the basis for geochronology. Both D and N are measurable quantities, and D_0 is a constant whose value can be determined.

It is common practice to express radioactive decay rates in terms of half-lives. The *half-life* ($t_{1/2}$) is defined as the time required for one-half of a given number of atoms of a nuclide to decay. Therefore, when $t = t_{1/2}$, it follows that $N = \frac{1}{2}N_0$. Substitution of these values into equation 14.3 gives:

$$\tfrac{1}{2}N_0/N_0 = e^{-\lambda t_{1/2}},$$

or

$$\ln \tfrac{1}{2} = -\lambda t_{1/2}.$$

This is equivalent to:

$$\ln 2 = \lambda t_{1/2}.$$

Solving for $t_{1/2}$ gives:

$$t_{1/2} = \ln 2/\lambda = 0.693/\lambda. \tag{14.7}$$

Equation 14.7 expresses the relationship between the half-life of a nuclide and its decay constant.

The equations above show that the disintegration of a parent radioactive isotope and the generation of daughter nuclides are both exponential functions of time. This is illustrated in the following worked problem.

Worked Problem 14.2

Follow the decay of radioactive parent ^{87}Rb and the growth of daughter ^{87}Sr in a granite sample over the course of six half-lives. Assume that the granite sample initially contains 1.2×10^{20} atoms of ^{87}Rb and 0.3×10^{20} atoms of ^{87}Sr.

The half-life of ^{87}Rb is 48.8×10^9 years. The decay constant for this radionuclide can be calculated by using equation 14.7:

$$\lambda = 0.693/(48.8 \times 10^9 \text{ yr}) = 1.42 \times 10^{-11} \text{ yr}^{-1}.$$

Using equation 14.3, we can determine the number of atoms (N) of parent ^{87}Rb remaining after one half-life:

$$N/(1.2 \times 10^{20}) = e^{-(1.42 \times 10^{-11})(48.8 \times 10^9)},$$

or

$$N = 6.00 \times 10^{19} \text{ atoms after one half-life.}$$

To calculate the number of atoms of the parent nuclide remaining after two half-lives, we substitute for N_0 in the same equation the value of N just calculated:

$$N/6.00 \times 10^{19} = e^{-(1.42 \times 10^{-11})(48.8 \times 10^9)},$$

or

$$N = 3.00 \times 10^{19} \text{ atoms after two half-lives,}$$

and so forth.

To calculate the number of atoms (D) of daughter ^{87}Sr after one half-life, we use equation 14.6:

$$D = 0.3 \times 10^{20} + 6.00 \times 10^{19}[e^{(1.42 \times 10^{-11})(48.8 \times 10^9)} - 1]$$

or

$$D = 9.00 \times 10^{19} \text{ atoms after one half-life.}$$

We follow a similar procedure for calculating daughter atoms after each successive half-life.

The easiest way to summarize these calculations is by means of a graph, plotting the number of atoms of parent (N) or daughter (D) versus time in units of half-life. This diagram is shown in figure 14.2. The exponential character of the radio-

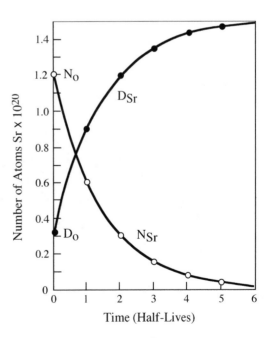

FIG. 14.2. The decay of radioactive ^{87}Sr (N) into its stable radiogenic daughter, ^{87}Rb (D), as a function of time measured in half-lives. N_0 and D_0 represent the initial number of atoms of parent and daughter, respectively.

active decay is obvious. After six half-lives, N approaches zero asymptotically and, for most practical purposes, the production of daughter atoms ceases.

Decay Series and Secular Equilibrium

The calculations above presume that a radioactive parent decays directly into a stable daughter. However, many radionuclides decay to a stable nuclide by way of transitory, unstable daughters; that is, a parent N_1 may decay to a radioactive daughter N_2, which in turn decays to a stable daughter N_3. Naturally occurring decay series arising from radioactive isotopes of uranium and thorium behave in this manner. For example, ^{238}U produces 13 separate radioactive daughters before finally arriving at stable ^{206}Pb, as illustrated in figure 14.3. Equations giving the number of atoms of any member of a decay series at any time t are presented in a text by Gunter Faure (1986), and will not be reproduced here. It is interesting, however, to examine the special case in which the half-life of the parent nuclide is very much longer than those of the radioactive daughters. In this situation, it can be shown that after some initial time interval has passed,

$$\lambda_1 N_1 = \lambda_2 N_2 = \lambda_3 N_3 = \lambda_n N_n,$$

where $\lambda_{1, 2, \dots, n}$ are the decay constants for each nuclide (N) in the series. This condition, in which the rate of decay of the daughters equals that of the parent, is known as *secular equilibrium*. It allows an important simplification of decay series calculations, because the system can be treated as if the original parent decayed directly to the stable daughter without intermediate steps. Secular equilibrium is assumed in uranium prospecting methods that utilize radiation measurements to calculate uranium abundances.

GEOCHRONOLOGY

One of the prime uses of radiogenic isotopes is, of course, to determine the ages of rocks. All radiometric dating systems assume that certain conditions are satisfied:

1. The system, which may be defined as the rock or as an individual mineral, must have remained closed so that neither parent nor daughter atoms were lost or gained except as a result of radioactive decay.
2. If atoms of the daughter nuclide were present before system closure, it must be possible to assign a value to this initial amount of material.

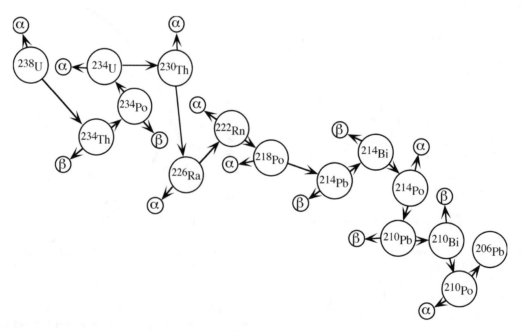

FIG. 14.3. Representation of the radioactive decay of ^{238}U to ^{206}Pb through a series of daughter isotopes. Alpha and beta particles produced at each step are labeled.

PROSPECTING AND WELL-LOGGING TECHNIQUES THAT USE RADIOACTIVITY

Uranium prospecting tools rely principally on detection of gamma radiation, because alpha and beta particles cannot penetrate an overburden cover of even a few cm thickness. The Geiger counter is commonly used for field surveys, although it is a rather inefficient detector. The instrument consists of a sealed glass tube containing a cathode and anode with a voltage applied. The tube is filled with a gas that is normally nonconducting, but when gamma radiation passes through the gas, it is ionized and the ions accelerate toward the electrodes. The resulting current pulses are recorded on a meter or heard as "clicks."

The scintillation counter is a more efficient tool for gamma ray detection. Certain kinds of crystals *scintillate*; that is, they emit tiny flashes of visible light when they absorb gamma radiation. These scintillations are detected by photomultiplier tubes and converted to electrical pulses that can be counted. This instrument can be used in either ground or airborne surveys.

A commonly used well-logging tool employs natural radioactivity to identify sedimentary lithologic boundaries in drill holes. The gamma ray log depends on scintillation detection of gamma rays produced by decay of ^{40}K and radioactive daughter products in the uranium and thorium decay series. These large ions are incompatible in most crystal structures, but are accommodated in clay minerals. Therefore, increases in gamma radiation are ascribed to clay concentrations (that is, shale formations) and, conversely, decreased radiation to cleaner sandstone or limestone units. An example of a portion of a gamma ray well log is shown in figure 14.4. Gamma rays from decay

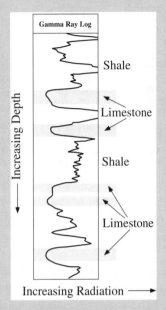

FIG. 14.4. An example of a gamma ray well log used in identifying sedimentary lithologic variations. The horizontal scale is radioactivity (increasing to the right) and the vertical scale is depth in the bore hole. Limestone horizons have low radioactivity relative to shale units because of lower contents of potassium, uranium, and thorium.

of any one nuclide have a particular energy, and the energy spectrum of this radiation is resolved in some gamma ray logs. This permits estimates to be made of the concentrations of potassium, uranium, and thorium (assuming secular equilibrium), and the K/U or K/Th ratios may serve as geochemical signatures for certain shale horizons, which may be useful in stratigraphic correlation.

3. The value of the decay constant must be known accurately. (This is particularly critical for nuclides with long half-lives, because a small error in the decay constant will translate into a large uncertainty in the age.)
4. The isotopic compositions of analyzed samples must be representative and must be measured accurately.

A number of naturally occurring radionuclides are in current use for geochronology; some of the most common are listed in table 14.1 and discussed below. Each

of these is employed for its special properties, such as rate of decay, response to heating and cooling, or concentration range in certain rocks.

Potassium-Argon System

In worked problem 14.1, we have seen that radioactive ^{40}K undergoes branching decay to produce two different daughter products. Its decay into ^{40}Ca is not a

TABLE 14.1. Radionuclides Commonly Used in Geochronology

Parent	Stable Daughter	Half-life (yr)	Decay Constant (yr^{-1})
Long-lived radionuclides			
^{40}K	^{40}Ar	1.25×10^9	5.54×10^{-10}
^{87}Rb	^{87}Sr	4.88×10^{10}	1.42×10^{-11}
^{147}Sm	^{143}Nd	1.06×10^{11}	6.54×10^{-13}
^{176}Lu	^{176}Hf	3.57×10^{10}	1.94×10^{-11}
^{187}Re	^{187}Os	4.56×10^{10}	1.52×10^{-11}
^{232}Th	^{208}Pb	1.40×10^{10}	4.95×10^{-11}
^{235}U	^{207}Pb	7.04×10^8	9.85×10^{-10}
^{238}U	^{206}Pb	4.47×10^9	1.55×10^{-10}
Short-lived and extinct radionuclides			
^{14}C	^{14}N	5.73×10^3	1.21×10^{-4}
^{10}Be	^{10}Be	1.50×10^6	4.62×10^{-7}
^{26}Al	^{26}Mg	7.20×10^5	9.60×10^{-7}
^{129}I	^{129}Xe	1.60×10^7	4.30×10^{-8}
^{182}Hf	^{182}W	9.0×10^{-6}	7.7×10^{-8}
^{244}Pu	$^{131-136}$Xe[1]	8.20×10^7	8.50×10^{-9}

[1] Many other fission products are also produced.

useful chronometer, because ^{40}Ca is also the most abundant stable isotope of calcium. As a result, the addition of radiogenic ^{40}Ca increases its abundance only slightly. Even though ^{40}Ar is the most abundant isotope of atmospheric argon, potassium-bearing minerals do not commonly contain argon. The radioactive decay of ^{40}K can therefore be monitored through time by measuring ^{40}Ar.

Each branch of the decay scheme has its own separate decay constant. The total decay constant λ is therefore the sum of these:

$$\lambda = \lambda_{Ca} + \lambda_{Ar}.$$

The commonly adopted value for the total decay constant and the corresponding half-life are given in table 14.1. The proportion of ^{40}K atoms that decay to ^{40}Ar is given by the ratio of decay constants, $\lambda_{Ar}/(\lambda_{Ar} + \lambda_{Ca})$, which has a value of 0.105.

We can use equation 14.6 to specify the increase in ^{40}Ar through time:

$$^{40}\text{Ar} = {}^{40}\text{Ar}_0 + 0.105 \ {}^{40}\text{K}(e^{\lambda t} - 1).$$

Notice that N, the number of atoms of ^{40}K, has been multiplied by $\lambda_{Ar}/(\lambda_{Ar} + \lambda_{Ca})$ to correct for the fact that most of these atoms will decay to another daughter nuclide. Because most minerals initially contain virtually no argon, $^{40}\text{Ar}_0 = 0$ and the equation reduces to:

$$^{40}\text{Ar} = 0.105 \ {}^{40}\text{K}(e^{\lambda t} - 1). \tag{14.8}$$

Application of equation 14.8 then requires only the measurement of ^{40}Ar and potassium in the sample; ^{40}K

is calculated from total potassium using its relative isotopic abundance of 0.01167%.

The K-Ar system is useful for dating igneous and, in a few cases, sedimentary rocks. Sediments containing glauconite and certain clay minerals that formed during diagenesis may be suitable for K-Ar chronology. However, because the daughter isotope is a gas, it may diffuse out of many minerals, especially if they have been buried deeply or have protracted thermal histories. Metamorphism appears to reset the system in most cases, and K-Ar dates may record the time of peak metamorphism in cases for which cooling was relatively rapid. During slow cooling, argon diffusion continues until the system reaches some critical temperature (the *blocking temperature*), at which it becomes closed to further diffusion. Different minerals in the same rock may have different blocking temperatures and thus yield slightly different ages.

Rubidium-Strontium System

^{87}Rb produces ^{87}Sr by beta decay, at a rate given in table 14.1. Unlike the K-Ar system, the radiogenic daughter is added to a system that already contains some ^{87}Sr. Thus, the difficulty in applying this geochronometer lies in determining how much of the ^{87}Sr measured in the sample was produced by decay of the parent isotope.

For the Rb-Sr system, equation 14.6 takes the form:

$$^{87}\text{Sr} = {}^{87}\text{Sr}_0 + {}^{87}\text{Rb}(e^{\lambda t} - 1).$$

Another isotope of strontium, ^{86}Sr, is stable and is not produced by decay of any naturally occurring isotope. If both sides of the equation are divided by the number of atoms of ^{86}Sr in the sample (a constant), we will not affect the equality:

$$^{87}\text{Sr}/^{86}\text{Sr} = {}^{87}\text{Sr}_0/^{86}\text{Sr} + {}^{87}\text{Rb}/^{86}\text{Sr}(e^{\lambda t} - 1). \tag{14.9}$$

Let's see how the initial ratio of $^{87}\text{Sr}_0/^{86}\text{Sr}$ in this equation, as well as the age, can be derived from graphical analysis of a suite of rock samples. Equation 14.9 is the expression for a straight line (of the form $y = b + mx$). We will construct a plot of $^{87}\text{Sr}/^{86}\text{Sr}$ versus $^{87}\text{Rb}/^{86}\text{Sr}$; that is, y versus x, as shown in figure 14.5. Both of these ratios are readily measurable in rocks. We can assume that, at the time of crystallization ($t = 0$), all minerals in a given rock will have the same $^{87}\text{Sr}/^{86}\text{Sr}$ value, because they cannot discriminate between these relatively heavy isotopes. The various minerals will have different con-

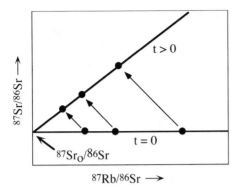

FIG. 14.5. Schematic Rb-Sr isochron diagram, illustrating the isotopic evolution of three samples with time. All samples have the same initial $^{87}Sr/^{86}Sr$ ratio but different $^{87}Rb/^{86}Sr$ ratios at time $t = 0$. The isotopic composition of each sample moves along a line of slope -1 as ^{87}Rb decays to ^{87}Sr. After some time has elapsed, the samples define a new isochron, the slope of which is $(e^{\lambda t} - 1)$. Extrapolation of this isochron to the abcissa gives the initial Sr isotopic composition.

tents of rubidium and strontium, however, and thus different values of $^{87}Rb/^{86}Sr$.

This situation is illustrated schematically by the line labeled $t = 0$ in figure 14.5. After some time has elapsed, a fraction of the ^{87}Rb atoms in each mineral have decayed to ^{87}Sr. Obviously, the mineral that had the greatest initial concentration of ^{87}Rb now has the greatest concentration of radiogenic ^{87}Sr. The position of each mineral in the rock shifts along a line with a slope of -1, as shown by arrows in figure 14.5. The slope of the resulting straight line is $(e^{\lambda t} - 1)$, and its intercept on the y axis is the initial strontium isotopic ratio, $^{87}Sr_0/^{86}Sr$. The diagonal line is called an *isochron*, and its slope increases with time. The fit of data points to a straight line is never perfect, and a linear regression is required. Analytical errors that lead to dispersion of data about the line give rise to corresponding uncertainties in ages.

Isochron diagrams such as figure 14.5 can be constructed by using data from coexisting minerals in the same rock (producing a *mineral isochron*) or from fractionated comagmatic rock samples that have concentrated different minerals to varying degrees (producing a *whole-rock isochron*). Metamorphism may rehomogenize rubidium and strontium isotopes, so that the timing of peak metamorphism may be recorded by this geochronometer. A mineral isochron is more likely than a whole-rock isochron to be reset by metamorphism, be-

cause it is easier to re-equilibrate adjacent minerals than different portions of a rock body. Diagenetic minerals in some sedimentary rocks may also have high enough rubidium contents that the age of burial and diagenesis can be determined using this system. The half-life of ^{87}Rb is long, so that this geochronometer complements the K-Ar system in terms of the accessible time range to which it can be applied.

Worked Problem 14.3

Larry Nyquist and his coworkers (1979) measured the following isotopic data for a whole-rock (WR) sample and for plagioclase (Plg), pyroxene (Px), and ilmenite (Ilm) mineral separates in an Apollo 12 lunar basalt (sample 12014).

Mineral	Rb (ppm)	Sr (ppm)	$^{87}Rb/^{86}Sr$	$^{87}Sr/^{86}Sr$
WR	0.926	90.4	0.0296	0.70096
Plg	0.599	323	0.00537	0.69989
Px	0.386	22.7	0.0492	0.70200
Ilm	3.76	96.5	0.1127	0.70490

What is the age of this rock?

We can determine the age and the initial Sr isotopic ratio, $^{87}Sr_0/^{86}Sr$, by plotting these data on an isochron diagram (fig. 14.6). A least-squares regression through the data can then be calculated, as illustrated by the diagonal line in the figure. The y intercept of this regression line corresponds to an initial isotopic ratio of 0.69964.

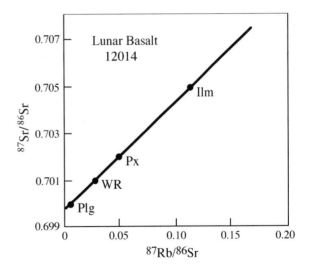

FIG. 14.6. Rb-Sr isochron for lunar basalt 12014. Isotopic data for the whole-rock (WR), pyroxene (Px), plagioclase (Plg), and ilmenite (Ilm) were obtained by Nyquist et al. (1979). The slope of this isochron corresponds to an age of 3.09 b.y.

Equation 14.9 can be solved for t:

$$t = 1/\lambda \ln([(^{87}Sr/^{86}Sr - ^{87}Sr_0/^{86}Sr)]/(^{87}Rb/^{86}Sr) + 1). \tag{14.10}$$

If we now insert the initial Sr isotopic ratio and the measured isotopic data for one sample (we use WR) into this expression, we find:

$$t = 1/(1.42 \times 10^{-11}) \ln([(0.70096 - 0.69964)/(0.0296)] + 1),$$

or

$$t = 3.09 \text{ b.y.}$$

This age corresponds to the time that the minerals in the basaltic magma crystallized. It differs slightly from the age published by Nyquist and coworkers, because they used a different decay constant for ^{87}Rb than that given in table 14.1. The estimated uncertainty in age based on fitting the regression line is $+0.11$ billion years.

Samarium-Neodymium System

The rare earth elements samarium and neodymium form the basis for another geochronometer. ^{147}Sm decays to ^{143}Nd at a rate indicated in table 14.1. The complication in this system is the same as that in the Rb-Sr system; that is, some ^{143}Nd is already present in rocks before radiogenic ^{143}Nd is added, so we must find a way to specify its initial abundance. This can be done by graphical methods after normalizing parent and daughter to a stable, nonradiogenic isotope of neodymium (^{144}Nd), in the same way we normalized to ^{86}Sr in the Rb-Sr system. An expression for the Sm-Nd system analogous to equation 14.9 can then be written:

$$^{143}Nd/^{144}Nd = ^{143}Nd_0/^{144}Nd + ^{147}Sm/^{144}Nd(e^{\lambda t} - 1), \tag{14.11}$$

where the subscript 0 indicates the initial isotopic composition of neodymium at the time of system closure. On an isochron plot of $^{143}Nd/^{144}Nd$ versus $^{147}Sm/^{144}Nd$, the y intercept is the initial isotopic ratio $^{143}Nd_0/^{144}Nd$ and the slope of the line becomes steeper with time.

The Sm-Nd radiometric clock is analogous in its mathematical form to that of the Rb-Sr system, but the two systems are applicable to different kinds of rocks. Basalts and gabbros usually cannot be dated precisely using the Rb-Sr system, because their contents of rubidium and strontium are so low. The Sm-Nd system is ideal for determination of the crystallization ages of mafic igneous rocks. This system may also be useful, in some cases, to "see through" a metamorphic event to the earlier igneous crystallization. Because both parent and daughter are relatively immobile elements, they are less likely to be disturbed by later thermal overprints.

Uranium-Thorium-Lead System

The U-Th-Pb system is actually a set of independent geochronometers that depend on the establishment of secular equilibrium. ^{238}U decays to ^{206}Pb, ^{235}U to ^{207}Pb, and ^{232}Th to ^{208}Pb, all through independent series of transient daughter isotopes. The half-lives of all three parent isotopes are much longer than those of their respective daughters, however, so that the prerequisite condition for secular equilibrium is present and the production rates of the stable daughters can be considered to be equal to the rates of decay of the parents at the beginning of the series. The half-lives and decay constants for the three parent isotopes are given in table 14.1. We refer to the decay constant for ^{238}U as λ_1, that for ^{235}U as λ_2, and that for ^{232}Th as λ_3.

In addition to these three radiogenic isotopes, lead also has a nonradiogenic isotope (^{204}Pb) that can be used for reference. (^{204}Pb is actually weakly radioactive, but it decays so slowly that it can be treated as a stable reference isotope.) We can express the isotopic composition of lead in minerals containing uranium and thorium by the following equations, analogous to equations 14.9 and 14.11:

$$^{206}Pb/^{204}Pb = ^{206}Pb_0/^{204}Pb + ^{238}U/^{204}Pb \, (e^{\lambda_1 t} - 1), \tag{14.12a}$$

$$^{207}Pb/^{204}Pb = ^{207}Pb_0/^{204}Pb + ^{235}U/^{204}Pb \, (e^{\lambda_2 t} - 1), \tag{14.12b}$$

$$^{208}Pb/^{204}Pb = ^{208}Pb_0/^{204}Pb + ^{232}Th/^{204}Pb \, (e^{\lambda_3 t} - 1). \tag{14.12c}$$

The subscript 0 in each case denotes an initial lead isotopic composition at the time the clock was set.

Each of these isotopic systems gives an independent age. In an ideal situation, the dates should all be the same. In many instances, though, these dates are not concordant, because the minerals do not remain completely closed to the diffusion of uranium, thorium, lead, or some of the intermediate daughter nuclides. To simplify this complicated situation, let's consider a system without thorium, so that we have to deal with only two radiogenic lead isotopes. From equation 14.5, we know that the amount of radiogenic ^{206}Pb at any time must be:

$$^{206}\text{Pb}^* = {}^{238}\text{U}(e^{\lambda_1 t} - 1),$$

or

$$^{206}\text{Pb}^*/{}^{238}\text{U} = e^{\lambda_1 t} - 1, \tag{14.13a}$$

where the superscript * denotes radiogenic lead; that is, $^{206}\text{Pb} - {}^{206}\text{Pb}_0$. Similarly, the amount of radiogenic ^{207}Pb at any time can be expressed as:

$$^{207}\text{Pb}^*/{}^{235}\text{U} = e^{\lambda_2 t} - 1. \tag{14.13b}$$

If a uranium-bearing mineral behaves as a closed system, we know that equations 14.13a and 14.13b should yield concordant ages (that is, the same value of t). Logically, then, we can reverse the procedure and calculate compatible values for $^{206}\text{Pb}^*/{}^{238}\text{U}$ and $^{207}\text{Pb}^*/{}^{235}\text{U}$ at any given time t. Figure 14.7 shows the results of such a calculation. The curve labeled *concordia* in this figure represents the locus of all concordant U-Pb systems.

At the time of system closure (t_0), no decay has occurred yet, so there is no radiogenic lead present. The sys-

tem's isotopic composition, therefore, plots at the origin of figure 14.7. At any time thereafter, the system is represented by some point along the concordia curve; numbers along the curve indicate the elapsed time in billions of years since formation of the system. If, instead, the system experiences loss of lead at time t_1, the system's composition shifts along a chord connecting t_1 to the origin. A complete loss of radiogenic lead would displace the isotopic composition all the way back to the origin, but this rarely happens. If some fraction of lead is lost, as is the more common situation, the system composition is represented by a point (such as x) along the chord. A series of related samples with discordant ages that have experienced varying degrees of lead loss will define the chord, called a *discordia* curve, more precisely. After the disturbing event, each point x on the discordia evolves along a path similar in form but not coincident with the concordia curve.

The result is that the discordia curve retains its linear form but changes slope, intersecting the concordia curve at a new set of points. The time of lead loss is translated along the concordia curve to t_3, and t_0 is similarly

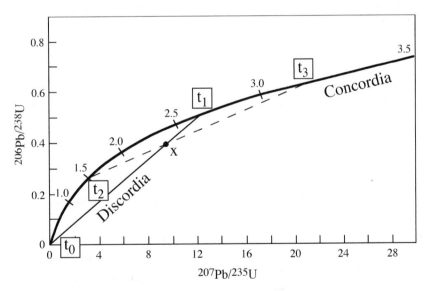

FIG. 14.7. Concordia diagram comparing isotopic data for two U-Pb systems. An undisturbed system formed at time t_0 will move along the concordia curve; the age at any time after formation is indicated by numbers (in billions of years) along the curve. If episodic lead loss occurs at time t_1, the position of the system will be displaced toward the origin by some amount, as illustrated by point x. Other samples that lose different amounts of lead are displaced along the same chord and give discordant dates. With further elapsed time, the discordia line will subsequently be rotated (dashed line) as t_0 moves to t_2 and t_1 moves to t_3.

translated to t_2, the overall effect being a rotation of discordia about point x. This is illustrated by the dashed discordia curve in figure 14.7.

If the discordia chord can be adequately defined, then this diagram may yield two geologically useful dates: the time of formation of the rock unit (t_2), and the time of a subsequent disturbance (possibly metamorphism or weathering) that caused lead loss (t_3). This interpretation assumes that loss of lead was instantaneous; if lead was lost by continuous diffusion over a long period of time, then t_3 is a fictitious date without geologic significance. Unfortunately, it is not possible to distinguish between these alternatives without other geological information.

It is possible to compensate for the effect of lead loss on U-Pb ages by calculating ages based on the $^{207}Pb/^{206}Pb$ ratio. This ratio is, of course, insensitive to lead mobilization, because it is practically impossible to fractionate isotopes of the same heavy element. By combining equations 14.12a and 14.12b, we obtain:

$$(^{207}Pb/^{204}Pb - {}^{207}Pb_0/^{204}Pb)/(^{206}Pb/^{204}Pb - {}^{206}Pb_0/^{204}Pb) = (^{235}U[e^{\lambda_2 t} - 1])/(^{238}U[e^{\lambda_1 t} - 1]).$$

$$(14.14a)$$

The left side of this equation is equivalent to the ratio of radiogenic ^{207}Pb to radiogenic ^{206}Pb; that is, to $(^{207}Pb/^{206}Pb)^*$. This is determined by subtracting initial $^{207}Pb/^{204}Pb$ and $^{206}Pb/^{204}Pb$ values from the measured values for these ratios. Consequently, ages can be determined solely on the basis of the isotopic ratios in the sample, without having to determine lead concentrations. Moreover, because the ratio $^{235}U/^{238}U$ is constant (1/137.8) for all natural materials at the present time, ages can be calculated without measuring uranium concentrations. Equation 14.14a can therefore be expressed as:

$$(^{207}Pb/^{206}Pb)^* = 1/137.8([e^{\lambda_2 t} - 1]/[e^{\lambda_1 t} - 1]).$$

$$(14.14b)$$

This is the expression for the $^{207}Pb/^{206}Pb$ age, also called the *lead-lead age*. The only difficulty in applying this method is that equation 14.14b cannot be solved for t by algebraic methods. However, the equation can be iteratively solved by a computer method of successive approximations until a solution is obtained within an acceptable level of precision.

The U-Th-Pb dating system is applicable to rocks containing such minerals as zircon, apatite, monazite, and sphene. These phases provide large structural sites for uranium and thorium. Igneous or metamorphic rocks of granitic composition are most suitable for dating by this method. Several generations of zircons, recognizable by distinct morphologies, have been found to give distinct ages in some rocks, and measurements of U-Th-Pb ages in zoned zircons using ion microprobes have shown overgrowths with younger ages. The possibility of minerals inherited from earlier events or affected by later events further complicates the interpretation of U-Th-Pb ages.

Worked Problem 14.4

John Aleinikoff and coworkers (1985) measured the following lead isotopic data for a zircon from a granitic pluton in Maine (units are atom %): $^{204}Pb = 0.010$, $^{206}Pb = 90.6$, $^{207}Pb = 4.91$, $^{208}Pb = 4.43$. The measured U concentration was 1396 ppm, and that for Pb was 62.2 ppm. What are the $^{206}Pb/^{238}U$, $^{207}Pb/^{235}U$, and $^{207}Pb/^{206}Pb$ ages for this rock, and are they concordant?

We begin by determining the proportions of ^{235}U and ^{238}U in this sample. All natural uranium has $^{235}U/^{238}U = 1/137.8$, corresponding to 99.27% ^{238}U. Aleinikoff and coworkers assumed that the nonradiogenic lead in this sample had isotopic ratios of 204:206:207:208 = 1:18.2:15.6:38, corresponding to an average atomic weight of 207.23. From these proportions, we can calculate the following ratios:

$$^{238}U/^{204}Pb = (1396/62.2)(207.23/238.03)(99.27/0.010)$$
$$= 193,970,$$

$$^{235}U/^{204}Pb = 193,970/137.8 = 1407.6,$$

$$^{206}Pb/^{204}Pb = (90.6/0.010) - 18.2 = 9041.8,$$

$$^{207}Pb/^{204}Pb = (4.91/0.010) - 15.6 = 475.4,$$

$$^{206}Pb_0/^{204}Pb = 18.2,$$

$$^{207}Pb_0/^{204}Pb = 15.6.$$

We can now insert values into equation 14.12 and solve for t. The expression for the $^{206}Pb/^{238}U$ age, corresponding to equation 14.12a, is:

$$t_{206} = 1/\lambda_1 \ln([[(^{206}Pb/^{204}Pb) - (^{206}Pb_0/^{204}Pb)]/[^{238}U/^{204}Pb] + 1).$$

Substitution of the ratios above into this equation gives:

$$t_{206} = 1/(1.55 \times 10^{-10}) \ln([9041.8 - 18.2]/[193,970] + 1) = 293 \text{ m.y.}$$

The $^{207}Pb/^{235}U$ age, calculated from the analogous expression derived from equation 14.12b, is:

$$t_{207} = 1/(9.85 \times 10^{-10}) \ln([475.4 - 15.6]/[1407.6] + 1)$$
$$= 287 \text{ m.y.}$$

Subtracting the initial ratios for the isotopic composition of non-radiogenic lead gives:

$$^{207}Pb/^{204}Pb = 491 - 15.6 = 475.4,$$

and

$$^{206}Pb/^{204}Pb = 9060 - 18.2 = 9041.8.$$

Dividing these results gives:

$$(^{207}Pb/^{206}Pb)^* = 475.4/9041.8 = 0.0526.$$

We substitute this ratio into equation 14.14b and solve for t to find that the $^{207}Pb/^{206}Pb$ age is 308 m.y.

These ages are similar and date the approximate time of crystallization of the pluton. This sample would plot on the concordia curve, because ages determined by both U-Pb systems are nearly identical.

Extinct Radionuclides

All of the naturally occurring radionuclides that we have considered so far have very long half-lives, so that most of the parent isotopes still occur naturally on the Earth. However, some terrestrial rock samples and meteorites contain evidence of the former presence of now extinct radionuclides that had short half-lives. The stable daughter products and rates of decay for ^{26}Al, ^{129}I, ^{182}Hf, and ^{244}Pu are given in table 14.1. That these isotopes occurred at all as "live" radionuclides demands that the samples that contain them formed very soon after the nuclides were created. In worked problem 14.5, we see how decay of now extinct ^{182}Hf can be used as a geochronological tool. Because of rapid decay rates, a few extinct radionuclides such as ^{26}Al may have been important sources of heat during the early stages of planet formation. The implications of the former presence of ^{26}Al and ^{182}Hf in meteorites will be considered in chapter 15.

Worked Problem 14.5

How can the now extinct radionuclide ^{182}Hf be used to constrain the time of core formation in the Earth? ^{182}Hf decays to ^{182}W with a half-life of only 9 million years, so any ^{182}Hf present in the early Earth has long since decayed. Hafnium is partitioned into silicates (that is, it is lithophile), and tungsten is concentrated in metal (it is siderophile). Consequently, tungsten was partitioned into the Earth's core when it separated from the mantle. If core separation happened after ^{182}Hf had decayed, then all of the daughter ^{182}W would have been sequestered in the core. If the core formed while ^{182}Hf was still alive, however,

this nuclide would have remained in the silicate mantle, and its subsequent decay would have produced an excess of radiogenic ^{182}W relative to nonradiogenic ^{184}W in the mantle. An anomalous $^{182}W/^{184}W$ ratio in the mantle would then be passed on to basaltic magmas derived from it.

Geochemists Alex Halliday and Der-Chuen Lee (1999) compared the tungsten isotopic composition in terrestrial silicate rocks with those measured in meteorites (fig. 14.8). In figure 14.8, tungsten isotopes are expressed as ε_W, defined as the deviation between $^{182}W/^{184}W$ in a sample relative to that ratio in a chondrite of identical age. The $^{182}W/^{184}W$ ratio in the Earth is indistinguishable from that of chondritic meteorites that have never experienced metal fractionation. However, iron meteorites, which are samples of the cores of differentiated asteroids, exhibit ^{182}W deficits. The absence of excess ^{182}W in the silicate

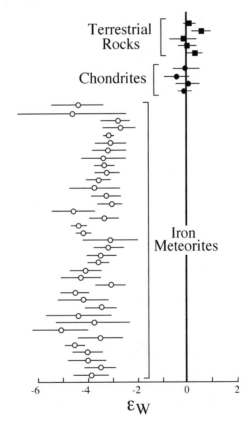

FIG. 14.8. The isotopic composition of tungsten ($^{182}W/^{184}W$), expressed as ε_W values, of terrestrial samples compared with those of meteorites. Terrestrial values are indistinguishable from those of carbonaceous chondrites, which have never experienced metal fractionation, but are distinct from iron meteorites, which formed as differentiated cores in asteroids. These data suggest that core formation in the Earth occurred after ^{182}Hf had decayed to ^{182}W, so that no excess radiogenic tungsten was produced in the mantle. (After Halliday and Lee 1999.)

portion of the Earth suggests that the Earth's core formation must have happened many ^{182}W half-lives after core formation in asteroids. Based on the elapsed time that would be necessary to resolve a ^{182}W/^{184}W ratio higher than in chondrites, Halliday and Lee estimated that core formation in the Earth occurred >50 m.y. after the formation of iron meteorites. A recent revision in the initial ratio of ^{182}Hf to stable ^{180}Hf (Yin et al. 2002) indicates a value half that assumed by Halliday and Chuen, which lowers the time of formation of the Earth's core to 29 m.y.

Fission Tracks

If you have examined micas or cordierite under the petrographic microscope, you may have noticed zones of discoloration around tiny inclusions of zircon and other uranium-bearing minerals. These pleochroic haloes are manifestations of radiation damage produced by alpha particles. The crystal lattices of minerals that contain dispersed uranium atoms also record the disruptive passage of alpha particles in the form of fission fragments. The fragments are very small, but they nevertheless produce *fission tracks* as they pass through the host mineral. Although the tracks are only ~10 microns long, they can be enlarged and made optically visible by etching with suitable solutions, because the damaged regions are more soluble than the unaffected regions. A microscopic view of fission tracks can be seen in figure 14.9. The density of fission tracks in a given area of sample can be determined by counting the tracks under a microscope.

These tracks, except in rare instances, are produced solely by spontaneous fission of ^{238}U. The observed track density is proportional to the concentration of ^{238}U in the sample and to the amount of time over which tracks have been accumulating; that is, the age of the host mineral. If a means can be found to measure the uranium concentration in the sample, fission tracks can then be used for geochronology. The uranium abundance can be determined by measuring the density of tracks induced by irradiating the sample in a reactor, where bombardment with neutrons causes more rapid decay (induced radioactivity is explained more fully in the next section). P. B. Price and Robert Walker (1963) formulated a solution for the fission track age equation (presented here without derivation):

$$t = \ln(1 + [\rho_s/\rho_I][\lambda_2 \phi \sigma U/\lambda_F]) 1/\lambda_2.$$

Explaining the terms in this equation illustrates how a fission track age is determined. λ_2 and λ_F are the decay

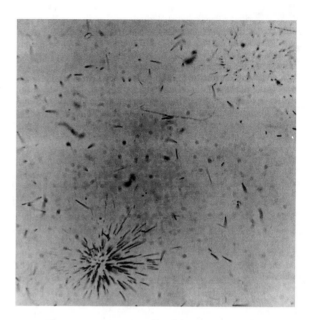

FIG. 14.9. Photomicrograph of fission tracks in mica. Radiating tracks emanate from point sources, such as small zircon inclusions; isolated tracks probably formed from single U or Th atoms. (Courtesy of G. Crozaz.)

constants for ^{238}U and spontaneous fission (8.42×10^{-17} yr^{-1}), respectively. The area density of fission tracks in the sample, ρ_s, is a function of age and uranium content. We noted earlier that the age can be determined uniquely if the uranium content can be measured. Uranium concentration is measured by placing the sample (after ρ_s has been visually counted) into a nuclear reactor, which induces rapid ^{238}U decay and increases the number of fission tracks in the sample. The remaining terms in the equation above are related to this irradiation process: ρ_i is the area density of induced tracks, ϕ is the flux of neutrons passed through the sample, σ is the target cross section for the induced fission reaction, and U is the (constant) atomic ratio ^{235}U/^{238}U.

Annealing of the sample at elevated temperatures causes fission tracks to fade, as the damage done to the crystal structure is healed. The annealing temperatures for most minerals used in fission track work are quite modest (several hundred degrees Centigrade or less), so the ages derived by this method must be interpreted as times the host rocks cooled through temperatures at which annealing ceased, commonly called *cooling ages*. Such phases as apatite, sphene, and epidote are useful for the interpretation of cooling histories, and they begin to retain fission tracks at temperatures that are different from the blocking temperatures for argon isotopes. Fission track annealing temperatures, like radiogenic isotope blocking temperatures, can be extrapolated from experimental data. The determination of cooling history using a combination of fission tracks and other isotopic methods is illustrated in the following worked problem.

Worked Problem 14.6

How can we determine cooling history from isotope blocking temperatures and fission track retention ages? The elucidation of cooling history usually requires integration of several different kinds of data, tempered with thoughtful geologic reasoning. D. L. Nielson and coworkers (1986) characterized the temperature history of a Cenozoic pluton in Utah in the this way. Critical observations for their interpretation of the cooling history are:

1. The K-Ar age of a hornblende sample from the pluton is 11.8 million years. The argon blocking temperature for hornblende is ~525° C.
2. The K-Ar age of a biotite sample from the pluton is 10.8 million years. The argon blocking temperature for biotite is ~275° C.
3. The fission track ages for zircons in these rocks range from 8.3 to 8.9 million years. The temperature at which zircon begins to retain tracks depends on the cooling rate, but is generally ~175° C.
4. The fission track ages for apatites range from 8.1 to 9.1 million years. Apatite begins to retain tracks at ~125° C.

Using these data, we can construct a temperature versus time plot that describes the cooling history of this pluton (fig. 14.10). We can also speculate about geologic controls on the cooling

path. The temperature-time curve in figure 14.10 implies that this body cooled from >500° C through ~100° C within a span of only 4 million years. This rapid cooling could not have taken place very deep in the crust and must indicate uplift and erosion. Nielson and coworkers calculated that uplift proceeded at a rate between 0.6 and 1.1 mm yr^{-1} for this interval, based on an assumed geothermal gradient of 30° km^{-1}.

GEOCHEMICAL APPLICATIONS OF INDUCED RADIOACTIVITY

In medieval times, alchemists toiled endlessly to turn various metals into gold. They were unsuccessful, but only because they did not know about nuclear irradiation. It is now possible to make gold, but it is still not economical, because the starting material must be platinum.

Nuclei that are bombarded with neutrons, protons, or other charged particles are transformed into different nuclides. In many cases, the nuclides produced by such irradiation are radioactive. Monitoring their decay provides useful analytical information. Nuclear irradiation can occur naturally or can be induced artificially. Here we consider some geochemical applications of both situations.

Neutron Activation Analysis

Nuclear reactions caused by neutron bombardment are commonly used to perform quantitative analyses of trace elements in geological samples. Fission of ^{235}U in a nuclear reactor produces large quantities of neutrons,

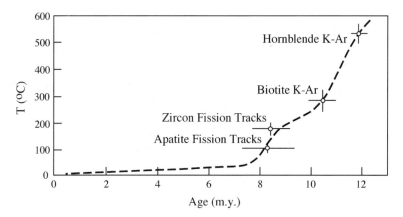

FIG. 14.10. The cooling history of an igneous pluton, inferred from argon blocking temperatures for hornblende and biotite and fission track ages for zircon and apatite. (Data reported by Nielson et al. 1986.)

which can be used to induce nuclear changes in mineral or rock samples. Slow neutrons are readily absorbed by the nuclei of most stable isotopes, transforming them into heavier elements. Because many of the irradiation products are radioactive, this procedure is called *neutron activation*.

In earlier sections, we described radioactivity in terms of N, the number of atoms remaining. In neutron activation analysis, radioactivity is monitored using activity (A), defined as:

$$A = c\lambda N, \tag{14.15}$$

where c is the detection coefficient (determined by calibrating the detector) and λN is the rate of decay. The easiest way to determine the concentration of an element by neutron activation is by irradiating a standard containing a known amount of that element at the same time as the unknown. After irradiation, the activities of both standard and unknown are measured at intervals using a scintillation counter to detect emitted gamma rays. (This step may be complicated for geological samples because they contain so many elements. It is necessary to screen out radiation at energies other than that corresponding to the decay of interest. This can normally be done by adjusting the detector to filter unwanted gamma radiation but, in some cases, chemical separations may be required to eliminate interfering radioactivities.) The activities are then plotted, as shown in figure 14.11; the exponential decay curves are transformed into straight lines by plotting $\ln A$. Extrapolation of these lines back to zero time gives values of A_0, the activities when the samples were first removed from the reactor. The amount of element X in the unknown can then be determined from the following relationship:

$$A_{0\,unkn}/A_{0\,stnd} = \text{amount of } X_{unkn}/$$
$$\text{amount of } X_{stnd}.$$

The concentration of element X in the unknown can be calculated by dividing its amount by the measured weight of sample.

It is also possible to calculate A_0 theoretically by using the following relationship:

$$A_0 = cN_0\phi\sigma(1 - e^{-\lambda t_i}), \tag{14.16}$$

where c is the detection coefficient, N_0 is the number of target nuclei, ϕ is the neutron flux in the reactor, σ is the neutron capture crosssection of target nuclei, and t_i is the time of irradiation. This computation, normally done

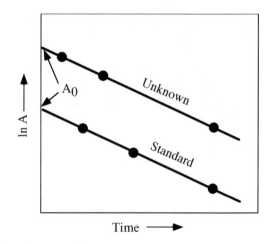

FIG. 14.11. Plot of the measured activity A of an unknown and a standard after neutron irradiation, as a function of time. Extrapolation of these straight lines gives A_0, the activity at the end of irradiation. These data are then used to determine the concentration of the irradiated element in the unknown.

by computer, makes it easier to handle many elements simultaneously in complex samples.

^{40}Argon-^{39}Argon Geochronology

One of the disadvantages of the K-Ar geochronometer is that gaseous argon tends to diffuse out of many minerals, even at modest temperatures. Loss of the radiogenic daughter can thus lead to an erroneously young radiometric age. ^{40}Ar-^{39}Ar measurements provide a way to get around the problem of argon loss. Let's examine how this technique works.

We might expect that the rims of crystals would lose argon more readily than the crystal interiors because of shorter diffusion distances. In this case, the ratio of radiogenic ^{40}Ar to ^{39}K, a stable isotope of potassium, will be highest in crystal centers and lower in crystal rims. Irradiating the sample with neutrons in a reactor causes some ^{39}K to be converted to ^{39}Ar. Consequently, the ratio ^{40}Ar/^{39}K becomes ^{40}Ar/^{39}Ar, which is readily analyzed using mass spectrometry.

The number of ^{39}Ar atoms produced during the irradiation is given by:

$$^{39}\text{Ar} = {}^{39}\text{K}t_i \int \phi\sigma \, d\varepsilon, \tag{14.17}$$

where t_i is the irradiation time, ϕ is the neutron flux in the reactor, σ is the neutron capture cross section, and the integration is carried out over the neutron energy spectrum $d\varepsilon$. The number of radiogenic ^{40}Ar atoms produced by decay of ^{40}K is given by equation 14.8. Dividing equation 14.8 by 14.17 gives the $^{40}Ar/^{39}Ar$ ratio in the sample after irradiation:

$$^{40}Ar/^{39}Ar = (0.124/t_i)(^{40}K/^{39}K)(e^{\lambda t} - 1)/\left(\int \phi\sigma\, d\varepsilon\right). \tag{14.18}$$

Because equation 14.18 can be solved for t, measurement of the $^{40}Ar/^{39}Ar$ ratio of a sample defines its age.

In practice, the sample is heated incrementally, and the isotopic composition of argon released at each temperature step is measured. Figure 14.12 illustrates a series of $^{40}Ar/^{39}Ar$ ages, calculated using equation 14.18, for samples of mica and amphibole in a blueschist sample from Alaska, as a function of total ^{39}Ar released. Each step represents a temperature increment of $50°$, starting at $550°$ C. If this sample had remained closed to argon loss since its formation, all of the ages would have been the same. The spectrum of ages that we see in figure 14.12 results from ^{40}Ar leakage over time. Argon from the cores of crystals is released at the higher temperature steps, and the plateau of ages at ~185 million years seen at high temperatures represents the time of metamorphism for this sample, because the mica blocking temperature is approximately equal to the peak metamorphic temperature.

Cosmic-Ray Exposure

In both of the examples we have just considered, the samples were irradiated intentionally by humans. Natural irradiation can also have useful geochemical applications. Cosmic rays, consisting mostly of high energy protons but also of neutrons and other charged particles generated in the Sun and outside the Solar System offer a natural irradiation source. Isotopes produced by interaction of matter with cosmic rays are called *cosmogenic* nuclides.

Radioactive ^{14}C is produced continuously in the atmosphere by the interaction of ^{14}N and cosmic rays. This isotope is then incorporated into carbon dioxide molecules, which in turn enter plant tissue through photosynthesis. The concentration of cosmogenic ^{14}C in living plants (and in the animals that eat them) is maintained at a constant level by atmospheric interaction and isotope decay. When the plant or animal dies, addition of ^{14}C from the atmosphere stops, so its concentration decreases due to decay. Measurement of the activity of ^{14}C in plant or animal remains thus provides a way to determine the time elapsed since death.

The activity A of carbon from a plant or animal that died t years ago is given by:

$$A = A_0 e^{-\lambda t}. \tag{14.19}$$

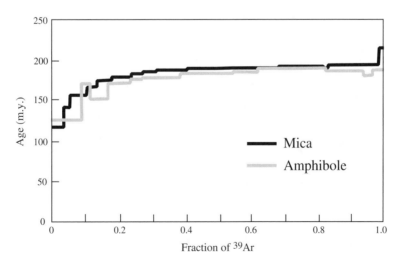

FIG. 14.12. ^{40}Ar-^{39}Ar age versus cumulative ^{39}Ar released for mica and amphibole samples from blueschist. The argon was released at successive temperature steps; each step in this figure corresponds to a 50° increment, starting at 550°C. The plateau of ages at 185 million years corresponds to radiogenic argon from the interiors of crystals. (After Sisson and Onstott 1986.)

where A_0 is the ^{14}C activity during life = 13.56 dpm g^{-1} (expressed as disintegrations per minute per gram of sample), and the decay constant λ is given in table 14.1. (Compare equations 14.19 and 14.15. Both express activity in terms of λN times a constant, c or A_0.) Radiocarbon ages have only a limited geologic range because of the short half-life of ^{14}C, but they are very useful for archaeological dating and for some environmental geochemistry applications.

Cosmic rays sometimes fragment other nuclides, a process called *spallation*. ^{10}Be and ^{26}Al are spallation products formed from oxygen and nitrogen in the atmosphere. Both are rapidly transferred to the oceans or to Earth's surface in rain or snow, where they can be incorporated into sediments on the ocean floor or into continental ice sheets. After deposition, these cosmogenic nuclides decay at rates given in table 14.1, so they can be used to date sediment or ice cores.

Rocks and soils at the Earth's surface are also exposed to cosmic rays, so the accumulation of cosmogenic nuclides in them can be used to measure the durations of sample exposure. These time intervals, called *cosmic ray exposure ages,* assess how long materials were situated on the Earth's surface, because a meter or more of overburden effectively shields out cosmic rays. Measured concentrations of cosmogenic nuclides are converted to exposure ages by knowing nuclide production rates, which can be determined experimentally. Quartz is an especially favorable target for measurement of cosmogenic ^{10}Be and ^{26}Al because of its simple chemistry, resistance to weathering, and common occurrence in many kinds of rocks. By measuring exposure ages, it is possible to quantify the rates of uplift and erosion. In a later section, we see how cosmogenic ^{10}Be can be employed as a tracer for the subduction of surface materials into the mantle and as a way of constraining subduction rates.

RADIONUCLIDES AS TRACERS OF GEOCHEMICAL PROCESSES

In biological research, certain organic molecules may be "tagged" with a radioactive isotope so that their participation in reactions or progress through an organism can be followed. Radionuclide tracers can be used in an analogous way to understand geochemical processes that have already taken place within the Earth. This usually requires a knowledge of the geochemical behavior of parent and daughter elements. In this section, we consider several examples of the use of these tracers.

Heterogeneity of the Earth's Mantle

We have already noted that most radioactive nuclides and their decay products cannot be fractionated by mass. However, radioactive isotopes generally have very different solid-liquid distribution coefficients from their stable daughter isotopes, so that partial melting can fractionate radioactive elements from their radiogenic daughters. As melts have been progressively extracted from the mantle to form the crust, the abundances of radioactive isotopes in these reservoirs have changed, and these differences have been magnified by subsequent decay of the fractionated radioactive isotopes. The resulting isotopic heterogeneity within the mantle can be monitored by analyzing magmas that are derived from different mantle depths.

Samarium and neodymium are partitioned differently between the continental crust and mantle. Although both elements are incompatible and thus are fractionated into the crust, neodymium is more incompatible and so is ~25 times more abundant in the crust than in the mantle, whereas samarium is only ~16 times more abundant. This means that the Sm/Nd ratio is higher in the mantle than in the crust. ^{147}Sm decays to ^{143}Nd over time, so the ratio of radiogenic ^{143}Nd to nonradiogenic ^{144}Nd changes in both crust and mantle. It is thought that the mantle originally contained samarium and neodymium in the same relative proportion as in chondritic meteorites. The deviation between ^{143}Nd/^{144}Nd measured in a mantle-derived rock and the same ratio in a chondrite of identical age is called ε_{Nd} (analogous to ε_{W}, defined in worked problem 14.5) and can be used to estimate the degree to which the mantle may have evolved from its original pristine state.

Similar arguments can be used to trace the evolution of the ^{87}Sr/^{86}Sr ratio, which is related to the Rb/Sr ratio through the decay of ^{87}Rb to ^{87}Sr. In this case, rubidium is more incompatible than strontium, so the Rb/Sr ratio is higher in the crust and the crust becomes more radiogenic over time—the opposite of the Sm-Nd system. The isotopic composition of strontium can also be expressed as ε_{Sr}.

If basalts are derived from melting of mantle material, their initial neodymium and strontium isotopic ratios

NATURAL NUCLEAR REACTORS

Certain heavy nuclides (^{235}U being the only naturally occurring example) are *fissile;* that is, on absorbing a neutron, they split into two nuclear fragments of unequal mass. Because fission is exothermic, it is the energy source for nuclear power reactors and for the atomic bomb. When ^{235}U fissions, it releases additional neutrons that collide with other ^{235}U nuclei, prompting further fission reactions. The nuclides produced by fission have higher ratios of N to Z than stable nuclides and are thus radioactive. This process is used to advantage in the design of breeder reactors, which produce plutonium and other transuranic elements. It also accounts for the radioactivity of nuclear wastes.

In 1972, workers at a nuclear-fuel processing plant in France found that some of the uranium ore shipped from the Oklo mine in the Republic of Gabon, Africa, was deficient in ^{235}U relative to ^{238}U. By exquisite analytical detective work, scientists at the French Atomic Energy Commission eventually traced this uranium isotopic anomaly to 17 individual pockets within the Oklo deposit and the surrounding area. They had discovered the intact remains of Precambrian natural nuclear reactors. Besides the depletion in ^{235}U in the spent nuclear fuel, these pockets contain fission-produced nuclides of Nd, Sm, Zr, Mo, Ru, Pd, Ag, Cd, Sn, Cs, Ba, and Te. These isotopes are also part of the radioactive waste from modern man-made nuclear reactors.

The Oklo reactors are small (usually 10–50 cm thick) seams of sandstone-hosted uraninite ore mantled by clay. Today, that ore would be unable to initiate self-sustaining neutron chain reactions, because the proportion of ^{235}U in natural uranium is too low. However, 1.9 billion years ago when these reactors were active, the proportion of ^{235}U was nearly five times greater. The high uranium concentration in quantities exceeding the critical mass was a necessary condition for the Oklo reactors. These ores contained few "pile poisons," which absorb neutrons and prevent a chain reaction. In modern reactors, neutrons are slowed (moderated) by water, which allows them to be absorbed by other nuclei. Geological observations show that water was also present as a moderator in the Oklo natural reactors. Moreover, the deposits also contain fission products of plutonium and other transuranic elements, which indicates that they functioned as fast breeder reactors. Breeding fissile material would have allowed the reactors to operate continuously for longer periods of time. You can read more about the Oklo reactors in a paper published by F. Gauthier-Lafaye and coworkers (1989).

Besides providing information on the physics of a fascinating natural phenomenon, these reactors are important for understanding radioactive waste containment. Oklo is the only occurrence in the world where actinides and fission products have been in a near-surface geological environment for an extremely long period of time. The fission products at these sites allow characterization of the geochemical mobility of various isotopes and thereby permit assessment of the effectiveness of radionuclide waste repositories.

---

should be the same as those of their mantle sources at the time of melting. Measured ε_{Nd} and ε_{Sr} values for young basalts from various tectonic settings are summarized in figure 14.13. Midocean ridge basalts (MORB) have positive ε_{Nd} and negative ε_{Sr} values; that is, they have high $^{143}Nd/^{144}Nd$ ratios and low $^{87}Sr/^{86}Sr$ ratios that are complementary to those of continental crust. From this, we infer that the upper mantle source regions for MORB have been depleted of materials used to form the continents. Ocean island basalts plot along an extension of the MORB trend in figure 14.13, defining what is commonly called the *mantle array*. The position of ocean island basalts in the diagram indicates that their source is less depleted than that of MORB. These basalts also show a wider range in isotopic composition. Ocean island basalts may be derived from the relatively undepleted lower mantle and are carried through the upper mantle in plumes, where they mix to varying degrees with depleted materials. Magmas in subduction zones (not shown in fig. 14.13) often plot off the mantle array,

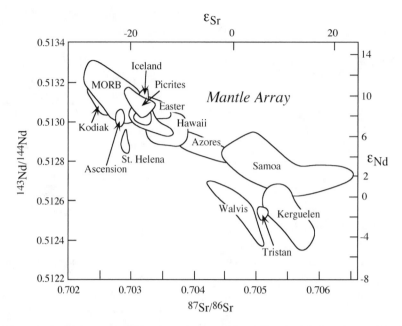

FIG. 14.13. Neodymium and strontium isotopic compositions of basalts. MORBs and ocean island basalts define a diagonal mantle array. MORBs are derived from a homogenized, depleted upper mantle, whereas ocean island basalts represent mixing of magmas derived from a less depleted lower mantle with upper mantle materials. (After Anderson 1994.)

reflecting the effects of mixing of seawater expelled from the subducted slab and crustal materials during ascent.

In terms of its radiogenic isotopes, the mantle can be visualized as consisting of two layers: the lower portion undifferentiated and the upper mantle highly fractionated. The lower mantle has presumably remained relatively undepleted since its formation, sampled only infrequently at hot spots to produce flood basalts and oceanic islands. However, periodic sinking of crustal plates into the lower mantle provides a mechanism for enriching it in incompatible elements. The upper mantle has been continuously cycled by the formation and subduction of oceanic crust, as well as the production of continental crust made in magmatic arcs at subduction zones. Its isotopic composition has become homogenized but remains relatively depleted in incompatible elements. This two-layer model of the mantle was previously presented in chapter 12, based on trace element abundances.

The estimated $+12\varepsilon_{Nd}$ value of the upper mantle is counterbalanced by the $-15\varepsilon_{Nd}$ value for continental crust. Knowing these values and the mass of the continental crust, it is possible to perform a mass balance

calculation to determine the proportion of the depleted upper mantle. The mass of the continental crust is ~2 × 10^{25} g and its neodymium abundance is ~50 times greater than the upper mantle. The mass of the upper mantle is thus $100 × 10^{25}$ g, which corresponds to a thickness on the order of 600–700 km. This depth range coincides with a prominent seismic discontinuity separating the upper and lower mantles.

If this view of a layered mantle is correct, other isotopic systems should be correlated with neodymium and strontium isotopic data. As an example, ε_{Pb}, the deviation in $^{206}Pb/^{204}Pb$ from chondritic evolution, varies systematically, so we should actually consider a Nd-Sr-Pb mantle array. Stan Hart and coworkers (1986) suggested that inclusion of additional isotopic systems may require that compositional models for the mantle be expanded to as many as five components.

Magmatic Assimilation

In some cases, ascending magmas are contaminated by assimilation of crustal materials, with the result that their compositions are altered. Radiogenic isotopes provide the least ambiguous way of recognizing this process. In uncontaminated basalts, the initial $^{87}Sr/^{86}Sr$ ratio is the same as that of the mantle source region. This ratio always has a relatively low value (ranging from ~0.699 to 0.704, depending on the age of the rock—the mantle slowly accumulates radiogenic ^{87}Sr over time), because the amount of parent rubidium relative to strontium in mantle rocks is low. Rb/Sr ratios in rocks of the continental crust are much higher; therefore, over time, crustal rocks evolve to contain more radiogenic ^{87}Sr. We can infer then that magmas derived from mantle sources should have low initial $^{87}Sr/^{86}Sr$ ratios. Conversely, those derived from, or contaminated with, continental crust should have high initial $^{87}Sr/^{86}Sr$ ratios. A corollary is that rocks related only through fractional crystallization of the same parental magma should have identical $^{87}Sr_0/^{86}Sr$ ratios.

Other isotopic systems can also be used to test for assimilation of crustal materials. Contaminated basalts should have lower $^{143}Nd/^{144}Nd$ ratios than uncontaminated rocks. It is not possible to predict the exact isotopic composition of crustal lead, but if Pb isotopes can be analyzed in plausible contaminating materials, its mixing effects can be assessed.

Assimilation is actually a rather complicated process, because it requires a great deal of heat. In chapter 12, we learned that melting requires enough heat to raise the temperature of the wall rocks to their melting points plus the heat necessary to fuse them. Most magmas are already below their liquidus temperatures, so the only plausible source of heat is exothermic crystallization of the magmas themselves. Consequently, we might expect assimilation and fractional crystallization to occur concurrently. An equation to model the change in isotopic composition of a magma experiencing both assimilation and fractionation is given below. Here we assume that a magma body of mass M^o is affected by two rate processes: the rate at which the magma assimilates a mass of rock, M_a; and the rate at which crystals separate during fractionation of the magma, M_c. At any time during the combined processes, the mass of magma remaining is M, and the parameter $F = M/M^o$ describes the mass of magma relative to the mass of the original magma. Any isotopic ratio in the magma I_m resulting from the combined effects of assimilation and fractional crystallization is given by:

$$I_m = \{(r/[r-1])(C_a/z)(1 - F^{-z})I_a + C_m^o F^{-z}I_m^o\}/$$
$$\{(r/[r-1])(C_a/z)(1 - F^{-z}) + C_m^o F^{-z}\}. \quad (14.20)$$

In this equation, r is defined as M_a/M_c, C_a is the concentration of the element of interest in the assimilated rock, C_m^o is its initial concentration in the original magma, and I_m^o and I_a are the initial isotopic ratios of the magma and assimilated rocks, respectively. The term z is defined as:

$$z = (r + D - 1)/(r - 1),$$

where D is the bulk distribution coefficient between the fractionating crystals and the magma. Equation 14.20 predicts the isotopic composition of a magma (I_m) resulting from combined assimilation and fractionation. I_m could be any isotope ratio, for example, $^{87}Sr/^{86}Sr$, or any normalized parameter describing such a ratio, such as ε_{Sr}.

Worked Problem 14.6

By considering the combined effects of assimilation and fractional crystallization, geochemist Don DePaolo (1981) showed that elevated $^{87}Sr/^{86}Sr$ ratios in modern andesitic lavas in the Andes resulted from crustal contamination. The inferred isotopic compositions of the initial magma and a plausible crustal contaminant (wallrock) are illustrated in figure 14.14. Measured data for most Andean volcanic rocks (open circles in fig. 14.14) do not lie on a straight mixing line connecting these compositions, which might suggest that mixing was not important in determining the isotopic compositions of these rocks. However, the combined effects of assimilation and fractional crystallization give a different result.

Models for the combined processes can be constructed using equation 14.20. To illustrate, we calculate the magma isotopic composition (I_m) using the following parameters: the original magma contained 400 ppm Sr and 10 ppm Rb and had an initial strontium isotopic ratio $^{87}Sr/^{86}Sr = 0.7030$. The contaminant had 500 ppm Sr and 40 ppm Rb, with an initial ratio $^{87}Sr/^{86}Sr = 0.71025$. We assume that $r = 0.8$, $D_{Sr} = 0.5$, and $F = 2$. The corresponding value for z is then:

$$z = (0.8 + 0.5 - 1)/(0.8 - 1) = -1.5.$$

Substitution of the above parameters into equation 14.20 gives:

$$I_m = \{(0.8/-0.2)(500/-1.5)(1 - 2^{1.5})(0.71025)$$
$$+ (400)(2^{1.5})(0.7030)\}/$$
$$\{(0.8/-0.2)(500/-1.5)(1 - 2^{1.5}) + (400)(2^{1.5})\}$$

$$= 0.7078.$$

This calculated value is only one point on a curve like those in figure 14.14, which illustrates how the strontium isotopic ratio changes during the combined processes. The curves actually shown in figure 14.14 are the values for Andean volcanics calculated by DePaolo. He assumed that D_{Sr} had a range of values as shown in figure 14.14 (we might expect that D_{Sr} would vary

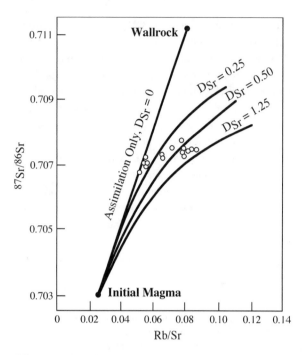

FIG. 14.14. Strontium isotopic data for volcanic rocks of the Peruvian Andes. The data points do not lie on a simple mixing line connecting the isotopic compositions of initial magma and wall-rock contaminant, but do lie near calculated mixing lines for an assimilation-fractional crystallization model with various values of D_{Sr}. It is expected that D_{Sr} increases in the latter stages of magma evolution. This model appears to explain the observed isotopic data fairly well. (After DePaolo 1981.)

during fractional crystallization and assimilation as different phases crystallized and the liquid composition changed). The measured strontium isotopic compositions and Rb/Sr ratios in the lavas fall within the calculated curves, thus supporting the model of assimilation and concurrent fractional crystallization.

Subduction of Sediments

Sediments deposited near oceanic trenches and subsequently subducted into the mantle could conceivably be melted and incorporated into arc volcanics. However, some geological and geophysical studies have stressed the mechanical difficulty of subducting low density sediments into the mantle. Radiogenic isotopes that are concentrated in such sediments may serve as tracers to follow their fate. These sediments characteristically have high $^{208}Pb/^{204}Pb$ and $^{207}Pb/^{204}Pb$ ratios. We have already seen that uranium and thorium are incompatible elements

that are concentrated in the crust, and the high radiogenic lead contents of these sediments are the result of their derivation from old U- and Th-enriched rocks on continents. The isotopic compositions of lead in some arc volcanics and in abyssal sediments are very similar, but are so distinct from those of oceanic basalts that recycling of crustal materials must occur. Mass balance calculations suggest that 1–2% sediment in the source could account for the observed lead isotope data.

This conclusion is supported by hafnium isotopic data on arc volcanics. We have not discussed the Lu-Hf system, but it is similar in most respects to the Sm-Nd system, with specifics given in table 14.1. Hafnium is partitioned strongly into the mineral zircon, which is commonly concentrated in continent-derived sands. Thus, there should be a major difference in Lu/Hf ratios between these sands and deep-sea muds. This behavior is distinct from Sm/Nd, which shows no fractionation during sedimentation. Therefore, the isotopic variations in hafnium and neodymium expected from mixing mantle with various types of sediment can be predicted and compared with data from arc volcanics. Jonathan Patchett and his collaborators (1984) estimated that <2% of nearly equal proportions of continental sand and oceanic mud could account for the measured Hf-Nd isotopic array.

The cosmogenic isotope ^{10}Be can also be used to shed light on this problem. The short half-life of this isotope, only 1.5 million years (table 14.1), is comparable to the time scale for subduction and magma generation. Young pelagic sediments contain appreciable amounts of ^{10}Be, so it is possible that this isotope might make the round trip from the Earth's surface into the mantle and back again if subduction and melting are fast enough. Analyses indicate that lavas that should not have incorporated crustal materials (MORB, ocean island basalts) do not contain significant amounts of ^{10}Be, but this isotope is enriched by factors of 10–100 in some arc volcanic rocks (e.g., Aleutians, Andes, Kurile arcs). ^{10}Be is absent in the lavas of some arcs, but its absence does not necessarily prove the absence of sediments, because the ^{10}Be signal depends on the rate of subduction and melting. Julie Morris (1991) provided a comprehensive review of the subject.

It might be argued that ^{10}Be in lavas reflects cosmic ray exposure on the surface after eruption, rather than incorporation of subducted sediments. However, analyses of modern lavas show a constant ratio of ^{10}Be to non-cosmogenic ^{9}Be in all minerals of the rocks, indicating that ^{10}Be must have been incorporated prior to igneous crystallization. Estimates of the amount of sediments incorporated into arc lavas are generally <3%. The time scales for subduction to mantle depths, transfer of material from the slab, and subsequent melting, ascent, and eruption are still debated. Constraints from decay of ^{10}Be can be coupled with U-Th disequilibrium systematics to give times ranging from a few hundred thousand years to as little as 10,000 years.

Isotopic Composition of the Oceans

The isotopic composition of ocean water depends on inputs from various sources. These are primarily (1) submarine weathering of and magmatic interaction with young basaltic rocks of the ocean floor, (2) recycling of marine carbonate rocks that now reside on continents, and (3) weathering of crystalline continental crust, which consists mostly of granitic igneous and metamorphic rocks. We briefly consider the behavior of strontium and neodymium isotopes in the oceanic system.

At the present time, all oceans have a ^{87}Sr/^{86}Sr ratio of 0.7090. In contrast, the measured ^{143}Nd/^{144}Nd ratios of ocean water indicate that each ocean has a distinct and characteristic compositional range, as summarized in figure 14.15 in terms of ε_{Nd}. What causes this difference in isotopic behavior?

It is thought that much of the strontium in seawater is derived from weathering of marine carbonates on continents. The ^{87}Sr/^{86}Sr ratio in these carbonates has varied during Phanerozoic time, decreasing to a minimum

of 0.7067 during the Jurassic and increasing again to the current high value of 0.7090. The isotopic composition of marine carbonates mimics that of the ocean in which they formed, so the oceans have varied in strontium isotopic composition over time. In fact, changes in the oceanic ^{87}Sr/^{86}Sr ratio have now been documented with sufficient precision that this isotope ratio can be used to date some fossil shells and carbonate sediments (the technique is most precise for intervals when the ^{87}Sr/^{86}Sr ratio is rapidly changing due to climatic or tectonic events). Different oceans drain continents with different carbonate sources, so the fact that oceans have the same strontium isotopic composition everywhere must relate to a long residence time (~4 million years) for this element in seawater. Interoceanic circulation patterns are rapid compared with strontium residence, so that a worldwide isotopic composition results at any given time.

Continental crystalline rocks probably supply most of the neodymium to the oceans. Differences in the ages of continental rocks whose weathering products are drained into the oceans produce distinct radiogenic isotopic signatures in river waters. We might expect that, as with strontium isotopes, these differences would be erased by global circulation, but they are not. The Atlantic is supplied by very old continental regions, with high ^{143}Nd/^{144}Nd, whereas the Pacific is ringed by younger regions with correspondingly lower values of neodymium isotope ratios. Waters in the two oceans retain these isotopic signatures. This can be explained only by assuming that the residence time for neodymium in seawater is short; that is, that neodymium entering the

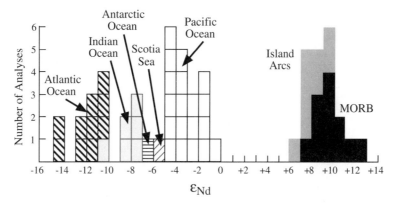

FIG. 14.15. The neodymium isotopic compositions of seawater and ferromagnesian nodules from different oceans, expressed as ε_{Nd}. Boxes represent individual analyses. These compositions are distinct from each other and from oceanic volcanic rocks. (After Piepgras et al. 1979.)

oceans is lost to the seafloor before interoceanic circulation patterns can homogenize the isotopic composition between oceans.

Radiogenic isotopic studies of seawater thus offer the possibility of shedding light on the rates at which oceans mix, provided that residence times can be fixed. Neodymium isotope measurements may even be able to identify the location of currents responsible for such mixing. Strontium isotopes also provide information on the kinds of rocks exposed to weathering on continents over geologic time.

Worked Problem 14.7

What are reasonable values for input into the modern oceans from various geochemical reservoirs? Gunter Faure (1986) presented a mass balance model based on the premise that the strontium isotopic composition of the oceans could be explained

by input from oceanic volcanics, marine carbonates, and crystalline continental crust. The $^{87}Sr/^{86}Sr$ ratio of seawater (sw) is given by:

$$(^{87}Sr/^{86}Sr)_{sw} = (^{87}Sr/^{86}Sr)_v V + (^{87}Sr/^{86}Sr)_m M + (^{87}Sr/^{86}Sr)_c C,$$

where $(^{87}Sr/^{86}Sr)_{v,m,c}$ are strontium isotopic ratios contributed by volcanic, marine carbonate, and continental crustal rocks, respectively, and the coefficients V, M, and C are the fractions of the total input contributed by these sources. By substituting present-day estimates for the strontium isotopic compositions of each of these sources, we obtain:

$$(^{87}Sr/^{86}Sr)_{sw} = 0.704V + 0.708M + 0.720C. \qquad (14.21)$$

If we then plot $(^{87}Sr/^{86}Sr)_{sw}$ versus V, as illustrated in figure 14.16, we can contour values of M and C from equation 14.21. Plausible solutions are limited by the restriction that no source can supply a negative amount of material; that is, V, M, and C must be ≥0. The isotopic composition of modern seawater is illustrated in figure 14.16 by a horizontal dashed line at $^{87}Sr/^{86}Sr$ = 0.7090. From this figure, we can see that V cannot be >0.68. However, V cannot be very close to this upper bound, because marine carbonates must make a substantial contribution. The high strontium contents and susceptibility to chemical weathering demand an important role for the volcanic rocks in this system. This model does not provide a unique solution to the problem unless the input from any one of the sources can be fixed unambiguously. However, it does allow us to assess the relative contributions from these sources if a reasonable value for one of them can be guessed. For example, if we assume that M has the value 0.7, then the contributions from other sources are V = 0.16 and C = 0.14.

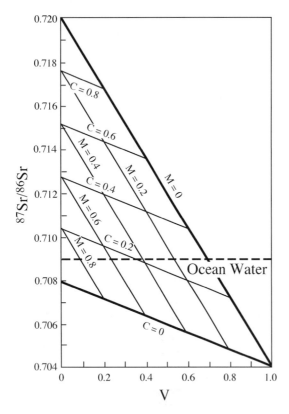

FIG. 14.16. A model to explain the strontium isotopic composition of seawater. Mixtures of various proportions of volcanic rocks (V), marine carbonates (M), and continental crustal rocks (C) can account for the $^{87}Sr/^{86}Sr$ ratio of modern ocean water (dashed horizontal line at 0.709). (After Faure 1986.)

Degassing of the Earth's Interior to Form the Atmosphere

The atmosphere is thought to have formed primarily by degassing of the interior of the Earth, and isotopic variations in gases dissolved in basalts provide information on this process. Noble gases are potentially very useful in understanding atmospheric evolution, because they are chemically inert and thus are not fractionated by chemical reactions, and because all except helium are too heavy to escape from the atmosphere to space. Helium and neon cannot be subducted but are continuously released at midocean ridges, demonstrating that the Earth's interior has not been completely degassed.

Radioactive decay results in the production of many noble gas isotopes. This is easily monitored by considering ratios of radiogenic and stable nuclides. For example, variations observed at the present time in $^3He/^4He$ ratios reflect the balance between primordial (stable) ^{3}He and radiogenic ^{4}He formed by decay of ^{232}Th, ^{235}U, and ^{238}U. (Why the radiogenic isotope is not the numerator in this ratio, as is conventional for other radiogenic isotope ratios, is a mystery.) Neon has one radiogenic isotope (^{21}Ne) and two nonradiogenic isotopes (^{20}Ne and ^{22}Ne), as does argon. The ^{40}Ar/^{36}Ar ratio is controlled by the progressive decay of ^{40}K to ^{40}Ar. The isotopic composition of xenon reflects several processes. Radiogenic ^{129}Xe was produced by decay of now extinct ^{129}I. Fission of ^{238}U also generates ^{134}Xe, ^{136}Xe, and several other xenon isotopes, creating variations in the ratios of these isotopes to nonradiogenic ^{130}Xe.

Two different geological processes also affect these isotopic ratios. Degassing occurs by volcanic and hydrothermal activity. Decreases in gas concentrations in the mantle reservoir affect both radiogenic and stable nuclides, but not the nongaseous parent nuclides. For example, degassing decreases the amount of ^{36}Ar (the isotope to which radiogenic ^{40}Ar is normalized) but does not affect ^{40}K (which generates ^{40}Ar). Fractionation of incompatible elements also affects these ratios. We have already learned that K, U, and Th can be fractionated into the crust by magmatic processes. Because some isotopes of these elements are parent nuclides for noble gas isotopes, this process leads to depletion of radiogenic gas isotopes in mantle source regions.

Degassing and crustal formation have occurred on different time scales, and the parent nuclides that produce the isotopic variations in noble gases have very different half-lives. Consequently, these isotopic systems may permit us to trace these geologic processes and determine their time dependence.

Claude Allegre and his coworkers (1986) modeled degassing of the mantle using mass balance calculations that employed argon and xenon isotopic ratios. Their model employs four noble gas reservoirs: the atmosphere, continental crust, upper mantle, and lower mantle. Their calculations indicate that approximately half the Earth's mantle is 99% degassed. Based on other isotopic evidence, such as neodymium isotopes discussed earlier, this is a larger mantle proportion than that corresponding to the depleted upper mantle. The disagreement might be rationalized by assuming that some of the primitive lower mantle has also lost a portion of its noble gases.

Helium was not used in the mass balance calculations because it has limited residence time in the atmosphere. However, this becomes an advantage if we want to cal-

culate the gas flux from the mantle. Assuming a steady state condition in which the mantle releases as much He as is lost from the atmosphere to space, Allegre and coworkers calculated He fluxes in the mantle-atmosphere system. Fluxes of Ar and Xe can also be estimated based on correlations of their isotopic ratios with ^{3}He/^{4}He. The model of Allegre and his collaborators is summarized in figure 14.17. From this diagram, we can see that the only recycling occurs at subduction zones, and that this recycling is temporary because gases in the subducted

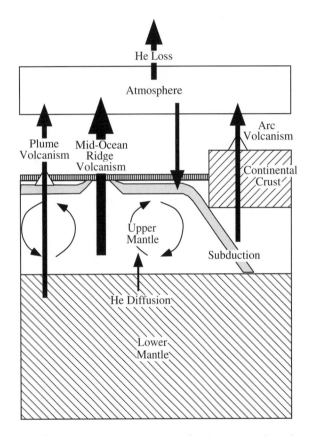

FIG. 14.17. Schematic representation of noble gas reservoirs and fluxes in the mantle-crust-atmosphere system. Degassing of the upper mantle occurs through volcanism at spreading centers, and degassing of a portion of the lower mantle results from hot spot volcanic activity. In this model, helium diffuses between mantle reservoirs, but they behave as closed systems to other noble gases. Subducted oceanic crust contains atmospheric gases, but these are only temporarily recycled, because arc volcanism returns them to the atmosphere. Continental crust is a significant reservoir for radiogenic gases, because their parent isotopes are fractionated into the crust and remain there for long periods of time. Helium is the only noble gas that escapes from the atmosphere to space. (Modified from Allegre et al. 1986.)

slabs are eventually returned to the atmosphere by arc volcanism.

However, the ratio $^{20}Ne/^{22}Ne$ in the mantle differs considerably from that in the atmosphere, and this difference cannot be attributed to nuclear processes (both are nonradiogenic isotopes). This implies that mantle degassing to form the atmosphere has not been a closed system process. Either some dynamic process caused loss of some early atmosphere (escape processes can result in significant fractionations), or some of the atmosphere has a different origin (perhaps carried in impacting comets). Although the neon isotopic data appear to require a more complex evolution of the Earth's atmosphere, simple degassing of the mantle as advocated by Allegre and coworkers is probably sufficient to explain the argon isotopes, because the mantle is such a huge reservoir of potassium. Noble gas isotopes in the Earth's mantle were reviewed by Ken Farley and Elizabeth Neroda (1998).

SUMMARY

In this chapter, we have seen that isotope stability is the exception rather than the rule. Radionuclides may decay by emission of beta particles, alpha particles, positrons, capture of electrons, or spontaneous fission. Branched decay involving several of these decay mechanisms occurs in some cases. The rate of radioactive decay is exponential and is most easily expressed in terms of half-life. We have derived equations that allow determination of the isotopic evolution of a system as a function of time. These equations permit the age dating of rocks, based on the decay of certain naturally occurring radionuclides, such as ^{40}K, ^{147}Sm, ^{87}Rb, ^{232}Th, ^{235}U, and ^{238}U. In such systems, it is possible to determine the age of rocks even though the daughter isotope may have been present in the system initially. Uranium and thorium isotopes decay through a series of intermediate radioactive daughters, but the half-lives of transitory daughters are short enough that each system can usually be treated as the decay of parent directly to the stable daughter. Independent U-Th-Pb geochronometers can give the same or different results, but useful information can be gained even in cases of discordancy. The interpretation of geochronological data can be complicated by the failure of rock systems to remain closed, as well as other factors. The earlier presence of now extinct radionuclides can, in some cases, be

determined, and their existence places constraints on early processes such as planetary differentiation.

The abundances of radiogenic isotopes can be determined directly from mass spectrometry or indirectly from fission track techniques. The production of radioactive isotopes can also be induced artificially in a nuclear reactor, which forms the basis for elemental quantitative analysis by neutron activation. The ^{40}Ar-^{39}Ar geochronometer also depends on induced radioactivity. Cosmogenic nuclides are produced naturally by cosmic ray exposure, which forms ^{14}C for dating recent events and other radionuclides useful in quantifying surface processes such as erosion.

Radionuclides also provide powerful tracers for geochemical processes. Uses we have explored here include understanding mantle heterogeneity, cycling of materials between crust and mantle reservoirs, global oceanic mixing, and degassing of the Earth's interior to form the atmosphere. The importance of radiogenic isotopes is considered further in the following chapter on the early solar system and processes at planetary scales.

SUGGESTED READINGS

There are an increasing number of textbooks that treat isotope geochemistry. The books by Faure and Dickin are required reading for anyone seriously interested in this subject, and the other references provide further elaboration on certain aspects of radiogenic isotopes.

Bowen, R. 1988. *Isotopes in the Earth Sciences*. London: Elsevier Applied Science. (Chapter 2 provides a thorough discussion of mass spectrometry, and later chapters deal primarily with radiometric dating techniques.)

Dickin, A. P. 1995. *Radiogenic Isotope Geology*. Cambridge: Cambridge University Press. (An up-to-date treatment of this field; everything you want to know about geochronology.)

Durrance, E. M. 1986. *Radioactivity in Geology: Principles and Applications*. New York: Wiley. (This unconventional book deals with many applications of radioactivity not treated here; examples include its use in prospecting for radioactive ores and petroleum, environmental applications, and the heat generated by radioactive decay.)

Farley, K. A., and E. Neroda. 1998. Noble gases in the Earth's mantle. *Annual Reviews of Earth and Planetary Sciences* 26: 189–218. (A lucid review of a particularly complex subject; noble gas geochemistry is complex and often understood only by its practitioners, but this paper really helps.)

Faure, G. 1986. *Principles of Isotope Geology*, 2nd ed. New York: Wiley. (One of the best available texts on isotope geochemistry. This is an excellent source for detailed information of radiogenic isotopes; geochronology is covered more exhaustively than other aspects of this subject.)

Jager, E., and J. C. Hunziker, eds. 1979. *Lectures in Isotope Geology.* Berlin: Springer-Verlag. (A series of lectures given in Switzerland by respected isotope geochemists; the first half of the book deals with geochronology.)

McLennan, S. M., S. Hemming, D. K. McDaniel, and G. N. Hanson. 1993. Geochemical approaches to sedimentation, provenance, and tectonics. *Geological Society of America Special Paper* 284:21–40. (This review paper is a primer on the use of radiogenic isotopes in deciphering the tectonic setting of clastic sediments.)

Philpotts, A. R. 1990. *Principles of Igneous and Metamorphic Petrology.* Englewood Cliffs: Prentice-Hall. (Chapter 21 gives a particularly good overview of isotope geochemistry, and especially of the use of radiogenic isotopes in understanding magma sources.)

Vertes, A., S. Nagy, and K. Suvegh. 1998. *Nuclear Methods in Mineralogy and Geology.* New York: Plenum. (This is an excellence source for learning about neutron activation analysis, and other subjects not covered here, such as dating groundwater.)

Walker, F. W., D. G. Miller, and F. Feiner. 1983. *Chart of the Nuclides.* San Jose: General Electric Company. (A real bargain that contains physical constants, conversion factors, and periodic table; order from General Electric Nuclear Energy Operations, 175 Curtner Ave., M/C 684, San Jose, California 95125.)

The following references were cited in this chapter. These provide more thorough treatments of the applications of radionuclides in solving geochemical problems.

Aleinikoff, J. N., R. H. Moench, and J. B. Lyons. 1985. Carboniferous U-Pb age of the Sebago batholith, southwestern Maine: Metamorphic and tectonic implications. *Geological Society of America Bulletin* 96:990–996.

Allegre, C. J., T. Staudacher, and P. Sarda. 1986. Rare gas systematics: Formation of the atmosphere, evolution and structure of the Earth's mantle. *Earth and Planetary Science Letters* 81:127–150.

Anderson, D. L. 1994. Komatiites and picrites: Evidence that the "plume" source is depleted. *Earth and Planetary Science Letters* 128:303–311.

DePaolo, D. J. 1981. Trace element and isotopic effects of combined wallrock assimilation and fractional crystallization. *Earth and Planetary Science Letters* 53:189–202.

Gauthier-Lafaye, F., F. Weber, and H. Ohomoto. 1989. Natural fission reactors of Oklo. *Economic Geology* 84:2286–2295.

Halliday, A. N., and D.-C. Lee. 1999. Tungsten isotopes and the early development of the Earth and Moon. *Geochimica et Cosmochimica Acta* 63:4157–4179.

Hart, S. R., D. C. Gerlach, and W. M. White. 1986. A possible Sr-Nd-Pb mantle array and consequences for mantle mixing. *Geochimica et Cosmochimica Acta* 50:1551–1557.

Morris, J. D. 1991. Applications of cosmogenic ^{10}Be to problems in the earth sciences. *Annual Reviews of Earth and Planetary Sciences* 19:313–350.

Nielson, D. L., S. H. Evans, and B. S. Sibbett. 1986. Magmatic, structural, and hydrothermal evolution of the Mineral Mountains intrusive complex, Utah. *Geological Society of America Bulletin* 97:765–777.

Nyquist, L. E., C. –Y. Shih, J. L. Wooden, B. M. Bansal, and H. Wiesmann. 1979. The Sr and Nd isotopic record of Apollo 12 basalts: Implications for lunar geochemical evolution. *Proceedings of the Lunar and Planetary Science Conference* 10:77–114.

Patchett, P. J., W. M. White, H. Feldman, S. Kielinszuk, and A. W. Hoffman. 1984. Hafnium rare-earth element fractionation in the sedimentary system and crustal recycling into the earth's mantle. *Earth and Planetary Science Letters* 69:365–378.

Piepgras, D. J., G. J. Wasserburg, and E. J. Dasch. 1979. The isotopic composition of Nd in different ocean masses. *Earth and Planetary Science Letters* 45:223–236.

Price, P. B., and R. M. Walker. 1963. Fossil tracks of charged particles in mica and the age of minerals. *Journal of Geophysical Research* 68:4847–4862.

Sisson, V. B., and T. C. Onstott. 1986. Dating blueschist metamorphism: A combined ^{40}Ar/^{39}Ar and electron microprobe approach. *Geochimica et Cosmochimica Acta* 50:2111–2117.

Yin, Q., S. B. Jacobsen, K. Yamashita, J. Blichert-Toft, P. Talouk, and F. Albarede. 2002. A short timescale for terrestrial planet formation from Hf-W chronometry of meteorites. *Nature* 418:949–952.

PROBLEMS

(14.1) Given the following Rb-Sr isotopic data for whole-rock samples of a granitic pluton, determine the initial strontium isotopic ratio and age of the pluton.

Sample No.	$^{87}Rb/^{86}Sr$	$^{87}Sr/^{86}Sr$
1	11.86	0.7718
2	7.66	0.7481
3	6.95	0.7436
4	9.68	0.7587
5	6.54	0.7413
6	9.69	0.7599
7	3.74	0.7259

(14.2) (a) Solve equation 14.8 for t. (b) Use the equation you just derived to determine the age of a biotite sample, for which the following data have been obtained: K = 7.10 wt. %, $^{40}Ar = 1.5 \times 10^{12}$ atoms g^{-1}.

(14.3) A zircon grain initially contained 1000 ppm ^{235}U and 100 ppm ^{207}Pb. How many atoms of ^{207}Pb will this zircon contain after 1 billion years?

(14.4) Using the data below and figure 14.7, construct a concordia diagram. Are these samples concordant? Interpret your results, explaining any assumptions that you make.

Sample No.	$^{206}Pb/^{238}U$	$^{207}Pb/^{235}U$
1	0.500	15.93
2	0.543	18.77
3	0.565	20.00

(14.5) You obtain the following geochronological data for a portion of the Appalachians. A gneiss unit gives whole-rock Rb-Sr and zircon U-Pb ages of 315 million years. Mineral geothermometers in this unit give a peak metamorphic temperature of 530° C. ^{40}Ar-^{39}Ar gas retention ages are 295 million years for hornblende and 283 million years for biotite. Fission track ages for apatites range from 130 to 150 million years. Using these data, construct a time-temperature path illustrating the late Paleozoic thermal history of this part of the Appalachian orogen.

(14.6) The measured activity of a sample of charcoal found at an archaeological site is 5.30 dpm g^{-1}. What is the age of the sample?

(14.7) How much heat energy would be generated in 100,000 years by a one cubic meter sample of rock containing live ^{26}Al? The half-life of ^{26}Al is given in table 14.1. Assume that the sample initially contains 100 g of aluminum with a $^{26}Al/^{27}Al$ atomic ratio of 0.1. Also assume that the energy of decay for ^{26}Al is 10^{-14} cal atom^{-1}.

(14.8) A magma containing 200 ppm strontium is contaminated by crustal rocks containing 1000 ppm Sr. The magma simultaneously undergoes fractional crystallization of phases with a bulk distribution coefficient $D_{Sr} = 0.75$. The initial Sr isotopic ratios of the magma and crustal rocks are 0.704 and 0.713, respectively. Assume that $r = 0.5$. Use equation 14.20 to calculate the isotopic composition of the resulting hybrid magma as a function of mass of magma relative to original mass (F), and illustrate the result with a graph.

CHAPTER 15

STRETCHING OUR HORIZONS

Cosmochemistry

OVERVIEW

In this chapter, we explore the rapidly emerging field of cosmochemistry. This subject involves the geochemical aspects of systems of planetary or solar system scale. We first consider nucleosynthesis processes in stars and use this foundation to understand the abundances of elements in the Sun and the solar system. Chondritic meteorites are discussed next, as samples of average solar system material stripped only of the lightest elements. Analyses of these meteorites provide information on the behavior of various elements in the early solar nebula; from these, we learn that elements were fractionated according to both geochemical affinity and volatility. Chondritic fractionation patterns are also important for understanding planetary compositions. We then consider evidence for proposed cosmochemical processes in the early solar nebula, such as condensation of solids from vapor and infusion of materials formed in and around other stars. The characteristics and origin of extraterrestrial organic molecules and ices are also explored; these, too, were important planetary building blocks. Finally, we consider how geochemical models for planets are formulated.

WHY STUDY COSMOCHEMISTRY?

In earlier chapters, we have sometimes used observations from extraterrestrial materials or other planetary bodies to illustrate processes or pathways in geochemistry. Perhaps these seemed exotic or even esoteric. From the geologist's uniformitarian perspective, however, the planets and other extraterrestrial bodies provide a larger laboratory in which to study geochemical behavior. Historically, the study of cosmochemistry has paralleled (and, in some cases, spurred) the development of geochemistry as a discipline, and has involved many of the same scientists. It is significant that the leading journal in geochemistry is also the premier journal in cosmochemistry and incorporates both fields in its title (*Geochimica et Cosmochimica Acta*). In this chapter, we examine several problems of huge (even astronomic) scale that attract geochemists to study the solar system and beyond, and we see how some of the answers improve our understanding of the world under our feet.

Our solar system consists of a single star surrounded by nine planets and a large retinue of smaller bodies, including their satellites, asteroids, comets, and a lot of small rocks and dust. Samples of the small rocks and dust are provided free of charge as meteorites and interplanetary dust particles, and in some cases, these can be related to the parent planets or asteroids on which they formed. Cosmochemists focus on chemical differences and similarities among these bodies. To what extent have they evolved separately, and what characteristics have they inherited from a common origin? Which processes are unique in a given body, and which processes are likely to have affected all of them? Do they contain evidence for how the solar system formed, and how has it evolved to its present state?

Most (>99%) of the mass of our solar system is concentrated in the Sun, so many questions regarding the bulk chemistry of the solar system and its early history must first address the behavior and composition of stars. Here, we rely on astrophysicists to construct models that infer the structure and evolutionary development of stars in general. These are based on observations of the Sun and distant stars and on the principles of thermodynamics and nuclear chemistry we have discussed in earlier chapters, although the language of astrophysics is different from that of geochemistry.

From the Sun, we turn to the planets, which are usually divided into two groups. Those closest to the Sun (Mercury, Venus, Earth, and Mars) are called the terrestrial planets, to distinguish them from the larger Jovian planets beyond (Jupiter, Saturn, Uranus, and Neptune). The Jovian worlds are part rock and part ice, now transformed into unusual forms by high pressures. Pluto and its satellite Charon, small icy bodies at the outer reaches of the solar system, have unusual orbits and may be related to comets. In this chapter, we focus on the terrestrial planets, as they are more likely to provide geochemical insights into the Earth. The presence and compositions of the Jovian planets, however, place important restrictions on any models for the history of the solar system as a whole.

The terrestrial bodies include not only the four inner planets, but also the moons of Earth and Mars and several thousand asteroids. Studies of these, particularly of our own Moon, have helped geochemists to determine how planetary size affects differentiation and other global-scale processes. Lunar samples brought to Earth by Apollo astronauts, by Soviet unmanned spacecraft, and as meteorites are an important resource for cosmochemists. A small group of meteorites are also thought to have come from Mars. Some of the most valuable information about terrestrial bodies, however, is based on meteorites that are fragments of asteroids. Some asteroids experienced differentiation and core formation, producing igneous rocks (achondrites) and iron meteorites, whereas others (chondrites) remained almost unchanged after their parent asteroids accreted. From meteorites, we obtain data that can be used to model the development of planetary cores, mantles, and crusts. We also use the nearly pristine nature of chondrites to deduce the chemistry of materials from which the larger terrestrial planets were assembled.

We begin a survey of cosmochemistry by considering the Sun as a star, to find chemical clues to its origin and the origin of other solar system materials. As we proceed, our focus will be drawn more to the development of planets, and finally back to the Earth itself. In this tour, you will recognize many familiar geochemical concerns and, we hope, see the place of cosmochemistry in understanding how the Earth works.

ORIGIN AND ABUNDANCE OF THE ELEMENTS

Nucleosynthesis in Stars

The universe is believed to have begun in a cataclysmic explosion—the Big Bang—and has been expanding ever since. At this beginning stage, matter existed presumably in the form of a stew of neutrons or as simple atoms (hydrogen and deuterium). The Big Bang itself may have produced some other nuclides, but only ^4He was formed in any abundance. In chemical terms, this was a pretty dull universe. How then did all of the other elements originate?

Local concentrations of matter periodically coalesce to form stars. In 1957, Margaret Burbidge and her husband, Geoffrey Burbidge, along with William Fowler and Fred Hoyle, teamed together to write a remarkable scientific paper that argued that other elements formed in stellar interiors by nuclear reactions with hydrogen as the sole starting material. (This paper is now rightfully considered a classic, often referred to as B^2FH from the initials of the authors' surnames.) When hydrogen atoms are heated to sufficiently high temperatures and held together by enormous pressures—such as occur in the deep interior of the Sun—fusion reactions occur. The proton-proton chain, the dominant energy-producing reaction in the Sun, is illustrated in figure 15.1. This fusion process, commonly called hydrogen burning, produces helium. When the hydrogen fuel in the interior begins to run low and the stellar core becomes dense enough to sustain reactions at higher pressures, another nuclear reaction may take place, in which helium atoms are fused to make carbon and oxygen. While helium burning proceeds in the core of a star, a hydrogen-burning shell works its way toward the surface; a star at this evolutionary stage expands into a red giant. This will be the fate of our Sun.

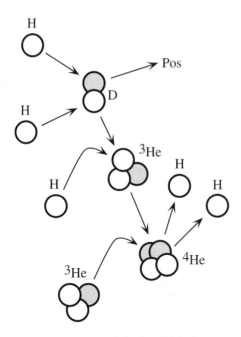

FIG. 15.1. The proton-proton chain, in which hydrogen atoms are fused into helium, is the predominant energy-producing reaction in the Sun. Open circles are protons, filled circles are neutrons, and Pos is a positron. For each gram of ^{4}He produced, 175,000 kwh of energy are released.

A star more massive than the Sun, however, can employ other fusion reactions, successively burning carbon, neon, oxygen, and silicon. The ashes from one burning stage provide the fuel for the next. The internal structure of such a massive, highly evolved star would then consist of many concentric shells, each of which produces fusion products that are then burned in the adjacent inward shell. The ultimate products of such fusion reactions are elements near iron in the periodic table (V, Cr, Mn, Fe, Co, and Ni). Fusion between nuclei cannot produce nuclides heavier than the iron group elements. For elements lighter than iron, the energy yield is higher for fusion reactions than for fission, but for elements heavier than iron, the energy yield for fission is greater. When a star reaches this evolutionary dead end, a series of both disintegrative and constructive nuclear reactions occurs among the iron group nuclei. After some time, a steady state, called *nuclear statistical equilibrium,* is reached; the relative abundances of iron group elements reflect this process.

Elements heavier than the iron group can be formed by addition of neutrons to iron group seed nuclei. We have already seen in chapter 14 that nuclei with neutron/proton ratios greater than the band of stability undergo beta decay to form more stable nuclei. Helium burning produces neutrons that are captured by iron group nuclei in a slow and rather orderly way, called the *slow* or *s process.* This process is slow enough that nuclei, if they are unstable, experience beta decay into stable nuclei before additional neutrons are added. This is illustrated in figure 15.2, in which stable and unstable nuclei are shown as white and gray boxes, respectively. A portion of the s process track is illustrated by the upper set of arrows. Neutrons are added (increasing N) until an unstable nuclide is produced; then beta decay occurs, resulting in a decrease of both N and Z by one. The resulting nuclide can then capture more neutrons until it beta decays again, and so forth. Notice that a helium burning star producing heavy elements by this process would have to be at least a second generation star, having somehow inherited iron group nuclei from an earlier star whose material had been recycled.

Fusion, nuclear statistical equilibrium, and s process neutron-capture reactions in massive stars thus provide mechanisms by which elements heavier than hydrogen can be produced. But how are these nuclides placed into interstellar space for later use in making planets, oceans, and people? Processed stellar matter is lost continuously from stars as fluxes of energetic ions, such as the solar wind. Other elements are liberated by supernovae—stellar explosions that scatter matter over vast distances. At the same time, supernovae generate other nuclides by a different nucleosynthetic pathway, the *rapid* or *r process.* Very rapid addition of neutrons to seed nuclei results in a chain of unstable nuclides.

One example of the r process is illustrated by the set of arrows at the bottom of figure 15.2. Reaching this part of the diagram, which is populated by unstable nuclei, happens only when neutrons are added more rapidly than the resulting nuclei can decay. The nuclides in crosshatched boxes decay much more rapidly than the seconds or minutes required for the decay of nuclides occupying gray boxes, so a nucleus shifted into a crosshatched box by r-process neutron capture immediately transforms by beta decay, as shown. When the supernova event ends, the r process nuclides transform more slowly by successive beta decays into stable nuclides. Many of these stable isotopes are the same as produced by the s process but some, such as ^{86}Kr, ^{87}Rb, and ^{96}Zr, can be reached only by the r process. In an analogous manner, such nuclides

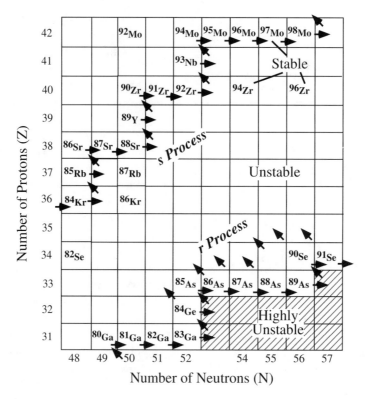

FIG. 15.2. A portion of the nuclide chart, illustrating how the s process (upper set of arrows) and the r process (lower set of arrows) produce new elements by neutron capture and subsequent decay. Shaded boxes represent stable nuclides, whereas white boxes are unstable and undergo beta decay, which shifts them up and to the left. Crosshatched boxes represent extremely unstable nuclides that transform rapidly by neutron capture.

as ^{92}Mo (fig. 15.2) on the proton-rich side of the s process band can be produced during supernovae by addition of protons to seed nuclei (the *p process*), followed by positron emission or electron capture.

Cosmic Abundance Patterns

The relative abundances of the elements in a star, then, are controlled by a combination of nucleosynthetic processes. For the moment, let's ignore how stellar or solar abundances are determined (we return to this problem shortly). Elemental abundances in the Sun, relative to 10^6 atoms of silicon, are given in table 15.1 and illustrated graphically in figure 15.3. (Normalization to silicon keeps these numbers from being astronomically large.) This abundance pattern is commonly called the *cosmic abundance of the elements*. This is a misnomer, of course; the composition of our Sun is not necessarily representative of that of the universe, but it does define the composition of our own solar system. Other stars have their own peculiar compositions because of different contributions from fusion, s and r processes, and so forth.

Let's inspect the cosmic abundance pattern in figure 15.3 to see why the Sun has this particular composition. Clearly, hydrogen and helium are the dominant elements. This is because only these two emerged from the Big Bang. Elements between helium and the iron group formed predominantly by fusion reactions. These show a rapid exponential decrease with increasing atomic number, reflecting decreasing production in the more advanced burning cycles. The Sun is presently burning only hydrogen, so these elements must have been inherited from an earlier generation of stars. Exceptions are the elements lithium, beryllium, and boron, which have abnormally low abundances. The abundances of these

TABLE 15.1. Cosmic Abundances of the Elements, Based on 10^6 Silicon Atoms

Atomic Number	Element	Symbol	Abundance	Atomic Number	Element	Symbol	Abundance
1	Hydrogen	H	2.79×10^{10}	45	Rhodium	Rh	0.344
2	Helium	He	2.72×10^9	46	Palladium	Pd	1.39
3	Lithium	Li	57.1	47	Silver	Ag	0.486
4	Beryllium	Be	0.73	48	Cadmium	Cd	1.61
5	Boron	B	21.2	49	Indium	In	0.184
6	Carbon	C	1.01×10^7	50	Tin	Sn	3.82
7	Nitrogen	N	3.13×10^6	51	Antimony	Sb	0.309
8	Oxygen	O	2.38×10^7	52	Tellurium	Te	4.81
9	Fluorine	F	843	53	Iodine	I	0.90
10	Neon	Ne	3.44×10^6	54	Xenon	Xe	4.7
11	Sodium	Na	5.74×10^4	55	Cesium	Cs	0.372
12	Magnesium	Mg	1.074×10^6	56	Barium	Ba	4.49
13	Aluminum	Al	8.49×10^4	57	Lanthanum	La	0.4460
14	Silicon	Si	1.00×10^6	58	Cerium	Ce	1.136
15	Phosphorus	P	1.04×10^4	59	Praseodymium	Pr	0.1669
16	Sulfur	S	5.15×10^5	60	Neodymium	Nd	0.8279
17	Chlorine	Cl	5240	61	Promethium[1]	Pm	—
18	Argon	Ar	1.01×10^5	62	Samarium	Sm	0.2582
19	Potassium	K	3770	63	Europium	Eu	0.0973
20	Calcium	Ca	6.11×10^4	64	Gadolinium	Gd	0.3300
21	Scandium	Sc	34.2	65	Terbium	Tb	0.0603
22	Titanium	Ti	2400	66	Dysprosium	Dy	0.3942
23	Vanadium	V	293	67	Holmium	Ho	0.0889
24	Chromium	Cr	1.35×10^4	68	Erbium	Er	0.2508
25	Manganese	Mn	9550	69	Thulium	Tm	0.0378
26	Iron	Fe	9.00×10^5	70	Ytterbium	Yb	0.2479
27	Cobalt	Co	2250	71	Lutetium	Lu	0.0367
28	Nickel	Ni	4.93×10^4	72	Hafnium	Hf	0.154
29	Copper	Cu	522	73	Tantalum	Ta	0.0207
30	Zinc	Zn	1260	74	Tungsten	W	0.133
31	Gallium	Ga	37.8	75	Rhenium	Re	0.0517
32	Germanium	Ge	119	76	Osmium	Os	0.675
33	Arsenic	As	6.56	77	Iridium	Ir	0.661
34	Selenium	Se	62.1	78	Platinum	Pt	1.34
35	Bromine	Br	11.8	79	Gold	Au	0.187
36	Krypton	Kr	45	80	Mercury	Hg	0.34
37	Rubidium	Rb	7.09	81	Thallium	Tl	0.184
38	Strontium	Sr	23.5	82	Lead	Pb	3.15
39	Yttrium	Y	4.64	83	Bismuth	Bi	0.144
40	Zirconium	Zr	11.4	90	Thorium	Th	0.0335
41	Niobium	Nb	0.698	91	Protactinium[1]	Pa	—
42	Molybdenum	Mo	2.55	92	Uranium	U	0.0090
43	Technetium[1]	Tc	—	84–89	Unstable elements[1]		—
44	Ruthenium	Ru	1.86				

Data from Anders and Grevesse (1989).

[1]Unstable element.

three elements may have been reduced in the Sun by bombardment with neutrons and protons, although the production processes just discussed tend to bypass these elements. The peak corresponding to the iron group elements represents nuclides formed by nuclear statistical equilibrium, also an addition from earlier stars. The abundances in the Sun of elements heavier than iron reflect additions of materials formed in supernovae.

Superimposed on the peaks and valleys in figure 15.3 is a peculiar sawtooth pattern, caused by higher abundances of elements with even rather than odd atomic numbers (Z). The even-numbered elements form a lopsided 98% of cosmic abundances, because nuclides with even atomic numbers are more likely to be stable, as already noted in chapter 14. This explanation, however, begs the question: Why are even-numbered atomic

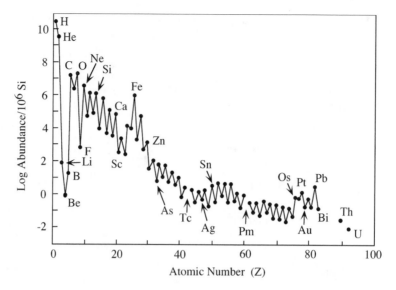

FIG. 15.3. Cosmic abundance of the elements, relative to 10^6 silicon atoms.

numbers more stable? Whenever nuclear particles, either protons or neutrons, can pair with spins in opposite directions, they can come closer together and the forces holding them together will be stronger. This stronger nuclear binding force results in higher stability.

The chemical composition of the Sun, then, can be understood primarily in terms of nucleosysthesis reactions from an earlier generation of stars whose materials have been recycled. These earlier generation stars also provided heavy elements for the formation of planets like the Earth. Some helium has been produced at the expense of hydrogen during the Sun's lifetime. Every second, 700 million tons of hydrogen is fused into 695 million tons of helium, with the missing 5 million tons converted into energy according to Einstein's famous equation. This is, of course, the source of the Sun's luminosity.

CHONDRITES AS SOURCES OF COSMOCHEMICAL DATA

Chondrites are the most common type of meteorite. These objects are basically ultramafic rocks, and petrographic studies suggest that they have never been geologically processed by melting and differentiation (although some components of chondrites have been melted prior to incorporation into the meteorites). Radiometric age determinations indicate that chondrites are the oldest samples surviving from the early solar system, approximately 4.56 billion years old.

Chondrites occupy a singularly important niche in cosmochemistry and geochemistry: they are the baseline from which the effects of most chemical processes can be gauged. The reason for this is quite simple. Of all known materials that can be studied directly, chondrites most closely match the composition of the Sun. This is illustrated in figure 15.5, a plot of the abundances of elements in the solar atmosphere against those in a chondrite, all relative to 10^6 atoms of silicon. A perfect correspondence is given by the diagonal line in this figure. The correspondence may even be better than indicated, because of some uncertainties in measuring the Sun's composition. To be honest, we should note that the scales in this figure are logarithmic but, with the exception of a few elements, the abundances are certainly within a factor of two of each other. Hydrogen, helium, and some other elements that occur commonly as gases—carbon, nitrogen, oxygen—are more abundant in the Sun, so it may be reasonable to visualize chondrites as a solar sludge from which gases have been distilled. In contrast, lithium, boron, and possibly beryllium have higher abundances in chondrites than in the Sun. Recall that these elements are systematically destroyed in the Sun by interactions with nuclear particles. In this regard, then, chondrites may record the original composition of the solar system even better than does the present day Sun.

It should now be obvious why in previous chapters we have repeatedly used chondrites as a basis for normalization of geochemical data: their nearly cosmic compo-

MEASURING THE COMPOSITION OF A STAR

The spectrum of light emitted by a star's *photosphere* (the region of the stellar atmosphere from which radiation escapes) provides a means for determining its composition. Superimposed on the continuous spectrum of a star are absorption bands, or missing wavelengths (fig. 15.4). These bands, produced by electron transitions in various atoms, appear as dark lines when photographed through a spectrograph. These are sometimes called *Fraunhofer lines,* after the German physicist who discovered them in 1817.

FIG. 15.4. The solar spectrum contains dark absorption bands (Fraunhofer lines) that can be used to estimate the Sun's composition. This figure shows only a small part of the spectrum.

Each Fraunhofer line corresponds to a particular element, and that element's abundance can be determined from the intensity of the absorption. What is actually measured is the width of the absorption band, because the Fraunhofer lines are broadened with increasing concentration. In practice, allowance must be made for the prevailing conditions of temperature and pressure in the star's photosphere before converting line width to element abundance. This is because the width of the absorption band depends not only on the elemental abundance, but also on the fraction of atoms that are in the right state of ionization and excitation to produce the band, which is in turn a function of temperature and pressure. From observations of the continuous spectrum, temperature and pressure can be modeled theoretically for various depths within the photosphere. It is then possible to combine the contributions from atoms at all depths throughout the photosphere to predict how line width varies with abundance.

Absorption bands corresponding to any one element are not necessarily visible in all stars. This is probably more a function of the varying temperatures and pressures of stars, which control the ability of a particular element to absorb radiation, than to the absence of that element. Measurements at certain wavelengths are also blocked by the Earth's atmosphere, although this problem has now been circumvented by spacecraft observations. Finally, note also that the composition of the photosphere may be quite different from that of the bulk of the star, unless convective mixing takes place.

To illustrate the importance of stellar spectral measurements, it is worth remembering that the second most abundant element was actually discovered in the nearest star before it was recognized on Earth. In 1869, Norman Lockyer, who founded the scientific journal *Nature,* noticed a line he could not identify in the solar spectrum. Based on this spectral evidence, Lockyer announced the existence of a new element which he called *helium,* because it had been discovered in the Sun—*helios* in Greek.

---

sitions provide the justification for this common practice. As an example, let's consider chondrite normalization of rare earth element (REE) patterns, which we introduced in chapter 13. REEs in both chondrites and basalts exhibit the sawtooth pattern of even-odd atomic numbers, as shown in figure 15.6. However, division of the abundance of each REE in basalt by the corresponding value in chondrites gives the smoothed, normalized values indicated by the open boxes. Without this normalizing step, the geochemical interpretation of REE patterns would be very cumbersome.

The ancient ages and nearly cosmic compositions of chondrites suggest that they are nearly pristine samples of early solar system matter (this is actually true only to a degree; see the accompanying box). There is ample evidence that chondrites are samples of asteroids—bodies generally too small to have sustained geologic processes. (However, we noted earlier that achondrites and iron

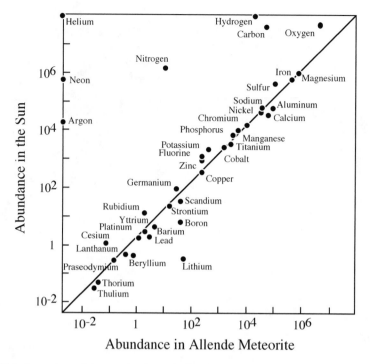

FIG. 15.5. Comparison of element abundances in the Allende chondritic meteorite with those in the solar photosphere. All elements are relative to 10^6 silicon atoms. This excellent correspondence indicates that chondrites have approximately cosmic compositions, except for the gaseous elements H, He, C, N, O, and noble gases.

meteorites are samples of melted and differentiated asteroids.) The terrestrial planets formed by accretion of smaller building blocks having chondritic composition. It is reasonable, therefore, to infer that the Earth itself must have a chondritic bulk composition.

COSMOCHEMICAL BEHAVIOR OF ELEMENTS

Controls on Cosmochemical Behavior

The cosmic abundance of an element is controlled by its nuclear structure, but its geochemical character is a function of its electron configuration. In chapter 2, we learned that Victor Goldschmidt first coined the terms *lithophile, chalcophile,* and *siderophile* to identify elements with affinity for silicate, sulfide, and metal, respectively. Geochemical affinity is actually a qualitative assessment of the relative magnitudes of free energies of formation for various compounds of the element under consideration. For example, suppose that metallic calcium

were exposed to an atmosphere containing oxygen and sulfur. Two potential reactions for calcium are:

$$Ca + \tfrac{1}{2}O_2 \rightarrow CaO,$$

and

$$Ca + S_2 \rightarrow CaS_2.$$

Of the two, the first is dominant, because ΔG_f^0 for the oxide is more negative than ΔG_f^0 of the sulfide. Therefore, calcium is lithophile. Of course, we must specify the conditions on the system for this conclusion to have meaning. For example, under extremely reducing conditions such as those under which enstatite chondrites were formed, calcium partitions more strongly into sulfide and thus is chalcophile.

There was little direct evidence on which to speculate about geochemical behavior when this idea was put forward, but Goldschmidt recognized that chondrites represent a natural experiment from which element behavior could be deduced. He and his coworkers measured the abundances of elements in coexisting silicate,

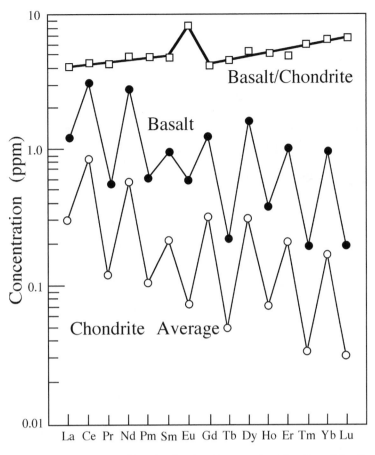

FIG. 15.6. Measured abundances of REEs in a basalt and average chondrites. Normalizing the basalt data to chondrites removes the zigzag pattern of odd-even abundances, revealing the pattern imposed during the igneous history of this rock.

sulfide, and metal from chondrites, and this pioneering work has been supplemented by studies of metal, sulfide, and silicate slag from smelters. Geochemical affinity is important in understanding element partitioning in any systems containing more than one of these phases—for example, the formation of magmatic sulfide ores or the differentiation of core and mantle.

This is not, however, the only factor controlling element behavior in cosmochemical systems. Another important factor is *volatility,* by which we mean the temperature range in which an element will condense from a gas of solar composition. Volatility depends on several factors, such as an element's vapor pressure, cosmic abundance (which controls partial pressure), total pressure of the system, and element reactivity (which determines whether the element in question occurs in pure form or in a compound). Refractory elements are those that

condense at temperatures >1400 K; volatile elements generally condense at temperatures <1200 K.

Chemical Fractionations Observed in Chondrites

In chondrites, groups of elements appear to move in unison. If one element of the group is enriched or depleted, all of the others are as well, and by nearly the same factor. At least four types of chemical fractionations, relative to cosmic (and C1 chondrite) composition, have been observed in chondrites:

1. Fractionation of refractory elements (for example, Ca, Al, Ti), resulting in high refractory element abundances in carbonaceous chondrites and lower abundances in other chondrite classes;

CHONDRITE PETROLOGY AND CLASSIFICATION

Chondrites take their name from *chondrules*: round, millimeter-sized, quenched droplets of silicate melt that are particularly abundant in these meteorites. Chondrules formed by flash melting at temperatures of 1700–2100 K and subsequent rapid cooling. They are typically composed of olivine, pyroxenes, and glass or feldspar. Irregularly shaped white objects called Ca,Al-rich inclusions (CAI) also occur in most chondrites. These are composed mostly of melilite, spinel, clinopyroxene, and anorthite. CAIs also formed at high temperatures of 1700–2400 K, and many were at least partly molten. Larger grains of iron-nickel metal and sulfides are also scattered throughout these meteorites. All of these objects are cemented together by a fine-grained matrix consisting of mixtures of olivine, pyroxene, feldspathoids, graphite, magnetite, and other minor phases. These components are illustrated in figure 15.7. These diverse kinds of materials have accreted together in space to form a kind of cosmic sediment. Despite extensive research, the origins of most of these components are unclear.

There are actually many classes of chondrites, differing slightly in petrography and chemical composition. The various classes of chondrites differ in their proportions of CAIs, chondrules, and metal. The *ordinary chondrites* are by far the most common extraterrestrial material that falls to Earth. The H, L, and LL subgroups of the ordinary chondrites contain varying proportions of iron, and differ in oxidation state (H chondrites contain the highest concentration of iron, mostly as metal; LL chondrites contain the least iron, mostly in oxidized form in silicates). Another class, *the enstatite chondrites* (E), are fully reduced, containing iron only as metal. The enstatite chondrites can also be subdivided by iron content into EH and EL groups. *Carbonaceous chondrites* (C) consist of many distinct chemical subgroups—CI, CM, CH, CO, CV, and CK, distinguished by variations in petrography and subtle differences in chemistry. Carbonaceous chondrites generally have high oxidation states relative to other chondrite types. The CI chondrites differ from other groups in that they contain no chondrules, but their chemical compositions are clearly chondritic. The *Rumaruti chondrites* (R) are also oxidized, but in other ways resemble ordinary chondrites.

Most ordinary and enstatite chondrites and a few carbonaceous chondrites have experienced thermal metamorphism. This has resulted in recrystallization, which blurs the distinctive chondrule textures. The major phases of chondrites (olivines and pyroxenes) are preserved, although their original scattered compositions have been homogenized, and glasses have been devitrified to form feldspars. The response of chondrites to heating is very different from that of terrestrial ultramafic rocks, because of the scarcity of fluids in chondrite parent bodies. Judging from the relative abundances of metamorphosed and unmetamorphosed chondrites in meteorite collections, asteroids must consist mostly of thermally sintered material. Many carbonaceous chondrites have been affected by a different process: aqueous alteration by fluids. The matrices of CM and CI chondrites have been transformed to complex mixtures of phyllosilicates, and veins of carbonates and sulfates permeate some samples.

Randy Van Schmus and John Wood (1967) formulated a classification scheme for chondrites that takes into account their compositions and thermal histories. Increasing metamorphic grade is indicated by num-

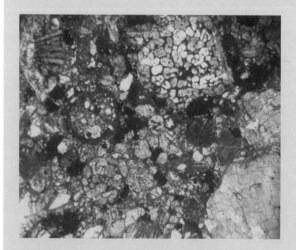

FIG. 15.7. Photomicrograph of the Tieschitz (Czech Republic) ordinary chondrite. The rounded objects are chondrules. The image is ~0.5 cm wide.

bers from 1 to 6; these are appended to the chemical group symbols, so that a particular chondrite may be classified, for example, as H3 or EL6. Van Schmus and Wood classified carbonaceous chondrites as metamorphic grades 1 and 2, inferring that they had experienced the least thermal effects. Grades 1 and 2 are now interpreted as reflecting varying degrees of aqueous alteration. Thus, we are left with a somewhat confusing situation, in which type 3 chondrites are the most primitive.

Many chondrites also show the effects of shock metamorphism, resulting from the impacts that dislodged them from the parent bodies. Despite the traumatizing effects of thermal and shock metamorphism and aqueous alteration in chondrites, their bulk chemical compositions seem little affected. In fact, CI1 chondrites (often abbreviated to C1, a convention we follow here), which have experienced pervasive mineralogic alteration, provide the best match for the solar composition. Some elements, however, were apparently mobilized by heating and were redistributed within ordinary chondrite parent bodies, and the stable isotopic compositions of CM2 and C1 chondrites were altered by exchange with fluids.

2. Fractionation of siderophile elements (for example, Fe, Ir, Ni, Au) to produce depletions in some ordinary and enstatite chondrites;

3. Fractionation of moderately volatile elements (for example, K, Sb, Ga, Zn), leading to depletions to varying degrees in carbonaceous, ordinary, and enstatite chondrites; and

4. Fractionation of highly volatile elements (for example, Pb, In, Tl) to produce severe depletions in ordinary and enstatite chondrites.

For three of the four of these fractionated groups, the only common property is volatility, which must have been a very important factor in controlling element behavior in the early Solar System. This is illustrated graphically in figure 15.8, in which elements are plotted in order of increasing volatility from left to right; degree of depletion in each chondrite type clearly increases with increasing volatility.

How could such element fractionations have been accomplished? The separation of siderophile elements may have resulted from the distinct densities of metal and silicates, possibly leading to differences in the rates at which grains of each accreted to form the meteorites. Fractionation of elements by volatility requires thermal processing in space. In the next section, we consider volatile behavior in a more quantitative fashion.

CONDENSATION OF THE ELEMENTS

Stars are so hot that all of the matter they contain is in the gaseous state, and any molecules are broken down into their constituent elements. However, as matter is expelled from stars, it cools and many elements may condense as solid metals or compounds. No liquids are produced, at least for a gas of solar composition cooling under equilibrium conditions, because the pressures in space are so low.

When interstellar dust and gas subsequently aggregate to form new stars, heating again occurs. Physical and dynamic conditions in the collapsing disk-shaped cloud that ultimately formed the Sun, called the solar nebula, have been modeled. Most astrophysical models suggest a hot nebula, with temperatures >1200 K near the center of the cloud but decreasing outward. These models have led cosmochemists to conclude that most of the presolar solid matter in the interior of the disk vaporized and then subsequently condensed as the nebula cooled after the Sun formed. Temperatures in the present-day asteroid belt are never this high, but outflows from the Sun's poles could have lofted condensed solids into the region where chondrites formed. Although the details of this process are sketchy, condensation is thought to have been an important process in establishing the chemical characteristics of the early Solar System.

How Equilibrium Condensation Works

The *condensation temperature* of an element is the temperature at which the most refractory solid phase containing that element first becomes stable relative to a gas of solar composition. The calculation of this temperature involves two steps. First, the distribution of a particular element among possible gaseous molecules as

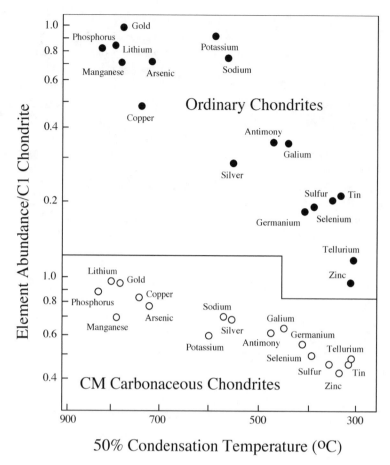

FIG. 15.8. The abundances of elements in two classes of chondrites show similar depletions related to element volatility, which increases from left to right. Serving as a volatility scale, 50% condensation temperature represents the temperature at which half of the element is not gaseous. All elements are normalized to cosmic abundances. (After Wasson 1985.)

a function of temperature must be determined. Next, the condensation temperatures of all potential solid phases that contain the element must be assessed. Comparison of these two sets of calculations yields the most stable phase—gas, solid, or some combination of the two. As with other types of thermodynamic calculations, it is necessary to consider all possible phases if the results are to be meaningful. Because such phases normally consist of several elements, the condensation temperature of any element may depend on the concentrations of other elements in the system. As a consequence, previously condensed phases may exert some control on the condensation temperatures of elements not contained in such phases.

Worked Problem 15.1

How does one calculate the condensation temperature of corundum from a gas of solar composition? For this exercise, we adopt a total pressure of 10^{-4} atm, which is appropriate for parts of the solar nebula. We have already seen that hydrogen has by far the highest cosmic abundance, and H_2 is much more abundant than any other hydrogen-containing molecule below 2000 K. Therefore,

$$P_{H_2} = P_{total},$$

where P_{total} is the pressure of some region of the nebula, in this case 10^{-4} atm. The ideal gas law, which is applicable at these low pressures, gives:

$$N_{H_2} = P_{H_2}/RT,$$

where N_{H_2} refers to the number of moles per liter of H_2. Because $2N_{H_2} = N_H$, we can express the concentration of any element X (in this case Al or O) in the gas as:

$$N_X = (A_X/A_H)N_H. \tag{15.1}$$

In this equation, A_X and A_H are the molar cosmic abundances of element X and hydrogen, respectively.

A mass balance equation can now be written for each element X that describes its partitioning into various gaseous molecules. For example, the total number of oxygen atoms can be expressed as:

$$N_O + N_{H_2O} + N_{CO} + 2N_{CO_2} + N_{MgO} + \ldots = N_{O\ total}. \tag{15.2}$$

All of these gaseous compounds are assumed to have formed by reaction of monatomic elements. In the example above, water formed by the reaction:

$$2H(g) + O(g) \rightleftarrows H_2O(g).$$

From the free energies of the species in this reaction at the temperature of interest, an equilibrium constant K can be calculated such that:

$$K = P_{H_2O}/(P_{H_2}P_O). \tag{15.3}$$

Equation 15.3 assumes, of course, that activities of these components are equal to partial pressures, probably a valid supposition at these low pressures. Substituting equation 15.3 into the ideal gas law and rearranging gives:

$$N_{H_2O} = KN_{H_2}N_O(RT)^2. \tag{15.4}$$

Equations of this form can be derived for each gaseous species and substituted into the mass balance equation 15.2. Other analogous mass balance equations for all elements of interest are then generated. The result is a series of n simultaneous equations in n unknowns, the amounts of monatomic gaseous elements, such as $N_{O\ total}$. These equations can then be solved, using a method of successive approximations, for the species composition of the gas phase at any temperature and a nebular pressure of 10^{-4} atm.

For the second step, we write an equation representing equilibrium between solid corundum and gas:

$$Al_2O_3(s) \rightleftarrows 2Al(g) + 3O(g),$$

for which:

$$\log K_{eq} = 2\log P_{Al} + 3\log P_O - \log a_{Al_2O_3}.$$

Because the activity of pure crystalline corundum is unity, this reduces to:

$$\log K_{eq} = 2\log P_{Al} + 3\log P_O. \tag{15.5}$$

Values for $\log K_{eq}$ at various temperatures can be determined from appropriate free energy data. (It is easy to make an error at this step, because tabulated values of free energy of formation from the elements for corundum do not use Al(g) and O(g)

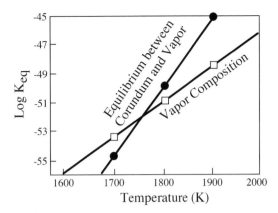

FIG. 15.9. Plot of log K_{eq} versus temperature, illustrating how the condensation temperature of corundum is obtained. Circles represent calculated conditions of equilibrium between corundum and aluminum vapor, and squares represent the evolving Al(g) concentration in the nebula as temperature changes, determined by solving the system of mass balance equations. The intersection of these lines marks the point at which corundum condenses from the cooling vapor.

as the standard states. JANAF [1971] tables give free energy data for the formation of gaseous monatomic species.) Log K_{eq} values for corundum calculated in this way are shown by circles in figure 15.9. The line connecting these points divides the figure into regions in which solid corundum and gas are stable. We then calculate log K_{eq} for the same reaction by using the partial pressures P_{Al} and P_O from the mass balance equations. These values are shown by squares in figure 15.9. The temperature at which the two values of log K_{eq} are the same (that is, the point at which the two lines in fig. 15.9 cross) is the temperature at which corundum condensation occurs, in this case 1758 K. For illustration purposes, we have treated this problem graphically, but it could also be solved numerically by finding the temperature at which the two values of log K_{eq} are equal.

Determining condensation temperatures for subsequently condensed phases is a little tricky. Once the condensation temperature for any element X is reached, the equation analogous to 15.5 must be solved again, this time adding terms for the concentration of the crystalline phase in the appropriate mass balance equations. If we require that the condensed phase be in equilibrium with the gas as it cools, the gas composition will have a lower P_X than would have been the case had not the solid phase appeared. Therefore, the affected mass balance equations must be corrected for the appearance of each new condensed phase.

As elegant as condensation theory is, it seems likely that kinetic factors may have been important in the condensation process. Complexities such as barriers to nucleation, the formation and persistence of metastable condensates, and supersaturation by condensable species in the gas phase are not currently understood and cannot be modeled. Treating the solar nebula as if it were in thermodynamic equilibrium provides great insight, but this may serve only as a boundary condition for the condensation process.

The Condensation Sequence

The equilibrium condensation sequence for a gas of solar composition at 10^{-4} atm, as calculated by cosmochemists Larry Grossman and John Larimer (1974), is illustrated schematically in figure 15.10. Some solid phases condense directly from the vapor. Others form by reaction of vapor with previously condensed phases. The latter situation is the cosmic equivalent of a peritectic reaction in solid-liquid systems.

Corundum (Al_2O_3) is the first solid phase to form that contains a major element, although such trace elements as osmium, zirconium, and rhenium may condense at even higher temperatures. Corundum is followed by perovskite $(CaTiO_3)$, which may contain uranium, thorium, tantalum, niobium, and REEs in solid solution.

Corundum then reacts with vapor to form spinel $(MgAl_2O_4)$ and melilite $(Ca_2[MgSi, Al_2]SiO_7)$, which in turn react to produce diopside at a lower temperature. Iron metal, containing nickel and cobalt, condenses by 1375 K. Pure magnesian forsterite appears shortly thereafter; it later reacts with vapor to form enstatite. Anorthite then forms by reaction of the previously condensed phases with vapor. All of these refractory phases appear above 1250 K.

Below this temperature, iron metal reacts with vapor and becomes enriched in germanium, gallium, and copper. Similarly, anorthite reacts to form solid solutions with the alkalis—potassium, sodium, and rubidium. Condensation of these moderately volatile elements occurs between 1200 and 600 K. These elements are somewhat depleted in chondrites relative to cosmic values (see fig. 15.8) and are sometimes called the *normally depleted elements*. Below 750 K, iron begins to oxidize, and the iron contents of olivine and pyroxene increase rapidly as the temperature drops further. This leads to the interesting conclusion that the oxidation state changes during condensation. Iron occurs only in metallic form at high temperatures, and only in oxidized form at the end of the condensation sequence. Metallic iron also

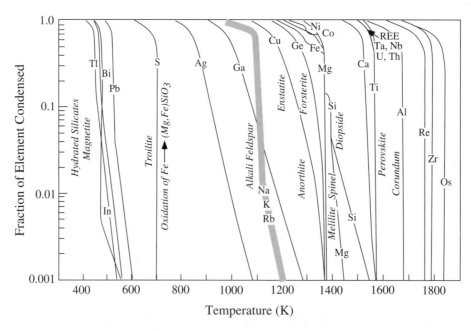

FIG. 15.10. Summary of the calculated condensation sequence for a gas of solar composition at 10^{-4} atm as a function of temperature and fraction of element condensed. The temperatures at which various minerals condense are indicated by the name of the mineral in italics. (Modified from Grossman and Larimer 1974.)

reacts with sulfur in the vapor to form troilite (FeS) below 700 K.

The highly volatile elements, which are strongly depleted in most chondrites, condense in the interval 600 to 400 K. Examples are lead, bismuth, and thalium. Magnetite becomes a stable phase at 405 K, and olivine and pyroxene react with vapor to form hydrated phyllosilicates at lower temperatures.

Evidence for Condensation in Chondrites

The mineralogy of chondrites is very similar to that predicted from condensation theory. The most refractory material occurs in the form of CAIs, one of which is shown in figure 15.11. These occur most frequently in carbonaceous chondrites and consist largely of melilite, spinel, diopside, anorthite, and perovskite—all phases predicted to condense at high temperatures. Hibonite ($CaO \cdot 9Al_2O_3$), not included in the original calculations because of the absence of appropriate thermodynamic data, appears to take the place of corundum as an early condensate. These inclusions sometimes contain tiny nuggets of refractory platinum group metals as well. Geochemical studies of CAIs demonstrate that they are markedly enriched in other refractory trace elements.

The evidence that CAIs are condensates is compelling, but they have not always survived intact. Many CAIs have been melted, and some appear to be residues from the evaporation of more volatile components.

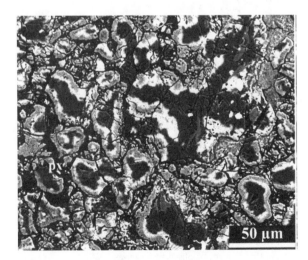

FIG. 15.11. CAI in a carbonaceous chondrite. The dark cores consist of spinel, hibonite, and perovskite (all early condensing phases), rimmed by later condensing melilite and diopside.

Olivine and pyroxene are the major constituents of chondrules and unaltered fine-grained matrix, and metallic iron occurs in abundance in most chondrites. Most silicate grains in chondrules are clearly not condensates, because they crystallized from melted droplets. However, the chondrules themselves may have formed by remelting of solid condensate material. A few relict olivine grains in chondrules appear to have survived the melting process; these have higher refractory trace element contents than other olivines and could be condensates. The phyllosilicate matrix phases of many carbonaceous chondrites, commonly mixed with magnetite, may possibly represent the low-temperature end of the condensation sequence, although there is considerable evidence that these phases formed by aqueous alteration in asteroids.

Thus, chondrites might be viewed as physical mixtures of material formed at various stages in the condensation sequence, although in many cases these materials were reheated after they condensed. These were aggregated in various proportions, accounting for the differences in abundances of refractory and volatile elements observed in ordinary, carbonaceous, and enstatite chondrite groups.

INFUSION OF MATTER FROM OUTSIDE THE SOLAR SYSTEM

Among the billions of stars in our galaxy, only two or three undergo supernova explosions each century. Yet supernova nucleosynthesis accounts for the abundances of many of the elements in the Solar System. Is supernovae debris flung so far and wide that it eventually gets incorporated into all stars, or are there other factors at work? Before we can answer this question, we must consider some isotopic evidence for the infusion of supernova products into the solar nebula.

Isotopic Diversity in Meteorites

Variability in isotopic composition has been demonstrated for many elements in most geological samples. In chapters 13 and 14, we learned that such diversity can arise from (1) isotope fractionation processes, (2) mixing, (3) radioactive decay, and (4) interaction with cosmic rays. Given this variability, how could we use isotopes to search for and recognize materials that formed outside the Solar System? The similarity between minerals in the condensation sequence and those that constitute CAIs suggests that CAIs are some of the oldest surviving relicts of the early solar nebula. This conclusion is corroborated by radiometric dating of these inclusions. For example, U-Th-Pb dating methods indicate ages of about 4.56×10^9 yr, and initial $^{87}Sr/^{86}Sr$ isotopic compositions are the lowest known of any kind of materials. If we wish to search for supernova products that have not been completely blended with other Solar System matter, these inclusions provide a good place to start.

The oxygen isotopic compositions of CAIs were first analyzed by Robert Clayton and his coworkers (1973). Figure 15.12 illustrates the relationship between $\delta^{17}O$ and $\delta^{18}O$ in these inclusions and terrestrial rocks. The terrestrial mass fractionation line has a slope of $+\frac{1}{2}$, because any separation of ^{17}O from ^{16}O (a difference of 1 atomic mass unit) will be half as effective as the separation of ^{18}O from ^{16}O (a difference of 2 atomic mass units). The trend of the inclusion data, clearly distinct from the mass fractionation line, must signify some other process at work. Clayton and coworkers interpreted this as a mixing line, joining "normal" solar system oxygen on the terrestrial mass fractionation line to pure ^{16}O (to which the mixing line in this figure extrapolates). Pure

^{16}O is produced by explosive carbon burning in supernovae. Thus, it was argued that CAIs must contain an admixture of interstellar grains containing supernova-generated oxygen. This exotic component may have served as a substrate for nucleation of other material during condensation, or it may have survived evaporation in refractory mineral sites. Another possibility is that oxygen isotope diversity occurred in the hottest part of the nebula by non-mass-dependent gas phase reactions that produced ^{16}O enrichments.

The exciting discovery of oxygen isotopic diversity initiated a flurry of research activity to find anomalies in the isotopic compositions of other elements. Nucleosynthesis theory leads us to expect anomalies in magnesium and silicon in samples with large concentrations of ^{16}O, but the results obtained so far are puzzling. Although nuclear anomalies in magnesium and silicon, as well as titanium, calcium, barium, and strontium, have been found, these occur in only a very few CAIs and are not generally correlated.

One of the most interesting isotopic anomalies in these objects indicates the former existence of now-extinct radionuclides. Excess amounts of ^{26}Mg, attributed to

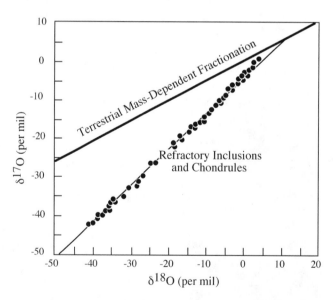

FIG. 15.12. The relationship between ^{16}O, ^{17}O, and ^{18}O in CAIs and chondrules in carbonaceous chondrites. Mass fractionation produces a line of slope $+\frac{1}{2}$ in this diagram, as observed for terrestrial materials. CAIs and chondrules define an apparent mixing line between "normal" solar system material, such as that along the terrestrial mass fractionation line, and pure ^{16}O. δ values are per mil relative to SMOW. (From Clayton et al. 1973.)

radioactive decay of ^{26}Al, have been found in CAIs. The half-life of ^{26}Al is only 0.7 Ma, so its incorporation as a "live" radionuclide indicates that formation of the solar nebula occurred very soon after nucleosynthesis. In fact, after just a few million years had elapsed, the abundance of this nuclide should have decreased to the point at which it would be undetectable. Like ^{16}O, ^{26}Al is produced by explosive carbon burning. The occurrence of this isotope provides a time scale for the addition of supernova material to the solar nebula. Besides ^{26}Al, other short-lived radionuclides formed in supernovae and found in chondrites include the following (half-life in parentheses): ^{41}Ca (0.1 Ma), ^{60}Fe (1.5 Ma), ^{53}Mn (3.7 Ma), ^{107}Pd (6.5 Ma), ^{182}Hf (9 Ma), ^{129}I (15.7 Ma), and ^{244}Pu (82 Ma).

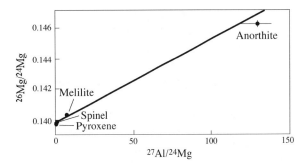

FIG. 15.13. Correlation between ^{26}Mg and ^{27}Al, both normalized to ^{24}Mg, in a CAI. This relationship suggests that the ^{26}Mg excess was produced by decay of the extinct radionuclide ^{26}Al. (From Lee et al. 1977.)

Worked Problem 15.2

How would one demonstrate that ^{26}Al was incorporated as a "live" radionuclide? First let's review the evidence that ^{26}Al, dead or alive, was added to CAIs. The decay product of this radionuclide is ^{26}Mg, so it is only necessary to look for an excess of this isotope. This turns out to be a rather difficult measurement, however, because ^{26}Mg is already an abundant isotope, constituting of ~11% of normal magnesium. Only in samples with a high original proportion of ^{26}Al would the addition of a relatively small amount of extra radiogenic daughter ^{26}Mg be detectable.

Caltech graduate student Typhoon Lee, along with Dimitri Papanastassiou and Gerry Wasserburg (1977), found enhancements of ^{26}Mg that changed the relative abundance of this isotope in inclusions by up to ~11.5%. Isotopic fractionation can be ruled out as a cause for this anomaly, because there should have been a similar but smaller effect in ^{25}Mg relative to ^{24}Mg, which is not observed. Thus, it is clear that the ^{26}Mg anomaly must be nuclear in origin; but there are nuclear reactions other than ^{26}Al decay that could give rise to this product.

The origin of the excess ^{26}Mg can be established by demonstrating that ^{26}Mg is correlated with aluminum in individual phases of the inclusions. Lee separated anorthite, melilite, spinel, and diopside, each of which contains different amounts of aluminum and magnesium. The results are illustrated in figure 15.13. (A word of caution is in order here. This diagram is not analogous to the plots utilized for Rb-Sr and Nd-Sm chronology in chapter 14. ^{26}Mg is not the radioactive parent, but a stable isotope of magnesium. If the ordinate were ^{26}Al/^{24}Mg, this would be an isochron diagram and the slope of the regression line would be a function of time. Such a diagram could only be constructed during the first few million years after nucleosynthesis, when ^{26}Al was still measurable.) Figure 15.13 shows that phases that initially had high total aluminum and low total magnesium contents incorporated more ^{26}Al and there-

fore now have the highest ^{26}Mg. Spinel and diopside, with very low Al/Mg ratios, have virtually no excess ^{26}Mg. Melilite, which contains both elements, has a 4% ^{26}Mg excess and anorthite, with a high Al/Mg ratio, has a 9% excess. Because the radiogenic ^{26}Mg now occurs in crystallographic sites occupied by aluminum, it must have been incorporated as ^{26}Al into these phases.

From the slope of the correlation line in figure 15.13, it is possible to calculate the initial isotopic ratio ^{26}Al/^{27}Al at the time the inclusion formed. This value is ~5×10^{-5}, a value so common in CAIs that it is normally taken as the canonical ^{26}Al/^{27}Al ratio for the early Solar System.

A Supernova Trigger?

The presence of freshly synthesized radionuclides in chondrites demands that these meteoritic materials formed soon after nucleosynthesis. Because of the long travel times across vast interstellar distances, these data further suggest that a supernova occurred in our cosmic backyard at about the time that the Solar System formed. The time from synthesis to injection into the collapsing solar nebula must have been on the order of 300,000 yr, consistent with the half-lives of the shortest lived radionuclides. Does this not seem highly unlikely? That would indeed be the case if there were no connection between the injection of short-lived radionuclides and solar nebula formation. Astrophysicists Al Cameron and J. W. Truran (1977) turned the argument around, proposing that a nearby supernova caused the formation of the solar nebula. They suggested that the advancing shock wave generated by a supernova explosion triggered the collapse of gas clouds surrounding it to form nebulae and ultimately new stars. The isotopic anomalies found in chondrites would be a natural consequence of the blending of small amounts of supernova products with matter indigenous to the gas-dust cloud. Any evidence of this addition of superova products in materials other than chondrites would have been subsequently obliterated by geological processing.

Astronomers have now observed the formation of new stars at the leading edges of expanding supernova remnants, providing independent evidence that such a process is possible. The idea of a supernova trigger neatly explains some of the most vexing questions in cosmochemistry. It is also an interesting demonstration of the utility of geochemistry in solving interdisciplinary problems.

The Discovery of Stardust in Chondrites

Cosmochemist Edward Anders recognized that solid grains formed from matter expelled from supernova explosions would likely be tagged with exotic r-process nuclides. In 1987, after working on and off for nearly two decades, he and his coworkers finally succeeded in isolating stardust from chondrites. Their method involved the stepwise dissolution of meteorites in acids while monitoring the presence of carriers of exotic nuclides (in this case, isotopes of neon and xenon) at each step in the residue. After numerous steps, less than a thousandth of the original of chondrite mass remained, and Anders finally arrived at his prize: a tiny amount of interstellar powder that proved to be miniature diamonds!

Buoyed by this success, Anders' University of Chicago group soon separated grains of another presolar mineral, silicon carbide. By measuring the isotopic composition of silicon and carbon with the ion microprobe, they demonstrated that these, too, were stardust. Subsequent work has isolated other interstellar phases, including graphite, corundum, and titanium carbide. The careful chemistry necessary to separate and characterize minute quantities of these grains tests the analytical skills of cosmochemists, but the results are astounding. Isotopic measurements of stardust provide ground truth for astrophysical models of nucleosynthesis in stars. With presolar grains, we can see the nuclear fusion processes at work. Who would have dreamed that the chemical analysis of tiny grains in meteorites would help us understand how the stars shine?

THE MOST VOLATILE MATERIALS: ORGANIC COMPOUNDS AND ICES

It is sobering to realize that most of the carbon atoms in our bodies were produced by nucleosynthesis in other stars. But how did they arrive at their present molecular complexity? Did all chemical evolution of organic compounds occur in the Earth's atmosphere and oceans, or was the process initiated at some other location? And for that matter, where did the gases that make up the atmosphere and oceans come from? We now turn to the cosmochemistry of the most volatile elements—hydrogen, carbon, nitrogen, and oxygen—which are depleted in chondrites relative to cosmic abundances.

Extraterrestrial Organic Compounds

Many organic molecules have been discovered in space by using microwave techniques. Detected organic compounds include most of the functional groups of biochemical compounds, except those involving phosphorus. Much of the interstellar medium is an inhospitable place for organic molecules, because any molecules that form are promptly dissociated by ultraviolet radiation. Dense interstellar clouds, however, are relatively opaque to such radiation, and organic compounds form readily within them. We can recognize these extraterrestrial organic molecules because they are enriched in deuterium. Under the frigid conditions within these clouds, deuterium is utilized in preference to hydrogen in chemical reactions. It combines with carbon, oxygen, nitrogen, and sometimes sulfur to produce an array of interstellar organic molecules.

When a molecular cloud collapses during star formation, the nebula inherits these elements already in some state of molecular complexity, although simpler compounds such as CO and CH_4 may predominate. These molecules probably occur as mantles on interstellar silicate dust grains.

When the solar nebula formed, organic compounds in what is now the inner Solar System were heated and partially decomposed into gases plus more refractory residues, a process called pyrolysis. Complex molecules could have survived in the outer portions of the nebula, and some of these were probably incorporated into comets prior to their ejection out of the vicinity of the Jovian planets. During subsequent cooling of the nebula, condensation of the vaporized fraction ensued. Some organic compounds are extremely volatile, so condensation occurred at low temperatures. As the temperature dropped, it became possible to stabilize and preserve organic compounds in regions where high temperatures had previously prevented their condensation, so we can envision these components as infiltrating the inner solar system late in the evolutionary history of the nebula. These compounds may have been relict interstellar materials transported from greater radial distances, local condensates, or both.

Let's first consider how organic compounds condense. Under equilibrium conditions, the dominant carbon and nitrogen species in the nebula should have been CO and N_2 at high temperatures, converting to CH_4 and NH_3 at low temperatures. Predicting the condensation of messy organic compounds is difficult because of inadequate thermodynamic data; in addition, kinetics must have played a major role.

One condensation model involves the Fischer-Tropsch type process, based on a commercial synthesis of hydrocarbon fuels from coal. As the nebula cooled below 600 K, adsorption of CO, NH_3, and H_2 onto previously condensed mineral grains led to the formation of a variety of compounds. The Fischer-Tropsch synthesis depends on catalytic activity on the mineral substrate. It has been proposed that complex organic compounds, as well as H_2O and CO_2, formed in this way. This is a very difficult model to test. Laboratory experiments indicate that the expected compounds can be produced, but only under very different conditions from those found in the nebula. Because the dominant nitrogen species at nebular temperatures is N_2 rather than NH_3, the Fischer-Tropsch systhesis would need to have been very efficient. The formation of complex organic molecules may also have been accomplished in the nebula by other processes, such as photochemical reactions driven by ultraviolet radiation.

Chondritic meteorites, especially carbonaceous chondrites, provide the only direct sampling of organic compounds from the nebula. Two categories of carbonaceous materials have been isolated from these samples. One consists of amorphous carbon mixed with macromolecular organic material similar to terrestrial kerogen. At least some fraction of this acid-insoluble component contains noble gases, whose elemental and isotopic compositions suggest that it may be relict interplanetary material. The smaller category consists of organic compounds that are readily soluble and thus easier to char-

TABLE 15.2. Distribution of Carbon in the Murchison Carbonaceous Chondrite

Species	Abundance
Acid-insoluble carbonaceous phase	1.3–1.8 wt. %
CO_2 and carbonate	0.1–0.5 wt. %
Hydrocarbons	
Aliphatic	12–35 ppm
Aromatic	15–28 ppm
Acids	
Monocarboxylic C_2-C_8)	170 ppm
Hydroxy (C_2-C_5)	6 ppm
Amino acids	10–20 ppm
Alcohols (C_1-C_4)	6 ppm
Aldehydes (C_2-C_4)	6 ppm
Ketones (C_3-C_5)	10 ppm
Ureas	2 ppm
Amines (C_1-C_4)	2 ppm
N-heterocycles	1.1–1.5 ppm

Data from Wood and Chang (1985)

acterize. Among these are many of the compounds present in living systems, such as carboxylic and amino acids. At one time, it was argued that such compounds were biologic in origin. However, the molecular structures of compounds in chondrites do not usually show the preferred chiral forms that characterize those produced biologically. Most amino acids, for example, are racemic mixtures of enantiomers, rather than being dominated by L-forms as in the amino acids in the terrestrial biosphere. The recent discovery of a slight excess of one enantiomer in several nonbiologic amino acids in chondrites suggests that organic synthesis in space may somehow introduce slight variations in D- and L- abundances. Table 15.2 lists the carbonaceous materials identified in one well-characterized carbonaceous chondrite.

These two carbonaceous components have very different stable isotopic compositions. For the kerogen component, measured mean values of δD, $\delta^{13}C$, and $\delta^{15}N$ are +800, −12, and +20, respectively; corresponding values for the soluble organic component are +250, +25, and +75. These major differences appear to require either more than one source or more than one production mechanism for meteoritic carbonaceous materials. Such data are at least consistent with the idea that some organic components may be relict interstellar phases.

Ices—The Only Thing Left

What happened to the simple gaseous molecules, such as CH_4, NH_3, and H_2O, that were not used in the production of more complex organic compounds? Under equilibrium nebular conditions, some of the water should have been incorporated into hydrous silicates, but there was no obvious sink for methane and ammonia. At temperatures below ~200 K, these compounds finally condensed into ices. This probably did not occur in the inner Solar System, but evidence of abundant ices is obvious at greater radial distances. The Jovian planets and their satellites consist of rocky cores surrounded by great thicknesses of these ices or their high-pressure equivalents. Comets are probably dirty snowballs containing considerable ice. It is generally thought that comets are relatively pristine samples of condensed matter, containing even the most volatile elements in their cosmic proportions.

It should come as no surprise that these ices are associated physically with organic compounds, because both condensed at low temperatures. We know this from spectroscopic observations of comet comas. It is not possible to reconstruct the primitive organic molecules that may have occurred in comet nuclei, however, because solar irradiation breaks down all but the simplest molecules before they are blown away into the tail.

Worked Problem 15.3

What are the relative weight proportions of ices to rock in a fully condensed solar nebula? Let's restrict this calculation to those elements with cosmic abundances >10^4 atoms per 10^6 atoms of silicon (table 15.1). Of these, we assume that the following elements condense as oxides to form rock: Na, Mg, Al, Si, Ca, Fe, and Ni. (Under reducing conditions appropriate for the nebula, some Fe and Ni condense as metals, but we can ignore this complication for now.) The following elements condense as ices: C, N, O, Ne, S, Ar, and Cl. We can model all of these ices as hydrides, except for the noble gases. A correction must be made to oxygen to adjust for the amount already partitioned into oxides. The element He does not condense, but remains in the gas phase; some H is consumed in hydride ices, and the remainder is gaseous.

Rock components are calculated in the following way. For sodium, convert atomic abundance to moles by dividing by Avogadro's number:

$$5.7 \times 10^4 \text{ atoms}/6.023 \times 10^{23} \text{ atoms mol}^{-1}$$
$$= 9.46 \times 10^{-20} \text{ mol.}$$

Now multiply half this number of moles by the gram formula weight of Na_2O:

$$(9.46 \times 10^{-20}/2 \text{ mol})(61.98 \text{ g mol}^{-1})$$
$$= 2.93 \times 10^{-18} \text{ g } Na_2O.$$

If we follow the same procedure for the each of the other elements condensable as rock, these sum to 3.04×10^{-16} g of oxides.

Now repeat the procedure for elements condensable as ices, using the gram formula weight of the hydride for all but the noble gases. For oxygen, we obtain:

$$2.01 \times 10^7 \text{ atoms}/6.023 \times 10^{23} \text{ atoms mol}^{-1}$$
$$= 3.34 \times 10^{-17} \text{ mol.}$$

In the case of oxygen, an adjustment must be made to correct for the amount already partitioned into oxides. Summing the amounts of oxygen needed for Na_2O, MgO, Al_2O_3, SiO_2, CaO, FeO, and NiO, we find that 7.03×10^{-18} mol of oxygen are required to combine with the condensable elements to make rock. Subtracting this from the total moles of oxygen available leaves 2.63×10^{-17} mol of oxygen that can be condensed as ice. Multiplying this result by the gram formula weight of the hydride gives:

$$(2.63 \times 10^{-17} \text{ mol})(18.016 \text{ g mol}^{-1})$$
$$= 4.73 \times 10^{-16} \text{ g } H_2O.$$

Repeating for other elements condensable as ice, we obtain a total of 1.02×10^{-15} g.

The ratio of these two numbers is:

$$\text{ice/rock} = 1.02 \times 10^{-15}/3.04 \times 10^{-16} = 3.38.$$

Thus, potential ices are greater than three times more abundant by weight than potential rock. Under reducing conditions, even more oxygen would be available for the production of ice. From this comparison, it is easy to see why the outer planets consist mostly of gases derived from ices.

A TIME SCALE FOR CREATION

The short-lived radionuclides in chondrites are potentially useful as a chronometer, if their initial relative abundances were uniform throughout the solar nebula. These isotopic systems give only relative time, but with very high resolution. To get absolute time scales, the short-lived nuclide time scales must be coupled with precise absolute ages, such as are determined from the long-lived U-Pb system. Consistency among all the chronometers would indicate an originally uniform nebula composition. These measurements are very challenging, but most evidence favors a homogeneous nebula.

Unaltered CAIs have the oldest measured U-Pb ages (determined very precisely to be 4.566 ± 0.002 b.y. by Claude Allegre and coworkers [1995]). This is normally taken to be the age of the solar system. CAIs also have the highest initial abundances of short-lived ^{26}Al, ^{41}Ca, and ^{53}Mn, as appropriate for the earliest formed nebular materials. Chondrule ages based on ^{26}Al decay are usually several million years younger than CAIs. The formation times of chondrites are impossible to date directly, because these rocks are assortments of components that formed at different times and the accretion process did not reset their radiometric clocks. After chondritic asteroids accreted, however, many experienced metamorphism (probably heated by ^{26}Al decay). As parts of these asteroids subsequently cooled through appropriate blocking temperatures, their radiometric ages were reset. The oldest measured chondrite, Ste. Marguerite (H4), has an absolute age of 4.563 ± 0.001 b.y., an ^{26}Al age of 5.6 ± 0.4 m.y. after CAIs, and a ^{182}Hf age of 4 ± 2 m.y.

The parent asteroids of achondrites and iron meteorites obviously had their radiometric systems reset by melting and differentiation. The oldest crystallization age for an achondrite is 4.558 ± 0.001 b.y. ^{182}Hf dating of iron meteorites also points to rapid differentiation within ~4 m.y. after CAI formation. These data indicate that asteroid-size objects (tens to hundreds of kilometers in diameter) took less than a few million years to form and differentiate. These planetesimals were rapidly heated to varying degrees, resulting in metamorphism or melting.

Determining when the Earth and other terrestrial planets formed is difficult because of their prolonged geologic evolution. In worked problem 14.5, we learned how the ^{182}Hf system can be used to constrain the timing of core formation. ^{182}Hf data for Martian meteorites indicate that Mars was differentiated within ~13 m.y. of Solar System formation (Yin et al. 2002). For the Earth, ^{182}Hf chronology suggests formation within ~29 m.y. of CAI formation. Mars is smaller than the Earth and its formation was probably completed before that of the Earth. The abundances of radiogenic ^{40}Ar and ^{129}Xe in the Earth's atmosphere indicate atmospheric retention beginning at 4.46 b.y. Thus, the terrestrial planets accreted after thermal processing of smaller asteroids. Because planetesimals in the inner Solar System were probably even more likely to have been melted, it is probable that they formed from already differentiated bodies. However, the bulk compositions of these asteroids were chondritic, so the Earth also has a bulk composition like chondrites.

ESTIMATING THE BULK COMPOSITIONS OF PLANETS

Some Constraints on Cosmochemical Models

Estimating the bulk compositions of planets is difficult, because differentiation ensures that no sample from anywhere within the body has the composition of the whole body. However, there are some constraints that allow us to test cosmochemical models.

Knowing a planet's mass (obtainable from its effects on the orbits of moons and nearby spacecraft) and volume (calculated from its measured diameter), we can calculate its mean density. The observed mean densities of the planets range from 3.9–5.5 g cm^{-3} for the terrestrial planets to 0.7–1.6 g cm^{-3} for the Jovian planets. Most of this difference can be explained by incorporation of varying amounts of rock and ice to form these bodies, due to radial temperature differences within the solar nebula. Within the terrestrial planet group, there are density variations, less dramatic but still of sufficient magnitude to suggest important chemical differences. To compare these density differences, we must correct mean densities for the effects of self-compression due to gravity. The uncompressed mean densities (in g cm^{-3}) for the terrestrial planets are: Mercury, 5.30; Venus, 4.00; Earth, 4.05; Mars 3.74; Moon, 3.34. Even after the effects of self-compression have been compensated for, significant density variations remain.

Several explanations have been put forward to explain these data. Cosmic abundance considerations indicate that planetary density variations must involve iron, the only abundant heavy element that occurs in minerals with different densities. Harold Urey (1952) proposed that varying proportions of silicate and metal, with densities of 3.3 and 7.9 g cm^{-3}, respectively, had been mixed to produce the inner planets. Density would then be a straightfoward function of metal/silicate ratio. Urey envisioned that metal-silicate fractionation occurred in the nebula, due to differences in the physical properties of these materials. A few years later, Ted Ringwood appealed to variations in oxidation state to explain density differences. The effect of redox state on chondrites illustrates the principle: if all the iron and nickel in these meteorites were converted to oxides, the density would be 3.78 g cm^{-3}. If FeO were instead fully reduced to metallic iron, the density would increase to 3.99 g cm^{-3}. This density range is not sufficient to encompass the mean densities of all the planets, so it is necessary to invoke some metal-silicate fractionation as well. Both of these proposals find some support in chondrites. The various chondrite groups can be distinguished on the basis of both Fe/Si and Fe^0/FeO ratios.

The value of its moment of inertia constrains how mass is distributed within a planet's interior. This can also be determined from seismic data, but aside from the Earth, we have almost no seismic information on the planets. The moment of inertia can be determined from the observed degree of flattening of a planet as it rotates, or from the precession of its rotational axis over time. Although moment of inertia is most affected by the presence and size of a metallic core, it also depends on the densities and thicknesses of mantle and crustal layers. Once a planet's bulk composition has been modeled and the proportions and mineralogies of core, mantle, and crust are estimated, its mean density and moment of inertia can be calculated. Comparisons between the calculated and observed mean density and moment of inertia constrain acceptable geochemical models for planet bulk compositions.

In the preceding section, we learned that the bulk composition of the Earth is chondritic, but what does that really mean? To cosmochemists, the term *chondritic* indicates that the relative abundances of elements with similar volatility are the same as in chondrites. We can clarify this statement by considering figure 15.14, which presents analyses of lanthanum, potassium, and uranium

in a variety of samples. Included in these diagrams are data from chondrites, terrestrial and lunar rocks, and meteorites from Mars and from the differentiated asteroid Vesta. (Martian meteorites are basalts and ultramafic rocks classified as shergottites, nakhlites, and chassignites, collectively called *SNCs*; Vesta meteorites are basalts and ultramafic rocks classified as howardites, eucrites, and diogenites, collectively called *HED achondrites*.) Lanthanum, potassium, and uranium all form lithophile cations having large ionic sizes, so they are incompatible elements. In other words, they are fractionated together during melting and crystallization, so that La/U or K/U *ratios* do not change from their original values, even though the absolute concentrations may be modified. Stated another way, these three elements exhibit similar geochemical behavior. Planetary differentiation has not significantly modified La/U or K/U ratios from their original values; thus, the measured ratios in any samples should give the bulk planet ratios.

However, these elements do exhibit differences in cosmochemical behavior. Uranium and lanthanum are refractory elements, whereas potassium is volatile. In figure 15.14a, we can see that all the bodies in the study, including the Earth, have a chondritic U/La ratio, because both U and La are refractory elements. In contrast, the K/La ratios vary from body to body (Fig. 15.14b), because elements with different volatilities were fractionated in the solar nebula. All the bodies shown are in fact depleted in volatile elements such as K, relative to chondrites. Recall that the chondrites themselves are

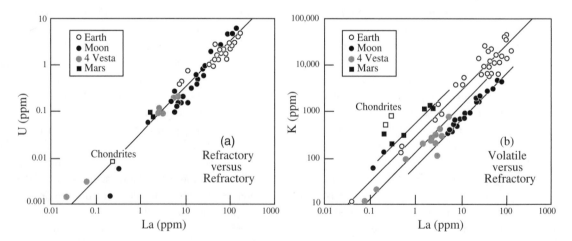

FIG. 15.14. Fractionation of volatile and refractory elements is shown by comparison of the ratios of uranium and potassium to lanthanum in planetary samples. All the bodies shown have chondritic ratios of refractory elements (U/La) (a), but differ in the ratio of volatile K to refractory La (b). (After Wänke and Dreibus 1988.)

depleted in volatile elements to varying degrees, relative to the Sun. This depletion of volatiles, although not well understood, is an important characteristic of planets that must be accounted for in cosmochemical models. Equally important, though, is the observation that planets have chondritic bulk compositions, in terms of the relative proportions of elements having similar volatility.

The Equilibrium Condensation Model

The condensation sequence already described defines a succession of equilibrium assemblages that form from a cooling gas of solar composition. The basis for the equilibrium condensation model for planet formation, as originally devised by John Lewis (1972), is that solids were thermally equilibrated with the surrounding nebula gas, and thus had compositions dictated by condensation theory. The temperature in the nebula is assumed to have been highest near the Sun and to have declined outward. Whatever solids had already condensed at particular radial distances were accreted to form the various planets and thereby isolated from further reaction with the gas. Any uncondensed materials were somehow flushed from the system.

At the location of Mercury, the temperature (1400 K) was such that refractory elements such as calcium and aluminum had completely condensed and all iron was metallic. Magnesium and silicon had only partly condensed at this point. The high uncompressed mean density of Mercury is thus explained by the occurrence of iron only in its most dense, metallic state, as well as the high iron abundance relative to magnesium and silicon. The temperature (900 K) at the orbital distance of Venus allowed complete condensation of magnesium and silicon, but no oxidation of iron. Earth formed in a somewhat cooler (600 K) region, in which some iron reacted to form sulfide and ferrous silicate. The addition of sulfur, with an atomic weight greater than the mean atomic weight of the other condensed elements, resulted in a mean density for Earth that is higher than that of Venus. The lower nebular temperature (450 K) for Mars permitted oxidation of all of the remaining iron and its incorporation into silicates. Its mean density was further reduced by hydration of some olivine and pyroxene to form phyllosilicates. Conditions in the asteroid belt were appropriate for the formation of carbonaceous chondrites and, further outward, ices were stable with silicates. These combined to form the Jovian planets.

Worked Problem 15.4

How can we estimate a planetary bulk composition using the equilibrium condensation model? This is a very long and complex problem, so we simply illustrate how it is done by examining how to set up the problem. It is first necessary to define the dimensions of a "feeding zone," from which each planet will draw in material during accretion. These annular zones can be the exclusive property of each planet or can be overlapping.

The condensation sequence represented in figure 15.10 was calculated at one pressure. However, pressure and temperature decreased outward in the solar nebula. Lewis (1972) generalized the condensation sequence by constructing a plot showing how some of these reactions varied with temperature and pressure, illustrated in figure 15.15. Lewis adopted a P-T gradient, also shown in this figure, that produced mineral assemblages at the locations of planets whose densities corresponded with those of the planets. The temperature at which a mineral first becomes stable was taken from the intersection of its reaction line with the P-T gradient in figure 15.15. The calculations of Grossman (1972) were then used to estimate the degree of condensation below that temperature. The composition of each

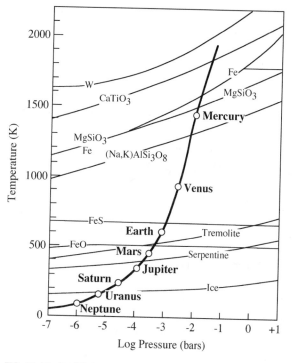

FIG. 15.15. Equilibrium condensation in the solar nebula can approximately account for the compositions of the terrestrial and Jovian planets, if temperature and pressure decreased radially away from the Sun, as shown by the heavy curved line. (Modified from Barshay and Lewis 1976.)

planet was defined by integrating, over equal increments of distance, the compositions of solid material within its feeding zone.

This model satisfactorily accounts for planetary mean densities, but a number of other problems are evident. First, it is not possible to explain the large difference in mean density of the Earth and Moon using equilibrium condensation. Also, it is unrealistic to assume that a large planet would accrete only condensates formed at a single temperature. Mixing would have been unavoidable and, in fact, orbital calculations suggest that the feeding zones for the terrestrial planets overlapped considerably. The kinetic barriers to gas-solid equilibrium must have also played a role, which would have resulted in further mixing of materials with different thermal histories. Finally, at the condensation temperatures for Venus, Earth, and Mars, no condensed volatiles are possible. The atmospheres of these planets and the Earth's oceans are thus an embarrassment to equilibrium condensation models, as is the presence of Fe^{3+} in terrestrial rocks. Despite these problems, the equilibrium condensation model is very useful as one of several end members of idealized scenarios for planet formation.

The Heterogeneous Accretion Model

Another model for planet formation is based on the premise that all solids were vaporized in a hot solar nebula. During cooling, condensation ensued. As each new phase formed, it was immediately accreted. The equilibrium condensation sequence was, in a sense, beheaded, because the removal of each solid phase prevented its subsequent reaction with the vapor. This is somewhat analogous to fractional crystallization, which we discussed in chapter 12, but in a cosmic setting. At some point, the process was interrupted as the nebula dissipated. Because temperatures in the nebula decreased radially outward, the progress of condensation was not everywhere the same. Mercury might have passed through only the stages at which Ca,Al-rich minerals, metallic iron, and some magnesium silicates condensed. Planets farther out may have acquired veneers of phyllosilicate- and organic-rich material before condensation was halted.

This model, originally developed by Karl Turekian and Sydney Clark (1969), is called *heterogeneous accretion*, because it results in layered planets, with the most refractory material in cores and more volatile phases at the surfaces. In practice, the model cannot be applied exclusively, because it leads to bizarre results. For example, the Earth should contain no FeO or FeS, because all of the iron was buried as metal in the core and isolated from further reaction through mantling by magnesian silicates. The advocates of this model do not practice undeviating loyalty, however; they ask only that equilibrium condensation be modified to allow some heterogeneous accretion to take place. The flexibility in this model makes it impossible to derive a unique composition for a planet.

Heterogeneous accretion does, however, suggest a source for volatile compounds in planetary atmospheres and the Earth's hydrosphere. Volatiles were accreted late in the condensation sequence onto planetary surfaces. The model also explains why rocks of the Earth's crust and mantle contain quantities of Fe^{2+} and Fe^{3+}, which are not stable in the presence of metallic iron. If reduced and oxidized iron had ever been mingled, they should have reacted to a uniform, intermediate oxidation state. A major difficulty with this model is that the requirement for total vaporization of dust in the nebula and for very rapid accretion are at odds with current conceptions of nebular conditions.

The Chondrite Mixing Model

If we accept the proposition that chondrites are left-over planetary building blocks, it is not necessary to rely on abstract concepts of nebular condensate material to construct models. In this case, we have actual samples that can be used in various proportions in cosmochemical models. If planets formed by accretion of already differentiated planetsimals, the chondrite mixing model may still be valid, because the bulk compositions of these bodies were chondritic.

Our task would be simple if the compositions of the planets matched those of individual chondrite groups. As appealing a model as this is, it doesn't quite work. We have already seen that planets show the same kinds of volatile element depletions that chondrite classes do, but planetary depletions are more extreme. Mixing various classes of chondrites provides better matches for planetary compositions, but even this is not enough to produce a perfect fit.

One example of a chondrite mixing model is based on the work of German cosmochemists Heinrich Wänke and Gerlind Dreibus (1988). Their model mixes two components, both having chondritic compositions, to produce the planets. Component A is refractory, free of all elements with equal or higher volatility than sodium, but containing all other elements in C1 abundance ratios. It is also reduced, with all iron present as metal; its degree of reduction is comparable to that of enstatite chondrites. This component, with its complete absence of highly volatile elements, does not correspond to any particular chondrite class. However, an end member extremely depleted in volatiles is necessary to explain the volatile depletions seen in planets. Component B is oxidized and contains all elements, including volatiles, in C1 abundances. In other words, it is C1 chondrite. Wänke and Dreibus were able to model the compositions of Earth and Mars by mixing these two chondrite components in differing amounts. The relative proportions of these components were constrained by element ratios in terrestrial mantle rocks and in Martian meteorites. They also argued that heterogeneous accretion (component A first, with only limited mixing of A and B later) best explained the composition of the Earth's mantle, whereas homogeneous accretion (thorough mixing of components A and B during accretion) was neces-

sary for Mars. The elemental abundances in the silicate (that is, noncore) parts of the Earth and Mars, based on the Wänke-Dreibus chondrite mixing model, are compared in figure 15.16.

Several other recent chondrite mixing models for Mars have been constructed in an attempt to duplicate the oxygen isotopic composition of SNC meteorites. One model mixes ordinary and carbonaceous chondrites, whereas another mixes ordinary and enstatite chondrites to arrive at the observed $\delta^{17}O$ and $\delta^{18}O$. The admixed materials obviously have very different oxidation states, as do components A and B in the Wänke-Dreibus model, so redox reactions involving metal, sulfide, carbon, water, and the FeO in silicates must be used to adjust these amounts to an equilibrium assemblage for the bulk planet.

Other chondrite mixing models use the components of chondrites, rather than bulk chondrites, to make planets. One formulation, by John Morgan and Edward Anders (1979), identifies components such as refractory material, metal, sulfide, silicates, and volatile material. Although these were actually defined as chemical

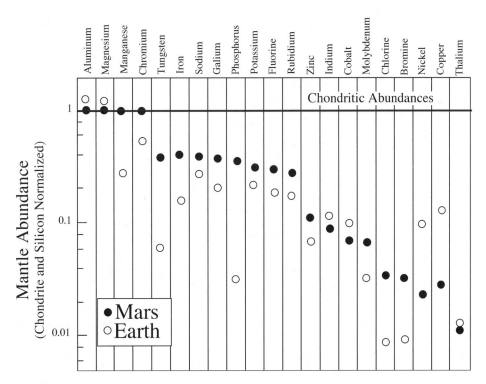

FIG. 15.16. Comparison of element abundances in the silicate portions of Earth and Mars, the latter calculated from SNC meteorites by Wänke and Dreibus (1988). All elements are normalized to chondrites and 10^6 silicon atoms.

components, they have physical counterparts in chondrites. For example, CAIs qualify as refractory material, chondrules are silicates, and matrix is volatile material. The chondrite model is based on the postulate that the materials that made the planets experienced the same fractionations as recorded by chondrites. Because the components can be mixed in different proportions to reproduce the chemistry of the various chondrite groups, they clearly were fractionated in the nebula. Thus, these same components can be blended to make planets, although more extreme proportions of some may be required. Each component carries its own suite of elements in approximately cosmic proportions, but the components have been fractionated from each other to varying degrees. This procedure seems to work reasonably well for most elements, but breaks down for elements that do not partition cleanly into one component. For example, significant errors in volatile element abundances may

GEOCHEMICAL CONSTRAINTS ON THE ORIGIN OF THE MOON

At various times, five principal hypotheses have been favored as possible explanations for the origin of the Earth's moon. These are:

1. Intact capture: the Moon formed elsewhere in the solar system and was gravitationally captured by the Earth.
2. Binary accretion: the Moon accreted as a companion to the Earth from materials captured in geocentric orbit.
3. Rotational fission: the Moon separated from an early Earth that was rotating so rapidly that it was unstable.
4. Collisional ejection: off-center collision of a very large projectile with the Earth ejected debris into geocentric orbit, from which the Moon ultimately accreted.
5. Disintegrative capture: a very large body that barely missed the Earth was tidally disrupted. Debris was retained in a geocentric orbit and accreted to form the Moon.

Aside from dynamical arguments, how can we choose among these possibilities?

Cosmochemical data do not provide a definitive answer, but they at least serve as important constraints for these hypotheses. Analyses of lunar rocks allow some general conclusions about the geochemistry of the bulk Moon. It is strongly depleted, by factors of several hundred relative to cosmic abundances, in highly volatile elements (for instance, Bi, Tl) and enriched in refractory elements (such as Ca, Al, Ti, REEs, U, Th). Its bulk FeO content of 13% is significantly lower than that of C1 chondrites and higher than that of the Earth. Trace siderophile elements are depleted in order of the metal-silicate partition coefficients, consistent with their removal into a small lunar core.

Let's first list some cosmochemical predictions that each hypothesis for its origin makes. Intact capture is largely untestable, but it seems likely that any body formed elsewhere in the Solar System might have a distinct composition, both elemental and isotopic, from the Earth. In contrast, binary accretion suggests that the compositions of Earth and Moon should be virtually identical, because they accreted from similar infalling materials at the same time. Rotational fission is normally envisioned as occurring after core formation in the Earth. Thus, in this model, the bulk Moon should be depleted in siderophile elements to the same degree as the Earth's mantle. Dynamical simulations of collisional ejection suggest that the incoming projectile provided most of the ejecta that was accreted to form the Moon. During the violent collision, most ejecta was vaporized and later recondensed, allowing for volatile element depletions in the Moon. Disintegrative capture is a variant of the intact capture hypothesis, from a cosmochemical perspective.

The oxygen isotopic compositions of lunar rocks lie along the same mass fractionation line as terrestrial rocks, as illustrated in figure 15.12. This is a very significant observation, because most other extraterrestrial samples define oxygen isotope fractionation lines displaced from the Earth-Moon line. The stable isotopic data are most compatible with hypotheses (2) and (3), although (4) is allowed if the Earth contributes a significant fraction of ejecta to the Moon.

The Mg/(Mg + Fe) ratio is an important indicator of the extent of geochemical differentiation. A value of 0.89 for this ratio for the Earth's mantle seems fairly well constrained, but that for the Moon is less certain. A ratio of 0.80 is consistent with Fe and Mg abundances in lunar basalts, which are more iron-rich than terrestrial basalts. If the Moon really does have a lower Mg/(Mg + Fe) ratio, this would argue against its derivation from the Earth by hypothesis (3).

The low uncompressed mean density of the Moon (3.34 g cm^{-3}) is very different from that of the Earth (4.05 g cm^{-3}). Variations in oxidation state cannot account for density differences of this magnitude, and it is generally accepted that the Moon is impoverished in iron metal relative to the Earth. Siderophile trace elements are depleted in both the Earth's mantle and the Moon. The Earth's situation is easy enough to understand. Elements such as nickel and cobalt are now sequestered in the core, but the Moon has only a small core. Hypotheses (3) and (4) can readily explain this constraint if fission or collision occurred after core formation. Preferential accretion of the Moon from the silicate parts of differentiated planetesimals could allow hypotheses (2) and (5) to be consistent. Hypothesis (1) must explain how two independent

bodies with such different bulk iron contents have such similar siderophile trace element abundances.

Volatile trace elements are more depleted, and refractory elements more enriched, in the Moon than in the Earth. This compositional distinction could be an inherent property of the materials that accreted to form these bodies, or it could result from some later, high-temperature process. This observation is difficult to explain if hypothesis (2) is correct. Volatile element differences are expected for hypothesis (1) and (5). Most ejecta formed during collision would be vaporized, and the observed volatile element depletion might result from recondensation in hypothesis (3). Because a rapidly rotating Earth would probably be partially molten, hypothesis (4) may also be consistent with this constraint.

These arguments obviously do not solve the problem, but they illustrate how cosmochemical data can be used to test possible solutions for complex problems of vast scale. The chemical data, plus the observation that the Earth-Moon system has very high angular momentum, suggest to many workers that a collisional ejection model is likely. Cosmochemical evidence bearing on the question of the Moon's origin has been explained more fully by Ross Taylor (1987).

result from this approach, because some of these may be only partly condensed.

Planetary Models: Cores and Mantles

The techniques we have just discussed provide ways to estimate the bulk chemical compositions of planets. It is also of obvious value to know how these elements are distributed within a planet; that is, which are partitioned into crust, mantle, and core. Because the crust contains such a minor portion of a planet's mass, it is common practice in constructing planetary models to combine mantle plus crust as one differentiated component. The problem then is reduced to that of how to estimate the compositions of the silicate part of the planet and its metallic core.

Worked problem 12.1 illustrated how the composition of the Earth's core can be estimated, based on chondritic relative abundances.

In chapter 12, we considered constraints on the composition of the Earth's crust, mantle, and core. Because the constraints are so meager, the problem is more difficult for other planets. Let's now consider the mantle and core of Mars, to illustrate how this problem is addressed for other planets.

Core composition can be estimated indirectly if the compositions of the bulk planet and mantle are known. The core composition is presumably complementary to that of the mantle from which it was extracted, so that depletions of siderophile and chalcophile elements relative to the bulk planet abundance can provide insight into the relative abundances of core-forming elements. This approach was used by Wänke and Dreibus (1988) to estimate that the core of Mars contains 61% FeNi metal and 39% FeS.

Connie Bertka and Yingwei Fei (1998) considered a variety of possible Martian core compositions, including iron as metal, sulfide, oxide, carbide, and hydride.

They also calculated the densities of these model cores, as shown in figure 15.17. The previous year, they performed high-pressure experiments, using the Wänke-Dreibus Mars composition, to determine the mineralogy of the Martian mantle as a function of depth. These data allowed Bertka and Fei to estimate a density profile for the mantle, also shown in figure 15.17. By assuming a 50 km-thick crust with the density of basalt, they then calculated the mean density and moment of inertia. Comparison with measured values suggested some inconsistencies with this Mars model. Although a firm estimate of the composition of Mars that meets all geophysical constraints is not yet available, this illustrates how cosmochemical models for planetary mantles and cores can be constructed and tested.

SUMMARY

Cosmochemistry is the application of geochemical principles to systems of vast scale. Fundamental to this discipline is the determination of the cosmic abundance of the elements. At the beginning of this chapter, we saw how elemental abundance patterns were controlled by nucleosynthesis in stars. Mixture of the products of fusion, nuclear statistical equilibrium, neutron capture, and proton capture reactions in earlier generations of stars accounts for the composition of the Sun and thus the cosmic abundance of the elements.

Chondritic meteorites are materials surviving from the earliest stages of solar system history. Their chemical compositions are essentially cosmic, although small fractionations occur in all chondrite classes. Element fractionations in the solar nebula, inferred from chondrites, were governed by volatility and by geochemical affinity, the latter reflected in siderophile, chalcophile, and lithophile behavior.

The equilibrium condensation sequence provides a conceptual framework, in which we can examine the effects of volatility. We have also seen that the isotopic compositions of the most refractory components of chondrites point to a supernova origin. The infusion of now extinct radionuclides indicates that a stellar explosion must have occurred in the vicinity of the solar nebula. It has been proposed that the shock wave from such a supernova triggered the collapse of gas and dust to form the solar system. Relict stardust, recognized by its isotopic fingerprints, has now been isolated from chondrites.

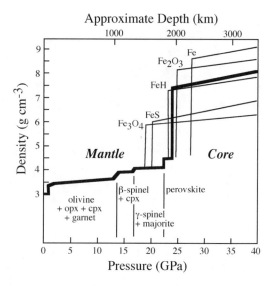

FIG. 15.17. Density profile for the Martian mantle and core (heavy line) and stability fields for mantle minerals, based on experiments using the Wänke-Dreibus Mars model by Bertka and Fei (1998). Also shown are density profiles and positions of the core-mantle boundary for a range of model core compositions (thin lines).

Some of the most volatile materials in chondrites are organic compounds, synthesized in cold molecular clouds and often modified by catalytic reactions in the nebula. Among these are most of the raw materials for life, suggesting that Earth may not have had to synthesize organic compounds from scratch. Ices were the last materials to condense in the nebula. Highly volatile materials are concentrated in the outer solar system, primarily in the Jovian planets and in comets.

The bulk compositions of the terrestrial planets can be modeled based on equilibrium condensation, assuming that a radial temperature and pressure distribution controlled mineral stability. The heterogeneous accretion model assumes that materials were aggregated and isolated from the nebula as soon as they condensed. The basis for a third planetary model is that fractionations in protoplanetary materials were the same as those observed in chondrites, so that chondritic comositions can be used to construct planets. The compositions of planetary mantles and cores can also be estimated from cosmochemical considerations. That these different techniques lead to similar results suggests that cosmochemical behavior offers a valid mechanism for understanding problems of planetary scale.

SUGGESTED READINGS

Because chondrites provide most of the data for cosmochemistry, much of the literature in this field is embedded in books and papers on meteorites. Many of the references below are technically demanding, but they provide a thorough background in this subject.

Anders, E., and N. Grevesse. 1989. Abundances of the elements: Meteoritic and solar. *Geochimica et Cosmochimica Acta* 53:197–214. (An important paper in which cosmic abundances are derived and discussed.)

Grossman, L., and J. W. Larimer. 1974. Early chemical history of the solar system. *Review of Geophysics and Space Physics* 12:71–101. (A comprehensive review of condensation and other nebular processes.)

McSween, H. Y., Jr. 1999. *Meteorites and Their Parent Planets,* 2nd ed. New York: Cambridge University Press. (A nontechnical introduction to meteorites and what can be inferred about their parent bodies.)

Newsom, H. E., and J. H. Jones, eds. 1990. *Origin of the Earth.* New York: Oxford University Press. (A series of papers dealing with geochemical and other kinds of constraints on the origin and early evolution of our planet.)

Suess, H. E. 1987. *Chemistry of the Solar System.* New York: Wiley. (A nice summary at a basic level. Chapter 1 gives the isotopic composition of the elements, and chapter 2 discusses nucleosynthesis.)

Taylor, S. R. 1992. *Solar System Evolution: A New Perspective.* Cambridge: Cambridge University Press. (A fascinating introduction to the solar system, as constrained by cosmochemistry, and arguing for the importance of impacts in its evolution.)

Wasson, J. T. 1985. *Meteorites, Their Record of Early Solar-System History.* New York: Freeman. (A meteorite monograph that is rich in cosmochemical information. Chapters 7 and 8 describe nebula fractionation in detail.)

Additional papers referenced in this chapter are the following:

Allegre, C. J., G. Manhes, and C. Gopel. 1995. The age of the Earth. *Geochimica et Cosmochimica Acta* 59:1445–1456.

Barshay, S. S., and J. S. Lewis. 1976. Chemistry of primitive solar material. *Annual Reviews of Astronomy and Astrophysics* 14:81–94.

Bertka, C. M., and Y. Fei. 1998. Implications of Mars Pathfinder data for the accretion history of the terrestrial planets. *Science* 281:1838–1840.

Burbidge, E. M., G. R. Burbidge, W. A. Fowler, and F. Hoyle. 1957. Synthesis of the elements in stars. *Reviews of Modern Physics* 29:547–650.

Cameron, A. G. W., and J. W. Truran. 1977. The supernova trigger for formation of the solar system. *Icarus* 30:447–461.

Clayton, R. N., L. Grossman, and T. K. Mayeda. 1973. A component of primitive nuclear composition in carbonaceous meteorites. *Science* 182:485–488.

Grossman, L. 1972. Condensation in the primitive solar nebula. *Geochimica et Cosmochimica Acta* 36:597–619.

JANAF 1971. *Thermochemical Tables,* 2nd ed. U.S. National Standards Reference Data Series 37. Washington, D.C.: National Bureau of Standards.

Lee, T., D. Papanastassiou, and G. J. Wasserburg. 1977. Aluminum-26 in the early solar system: Fossil or fuel? *Astrophysical Journal* 211:L107–110.

Lewis, J. S. 1972. Low temperature condensation in the solar nebula. *Icarus* 16:241–252.

Morgan, J.W., and E. Anders. 1979. Chemical composition of Mars. *Geochimica et Cosmochimica Acta* 43:1601–1610.

Taylor, S. R. 1987. The unique lunar composition and its bearing on the origin of the Moon. *Geochimica et Cosmochimica Acta* 31:1297–1306.

Turekian, K., and S. P. Clark. 1969. Inhomogeneous accretion of the earth from the primitive solar nebula. *Earth and Planetary Science Letters* 6:346–348.

Urey, H. C. 1952. *The Planets: Their Origin and Development.* New Haven: Yale University Press.

Van Schmus, W. R., and J. A. Wood. 1967. A chemical-petrologic classification for the chondritic meteorites. *Geochimica et Cosmochimica Acta* 31:747–765.

Wänke H., and G. Dreibus. 1988. Chemical composition and accretion history of terrestrial planets. *Philosophical Transactions of the Royal Society (London)* A 325:545–557.

Wood, J. A., and S. Chang, eds. 1985. *The Cosmic History of the Biogenic Elements and Compounds.* Washington, D.C.: NASA.

Yin, Q., S. B. Jacobsen, K. Yamashita, J. Blichert-Toft, P. Telouk, and F. Albarede. 2002. A short timescale for terrestrial planet formation from Hf-W chronometry of meteorites. *Nature* 418:949–952.

PROBLEMS

(15.1) Discuss why the comparison of terrestrial Nd-Sm isotopic data with a chondritic evolution curve (ε_{Nd} versus ε_{Sr}) provides useful information.

(15.2) Using figure 15.10 and the cosmic abundances in table 15.1, estimate the bulk composition of the refractory nebular condensate that forms in the temperature interval from 1900 to 1400 K.

(15.3) What are the relative weight proportions of condensable matter (rock plus ices) versus non-condensable gases (He and leftover H) in a gas of solar composition? (Part of the solution has already been calculated in worked problem 15.3.)

(15.4) Calculate the composition of the Earth's core in terms of Fe, Ni, Co, P, S, and O, assuming that half of the light element is sulfur and half is oxygen. (See worked problem 12.1.)

(15.5) List the chemical fractionations observed in chondrites, and suggest a plausible physical mechanism for each.

(15.6) Using figure 15.12, calculate the proportion of an exotic oxygen component consisting of pure ^{16}O that would have to be added to "normal" solar system oxygen (lying along the terrestrial mass fractionation line) to produce a CAI plotting along the mixing line at $\delta^{17}O = -25$ per mil.

APPENDICES

APPENDIX A: MATHEMATICAL METHODS

Most of the problems in this book require mathematics that is no more advanced than standard algebra or first-year calculus. We have found, however, that students often need a refresher in some of these basics to boost their confidence. We strongly urge you to use the problems in this text as an excuse to dust off your old math books and get some practice. If you do not already own a handbook of standard functions or a guide to mathematical methods in the sciences, you should probably add one to your shelf. We have found Burington (1973), Dence (1975), Boas (1984), and Potter and Goldberg (1997) to be particularly useful.

This short appendix is intended to introduce a few concepts not generally included in math courses required for the geology curriculum, but that are very useful in geochemistry. We do not pretend that it is an in-depth presentation; our goal is purely pragmatic. Each of these topics appears in one or more of the problems in this text, and should become familiar as you advance in geochemistry.

Partial Differentiation

If a function f has values that depend on only one physical parameter, x, then you know from your first calculus course that the derivative, $df(x)/dx$, represents the rate of change of $f(x)$ with respect to x. In graphical terms, $df(x)/dx$ (if it exists in the range of interest) is the instantaneous slope of the curve defined by $y = f(x)$. Strictly, $df(x)/dx$ is given by:

$$df(x)/dx = \lim_{h \to 0}[(f(x + h) - f(x))/h].$$

Equations involving rates of change in the physical world are extremely common, but it is rare to find one in which the value of the function f depends on only one parameter. More commonly, we deal with $f(x, y, z, \ldots)$.

Still, it is useful to know how the function f varies if we change the value of only one of its controlling parameters at a time. For a function $f(x, y)$, for example, we define *partial derivatives* $(\partial f(x, y)/\partial x)_y$ and $(\partial f(x, y)/\partial y)_x$ by:

$$(\partial f(x, y)/\partial x)_y = \lim_{h \to 0}[(f(x + h, y) - f(x, y))/h],$$

and

$$(\partial f(x, y)/\partial y)_x = \lim_{h \to 0}[(f(x, y + h) - f(x, y))/h].$$

When we write partial derivatives, we use the Greek symbol ∂, rather than d, to remind ourselves that f is a function of more than one variable, and we indicate the variables that are being held fixed by means of subscripts. In chapter 3, for example, we find that the Gibbs free energy (G) of a phase is a function of temperature (T), pressure (P), and the number of moles of each of the components $(n_1, n_2, \ldots, n_i)$ in the phase. To consider how G changes as a function of T alone, we look for the value of $(\partial G/\partial T)_{P,n_1,n_2,\ldots,n_i}$.

Because there can be many interrelationships among variables that describe a system, we can write many different partial derivatives. If pressure, temperature, and volume all influence the state of a system, for example, we may be interested in $(\partial P/\partial T)_V$, $(\partial V/\partial P)_T$, $(\partial T/\partial V)_P$, or any of their reciprocals. These various partial derivatives are not independent of each other. This should be evident from our discussion of the Maxwell relations in chapter 3, in which we investigated some of the relationships among partial derivatives of thermodynamic functions. If that discussion was unfamiliar ground for you, try learning four simple rules for working with partial derivatives. These follow from the standard Chain Rule, which states that if $w = f(x, y, z)$ and if x, y, and z are each functions of u and v, then:

$$(\partial w/\partial u)_v = (\partial w/\partial x)_{y,z}(\partial x/\partial u)_v + (\partial w/\partial y)_{x,z}(\partial y/\partial u)_v$$
$$+ (\partial w/\partial z)_{x,y}(\partial z/\partial u)_v, \qquad (A.1)$$

and

$$(\partial w/\partial v)_u = (\partial w/\partial x)_{y,z}(\partial x/\partial v)_u + (\partial w/\partial y)_{x,z}(\partial y/\partial v)_u$$
$$+ (\partial w/\partial z)_{x,y}(\partial z/\partial v)_u. \qquad (A.2)$$

The symmetry of the Chain Rule makes it easy to remember. Consider, now, what would happen if we applied the Chain Rule to a function $w = f(x, y)$, in which x was itself a function of y and another variable, z; that is, $x = g(y, z)$. Then:

$$(\partial w/\partial z)_y = (\partial w/\partial x)_y(\partial x/\partial z)_y + (\partial w/\partial y)_x(\partial y/\partial z)_y,$$

but

$$(\partial y/\partial z)_y = 0,$$

so that:

Rule 1: $(\partial w/\partial x)_y(\partial x/\partial z)_y = (\partial w/\partial z)_y. \qquad (A.3)$

It can also be shown that:

Rule 2: $(\partial w/\partial x)_y = 1/(\partial x/\partial w)_y, \qquad (A.4)$

Rule 3: $(\partial x/\partial y)_z(\partial y/\partial z)_x(\partial z/\partial x)_y = -1, \qquad (A.5)$

and

Rule 4: $(\partial w/\partial y)_x = (\partial w/\partial y)_z + (\partial w/\partial z)_y(\partial z/\partial y)_x. \qquad (A.6)$

We commonly describe changes in some property of a system by writing *differential equations*, in which an infinitesimal change df is related to infinitesimal changes in one or more of the variables that control the property. For a function of a single variable, $f(x)$, we can write:

$$df = (df/dx)dx.$$

If more than one variable $(x, y, z, \ldots)$ can potentially influence the value of f, the equivalent expression is:

$$df = (\partial f/\partial x)_{y,z,\ldots}\, dx + (\partial f/\partial y)_{x,z,\ldots}\, dy$$
$$+ (\partial f/\partial z)_{x,y\ldots}\, dz + \ldots.$$

The expression on the right side of this equation is called a *total differential*, because it accounts for the total change df by summing the changes due to each independent variable separately. Consider land elevation, for example, which varies as a function of both latitude and longitude on the Earth's surface. If you were to move a short distance in some random direction, your total change in elevation would be equal to the slope of the land surface in a north-south direction times the distance you actually moved in that direction, plus the slope in an east-west direction times the distance you moved in that direction.

In many geochemical applications, we find ourselves working with a special type of total differential called an *exact* or *perfect differential*. The differentials of each of the energy functions (dE, dH, dF, dG) introduced in chapter 3, for example, belong in this category. To define what we mean, suppose that we know of three properties of a system, each of which is a function of the same set of three independent variables. Identify these properties as $P(x, y, z)$, $Q(x, y, z)$, and $R(x, y, z)$. Can these three properties be related by some function $f(x, y, z)$ in such a way that $df = Pdx + Qdy + Rdz$? If so, then the expression on the right side is not only the total differential of f, it is also an exact differential. We then say that f is a *function of state*, with the following particularly useful characteristics:

1. Not only are P, Q, and R the partial derivatives of f with respect to x, y, and z, but their own partial derivatives (second derivatives of f) are also interrelated. The following *cross-partial reciprocity expressions* are true:

$$(\partial P/\partial y)_{x,z} = (\partial Q/\partial x)_{y,z}; \ (\partial P/\partial z)_{x,y} = (\partial R/\partial x)_{y,z};$$
$$(\partial Q/\partial z)_{x,y} = (\partial R/\partial y)_{x,z}.$$

This property is evident in the Maxwell relations, which derive from the fact that each of the energy functions (E, F, H, and G) is function of state. For example, $G = G(P, T, n) = -SdT + VdP + \mu dn$. $S(P, T, n)$, $V(P, T, n)$, and $\mu(P, T, n)$, therefore, are like the functions $P(x, y, z)$, $Q(x, y, z)$, and $R(x, y, z)$. Each is a partial differential of G (that is, $S = -(\partial G/\partial T)_{P,n}$, $V = (\partial G/\partial P)_{T,n}$, and $\mu = (\partial G/\partial n)_{T,P}$) and the Maxwell relations in this case,

$$-(\partial S/\partial P)_{T,n} = (\partial V/\partial T)_{P,n} = (\partial \mu/\partial n)_{P,T},$$

are the cross-partial reciprocity expressions we expect, because G is an exact differential.

2. We can determine the total change in f between two states of the system by integrating df between (x_1, y_1, z_1) and (x_2, y_2, z_2) along some reaction pathway, C. An integral of this type is known as a *line integral*.

For an illustration of how this operation is performed, see worked problem 3.1. When we do this integration for a function of state, we discover that:

$$\oint_{x_1,y_1,z_1}^{x_2,y_2,z_2} df = \oint_{x_1,y_1,z_1}^{x_2,y_2,z_2} (Pdx + Qdy + Rdz)$$
$$= f(x_2, y_2, z_2) - f(x_1, y_1, z_1).$$

Put simply, this means that we always get the same answer, regardless of how we get from state 1 to state 2. The integral of df is said to be *independent of path*.

3. Path independence implies that if we integrate df around a closed loop from state 1 to state 2 and back again, the net change in f will be zero. Again, see worked problem 3.1 to verify that the change in internal energy (an exact differential) around a closed path is zero, whereas the change in either work or heat is not.

Root Finding

It is easy to find the roots of sets of first-order equations by simple algebraic substitution. It is very common, however, to encounter problems in the physical sciences that yield equations in which a key variable x is raised to a higher power or appears in a transcendental function like sin x. A familiar equation from which you learned to extract x when you were in high school is the polynomial:

$$0 = ax^2 + bx + c.$$

In this example, x can be determined by applying the quadratic formula:

$$x = \left(-b \pm \sqrt{(b^2 - 4ac)}\right)/2a.$$

Unfortunately, there are no simple solutions of this type for most equations you will encounter. Adding a cubic term to this polynomial, for example, would make the task of root finding considerably more difficult. There are many ways around this problem, most of which involve a numerical approach instead of a closed-form or analytical one. They are thus well suited to analysis by computer or a pocket calculator. We briefly discuss one of these approaches, the *Newton-Raphson method*, which is mentioned as a means for solving several problems in chapters 4, 7, and 14.

Consider the curve for the function $y = f(x)$ in figure A.1. Suppose that $f(x)$ has a real root at $x = a_0$ that we want to find. One way to do this would be simply to

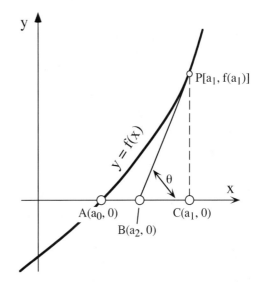

FIG. A.1. Geometric basis for the Newton-Raphson method for finding a root of $f(x)$. Line BP is the tangent to $f(x)$ at point P.

guess at random successive values of x. If we are patient and watch to see that $f(x)$ is closer to zero with each guess, we will eventually find the value for which $f(x) = 0$. By examining figure A.1, however, we can see a way to make guesses in a more sophisticated way. Suppose that our first guess is that $x = a_1$, reasonably close to a_0 but not correct. The error in this guess is $|a_1 - a_0|$, shown by the distance AC in the figure. Assuming that this error is unacceptably large, we now make a better guess by drawing the tangent BP to the curve at point P, and see that the tangent crosses the x axis at $x = a_2$. The error is now only $|a_2 - a_0|$, or the distance AB. We repeat the process, each time estimating the improved value of a_{i+1} by drawing a tangent to $y = f(x)$ at the point corresponding to a_i, until the error $|a_i - a_0|$ is acceptably small.

We can write an algorithm to express this method in a way that can be used in a computer program. Comparing the first two guesses geometrically, we see that:

$$AB = AC - BC,$$

or

$$(a_2 - a_0) = (a_1 - a_0) - PC/\tan \theta,$$

from which $a_2 = a_1 - f(a_1)/f'(a_1)$, where $f'(a_1)$ is the instantaneous slope of $f(x)$ at a_1 (that is, $f'(a_1) = dy/dx$ at a_1). This simple rule gives us a powerful means for making consecutive guesses, provided that (1) $f(x)$ has a derivative in the vicinity of a_0, (2) a_1 is reasonably close to a_0, so the method doesn't converge on some other root, and (3) $f'(a_{i+1})$ is not very close to zero, so that we avoid introducing a large numerical error in a_{i+1}. The algorithm, then, is:

1. Make an initial guess, a_1, of the root.
2. Calculate $f(a_1)$ and $f'(a_1)$.
3. Calculate a better value for the root, $a_2 = a_1 - f'(a_1)/f(a_1)$.
4. Calculate $f(a_2)$.
5. If $f(a_2)$ is acceptably close to zero, stop. Otherwise, repeat steps (2)–(5) until the root *is* acceptably close.

You may define "acceptably close" with a specific numerical value, or by noting when $f(a_i)$ does not change much from one iteration to the next.

Most texts that discuss the Newton-Raphson method fail to point out how easily it can be applied to systems of equations. Because geochemical problems commonly involve functions of more than one variable, you may find it useful to know how this is done. Suppose that several variables $(x_1, x_2, \ldots, x_n)$, related by a set of n independent equations, describe a system. To make life frustrating, each of the equations is a complicated function in which the variables appear in several terms, so that finding the values of $x_1 - x_n$ by substitution or simple matrix methods is impossible. If each of the functions can be differentiated with respect to each of the variables, however, there is hope for use of the Newton-Raphson method.

As in the simple case above, begin by guessing initial values for x_1 through x_n. Using these, calculate the numerical values of each of the functions $(f_1, f_2, \ldots, f_n)$ and place them in a column vector F. Also, calculate each of the partial derivatives $(\partial f/\partial x)$ and build a matrix J that looks like this:

$$J = \begin{pmatrix} (\partial f_1/\partial x_1) & (\partial f_1/\partial x_2) & \cdots & (\partial f_1/\partial x_n) \\ (\partial f_2/\partial x_1) & (\partial f_2/\partial x_2) & \cdots & (\partial f_2/\partial x_n) \\ (\partial f_3/\partial x_1) & (\partial f_3/\partial x_2) & \cdots & (\partial f_3/\partial x_n) \\ \vdots & \vdots & \vdots & \vdots \\ (\partial f_n/\partial x_1) & (\partial f_n/\partial x_2) & \cdots & (\partial f_n/\partial x_n) \end{pmatrix}.$$

The column vector $X = (x_1, x_2, \ldots, x_n)$ can then be found by successive approximations from the matrix version of the Newton-Raphson formula, which we present without proof:

$$X_{k+1} = X_k - (J_k)^{-1} F_k.$$

This is too cumbersome for a pocket calculator, because it involves inverting the matrix J at each iteration, but it is quite easily done with standard software, such as Microsoft Excel™. An algorithm based on this matrix method tests F_k at each step to see whether each of its elements is acceptably close to zero.

Fitting a Function to Data

In a typical research problem, a geochemist gathers observations on a natural or experimental system and then tries to make sense of the results by looking for functional relationships between the data and system variables that control them. Sometimes the function that best describes the data has a form based on theoretical considerations. At other times, you may choose an empirical function—a polynomial or exponential equation, for example. In any case, the task of fitting that function to your data set involves finding the values of one or more coefficients in the equation so that it gives results that are as close to the observed data as possible.

We demonstrate a common approach to the problem by considering the case in which measured values of some property y appear to depend on some other property x in such a way that a graph of y against x is a straight line. That is, $y = f(x)$ has the form:

$$y = a_0 + a_1 x.$$

What values of a_0 and a_1 are most appropriate for your data? If all of the observed values of y lie precisely on a line, the answer would be easy to find. Unfortunately, however, there are always random errors in any data set, due to sampling technique or undiagnosed complexities in the system (fig. A.2.) The task, then, is to find values of a_0 and a_1 such that the distance between $f(x)$ and each of the measured values of y is as small as possible. In practice, we generally go one step further and look for a_0 and a_1 such that the sum of the squares of the deviations from $f(x)$ is minimized. This overall approach, therefore, is known as the *method of least squares*.

Mathematically, the problem involves minimizing the function:

$$Q(a_0, a_1) = \sum_i (y_i - a_0 - a_1 x_i)^2,$$

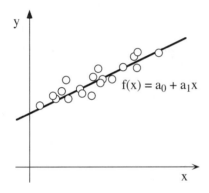

FIG. A.2. The measured values of property *x* (open circles), in this example, are apparently related to values of property *y* by some function $y = f(x)$. Because of random errors in the data set, however, the observed values are scattered around the most probable linear function (line).

or, if we have some reason to trust some observations more than others,

$$Q(a_0, a_1) = \sum_i w_i(y_i - a_0 - a_1 x_i)^2,$$

where w_i is a weighting factor for observation *i*. The summations are each taken over all values of *i* from 1 to *N*, where *N* is the total number of observations. To minimize *Q*, we first find expressions for the two partial derivatives of *Q* with respect to a_0 and a_1 and set each equal to zero:

$$(\partial Q/\partial a_0)_{a_1} = 2\left(\sum w_i y_i - a_0 \sum w_i - a_1 \sum w_i x_i\right) = 0,$$

and

$$(\partial Q/\partial a_1)_{a_0} = 2\left(\sum w_i x_i y_i - a_0 \sum w_i x_i - a_1 \sum w_i x_i^2\right)$$
$$= 0.$$

These can be solved simultaneously to yield:

$$a_0 = \left[\left(\sum w_i x_i^2\right)\left(\sum w_i y_i\right) - \left(\sum w_i x_i\right)\left(\sum w_i x_i y_i\right)\right]/$$
$$\left[\left(\sum w_i\right)\left(\sum w_i x_i^2\right) - \left(\sum w_i x_i\right)^2\right],$$

$$a_1 = \left[\left(\sum w_i\right)\left(\sum w_i x_i y_i\right) - \left(\sum w_i x_i\right)\left(\sum w_i y_i\right)\right]/$$
$$\left[\left(\sum w_i\right)\left(\sum w_i x_i^2\right) - \left(\sum w_i x_i\right)^2\right].$$

The same procedure can be followed for any polynomial expression.

The Error Function

You are probably somewhat familiar with the function $f(x) = (1/\sqrt{[2\pi]})\, e^{-t^2/2}$, where $t = (x - \bar{x})/\sigma$. This is the standard form of the normal frequency function, which describes the frequency distribution of events (*x*) around a mean, $\bar{x}$. The quantity σ is the *standard deviation* of *x*, a measure of the spread of values around the mean. This is the "curve" that teachers once used for calculating the grade distribution in classes. It describes the expected distribution of random variations ("errors") in many other natural situations. In chapter 5 and elsewhere, we have used it indirectly as a means of finding the distribution of mobile ions diffusing across a boundary between adjacent phases.

It can be shown that the area under this curve between zero and some penetration distance *x* is equal to the probability that an ion lies within that range. This, in fact, is the context in which the normal distribution appears in chapter 5. The error function, erf(*x*), can be defined as:

$$\mathrm{erf}(x) = (1/\sqrt{[2\pi]}) \int_0^x e^{-t^2/2} dt,$$

and is used to calculate this probability. This form of the error function is commonly used by statisticians. Unfortunately, several other forms are also used by physical scientists. The definition most commonly found in discussions of transport equations is the one we have applied in chapter 5:

$$\mathrm{erf}(x) = (2/\sqrt{\pi}) \int_0^x e^{-t^2} dt.$$

It is a simple matter to relate one form of the error function to another by making an appropriate substitution for *t* and adjusting the integration limits accordingly. Because several subtly different versions exist, you should always check to see how erf(*x*) is defined for the particular application you have in mind.

Unfortunately, the integration of e^{-t^2} cannot be performed analytically (that is, in neat, closed form). Because erf(*x*) is a widely used function, however, there are many popular ways to evaluate it. The easiest is simply to look it up in a table. The error function and its complement, erfc(*x*) = 1 − erf(*x*), are tabulated in most standard volumes of mathematical functions (for example, Burington 1973). There are also many approximate solutions that are reliable for calculations with pencil and paper. One that yields <0.7% error is the function erf(*x*) = $\sqrt{[1 - \exp(-4x2/\pi)]}/2$. In most applications today, however, erf(*x*) is calculated by numerical integration.

REFERENCES

Boas, M. L. 1984. *Mathematical Methods in the Physical Sciences: Solutions of Selected Problems,* 2nd ed. New York: Wiley.

Burington, R. S. 1973. *Handbook of Mathematical Tables and Formulas,* 5th ed. New York: McGraw-Hill.

Dence, J. B. 1975. *Mathematical Techniques in Chemistry.* New York: Wiley.

Potter, M. C., and J. Goldberg. 1997. *Mathematical Methods for Engineers,* 2nd ed. Wildwood: Great Lakes Press.

APPENDIX B: FINDING AND EVALUATING GEOCHEMICAL DATA

Selected Data Sources

The data of geochemical interest are scattered throughout a large number of journals and research reports that focus on the results of narrowly defined studies. In most cases, a search for data on specific systems should begin with the major abstracting journals, *Chemical Abstracts* and *Mineralogical Abstracts,* which regularly review the major journals and provide a topical summary of their contents. Limited literature surveys, such as the excellent summary of thermodynamic data sources in Nordstrom and Muñoz (1986), are also useful guides to the primary literature. Rather than provide another survey of this type, here we have compiled a short list of general references that are widely used by geochemists. In almost all cases, these are tabulations or databases built of information gathered from the primary research literature. They offer the advantage of being a quick overview of values that might otherwise be hard to find except in obscure corners of the library. More importantly, a large number of these (but not all!) are critical evaluations of the data, checked for internal consistency.

Using the Data

Before you use the data in any tabulation, be sure to read the text material that accompanies it, describing the criteria for data selection, the reactions that were investigated experimentally to obtain the data, the choice of standard and reference states, and the methods that were used to calculate interpolated, or "derived," values. Because data are gathered for many purposes, these criteria may not be the same from one source to another. Also, tests for the internal consistency of data compilations may be applied in different ways by various teams of compilers. Therefore, you should not assume that you can safely mix values garnered from different sources in a single problem. For a very good discussion of the potential problems of dealing with multiple data sources, consult chapter 12 of Nordstrom and Muñoz (1986).

Despite these problems, it is usually advisable and often necessary to consult several sources as you compile data for use in your calculations. You will find that teams of researchers have made idiosyncratic choices that emphasize different types of information for the same substances. The JANAF tables (Stull and Prophet 1971 et seq.), for example, indicate the temperatures at which phase changes take place in the substance being reported, but do not indicate phase changes in the reference state elements used to determine $\Delta \bar{H}_f^0$ and $\Delta \bar{G}_f^0$. Robie and coworkers (1978) report both. In addition, Robie and coworkers provide pairs of tables for silicate minerals that report free energies and enthalpies of formation from both the elements and from the oxides. The JANAF tables, however, include a short summary of the data sources consulted during compilation and, in most cases, a critical justification for the particular choices that were made in preparing their tables. This is a useful feature rarely found in other compendia. Helgeson and colleagues (1978) carry this approach to an extreme by producing a document in which evaluation of the data is the major concern. The data tables are only a few pages of the text and are in a compressed form. It is left to the reader, for example, to calculate thermodynamic values above 298 K by laboriously integrating heat capacity power functions.

Most of the data in standard tabulations should be familiar to students who have read chapters 3, 4, and 9 of this text. The lone exception is the *free energy function,* defined as $(\Delta \bar{G}^0 - \Delta \bar{H}_{298}^0)/T$. Values of this function do not change very much with temperature, so that it is usually very safe to interpolate linearly to find values between tabulated temperatures. For reactions among various substances, then, it can be shown that:

$$(\Delta \bar{G}^0 - \Delta \bar{H}_{298}^0)/T = \Delta \bar{G}_r^0/T - \Delta \bar{H}_{r,298}^0/T,$$

so that

$$\Delta \bar{G}_{r,T}^0 = T([\Delta \bar{G}^0 - \Delta \bar{H}_{298}^0]/T)_r + \Delta \bar{H}_{r,298}^0,$$

where we have followed our convention that quantities with the subscript r are the stoichiometric sum of values for the product phases minus the reactant phases. When the free energy function is available, it serves as a labor-saving alternative to calculating $\Delta \bar{G}^0_{r,T}$ from heat capacity data and reference state values for enthalpy and entropy.

For those instances when it is preferable to use heat capacity data, it is convenient to have them in the form of a power law. A commonly used expression is:

$$C_P = R(a + bT + cT^2 + dT^3 + eT^4),$$

in which R ($= 1.98726$ cal mole^{-1} K^{-1}) is the gas constant. Robie and coworkers (1978), for example, tabulate values of the constants a, b, c, d, and e for each of the substances in their compendium.

GENERAL REFERENCE

Carmichael, I.S.E., and H. P. Eugster, eds. 1987. *Thermodynamic Modeling of Geological Materials: Minerals, Fluids and Melts.* Reviews in Mineralogy 17. Washington, D.C.: Mineralogical Society of America.

DeBievre, P., M Gallet, N. E. Holden, and I. L. Barnes. 1984. Isotopic abundances and atomic weights of the elements. *Journal of Physical and Chemical Reference Data* 13:809–891.

Greenwood, N. N., and A. Earnshaw. 1997. *Chemistry of the Elements,* 2nd ed. Oxford: Pergamon.

Li, Y.-H. 2000. *A Compendium of Geochemistry.* Princeton: Princeton University Press.

Lide, D. R., ed. 2001. *CRC Handbook of Chemistry and Physics,* 81st ed. Boca Raton: CRC.

Nordstrom, D. K., and J. L. Muñoz. 1986. *Geochemical Thermodynamics.* Malden: Blackwell.

Ronov, A. B., and A. A. Yaroshevsky. 1969. Chemical composition of the earth's crust. In P. J. Hart, ed., *The Earth's Crust and Upper Mantle.* American Geophysical Union Monograph 13. Washington, D.C.: American Geophysical Union, pp. 35–57.

Wedepohl, K. H., ed. 1969. *Handbook of Geochemistry,* 2 vols. New York: Springer-Verlag.

ELEMENTS AND INORGANIC COMPOUNDS

Chase, M. W., Jr. 1998. *NIST-JANAF Thermochemical Tables,* 4th ed. Monograph 9. *Journal of Physical and Chemical Reference Data.* Washington, D.C.: National Institute of Standards and Technology.

CODATA. 1978. *CODATA Recommended Key Values for Thermodynamics 1977.* CODATA Bulletin 28. Oxford: Pergamon.

CODATA. 1987. *1986 Adjustment of the Fundamental Physical Constants.* CODATA Bulletin 63. Oxford: Pergamon.

Cox, J. D., D. D. Wagman, and V. A. Medvedev, eds. 1989. *CODATA Key Values for Thermodynamics.* New York: Hemisphere.

Hultgren, R., P. D. Desai, D. T. Hawkins, M. Gleiser, K. K. Kelly, and D. D. Wagman. 1973. *Selected Values of the Thermodynamic Properties of Binary Alloys.* Metals Park: American Society for Metals.

Merrill, L. 1982. Behavior of the AB$_2$-type compounds at high pressures and high temperatures. *Journal of Physical and Chemical Reference Data* 11:1005–1064.

Naunov, G. B., B. N. Ryzhenko, and I. L. Khodakovskii. 1974. *Handbook of Thermodynamic Data.* NTIS Document Pb-226, 722/7CxA. Washington, D.C.: U.S. Department of Commerce.

Parker, V. B., D. D Wagman, and W. H. Evans. 1971. *Selected Values of Chemical Thermodynamic Properties: Tables for the Alkaline Earth Elements (Elements 92 through 97 in the Standard Order of Arrangement).* U.S. National Bureau of Standards Technical Note 270-6. Washington, D.C.: U.S. Department of Commerce.

Schumm, R. H., D. D. Wagman, S. Bailey, W. H. Evans, and V. B. Parker 1973. *Selected Values of Chemical Thermodynamic Properties: Tables for the Lanthanide (Rare Earth) Elements (Elements 62 through 76 in the Standard Order of Arrangement).* U.S. National Bureau of Standards Technical Note 270-7. Washington, D.C.: U.S. Department of Commerce.

Stull, D. R., and H. Prophet. 1971. *JANAF Thermochemical Tables.* U.S. National Bureau of Standards NSRDS-NBS 37. Washington, D.C.: U.S. Department of Commerce.

Supplement 1974 by Chase, M. W., Jr., J. L. Curnutt, A. T. Hu, H. Prophet, and L. C. Walker. *Journal of Physical and Chemical Reference Data* 3:311–480.

Supplement 1975 by Chase, M. W., Jr., J. L. Curnutt, H. Prophet, R. A. McDonald, and A. N. Syverud. *Journal of Physical and Chemical Reference Data* 4:1–175.

Supplement 1978 by Chase, M. W., Jr., J. L. Curnutt, R. A. McDonald, and A. N. Syverud. *Journal of Physical and Chemical Reference Data* 7:793–940.

Supplement 1982 by Chase, M. W., Jr., J. L. Curnutt, J. R. Downey Jr., R. A. McDonald, A. N. Syverud, and E. A. Valenzuela. *Journal of Physical and Chemical Reference Data* 11:695–940.

Wagman, D. D., W. H. Evans, V. B. Parker, I. Halow, S. M. Bailey, and R. H. Schumm. 1968. *Selected Values of Chemical Thermodynamic Properties: Tables for the First Thirty-Four Elements in the Standard Order of Arrangement.* U.S. National Bureau of Standards Technical Note 270-3. Washington, D.C.: U.S. Department of Commerce.

Wagman, D. D., W. H. Evans, V. B. Parker, I. Halow, S. M. Bailey, and R. H. Schumm. 1969. *Selected Values of Chemical Thermodynamic Properties: Tables for Elements 35 through 53 in the Standard Order of Arrangement.* U.S. National Bureau of Standards Technical Note 270-4. Washington, D.C.: U.S. Department of Commerce.

Wagman, D. D., W. H. Evans, V. B. Parker, I. Halow, S. M. Bailey, R. H. Schumm, and K. L. Churney. 1971. *Selected Values of Chemical Thermodynamic Properties: Tables for Elements 54 through 61 in the Standard Order of Arrangement.* U.S. National Bureau of Standards Technical Note 270-5. Washington, D.C.: U.S. Department of Commerce.

Wagman, D. D., W. H. Evans, V. B. Parker, and R. H. Schumm. 1976. *Chemical Thermodynamic Properties of Sodium, Potassium, and Rubidium: An Interim Tabulation of Selected Material.* U.S. National Bureau of Standards Interim Report NBSIR 76-1034. Washington, D.C.: U.S. Department of Commerce.

Wagman, D. D., W. H. Evans, V. B. Parker, R. H. Schumm, and R. L. Nuttall. 1981. *Selected Values of Chemical Thermodynamic Properties: Compounds of Uranium, Protactinium, Thorium, Actinium, and the Alkaline Metals.* U.S. National Bureau of Standards Technical Note 270-8. Washington, D.C.: U.S. Department of Commerce.

Wagman, D. D., W. H. Evans, V. B. Parker, R. H. Schumm, I. Halow, S. M. Bailey, K. L. Churney, and R. L. Nuttall. 1982. The NBS tables of chemical thermodynamic properties: Selected values for inorganic and C_1 and C_2 organic substances in SI units. *Journal of Physical and Chemical Reference Data* 11(Suppl. 2):1–392.

MINERALS

Haas, J. L., Jr., G. R. Robinson, and B. S. Hemingway. 1981. Thermodynamic tabulations for selected phases in the system CaO-Al_2O_3-SiO_2-H_2O at 101.325 kPa (1 atm) between 273.15 and 1800 K. *Journal of Physical and Chemical Reference Data* 10:575–669.

Helgeson, H. C., J. M. Delany, H. W. Nesbitt, and D. K. Bird. 1978. Summary and critique of the thermodynamic properties of rock-forming minerals. *American Journal of Science* 278A:1–229.

Robie, R. A., B. S. Hemingway, and J. R. Fisher. 1978. *Thermodynamic Properties of Minerals and Related Substances at 298.15 K and One Atmosphere Pressure and at Higher Temperatures.* U.S. Geological Survey Bulletin 1259. Washington, D.C.: U.S. Geological Survey.

Robie, R. A., B. S. Hemingway, and H. T. Haselton. 1983. *Thermodynamic Properties of Minerals.* U.S. Geological Survey Professional Paper 1375. Washington, D.C.: U.S. Geological Survey.

Robinson, G. R., J. L. Haas Jr., C. M. Schafer, and H. T. Hazelton Jr. 1982. *Thermodynamic and Thermophysical Properties of Selected Phases in the MgO-SiO_2-H_2O-CO_2, CaO-Al_2O_3-SiO_2-H_2O-CO_2, and Fe-FeO-Fe_2O_3-SiO_2 Chemical Systems, with Special Emphasis on the Properties of Basalts and Their Mineral Components.* U.S. Geological Survey Open-File Report 83-79. Washington, D.C.: U.S. Geological Survey.

Woods, T. L., and R. M. Garrells. 1987. *Thermodynamic Values at Low Temperature for Natural Inorganic Materials.* New York: Oxford University Press.

AQUEOUS SPECIES

Burnham, C. W., J. R. Holoway, and N. F. Davis. 1969. *Thermodynamic Properties of Water to 1000°C and 10,000 bars.* Geological Society of America Special Paper 132. Boulder: Geological Society of America.

Criss, C. M., and J. W. Cobble. 1964. The thermodynamic properties of high temperature aqueous solutions. IV. Entropies of the ions up to 200° and the correspondence principle. *Journal of the American Chemical Society* 86:5385–5390.

Criss, C. M., and J. W. Cobble. 1964. The thermodynamic properties of high temperature aqueous solutions. V. The calculation of ionic heat capacities up to 200°. Entropies and heat capacities above 200°. *Journal of the American Chemical Society* 86:5390–5393.

Hamer, W. J. 1968. *Theoretical Mean Activity Coefficients of Strong Electrolytes in Aqueous Solutions from 0 to 100°C.* U.S. National Bureau of Standards NSRDS-NBS 24. Washington, D.C.: U.S. Department of Commerce.

Helgeson, H. C. 1967. Thermodynamics of complex dissociation in aqueous solution at elevated temperatures. *Journal of Physical Chemistry* 71:3121–3136.

Helgeson, H. C. 1969. Thermodynamics of hydrothermal systems at elevated temperatures and pressure. *American Journal of Science* 267:729–804.

Helgeson, H. C. 1982. Errata: Thermodynamics of minerals, reactions, and aqueous solutions at high temperatures and pressures. *American Journal of Science* 282:1144–1149.

Helgeson, H. C. 1985. Errata II: Thermodynamics of minerals, reactions, and aqueous solutions at high pressures and temperatures. *American Journal of Science* 285:845–855.

Helgeson, H. C., and D. H. Kirkham. 1974. Theoretical prediction of the thermodynamic behavior of aqueous electrolytes at high pressures and temperatures: I. Summary of the thermodynamic/electrostatic properties of the solvent. *American Journal of Science* 274:1089–1198.

Helgeson, H. C., and D. H. Kirkham. 1974. Theoretical prediction of the thermodynamic behavior of aqueous electrolytes at high pressures and temperatures. II. Debye-Hückel parameters for activity coefficients and relative partial molal properties. *American Journal of Science* 274:1199–1261.

Helgeson, H. C., and D. H. Kirkham. 1976. Theoretical prediction of the thermodynamic behavior of aqueous electrolytes at high pressures and temperatures. III. Equation of state for aqueous species at infinite dilution. *American Journal of Science* 276:97–240.

Helgeson, H. C., D. H. Kirkham, and G. C. Flowers. 1982. Theoretical prediction of the thermodynamic behavior of aqueous electrolytes at high pressures and temperatures. IV. Calculation of activity coefficients, osmotic coefficients, and partial molal and standard and relative partial molal properties to 600°C and 5 kb. *American Journal of Science* 281:1249–1516.

Hogfeldt, E. 1982. *Stability Constants of Metal Ion Complexes: Part A. Inorganic Ligands.* IUPAC Chemical Data Series no. 21. Oxford: Pergamon.

Horne, R. A., ed. 1972. *Water and Aqueous Solutions: Structure, Thermodynamics, and Transport Properties.* New York: Wiley Interscience.

Sillén, L. G., and A. E. Martell. 1964. *Stability Constants of Metal-Ion Complexes.* The Chemical Society Special Publication no. 17. London: The Chemical Society. (Supplement 1971.)

REACTION KINETICS

Cohen, N., and K. R. Westberg. 1983. Chemical kinetic data sheets for high-temperature chemical reactions. *Journal of Physical and Chemical Reference Data* 12:531–590.

PHASE RELATIONS

Brookins, D. G. 1988. *Eh-pH Diagrams for Geochemistry.* New York: Springer-Verlag.

Ehlers, E. G. 1987. *The Interpretation of Geological Phase Diagrams.* Mineola: Dover.

Levin, E. M., C. R. Robbins, and H. F. McMurdie. 1964. *Phase Diagrams for Ceramists,* vol. 1. Columbus: American Ceramic Society.

APPENDIX C: NUMERICAL VALUES OF GEOCHEMICAL INTEREST

Dimensions of the Earth

Mass of the Earth	5.973×10^{24} kg
Mass of the atmosphere	5.1×10^{18} kg
Mass of the oceans	1.4×10^{21} kg
Mass of the crust	2.6×10^{22} kg
Mass of the mantle	4.0×10^{24} kg
Mass of the outer core	1.85×10^{24} kg
Mass of the inner core	9.7×10^{22} kg
Equatorial radius	6.378139×10^{6} m
Polar radius	6.35675×10^{6} m
Surface area of the Earth	5.1×10^{14} m^2
Surface area of the oceans	3.62×10^{14} m^2
Surface area of the continents	1.48×10^{14} m^2
Volume of the oceans	1.37×10^{21} l
Mean depth of the oceans	3.8×10^{3} m
Continental runoff rate	3.6×10^{16} l yr^{-1}

Physical Constants

Avogadro's number	$N = 6.022094 \times 10^{13}$ mol^{-1}
Gas constant	$R = 1.98717$ cal mol^{-1} K^{-1}
	$= 8.31433$ J mol^{-1} K^{-1}
	$= 82.06$ cm^3 atm mol^{-1} K^{-1}
Faraday's constant	$F = 96,487.0$ coulomb equiv^{-1}
	$= 23,060.9$ cal volt^{-1} equiv^{-1}
Gravitational constant	$G = 6.6732 \times 10^{-11}$ m^3 kg^{-1} sec^{-2}
	$= 6.6732 \times 10^{-11}$ nt m^2 kg^{-2}
Boltzmann's constant	$k = 1.380622 \times 10^{-23}$ J K^{-1}
Planck's constant	$h = 6.626176 \times 10^{-34}$ J sec
Base of natural logarithms	$e = 2.71828$

Conversion Factors

Distance:

1 centimeter (cm)	$= 10^8$ Ångström (Å)
	$= 0.3937$ inch

Time:

1 year	$= 3.154 \times 10^7$ sec

Mass:

1 gram (g)	$= 2.20462 \times 10^{-3}$ lb
1 atomic mass unit (amu)	$= 1.66054 \times 10^{-24}$ g

Temperature:

Kelvins (K)	$= °C + 273.15$
	$= 5(°F - 32)/9 + 273.15$

Pressure (mass length^{-1} time^{-1}):

1 atm	$= 1.013250 \times 10^6$ dyne cm^{-2}
	$= 1.013250$ bar
	$= 1.013250 \times 10^5$ pascal (Pa)
	$= 1.013250 \times 10^5$ nt m^2
	$= 14.696$ lb in^{-2}

Energy (mass length^2 time^{-1}):

   1 joule (J) $= 10^7$ erg
                        $= 2.389 \times 10^{-1}$ cal
                        $= 9.868 \times 10^{-1}$ liter atm
                        $= 6.242 \times 10^{11}$ eV
                        $= 2.778 \times 10^{-7}$ kWh
                        $= 9.482 \times 10^{-4}$ Btu

Entropy (mass length^2 time^{-2} deg^{-1}):

   1 Gibbs $= 1$ cal K^{-1}

Viscosity (mass length^{-1} time^{-1}):

   1 poise $= 1$ g sec^{-1} cm^{-1}
                        $= 1$ dyne sec cm^{-1}
                        $= 0.1$ Pa sec

Radioactive decay:

   1 Curie (Ci) $= 3.7 \times 10^{10}$ sec^{-1}

Logarithmic values: $\ln x = 2.303 \log x$

Units of concentration in solutions:

   Molarity (M)   = moles liter^{-1} of solution
   Normality (N)   = equiv. liter^{-1} of solution
   Molality (m)   = moles kg^{-1} of H$_2$O*
   Parts per million   = g/10^6 g (= mg/kg)
      (ppm)

*Concentrations are commonly reported in the geochemical literature as millimoles kg^{-1} of solution. To convert this measure to molal units requires that the density of the solution be known. Except in highly concentrated brines, however, little error is introduced if this difference is ignored.

GLOSSARY

Accretion. Accumulation of materials in space to form larger objects. This process may occur homogeneously (producing bodies with uniform composition) or heterogeneously (producing layered bodies).

Activation energy. The amount of energy that must be provided to overcome some kinetic barrier.

Activity. Concentration of a component, adjusted for any effects of nonideality; also, a measurement of the number of radioactive decay events per unit of time.

Activity coefficient. Ratio of the activity of a species to its concentration; a measure of the degree of nonideal behavior of a chemical species in solution.

Adiabatic process. A process that occurs without exchange of heat with the surroundings.

Advection. Transport of ions or molecules within a moving medium.

Alkalinity. Charge deficit between the sum of dissolved conservative cations and anions in an electrolyte solution.

Alkane. Saturated aliphatic hydrocarbon with general formula C_nH_{2n+2}; alkanes with at least 16 carbons are solids, usually found in primary producers (photosynthesizing organisms).

Alloy. A nonstoichiometric combination of metals.

Amino acid. Organic compound containing an amino (NH_2) and a carboxyl (COOH) group. There are 20 amino acids used in protein synthesis.

Assimilation. Incorporation of solid material into a magma.

Atomic number. The number of protons in an atomic nucleus.

Authigenesis. Mineral formation from dissolved or solid constituents already present at the site, as opposed to constituents transported from elsewhere.

Biomolecule. Organic compound, such as a protein or carbohydrate, that is a component of an organism.

Biopolymer. Complex organic molecule synthesized by plants or animals. Examples of these molecules include carbohydrates, proteins, lignin, and lipids.

Bitumin. Generic term applied to naturally occurring, flammable organic substances consisting of hydrocarbons. Bitumin is soluble in organic solvents.

Buffer. Assemblage of chemical species whose coexistence allows a system to resist change in some intensive property, such as pH or oxygen fugacity.

Calorimetry. Experimental measurement of heat evolved or absorbed during a specific reaction.

Carbohydrate. Organic compound present in living cells with general formula CH_2O. It is the most abundant class of organic compounds and includes sugars, starches, and cellulose.

Carbonate compensation depth (CCD). Depth in the oceans at which the downward flux of carbonate minerals is balanced by their rate of dissolution.

Chalcophile. An element with an affinity for sulfide phases.

Chelate. Large molecule or complex that can enclose weakly bound atoms or ions.

Chemical potential (μ). Partial molar free energy, which describes the way in which total free energy for a phase responds to a change in the amount of component i in the phase $[=(\partial G/\partial n_i)_{P,T,n_{j\neq i}}]$.

Chondrite. A common type of meteorite, with nearly cosmic elemental abundances, thought to be a sample of the earliest Solar System material.

Closed system. A system that can exchange energy, but not matter, with its surroundings. Compare: *Isolated system, Open system.*

Colloid. Stable electrostatic suspension of small particles in a liquid.

Components. Abstract chemical entities, independently variable within a system, which collectively describe all of the potential compositional variations within it.

Compressibility. Measure of the relationship between phase volume and lithostatic pressure, $[\beta_T = -1/V(\partial V/\partial P)_T; \beta_s = -1/V(\partial V/\partial P)_S]$.

Condensation. The formation of solids or liquids from a gas phase during cooling.

Congruent reaction. Reaction in which one phase melts or dissolves to form another phase of the same composition. Compare: *Incongruent reaction.*

Conservative species. Any dissolved species whose concentrations are not affected in solution in any way, other than dilution, by variations in the abundance of other species in solution.

Cosmic abundance of the elements. Relative elemental abundance pattern in the Sun.

Cosmogenic nuclide. Any isotope formed by interaction of matter with cosmic rays.

Cotectic. Boundary curve on an *n*-component phase diagram, along which two or more phases crystallize simultaneously. Compare: *Peritectic.*

Delta notation. Notation used for stable isotope data, equal to $[(R_{sample} - R_{standard})/R_{standard}] \times 1000$, where R is the ratio of heavy to light isotope of the element of interest.

Diagenesis. The set of processes, other than weathering or metamorphism, that change the texture or mineral composition of sediments; these include compaction, cementation, recrystallization, and authigenesis.

Differentiation. Separation within a homogeneous planet of crust, mantle, and core; also, any process by which magma can give rise to rocks of contrasting composition.

Diffusion. Dispersion of ions or molecules through a medium that is not moving, due to a gradient in some intensive property across the system.

Diffusion coefficient. Proportionality constant relating flux to gradient in a diffusion equation.

Distribution coefficient (K_D). Ratio of concentrations of a component in two coexisting phases.

e-folding time. Time required for a compositional variable to return to within a factor of $1/e$ of its steady state value following some perturbing event, numerically equal to the inverse of the kinetic rate constant.

E_h. Redox potential of a cell, expressed in terms of voltage.

Electrolyte. Any substance that dissociates into ions when dissolved in an appropriate medium.

Electron. Subatomic particle, found in a region surrounding the nucleus of an atom, with a unit negative electrical charge.

Electron affinity (EA). Amount of energy released when an electron is added to an atom, creating an anion.

Elementary reaction. Any chemical reaction that occurs at the molecular level among species as they appear in the written representation. Compare: *Overall reaction.*

Enantiomer. One of two compounds with the same composition but having structures that are mirror images of each other.

Endothermic reaction. Reaction that absorbs heat, resulting in a decrease in enthalpy. Compare: *Exothermic reaction.*

Enthalpy. Thermodynamic function of state, H; a measure of the energy that is irretrievably converted to heat during any natural process.

Entropy. Thermodynamic function, S, the change of which is defined as the change in heat gained by a body at a certain temperature in a reversible process $[dS = (dQ/T)_{rev}]$.

Equation of state. Function that defines interrelationships among intensive properties of a system.

Equilibrium. Condition in which the properties of a system do not change with time.

Equilibrium constant (K_{eq}). Constant relating the composition of a system at equilibrium to its Gibbs free energy.

Eutectic. Point on a phase diagram at which the liquidus touches the solidus, where two or more solids crystallize simultaneously from a melt.

Exothermic reaction. Reaction that evolves heat, resulting in an increase in enthalpy. Compare: *Endothermic reaction.*

Exsolution. Physical unmixing of two phases.

Extended defects. Dislocations and planar defects in crystals that provide effective pathways for diffusion.

Extensive property. Variable whose value is a measure of the size of the system. Compare: *Intensive property.*

Extinct radionuclide. Any unstable nuclide that decays so rapidly that virtually none remains at the present time.

Fluid inclusion. Fluid trapped within crystals. These may be primary (trapped as the crystal grew) or secondary (introduced at some later time).

Flux. Amount of material moving from one location to another per unit time.

Fractional crystallization. Physical separation of crystals and liquid, preventing equilibrium between phases.

Fractionation. Separation of any two entities, such as isotopes (by mass), elements (by geochemical or cosmochemical properties), or crystals and melt (by fractional crystallization or fractional melting).

Fractionation factor. Ratio of a heavy to a light isotope of an element in one phase divided by the same ratio in another coexisting phase.

Fugacity. Gas partial pressure that has been adjusted for nonideality.

Geobarometer. Calibrated mineral exchange reaction that is a strong function of pressure but not temperature.

Geochronology. Determination of the ages of rocks using radiometric dating techniques.

Geopolymer. Any of several complex organic compounds of high molecular weight, formed from biopolymers during diagenesis; these include kerogen and bitumen.

Geothermal gradient. Rate of change of temperature with depth in the earth; also called a *geotherm.*

Geothermometer. Calibrated mineral exchange reaction that is a strong function of temperature but not pressure.

Gibbs free energy. Thermodynamic function of state, G, which is a measure of the energy available for changing chemical bonds in a system.

Half-life. Time required for one-half of a given number of atoms of a radionuclide to decay.

Half-reaction. Hypothetical reaction that illustrates either gain or loss of electrons, useful for understanding redox equilibria.

Heat. Transfer of energy that results in an increase in temperature.

Heat capacity. Functional relationship between enthalpy and temperature [$C_P = (\partial H/\partial T)_P$; $C_V = (\partial H/\partial T)_V$].

Helmholtz free energy. Thermodynamic function of state, F, which, for an isothermal process, is a measure of the energy transferred as work; also called the *work function.*

Henry's law. Expression of solution behavior in which the activities of dissolved species are directly proportional to their concentrations at high dilution.

Humic substance. Altered complex organic substance derived from the remains of terrestrial plants and phytoplankton.

Ideal solution. Any solution in which the end-member constituents behave as if they were independent.

Immiscibility. Condition that results in spontaneous phase separation.

Incongruent reaction. Reaction in which one phase melts or dissolves to form phases of different composition. Compare: *Congruent reaction.*

Intensive property. Variable whose value is independent of the size of the system. Compare: *Extensive property.*

Internal energy. Thermodynamic function of state, E, the change of which is a measure of the energy transferred as heat or work between the system and its surroundings.

Ionization potential (IP). Amount of energy needed to remove an electron from an atom, creating a cation.

Isochron. Line on an isotope plot whose slope determines the age of a rock system. Isochrons may be defined by the isotopic compositions of constituent minerals or whole rocks.

Isolated system. A system that cannot exchange either matter or energy with the universe beyond the system's limits. Compare: *Closed system, Open system.*

Isotopes. Atoms of the same element but with different numbers of neutrons and hence different atomic weights. Isotopes may be stable or unstable; also called *nuclides.*

Kerogen. Insoluble, high-molecular-weight organic matter derived from algae and woody plant material and found in sedimentary rocks. This material may yield petroleum when heated.

Kinetics. Description of a system's behavior in terms of its rates of change. Compare: *Thermodynamics.*

Lignin. A complex polysaccharide that provides structural support for many plants and is one of the chief organic substances in wood.

Lipid. Class of aliphatic hydrocarbons containing the fatty acids, fats, waxes, and steroids. These compounds are energy sources for the host organism.

Liquidus. Boundary curve on a phase diagram representing the onset of crystallization of a liquid with lowering temperature. Compare: *Solidus.*

Lithophile. Element with an affinity for silicate phases.

Local equilibrium. Equilibrium that is attained on a small scale, but not on a larger scale; also called *mosaic equilibrium.*

Lysocline. Depth in the oceans at which the effects of carbonate dissolution are first discernable.

Mass number. The number of protons plus neutrons in a nucleus.

Mean residence time. Average residence time for a chemical species in a system before it is removed by some loss process.

Metamorphism. Collection of processes that change the texture or phase composition of a rock through the action of temperature, pressure, and fluids.

Metasomatism. Metamorphism with accompanying change in chemical composition, usually as a result of fluid migration.

Metastable system. A system that appears to be stable because it is observed over a short time compared to the rates of reactions that alter it.

Mobile component. A component free to migrate into or out of a system.

Monolayer. An organic coating with a thickness of only one molecule.

Monolayer equivalent. The mass of organic matter that is theoretically present over a square meter if all surfaces of the sediment have a monolayer coating.

Neutron. Subatomic particle, found in the nucleus of an atom, with no electrical charge.

Nuclear statistical equilibrium. The steady state in which disintegrative and constructive nuclear reactions balance in an evolved star.

Nucleation. The initiation of a small volume of a new phase. This process may occur homogeneously (without a substrate) or heterogeneously (on a substrate).

Nucleosynthesis. The production of nuclides, primarily in stars.

Nuclides. See *Isotopes.*

Open system. A system that can exchange matter and energy with its surroundings. Compare: *Closed system, Isolated system.*

Overall reaction. Chemical reaction that describes a net change involving several intermediate steps and competing pathways. Compare: *Elementary reaction.*

Partial melting. Fusion of something less than an entire rock, a process that may occur under equilibrium or fractional (continuous separation of melt and crystals) conditions.

Periodic table. A tabular arrangement of the elements, in order of their increasing mean atomic weight, in such a way that elements with similar chemical properties related to their electronic configuration are grouped together.

Peritectic. Point on a *phase diagram* at which an incongruent reaction occurs. Compare: *Cotectic.*

pH. Measure of acidity in solution [$pH = -\log_{10} a_{H^+}$].

Phase. Substance with continuous physical properties.

Phase diagram. Graphical summary of how a system reacts to changing conditions or composition.

Point defects. Crystal imperfections caused by the absence of atoms at lattice sites; such defects may be intrinsic or extrinsic.

Proton. Subatomic particle, found in the nucleus of an atom, with a unit positive electrical charge.

Quantum. A discrete quantity or level of energy.

Racemic. An equal mixture of D- and L- enantiomers.

Racemization. Transformation of optically active compounds, such as amino acids, from the D-form to the L-form or vice versa.

Radioactive decay. Spontaneous transformation of one unstable nuclide into another isotope; this occurs by alpha or beta decay, positron decay, electron capture, or spontaneous fission.

Radiogenic nuclide. Isotope formed by decay of some parent radionuclide.

Raoult's law. Expression of solution behavior in which the activities of dissolved species are equal to their concentrations.

Refractory. Having a high melting temperature or condensing from a gas at a high temperature.

Salinity. The total dissolved salt content.

Saturated hydrocarbons. Organic molecules consisting of carbon and hydrogen atoms linked only by single bonds; these may form chain structures called aliphatic compounds, or cyclical structures called *alicyclic* compounds.

Secular equilibrium. Condition in which the decay rate for daughter radionuclides in a decay series equals that of the parent.

Siderophile. An element with an affinity for metallic phases.

Solidus. Boundary curve on a phase diagram representing complete solidification of a system with lowering temperature. Compare: *Liquidus.*

Solubility product constant (K_{sp}). Equilibrium constant for a dissolution reaction.

Solute. The dissolved species in a solution.

Solvent. The host species in a solution.

Solvus. Region of unmixing on a phase diagram.

Spinodal. Region of unmixing inside the *solvus* on a phase diagram; such unmixing involves no nucleation barrier.

Standard state. Arbitrarily selected state of a system used as a reference against which changes in thermodynamic properties can be compared.

Steady state. Condition in which rates of change within a system balance one another, so that there is no net change in its appearance for an indefinite time period, even though various parts of the system are not in thermodynamic equilibrium.

Sterol. A group of solid, mainly unsaturated steroid alcohols, including cholesterol and ergosterol, present in plant and animal tissues.

Stoichiometric. Summing to the whole, as in stoichiometric equations or coefficients.

Supernova. Massive stellar explosion, important for nucleosynthesis.

Surface tension. Tensional force applied perpendicular to any line on a droplet surface.

System. That portion of the universe that is of interest for a particular problem; a system can be open, closed, or isolated.

Thermal expansion. Measure of the relationship between phase volume and temperature [$\alpha_p = -V(\partial V/\partial T)_p$].

Thermocline. Transitional range between the warm surface zone and the cold deeper zone in the oceans.

Thermodynamics. Set of laws that predict the equilibrium configuration of a system and how it will change if its environmental parameters are changed. Compare: *Kinetics.*

Titration. Experiment in which a reaction is allowed to proceed incrementally, so that proportions of reactants can be determined.

Trace elements. Elements that occur in rocks with concentrations of a few tenths of a percent or less by weight; these may exhibit compatible or incompatible behavior.

Transition elements. Elements with inner orbitals incompletely filled by electrons.

Troposphere. Lower portion of the atmosphere that contains the bulk of its mass.

Uncompressed mean density. Mean density of a planet, corrected for the effects of gravitational self-compression.

Undercooling. Difference between the equilibrium temperature for the appearance of a phase and the temperature at which it actually appears.

Unsaturated hydrocarbons. Organic molecules with double or triple bonds between carbon atoms; these may form chain structures, called *aliphatic* compounds, or cyclic structures, called *aromatic* compounds.

Variance. Number of independent parameters that must be defined in order to specify the state of a system at equilibrium; also called *degrees of freedom.*

Volatile. Term indicating that a given element or compound commonly occurs in the gaseous state at the temperature of interest, or indicating that a given element condenses from a gas at low temperature.

Weathering. Those processes occurring at the Earth's surface that cause decomposition of rocks.

Work. Transfer of energy that causes a mechanical change in a system or its surroundings; the integral of force × displacement.

INDEX